Aerodynamics for Engineering Students

Frontispiece *(see overleaf)*

Aircraft wake (photo courtesy of Cessna Aircraft Company).

This photograph first appeared in the Gallery of Fluid Motion, *Physics of Fluids* (published by the American Institute of Physics), Vol. 5, No. 9, Sept. 1993, p. S5, and was submitted by Professor Hiroshi Higuchi (Syracuse University). It shows the wake created by a Cessna Citation VI flown immediately above the fog bank over Lake Tahoe at approximately 313 km/h. Aircraft altitude was about 122 m above the lake, and its mass was approximately 8400 kg. The downwash caused the trailing vortices to descend over the fog layer and disturb it to make the flow field in the wake visible. The photograph was taken by P. Bowen for the Cessna Aircraft Company from the tail gunner's position in a B-25 flying slightly above and ahead of the Cessna.

Aerodynamics for Engineering Students

Fifth Edition

E.L. Houghton

and

P.W. Carpenter

Professor of Mechanical Engineering,
The University of Warwick

ELSEVIER
BUTTERWORTH
HEINEMANN

AMSTERDAM · BOSTON · HEIDELBERG · LONDON · NEW YORK · OXFORD
PARIS · SAN DIEGO · SAN FRANCISCO · SINGAPORE · SYDNEY · TOKYO

Elsevier Butterworth-Heinemann
Linacre House, Jordan Hill, Oxford OX2 8DP
30 Corporate Drive, Burlington, MA 01803

First published in Great Britain 1960
Fourth edition published in 1993 by Edward Arnold
Fifth edition published by Butterworth-Heinemann 2003, 2004, 2005

British Library Cataloguing in Publication Data
Houghton, E.L. (Edward Lewis)
    Aerodynamics for engineering students. – 5th ed.
    1 Aerodynamics
    I Title   II Carpenter, P.W.
    629.1′323

Library of Congress Cataloguing in Publication Data
Houghton, E.L. (Edward, Lewis)
    Aerodynamics for engineering students / E.L. Houghton and P.W. Carpenter. – 5th ed.
    p. cm.
    Includes index.
    ISBN 0 7506 5111 3
    1 Aerodynamics 2 Airplanes-Design and construction 1 Carpenter, P.W.
    (Peter William), 1942–   II Title.
    TL570 .H587 2002
    620.132′3–dc21   2002029945

ISBN 0 7506 5111 3

For information on all Elsevier Butterworth-Heinemann publications
visit our website at www.bh.com

Working together to grow
libraries in developing countries

www.elsevier.com | www.bookaid.org | www.sabre.org

ELSEVIER   BOOK AID International   Sabre Foundation

Typeset in 10/11pt Times by Integra Software Services Pvt. Ltd, Pondicherry, India
Printed and bound in Great Britain by Biddles Ltd, King's Lynn, Norfolk

Contents

Preface

This volume is intended for students of engineering on courses or programmes of study to graduate level.

The sequence of subject development in this edition commences with definitions and concepts and goes on to cover incompressible flow, low speed aerofoil and wing theory, compressible flow, high speed wing theory, viscous flow, boundary layers, transition and turbulence, wing design, propellers and propulsion.

Accordingly the work deals first with the units, dimensions and properties of the physical quantities used in aerodynamics then introduces common aeronautical definitions before explaining the aerodynamic forces involved and the basics of aerofoil characteristics. The fundamental fluid dynamics required for the development of aerodynamics and the analysis of flows within and around solid boundaries for air at subsonic speeds is explored in depth in the next two chapters, which continue with those immediately following to use these and other methods to develop aerofoil and wing theories for the estimation of aerodynamic characteristics in these regimes. Attention is then turned to the aerodynamics of high speed air flows. The laws governing the behaviour of the physical properties of air are applied to the transonic and supersonic regimes and the aerodynamics of the abrupt changes in the flow characteristics at these speeds are explained. The exploitation of these and other theories is then used to explain the significant effects on wings in transonic and supersonic flight respectively, and to develop appropriate aerodynamic characteristics. Viscosity is a key physical quantity of air and its significance in aerodynamic situations is next considered in depth. The useful concept of the boundary layer and the development of properties of various flows when adjacent to solid boundaries, build to a body of reliable methods for estimating the fluid forces due to viscosity and notably, in aerodynamics, of skin friction and profile drag. Finally the two chapters on wing design and flow control, and propellers and propulsion respectively, bring together disparate aspects of the previous chapters as appropriate, to some practical and individual applications of aerodynamics.

It is recognized that aerodynamic design makes extensive use of computational aids. This is reflected in part in this volume by the introduction, where appropriate, of descriptions and discussions of relevant computational techniques. However, no comprehensive cover of computational methods is intended, and experience in computational techniques is not required for a complete understanding of the aerodynamics in this book.

Equally, although experimental data have been quoted no attempt has been made to describe techniques or apparatus, as we feel that experimental aerodynamics demands its own considered and separate treatment.

We are indebted to the Senates of the Universities and other institutions referred to within for kindly giving permission for the use of past examination questions. Any answers and worked examples are the responsibility of the authors, and the authorities referred to are in no way committed to approval of such answers and examples.

This preface would be incomplete without reference to the many authors of classical and popular texts and of learned papers, whose works have formed the framework and guided the acquisitions of our own knowledge. A selection of these is given in the bibliography if not referred to in the text and we apologize if due recognition of a source has been inadvertently omitted in any particular in this volume.

ELH/PWC
2002

1

Basic concepts and definitions

Preamble

The study of aerodynamics, as is the case with that of all physical sciences and technologies, requires the common acceptance of a number of basic definitions including an unambiguous nomenclature and an understanding of the relevant physical properties, the related mechanics and the appropriate mathematics.

Of course, many of these are common to other disciplines and it is the purpose of this chapter to identify and explain those that are basic and pertinent to aerodynamics and which are to be used in the remainder of the volume.

The units and dimensions of all physical properties and the relevant properties of fluids are recalled, and after a review of the aeronautical definitions of wing and aerofoil geometry, the remainder of the chapter introduces aerodynamic force.

The origins of aerodynamic force and how it is manifest on wings and other aeronautical bodies and the theories that permit its evaluation and design are to be found in the remainder of the volume, but in this chapter the lift, drag, side-wind components and associated moments of aerodynamic force are conventionally identified, the application of dimensional theory establishing their coefficient form. The significance of the pressure distribution around an aerodynamic body and the estimation of lift, drag and pitching moment on it in flight, completes this chapter of basic concepts and definitions.

1.1   Units and dimensions

A study in any science must include measurement and calculation, which presupposes an agreed system of units in terms of which quantities can be measured and expressed. There is one system that has come to be accepted for most branches of science and engineering, and for aerodynamics in particular, in most parts of the world. That system is the Système International d'Unités, commonly abbreviated to SI units, and it is used throughout this book, except in a very few places as specially noted.

It is essential to distinguish between the terms 'dimension' and 'unit'. For example, the dimension 'length' expresses the *qualitative* concept of linear displacement, or distance between two points, as an abstract idea, without reference to actual quantitative measurement. The term 'unit' indicates a specified amount of the quantity. Thus a metre is a unit of length, being an actual 'amount' of linear displacement, and so also is a mile. The metre and mile are different *units*, since each contains a different *amount* of length, but both describe length and therefore are identical *dimensions*.*

Expressing this in symbolic form:

x metres $= [L]$ (a quantity of x metres has the dimension of length)
x miles $= [L]$ (a quantity of x miles has the dimension of length)
x metres $\neq x$ miles (x miles and x metres are unequal quantities of length)
$[x$ metres$] = [x$ miles$]$ (the dimension of x metres is the same as the dimension of x miles).

1.1.1 Fundamental dimensions and units

There are four fundamental dimensions in terms of which the dimensions of all other physical quantities may be expressed. They are mass $[M]$, length $[L]$, time $[T]$ and temperature $[\theta]$.[†] A consistent set of units is formed by specifying a unit of particular value for each of these dimensions. In aeronautical engineering the accepted units are respectively the kilogram, the metre, the second and the Kelvin or degree Celsius (see below). These are identical with the units of the same names in common use, and are defined by international agreement.

It is convenient and conventional to represent the names of these units by abbreviations:

kg for kilogram
m for metre
s for second
°C for degree Celsius
K for Kelvin

The degree Celsius is one one-hundredth part of the temperature rise involved when pure water at freezing temperature is heated to boiling temperature at standard pressure. In the Celsius scale, pure water at standard pressure freezes at 0 °C and boils at 100 °C.

The unit Kelvin (K) is identical in size with the degree Celsius (°C), but the Kelvin scale of temperature is measured from the absolute zero of temperature, which is approximately −273 °C. Thus a temperature in K is equal to the temperature in °C plus 273 (approximately).

1.1.2 Fractions and multiples

Sometimes, the fundamental units defined above are inconveniently large or inconveniently small for a particular case. In such cases, the quantity can be expressed in terms of some fraction or multiple of the fundamental unit. Such multiples and fractions are denoted by appending a prefix to the symbol denoting the fundamental unit. The prefixes most used in aerodynamics are:

---

* Quite often 'dimension' appears in the form 'a dimension of 8 metres' and thus means a specified length. This meaning of the word is thus closely related to the engineer's 'unit', and implies linear extension only. Another common example of its use is in 'three-dimensional geometry', implying three linear extensions in different directions. References in later chapters to two-dimensional flow, for example, illustrate this. The meaning above must not be confused with either of these uses.

† Some authorities express temperature in terms of length and time. This introduces complications that are briefly considered in Section 1.2.8.

M (mega) – denoting one million
k (kilo) – denoting one thousand
m (milli) – denoting one one-thousandth part
μ (micro) – denoting one-millionth part

Thus

1 MW = 1 000 000 W
1 mm = 0.001 m
1 μm = 0.001 mm

A prefix attached to a unit makes a new unit. For example,

$$1\,\text{mm}^2 = 1\,(\text{mm})^2 = 10^{-6}\,\text{m}^2,\ \text{not}\ 10^{-3}\,\text{m}^2$$

For some purposes, the hour or the minute can be used as the unit of time.

1.1.3   Units of other physical quantities

Having defined the four fundamental dimensions and their units, it is possible to establish units of all other physical quantities (see Table 1.1). Speed, for example, is defined as the distance travelled in unit time. It therefore has the dimension LT^{-1} and is measured in metres per second ($m\,s^{-1}$). It is sometimes desirable and permissible to use kilometres per hour or knots (nautical miles per hour, see Appendix 4) as units of speed, and care must then be exercised to avoid errors of inconsistency.

To find the dimensions and units of more complex quantities, appeal is made to the principle of *dimensional homogeneity*. This means simply that, in any valid physical equation, the dimensions of both sides must be the same. Thus if, for example, (mass)^n appears on the left-hand side of the equation, (mass)^n must also appear on the right-hand side, and similarly this applies to length, time and temperature.

Thus, to find the dimensions of force, use is made of Newton's second law of motion

$$\text{Force} = \text{mass} \times \text{acceleration}$$

while acceleration is speed $\div$ time.

Expressed dimensionally, this is

$$\text{Force} = [M] \times \left[\frac{L}{T} \div T\right] = [MLT^{-2}]$$

Writing in the appropriate units, it is seen that a force is measured in units of $kg\,m\,s^{-2}$. Since, however, the unit of force is given the name Newton (abbreviated usually to N), it follows that

$$1\,N = 1\,kg\,m\,s^{-2}$$

It should be noted that there could be confusion between the use of m for milli and its use for metre. This is avoided by use of spacing. Thus ms denotes millisecond while m s denotes the product of metre and second.

The concept of the dimension forms the basis of dimensional analysis. This is used to develop important and fundamental physical laws. Its treatment is postponed to Section 1.4 later in the current chapter.

Table 1.1 Units and dimensions

Quantity	Dimension	Unit (name and abbreviation)
Length	L	Metre (m)
Mass	M	Kilogram (kg)
Time	T	Second (s)
Temperature	θ	Degree Celsius (°C), Kelvin (K)
Area	L^2	Square metre (m^2)
Volume	L^3	Cubic metre (m^3)
Speed	LT^{-1}	Metres per second ($m\,s^{-1}$)
Acceleration	LT^{-2}	Metres per second per second ($m\,s^{-2}$)
Angle	1	Radian or degree (°)
		(The radian is expressed as a ratio and is therefore dimensionless)
Angular velocity	T^{-1}	Radians per second (s^{-1})
Angular acceleration	T^{-2}	Radians per second per second (s^{-2})
Frequency	T^{-1}	Cycles per second, Hertz (s^{-1} Hz)
Density	ML^{-3}	Kilograms per cubic metre ($kg\,m^{-3}$)
Force	MLT^{-2}	Newton (N)
Stress	$ML^{-1}T^{-2}$	Newtons per square metre *or* Pascal ($N\,m^{-2}$ *or* Pa)
Strain	1	None (expressed as %)
Pressure	$ML^{-1}T^{-2}$	Newtons per square metre *or* Pascal ($N\,m^{-2}$ *or* Pa)
Energy work	ML^2T^{-2}	Joule (J)
Power	ML^2T^{-3}	Watt (W)
Moment	ML^2T^{-2}	Newton metre (Nm)
Absolute viscosity	$ML^{-1}T^{-1}$	Kilogram per metre second *or* Poiseuille ($kg\,m^{-1}\,s^{-1}$ *or* PI)
Kinematic viscosity	L^2T^{-1}	Metre squared per second ($m^2\,s^{-1}$)
Bulk elasticity	$ML^{-1}T^{-2}$	Newtons per square metre *or* Pascal ($N\,m^{-2}$ *or* Pa)

1.1.4   Imperial units[‡]

Until about 1968, aeronautical engineers in some parts of the world, the United Kingdom in particular, used a set of units based on the Imperial set of units. In this system, the fundamental units were:

mass – the slug
length – the foot
time – the second
temperature – the degree Centigrade or Kelvin.

1.2   Relevant properties

1.2.1   Forms of matter

Matter may exist in three principal forms, solid, liquid or gas, corresponding in that order to decreasing rigidity of the bonds between the molecules of which the matter is composed. A special form of a gas, known as a *plasma*, has properties different from

---

[‡] Since many valuable texts and papers exist using those units, this book contains, as Appendix 4, a table of factors for converting from the Imperial system to the SI system.

those of a normal gas and, although belonging to the third group, can be regarded justifiably as a separate, distinct form of matter.

In a solid the intermolecular bonds are very rigid, maintaining the molecules in what is virtually a fixed spatial relationship. Thus a solid has a fixed volume and shape. This is seen particularly clearly in crystals, in which the molecules or atoms are arranged in a definite, uniform pattern, giving all crystals of that substance the same geometric shape.

A liquid has weaker bonds between the molecules. The distances between the molecules are fairly rigidly controlled but the arrangement in space is free. A liquid, therefore, has a closely defined volume but no definite shape, and may accommodate itself to the shape of its container within the limits imposed by its volume.

A gas has very weak bonding between the molecules and therefore has neither a definite shape nor a definite volume, but will always fill the whole of the vessel containing it.

A plasma is a special form of gas in which the atoms are ionized, i.e. they have lost one or more electrons and therefore have a net positive electrical charge. The electrons that have been stripped from the atoms are wandering free within the gas and have a negative electrical charge. If the numbers of ionized atoms and free electrons are such that the total positive and negative charges are approximately equal, so that the gas as a whole has little or no charge, it is termed a plasma. In astronautics the plasma is usually met as a jet of ionized gas produced by passing a stream of normal gas through an electric arc. It is of particular interest for the re-entry of rockets, satellites and space vehicles into the atmosphere.

1.2.2 Fluids

The basic feature of a fluid is that it can flow, and this is the essence of any definition of it. This feature, however, applies to substances that are not true fluids, e.g. a fine powder piled on a sloping surface will also flow. Fine powder, such as flour, poured in a column on to a flat surface will form a roughly conical pile, with a large angle of repose, whereas water, which is a true fluid, poured on to a fully wetted surface will spread uniformly over the whole surface. Equally, a powder may be heaped in a spoon or bowl, whereas a liquid will always form a level surface. A definition of a fluid must allow for these facts. Thus a fluid may be defined as 'matter capable of flowing, and either finding its own level (if a liquid), or filling the whole of its container (if a gas)'.

Experiment shows that an extremely fine powder, in which the particles are not much larger than molecular size, will also find its own level and may thus come under the common definition of a liquid. Also a phenomenon well known in the transport of sands, gravels, etc. is that they will find their own level if they are agitated by vibration, or the passage of air jets through the particles. These, however, are special cases and do not detract from the authority of the definition of a fluid as a substance that flows or (tautologically) that possesses fluidity.

1.2.3 Pressure

At any point in a fluid, whether liquid or gas, there is a pressure. If a body is placed in a fluid, its surface is bombarded by a large number of molecules moving at random. Under normal conditions the collisions on a small area of surface are so frequent that they cannot be distinguished as individual impacts. They appear as a steady force on the area. The *intensity* of this 'molecular bombardment' force is the *static pressure*.

Very frequently the static pressure is referred to simply as pressure. The term *static* is rather misleading. Note that its use does not imply the fluid is at rest.

For large bodies moving or at rest in the fluid, e.g. air, the pressure is not uniform over the surface and this gives rise to *aerodynamic force* or *aerostatic force* respectively.

Since a pressure is force per unit area, it has the dimensions

$$[\text{Force}] \div [\text{area}] = [\text{MLT}^{-2}] \div [\text{L}^2] = [\text{ML}^{-1}\text{T}^{-2}]$$

and is expressed in the units of Newtons per square metre or Pascals (N m^{-2} or Pa).

Pressure in fluid at rest

Consider a small cubic element containing fluid at rest in a larger bulk of fluid also at rest. The faces of the cube, assumed conceptually to be made of some thin flexible material, are subject to continual bombardment by the molecules of the fluid, and thus experience a *force*. The force on any face may be resolved into two components, one acting perpendicular to the face and the other along it, i.e. tangential to it. Consider for the moment the tangential components only; there are three significantly different arrangements possible (Fig. 1.1). The system (a) would cause the element to rotate and thus the fluid would not be at rest. System (b) would cause the element to move (upwards and to the right for the case shown) and once more, the fluid would not be at rest. Since a fluid cannot resist shear stress, but only rate of change of shear strain (Sections 1.2.6 and 2.7.2) the system (c) would cause the element to distort, the degree of distortion increasing with time, and the fluid would not remain at rest.

The conclusion is that a fluid at rest cannot sustain tangential stresses, or conversely, that in a fluid at rest the pressure on a surface must act in the direction perpendicular to that surface.

Pascal's law

Consider the right prism of length δz into the paper and cross-section ABC, the angle ABC being a right-angle (Fig. 1.2). The prism is constructed of material of the same density as a bulk of fluid in which the prism floats at rest with the face BC horizontal.

Pressures p_1, p_2 and p_3 act on the faces shown and, as proved above, these pressures act in the direction perpendicular to the respective face. Other pressures act on the end faces of the prism but are ignored in the present problem. In addition to these pressures, the weight W of the prism acts vertically downwards. Consider the forces acting on the wedge which is in equilibrium and at rest.

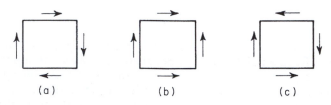

(a)   (b)   (c)

Fig. 1.1 Fictitious systems of tangential forces in static fluid

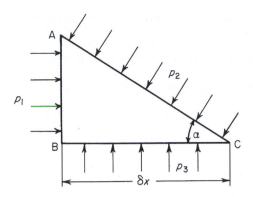

Fig. 1.2 The prism for Pascal's Law

Resolving forces horizontally,

$$p_1(\delta x \tan \alpha)\delta z - p_2(\delta x \sec \alpha) \, \delta z \sin \alpha = 0$$

Dividing by $\delta x \, \delta z \tan \alpha$, this becomes

$$p_1 - p_2 = 0$$

i.e.

$$p_1 = p_2 \tag{1.1}$$

Resolving forces vertically,

$$p_3 \delta x \, \delta z - p_2(\delta x \sec \alpha) \, \delta z \cos \alpha - W = 0 \tag{1.2}$$

Now

$$W = \rho g(\delta x)^2 \tan \alpha \, \delta z / 2$$

therefore, substituting this in Eqn (1.2) and dividing by $\delta x \, \delta z$,

$$p_3 - p_2 - \frac{1}{2}\rho g \, \tan \alpha \, \delta z = 0$$

If now the prism is imagined to become infinitely small, so that $\delta x \to 0$ and $\delta z \to 0$, then the third term tends to zero leaving

$$p_3 - p_2 = 0$$

Thus, finally,

$$p_1 = p_2 = p_3 \tag{1.3}$$

Having become infinitely small, the prism is in effect a point and thus the above analysis shows that, at a point, the three pressures considered are equal. In addition, the angle α is purely arbitrary and can take any value, while the whole prism could be rotated through a complete circle about a vertical axis without affecting the result. Consequently, it may be concluded that the pressure acting at a point in a fluid at rest is the same in all directions.

1.2.4   Temperature

In any form of matter the molecules are in motion relative to each other. In gases the motion is random movement of appreciable amplitude ranging from about 76×10^{-9} metres under normal conditions to some tens of millimetres at very low pressures. The distance of free movement of a molecule of gas is the distance it can travel before colliding with another molecule or the walls of the container. The mean value of this distance for all the molecules in a gas is called the length of mean molecular free path.

By virtue of this motion the molecules possess kinetic energy, and this energy is sensed as the *temperature* of the solid, liquid or gas. In the case of a gas in motion it is called the static temperature or more usually just the temperature. Temperature has the dimension $[\theta]$ and the units K or °C (Section 1.1). In practically all calculations in aerodynamics, temperature is measured in K, i.e. from absolute zero.

1.2.5   Density

The density of a material is a measure of the amount of the material contained in a given volume. In a fluid the density may vary from point to point. Consider the fluid contained within a small spherical region of volume δV centred at some point in the fluid, and let the mass of fluid within this spherical region be δm. Then the density of the fluid at the point on which the sphere is centred is defined by

$$\text{Density } \rho = \lim_{\delta v \to 0} \frac{\delta m}{\delta V} \qquad (1.4)$$

The dimensions of density are thus ML^{-3}, and it is measured in units of kilogram per cubic metre (kg m^{-3}). At standard temperature and pressure ($288\,\text{K}$, $101\,325\,\text{N m}^{-2}$) the density of dry air is $1.2256\,\text{kg m}^{-3}$.

Difficulties arise in applying the above definition rigorously to a real fluid composed of discrete molecules, since the sphere, when taken to the limit, either will or will not contain part of a molecule. If it does contain a molecule the value obtained for the density will be fictitiously high. If it does not contain a molecule the resultant value for the density will be zero. This difficulty can be avoided in two ways over the range of temperatures and pressures normally encountered in aerodynamics:

(i) The molecular nature of a gas may for many purposes be ignored, and the assumption made that the fluid is a continuum, i.e. does not consist of discrete particles.

(ii) The decrease in size of the imaginary sphere may be supposed to be carried to a limiting minimum size. This limiting size is such that, although the sphere is small compared with the dimensions of any physical body, e.g. an aeroplane, placed in the fluid, it is large compared with the fluid molecules and, therefore, contains a reasonable number of whole molecules.

1.2.6   Viscosity

Viscosity is regarded as the tendency of a fluid to resist sliding between layers or, more rigorously, a rate of change of shear strain. There is very little resistance to the movement of a knife-blade edge-on through air, but to produce the same motion through thick oil needs much more effort. This is because the viscosity of oil is high compared with that of air.

Dynamic viscosity

The dynamic (more properly called the coefficient of dynamic, or absolute, viscosity) viscosity is a direct measure of the viscosity of a fluid. Consider two parallel flat plates placed a distance h apart, the space between them being filled with fluid. One plate is held fixed and the other is moved in its own plane at a speed V (see Fig. 1.3). The fluid immediately adjacent to each plate will move with that plate, i.e. there is no slip. Thus the fluid in contact with the lower plate will be at rest, while that in contact with the upper plate will be moving with speed V. Between the plates the speed of the fluid will vary linearly as shown in Fig. 1.3, in the absence of other influences. As a direct result of viscosity a force F has to be applied to each plate to maintain the motion, the fluid tending to retard the moving plate and to drag the fixed plate to the right. If the area of fluid in contact with each plate is A, the shear stress is F/A. The rate of shear strain caused by the upper plate sliding over the lower is V/h.

These quantities are connected by Newton's equation, which serves to define the dynamic viscosity μ. This equation is

$$\frac{F}{A} = \mu\left(\frac{V}{h}\right) \tag{1.5}$$

Hence

$$[\mathrm{ML^{-1}T^{-2}}] = [\mu][\mathrm{LT^{-1}L^{-1}}] = [\mu][\mathrm{T^{-1}}]$$

Thus

$$[\mu] = [\mathrm{ML^{-1}T^{-1}}]$$

and the units of μ are therefore $\mathrm{kg\,m^{-1}\,s^{-1}}$; in the SI system the name Poiseuille (Pl) has been given to this combination of fundamental units. At $0\,°\mathrm{C}$ ($273\,\mathrm{K}$) the dynamic viscosity for dry air is $1.714 \times 10^{-5}\,\mathrm{kg\,m^{-1}\,s^{-1}}$.

The relationship of Eqn (1.5) with μ constant does not apply for all fluids. For an important class of fluids, which includes blood, some oils and some paints, μ is not constant but is a function of V/h, i.e. the rate at which the fluid is shearing.

Kinematic viscosity

The kinematic viscosity (or, more properly, coefficient of kinematic viscosity) is a convenient form in which the viscosity of a fluid may be expressed. It is formed

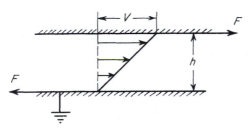

Fig. 1.3

by combining the density ρ and the dynamic viscosity μ according to the equation

$$v = \frac{\mu}{\rho}$$

and has the dimensions L^2T^{-1} and the units $m^2\,s^{-1}$.

It may be regarded as a measure of the relative magnitudes of viscosity and inertia of the fluid and has the practical advantage, in calculations, of replacing two values representing μ and ρ by a single value.

1.2.7   Speed of sound and bulk elasticity

The bulk elasticity is a measure of how much a fluid (or solid) will be compressed by the application of external pressure. If a certain small volume, V, of fluid is subjected to a rise in pressure, δp, this reduces the volume by an amount $-\delta V$, i.e. it produces a volumetric strain of $-\delta V/V$. Accordingly, the bulk elasticity is defined as

$$K = -\frac{\delta p}{\delta V/V} = -\frac{1}{V}\frac{dp}{dV} \qquad (1.6a)$$

The volumetric strain is the ratio of two volumes and evidently dimensionless, so the dimensions of K are the same as those for pressure, namely $ML^{-1}T^{-2}$. The SI units are Nm^{-2} (or Pa).

The propagation of sound waves involves alternating compression and expansion of the medium. Accordingly, the bulk elasticity is closely related to the speed of sound, a, as follows:

$$a = \sqrt{\frac{K}{\rho}} \qquad (1.6b)$$

Let the mass of the small volume of fluid be M, then by definition the density, $\rho = M/V$. By differentiating this definition keeping M constant, we obtain

$$d\rho = -\frac{M}{V^2}dV = -\rho\frac{dV}{V}$$

Therefore, combining this with Eqns (1.6a,b), it can be seen that

$$a = \sqrt{\frac{dp}{d\rho}} \qquad (1.6c)$$

The propagation of sound in a perfect gas is regarded as an isentropic process. Accordingly, (see the passage below on *Entropy*) the pressure and density are related by Eqn (1.24), so that for a perfect gas

$$a = \sqrt{\frac{\gamma p}{\rho}} \qquad (1.6d)$$

where γ is the ratio of the specific heats. Equation (1.6d) is the formula usually used to determine the speed of sound in gases for applications in aerodynamics.

1.2.8 Thermodynamic properties

Heat, like work, is a form of energy transfer. Consequently, it has the same dimensions as energy, i.e. ML^2T^{-2}, and is measured in units of Joules (J).

Specific heat

The specific heat of a material is the amount of heat necessary to raise the temperature of unit mass of the material by one degree. Thus it has the dimensions $L^2T^{-2}\theta^{-1}$ and is measured in units of $J\,kg^{-1}\,{}°C^{-1}$ or $J\,kg^{-1}\,K^{-1}$.

With a gas there are two distinct ways in which the heating operation may be performed: at constant volume and at constant pressure, and in turn these define important thermodynamic properties.

Specific heat at constant volume If unit mass of the gas is enclosed in a cylinder sealed by a piston, and the piston is locked in position, the volume of the gas cannot change, and any heat added is used solely to raise the temperature of the gas, i.e. the head added goes to increase the *internal energy* of the gas. It is assumed that the cylinder and piston do not receive any of the heat. The specific heat of the gas under these conditions is the specific heat at constant volume, c_V. For dry air at normal aerodynamic temperatures, $c_V = 718\,J\,kg^{-1}\,K^{-1}$.

Internal energy (E) is a measure of the kinetic energy of the molecules comprising the gas. Thus

internal energy per unit mass $E = c_V T$

or, more generally,

$$c_V = \left[\frac{\partial E}{\partial T}\right]_\rho \tag{1.7}$$

Specific heat at constant pressure Assume that the piston referred to above is now freed and acted on by a constant force. The pressure of the gas is that necessary to resist the force and is therefore constant. The application of heat to the gas causes its temperature to rise, which leads to an increase in the volume of the gas, in order to maintain the constant pressure. Thus the gas does mechanical work against the force. It is therefore necessary to supply the heat required to increase the temperature of the gas (as in the case at constant volume) and in addition the amount of heat equivalent to the mechanical work done against the force. This total amount of heat is called the specific heat at constant pressure, c_p, and is defined as that amount of heat required to raise the temperature of unit mass of the gas by one degree, the pressure of the gas being kept constant while heating. Therefore, c_p is always greater than c_V. For dry air at normal aerodynamic temperatures, $c_p = 1005\,J\,kg^{-1}\,K^{-1}$.

Now the sum of the internal energy and pressure energy is known as the *enthalpy* (h per unit mass) (see below). Thus

$$h = c_p T$$

or, more generally,

$$c_p = \left[\frac{\partial h}{\partial T}\right]_p \tag{1.8}$$

The ratio of specific heats

This is a property important in high-speed flows and is defined by the equation

$$\gamma = \frac{c_p}{c_V} \tag{1.9}$$

(The value of γ for air depends on the temperature, but for much of practical aerodynamics it may be regarded as constant at about 1.403. This value in turn is often approximated to $\gamma = 1.4$, which is, in fact, the theoretical value for an ideal diatomic gas.)

Enthalpy

The enthalpy h of a unit mass of gas is the sum of the internal energy E and pressure energy $p \times 1/\rho$. Thus,

$$h = E + p/\rho \tag{1.10}$$

But, from the definition of specific heat at constant volume, Eqn (1.7), Eqn (1.10) becomes

$$h = c_V T + p/\rho$$

Again from the definition, Eqn (1.8), Eqn (1.10) gives

$$c_p T = c_V T + p/\rho \tag{1.11}$$

Now the pressure, density and temperature are related in the equation of state, which for perfect gases takes the form

$$p/(\rho T) = \text{constant} = R \tag{1.12}$$

Substituting for p/ρ in Eqn (1.11) yields the relationship

$$c_p - c_V = R \tag{1.13}$$

The gas constant, R, is thus the amount of mechanical work that is obtained by heating unit mass of a gas through unit temperature rise at constant pressure. It follows that R is measured in units of $\text{J kg}^{-1}\text{K}^{-1}$ or $\text{J kg}^{-1}\,^\circ\text{C}^{-1}$. For air over the range of temperatures and pressures normally encountered in aerodynamics, R has the value $287.26\,\text{J kg}^{-1}\text{K}^{-1}$.

Introducing the ratio of specific heats (Eqn (1.9)) the following expressions are obtained:

$$c_p = \frac{\gamma}{\gamma - 1}R \quad \text{and} \quad c_V = \frac{R}{\gamma - 1} \tag{1.14}$$

Replacing $c_V T$ by $[1/(\gamma - 1)]p/\rho$ in Eqn (1.11) readily gives the enthalpy as

$$c_p T = \frac{\gamma}{\gamma - 1}\frac{p}{\rho} \tag{1.15}$$

It is often convenient to link the enthalpy or total heat above to the other energy of motion, the kinetic energy $\bar{K}$; that for unit mass of gas moving with mean velocity V is

$$\bar{K} = \frac{V^2}{2} \tag{1.16}$$

Thus the total energy flux in the absence of external, tangential surface forces and heat conduction becomes

$$\frac{V^2}{2} + c_p T = c_p T_0 = \text{constant} \tag{1.17}$$

where, with c_p invariant, T_0 is the absolute temperature when the gas is at rest. The quantity $c_p T_0$ is referred to as the *total* or *stagnation enthalpy*. This quantity is an important parameter of the equation of the conservation of energy.

Applying the *first law of thermodynamics* to the flow of non-heat-conducting inviscid fluids gives

$$\frac{\mathrm{d}(c_V T)}{\mathrm{d}t} + p\frac{\mathrm{d}(1/\rho)}{\mathrm{d}t} = 0 \tag{1.18}$$

Further, if the flow is unidirectional and $c_V T = E$, Eqn (1.18) becomes, on cancelling dt,

$$\mathrm{d}E + p\mathrm{d}\left(\frac{1}{\rho}\right) = 0 \tag{1.19}$$

but differentiating Eqn (1.10) gives

$$\mathrm{d}h = \mathrm{d}E + p\mathrm{d}\left(\frac{1}{\rho}\right) + \frac{1}{\rho}\mathrm{d}p \tag{1.20}$$

Combining Eqns (1.19) and (1.20)

$$\mathrm{d}h = \frac{1}{\rho}\mathrm{d}p \tag{1.21}$$

but

$$\mathrm{d}h = c_p\mathrm{d}T = \frac{c_p}{R}\mathrm{d}\left(\frac{p}{\rho}\right) = \frac{\gamma}{\gamma - 1}\left[\frac{1}{\rho}\mathrm{d}p + p\mathrm{d}\left(\frac{1}{\rho}\right)\right] \tag{1.22}$$

which, together with Eqn (1.21), gives the identity

$$\frac{\mathrm{d}p}{p} + \gamma\rho\mathrm{d}\left(\frac{1}{\rho}\right) = 0 \tag{1.23}$$

Integrating gives

$$\ln p + \gamma\ln\left(\frac{1}{\rho}\right) = \text{constant}$$

or

$$p = k\rho^\gamma \tag{1.24}$$

which is the isentropic relationship between pressure and density.

It should be remembered that this result is obtained from the equation of state for a perfect gas and the equation of conservation of energy of the flow of a non-heat-conducting inviscid fluid. Such a flow behaves isentropically and, notwithstanding the apparently restrictive nature of the assumptions made above, it can be used as a model for a great many aerodynamic applications.

Entropy

Entropy is a function of state that follows from, and indicates the working of, the *second law of thermodynamics*, that is concerned with the *direction* of any process involving heat and energy. Entropy is a function the positive increase of which during an adiabatic process indicates the consequences of the second law, i.e. a *reduction* in entropy under these circumstances contravenes the second law. Zero entropy change indicates an ideal or completely reversible process.

By definition, specific entropy (S)* (Joules per kilogram per Kelvin) is given by the integral

$$S = \int \frac{\mathrm{d}Q}{T} \tag{1.25}$$

for any reversible process, the integration extending from some datum condition; but, as seen above, it is the change in entropy that is important, i.e.

$$\mathrm{d}S = \frac{\mathrm{d}Q}{T} \tag{1.26}$$

In this and the previous equation $\mathrm{d}Q$ is a heat transfer to a unit mass of gas from an external source. This addition will go to changing the internal energy and will do work.

Thus, for a reversible process,

$$\mathrm{d}Q = \mathrm{d}E + p\mathrm{d}\left(\frac{1}{\rho}\right)$$

$$\mathrm{d}S = \frac{\mathrm{d}Q}{T} = \frac{c_V \mathrm{d}T}{T} + \frac{p\mathrm{d}(1/\rho)}{T} \tag{1.27}$$

but $p/T = R_\rho$, therefore

$$\mathrm{d}S = \frac{c_V \mathrm{d}T}{T} + \frac{R\mathrm{d}(1/\rho)}{1/\rho} \tag{1.28}$$

Integrating Eqn (1.28) from datum conditions to conditions given by suffix 1,

$$S_1 = c_V \ln \frac{T_1}{T_\mathrm{D}} + R \ln \frac{\rho_\mathrm{D}}{\rho_1}$$

Likewise,

$$S_2 = c_V \ln \frac{T_2}{T_\mathrm{D}} + R \ln \frac{\rho_\mathrm{D}}{\rho_2}$$

---

* Note that in this passage the unconventional symbol S is used for specific entropy to avoid confusion with the length symbols.

and the entropy change from conditions 1 to 2 is given by

$$\Delta S = S_2 - S_1 = c_V \ln \frac{T_2}{T_1} + R \ln \frac{\rho_1}{\rho_2} \tag{1.29}$$

With the use of Eqn (1.14) this is more usually rearranged to be

$$\frac{\Delta S}{c_V} = \ln \frac{T_2}{T_1} + (\gamma - 1) \ln \frac{\rho_1}{\rho_2} \tag{1.30}$$

or in the exponential form

$$e^{\Delta S/c_V} = \frac{T_2}{T_1} \left(\frac{\rho_1}{\rho_2} \right)^{\gamma - 1} \tag{1.31}$$

Alternatively, for example, by using the equation of state,

$$e^{\Delta S/c_V} = \left(\frac{T_2}{T_1} \right)^{\gamma} \left(\frac{p_1}{p_2} \right)^{\gamma - 1} \tag{1.32}$$

These latter expressions find use in particular problems.

1.3 Aeronautical definitions

1.3.1 Wing geometry

The *planform* of a wing is the shape of the wing seen on a plan view of the aircraft. Figure 1.4 illustrates this and includes the names of symbols of the various parameters of the planform geometry. Note that the root ends of the leading and trailing edges have been connected across the fuselage by straight lines. An alternative to this convention is that the leading and trailing edges, if straight, are produced to the aircraft centre-line.

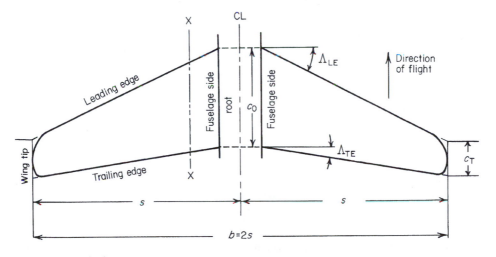

Fig. 1.4 Wing planform geometry

Wing span

The wing span is the dimension b, the distance between the extreme wingtips. The distance, s, from each tip to the centre-line, is the wing semi-span.

Chords

The two lengths c_T and c_0 are the tip and root chords respectively; with the alternative convention, the root chord is the distance between the intersections with the fuselage centre-line of the leading and trailing edges produced. The ratio c_T/c_0 is the taper ratio λ. Sometimes the reciprocal of this, namely c_0/c_T, is taken as the taper ratio. For most wings $c_T/c_0 < 1$.

Wing area

The plan-area of the wing including the continuation within the fuselage is the gross wing area, S_G. The unqualified term wing area S is usually intended to mean this gross wing area. The plan-area of the exposed wing, i.e. excluding the continuation within the fuselage, is the net wing area, S_N.

Mean chords

A useful parameter, the standard mean chord or the geometric mean chord, is denoted by $\bar{c}$, defined by $\bar{c} = S_G/b$ or S_N/b. It should be stated whether S_G or S_N is used. This definition may also be written as

$$\bar{c} = \frac{\int_{-s}^{+s} c \, dy}{\int_{-s}^{+s} dy}$$

where y is distance measured from the centre-line towards the starboard (right-hand to the pilot) tip. This standard mean chord is often abbreviated to SMC.

Another mean chord is the aerodynamic mean chord (AMC), denoted by $\bar{c}_A$ or $\bar{\bar{c}}$, and is defined by

$$\bar{c}_A = \frac{\int_{-s}^{+s} c^2 \, dy}{\int_{-s}^{+s} c \, dy}$$

Aspect ratio

The aspect ratio is a measure of the narrowness of the wing planform. It is denoted by A, or sometimes by (AR), and is given by

$$A = \frac{\text{span}}{\text{SMC}} = \frac{b}{\bar{c}}$$

If both top and bottom of this expression are multiplied by the wing span, b, it becomes:

$$A = \frac{b^2}{b\bar{c}} = \frac{(\text{span})^2}{\text{area}}$$

a form which is often more convenient.

Sweep-back

The sweep-back angle of a wing is the angle between a line drawn along the span at a constant fraction of the chord from the leading edge, and a line perpendicular to the centre-line. It is usually denoted by either Λ or ϕ. Sweep-back is commonly measured on the leading edge (Λ_{LE} or ϕ_{LE}), on the quarter-chord line, i.e. the line $\frac{1}{4}$ of the chord behind the leading edge ($\Lambda_{1/4}$ or $\phi_{1/4}$), or on the trailing edge (Λ_{TE} or ϕ_{TE}).

Dihedral angle

If an aeroplane is looked at from directly ahead, it is seen that the wings are not, in general, in a single plane (in the geometric sense), but are instead inclined to each other at a small angle. Imagine lines drawn on the wings along the locus of the intersections between the chord lines and the section noses, as in Fig. 1.5. Then the angle 2Γ is the dihedral angle of the wings. If the wings are inclined upwards, they are said to have *dihedral*, if inclined downwards they have *anhedral*.

Incidence, twist, wash-out and wash-in

When an aeroplane is in flight the chord lines of the various wing sections are not normally parallel to the direction of flight. The angle between the chord line of a given aerofoil section and the direction of flight or of the undisturbed stream is called the geometric angle of incidence, α.

Carrying this concept of incidence to the twist of a wing, it may be said that, if the geometric angles of incidence of all sections are not the same, the wing is twisted. If the incidence increases towards the tip, the wing has *wash-in*, whereas if the incidence decreases towards the tip the wing has *wash-out*.

1.3.2 Aerofoil geometry

If a horizontal wing is cut by a vertical plane parallel to the centre-line, such as X–X in Fig. 1.4, the shape of the resulting section is usually of a type shown in Fig. 1.6c.

Fig. 1.5 Illustrating the dihedral angle

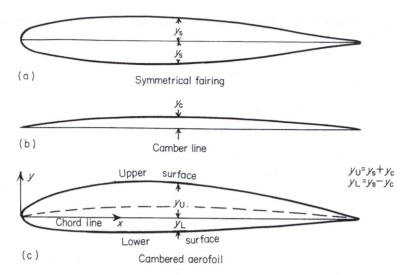

Fig. 1.6 Wing section geometry

This is an aerofoil section. For subsonic use, the aerofoil section has a rounded leading edge. The depth increases smoothly to a maximum that usually occurs between $\frac{1}{4}$ and $\frac{1}{2}$ way along the profile, and thereafter tapers off towards the rear of the section.

If the leading edge is rounded it has a definite radius of curvature. It is therefore possible to draw a circle of this radius that coincides with a very short arc of the section where the curvature is greatest. The trailing edge may be sharp or it, too, may have a radius of curvature, although this is normally much smaller than for the leading edge. Thus a small circle may be drawn to coincide with the arc of maximum curvature of the trailing edge, and a line may be drawn passing through the centres of maximum curvature of the leading and trailing edges. This line, when produced to intersect the section at each end, is called the chord line. The length of the chord line is the aerofoil chord, denoted by c.

The point where the chord line intersects the front (or nose) of the section is used as the origin of a pair of axes, the x-axis being the chord line and the y-axis being perpendicular to the chord line, positive in the upward direction. The shape of the section is then usually given as a table of values of x and the corresponding values of y. These section ordinates are usually expressed as percentages of the chord, $(100x/c)\%$ and $(100y/c)\%$.

Camber

At any distance along the chord from the nose, a point may be marked mid-way between the upper and lower surfaces. The locus of all such points, usually curved, is the median line of the section, usually called the camber line. The maximum height of the camber line above the chord line is denoted by δ and the quantity $100\delta/c\%$ is called the percentage camber of the section. Aerofoil sections have cambers that are usually in the range from zero (a symmetrical section) to 5%, although much larger cambers are used in cascades, e.g. turbine blading.

It is seldom that a camber line can be expressed in simple geometric or algebraic forms, although a few simple curves, such as circular arcs or parabolas, have been used.

Thickness distribution

Having found the median, or camber, line, the distances from this line to the upper and lower surfaces may be measured at any value of x. These are, by the definition of the camber line, equal. These distances may be measured at all points along the chord and then plotted against x from a straight line. The result is a symmetrical shape, called the thickness distribution or symmetrical fairing.

An important parameter of the thickness distribution is the maximum thickness, or depth, t. This, when expressed as a fraction of the chord, is called the thickness/chord ratio. It is commonly expressed as a percentage $100t/c\%$. Current values in use range from 13% to 18% for subsonic aircraft down to 3% or so for supersonic aircraft.

The position along the chord at which this maximum thickness occurs is another important parameter of the thickness distribution. Values usually lie between 30% and 60% of the chord from the leading edge. Some older sections had the maximum thickness at about 25% chord, whereas some more extreme sections have the maximum thickness more than 60% of the chord behind the leading edge.

It will be realized that any aerofoil section may be regarded as a thickness distribution plotted round a camber line. American and British conventions differ in the exact method of derivation of an aerofoil section from a given camber line and thickness distribution. In the British convention, the camber line is plotted, and the thickness ordinates are then plotted from this, perpendicular to the chord line. Thus the thickness distribution is, in effect, sheared until its median line, initially straight, has been distorted to coincide with the given camber line. The American convention is that the thickness ordinates are plotted perpendicular to the curved camber line. The thickness distribution is, therefore, regarded as being bent until its median line coincides with the given camber line.

Since the camber-line curvature is generally very small the difference in aerofoil section shape given by these two conventions is very small.

1.4 Dimensional analysis

1.4.1 Fundamental principles

The theory of dimensional homogeneity has additional uses to that described above. By predicting how one variable may depend on a number of others, it may be used to direct the course of an experiment or the analysis of experimental results. For example, when fluid flows past a circular cylinder the axis of which is perpendicular to the stream, eddies are formed behind the cylinder at a frequency that depends on a number of factors, such as the size of the cylinder, the speed of the stream, etc.

In an experiment to investigate the variation of eddy frequency the obvious procedure is to take several sizes of cylinder, place them in streams of various fluids at a number of different speeds and count the frequency of the eddies in each case. No matter how detailed, the results apply directly only to the cases tested, and it is necessary to find some pattern underlying the results. A theoretical guide is helpful in achieving this end, and it is in this direction that dimensional analysis is of use.

In the above problem the frequency of eddies, n, will depend primarily on:

(i) the size of the cylinder, represented by its diameter, d
(ii) the speed of the stream, V
(iii) the density of the fluid, ρ
(iv) the kinematic viscosity of the fluid, ν.

It should be noted that either μ or ν may be used to represent the viscosity of the fluid.

The factors should also include the geometric shape of the body. Since the problem here is concerned only with long circular cylinders with their axes perpendicular to the stream, this factor will be common to all readings and may be ignored in this analysis. It is also assumed that the speed is low compared to the speed of sound in the fluid, so that compressibility (represented by the modulus of bulk elasticity) may be ignored. Gravitational effects are also excluded.

Then

$$n = f(d, V, \rho, \nu)$$

and, assuming that this function (. . .) may be put in the form

$$n = \sum Cd^a V^b \rho^e \nu^f \tag{1.33}$$

where C is a constant and a, b, e and f are some unknown indices; putting Eqn (1.33) in dimensional form leads to

$$[\mathrm{T}^{-1}] = [\mathrm{L}^a (\mathrm{LT}^{-1})^b (\mathrm{ML}^{-3})^e (\mathrm{L}^2 \mathrm{T}^{-1})^f] \tag{1.34}$$

where each factor has been replaced by its dimensions. Now the dimensions of both sides must be the same and therefore the indices of M, L and T on the two sides of the equation may be equated as follows:

Mass (M) $0 = e$ $\qquad\qquad\qquad\qquad\qquad\qquad$ (1.35a)

Length (L) $0 = a + b - 3e + 2f$ $\qquad\qquad\qquad$ (1.35b)

Time (T) $-1 = -b - f$ $\qquad\qquad\qquad\qquad\quad$ (1.35c)

Here are three equations in four unknowns. One unknown must therefore be left undetermined: f, the index of ν, is selected for this role and the equations are solved for a, b and e in terms of f.

The solution is, therefore,

$$b = 1 - f \tag{1.35d}$$

$$e = 0 \tag{1.35e}$$

$$a = -1 - f \tag{1.35f}$$

Substituting these values in Eqn (1.33),

$$n = \sum Cd^{-1-f} V^{1-f} \rho^0 \nu^f \tag{1.36}$$

Rearranging Eqn (1.36), it becomes

$$n = \sum C \frac{V}{d} \left(\frac{Vd}{V} \right)^{-f} \tag{1.37}$$

or, alternatively,

$$\left(\frac{nd}{V} \right) = g \left(\frac{Vd}{\nu} \right) \tag{1.38}$$

where g represents some function which, as it includes the undetermined constant C and index f, is unknown from the present analysis.

Although it may not appear so at first sight, Eqn (1.38) is extremely valuable, as it shows that the values of nd/V should depend only on the corresponding value of Vd/ν, regardless of the actual values of the original variables. This means that if, for each observation, the values of nd/V and Vd/ν are calculated and plotted as a graph, all the results should lie on a single curve, this curve representing the unknown function g. A person wishing to estimate the eddy frequency for some given cylinder, fluid and speed need only calculate the value of Vd/ν, read from the curve the corresponding value of nd/V and convert this to eddy frequency n. Thus the results of the series of observations are now in a usable form.

Consider for a moment the two compound variables derived above:

(a) nd/V. The dimensions of this are given by

$$\frac{nd}{V} = [T^{-1} \times L \times (LT^{-1})^{-1}] = [L^0 T^0] = [1]$$

(b) Vd/ν. The dimensions of this are given by

$$\frac{Vd}{\nu} = [(LT^{-1}) \times L \times (L^2 T^{-1})^{-1}] = [1]$$

Thus the above analysis has collapsed the five original variables n, d, V, ρ and ν into two compound variables, both of which are non-dimensional. This has two advantages: (i) that the values obtained for these two quantities are independent of the consistent system of units used; and (ii) that the influence of four variables on a fifth term can be shown on a single graph instead of an extensive range of graphs.

It can now be seen why the index f was left unresolved. The variables with indices that were resolved appear in both dimensionless groups, although in the group nd/V the density ρ is to the power zero. These repeated variables have been combined in turn with each of the other variables to form dimensionless groups.

There are certain problems, e.g. the frequency of vibration of a stretched string, in which all the indices may be determined, leaving only the constant C undetermined. It is, however, usual to have more indices than equations, requiring one index or more to be left undetermined as above.

It must be noted that, while dimensional analysis will show which factors are not relevant to a given problem, the method cannot indicate which relevant factors, if any, have been left out. It is, therefore, advisable to include all factors likely to have any bearing on a given problem, leaving out only those factors which, on *a priori* considerations, can be shown to have little or no relevance.

1.4.2 Dimensional analysis applied to aerodynamic force

In discussing aerodynamic force it is necessary to know how the dependent variables, aerodynamic force and moment, vary with the independent variables thought to be relevant.

Assume, then, that the aerodynamic force, or one of its components, is denoted by F and when fully immersed depends on the following quantities: fluid density ρ, fluid kinematic viscosity ν, stream speed V, and fluid bulk elasticity K. The force and moment will also depend on the shape and size of the body, and its orientation to the stream. If, however, attention is confined to geometrically similar bodies, e.g. spheres, or models of a given aeroplane to different scales, the effects of shape as such will be eliminated, and the size of the body can be represented by a single typical dimension; e.g. the sphere diameter, or the wing span of the model aeroplane, denoted by D. Then, following the method above

$$F = \mathrm{f}(V,\ D,\ \rho,\ \nu,\ K)$$
$$= \Sigma C\, V^a\, D^b\, \rho^c \nu^d\, K^e \tag{1.39}$$

In dimensional form this becomes

$$\left[\frac{ML}{T^2}\right] = \left[\left(\frac{L}{T}\right)^a (L)^b \left(\frac{M}{L^3}\right)^c \left(\frac{L^2}{T}\right)^d \left(\frac{M}{LT^2}\right)^e\right]$$

Equating indices of mass, length and time separately leads to the three equations:

(Mass)	$1 = c + e$	(1.40a)
(Length)	$1 = a + b - 3c + 2d - e$	(1.40b)
(Time)	$-2 = -a - d - 2e$	(1.40c)

With five unknowns and three equations it is impossible to determine completely all unknowns, and two must be left undetermined. These will be d and e. The variables whose indices are solved here represent the most important characteristic of the body (the diameter), the most important characteristic of the fluid (the density), and the speed. These variables are known as repeated variables because they appear in each dimensionless group formed.

The Eqns (1.40) may then be solved for a, b and c in terms of d and e giving

$$a = 2 - d - 2e$$
$$b = 2 - d$$
$$c = 1 - e$$

Substituting these in Eqn (1.39) gives

$$F = V^{2-d-2e} D^{2-d} \rho^{1-e} \nu^d K^e$$
$$= \rho V^2 D^2 \left(\frac{\nu}{VD}\right)^d \left(\frac{K}{\rho V^2}\right)^e \tag{1.41}$$

The speed of sound is given by Eqns (1.6b,d) namely,

$$a^2 = \frac{\gamma p}{\rho} = \frac{K}{\rho}$$

Then

$$\frac{K}{\rho V^2} = \frac{\rho a^2}{\rho V^2} = \left(\frac{a}{V}\right)^2$$

and V/a is the Mach number, M, of the free stream. Therefore Eqn (1.41) may be written as

$$F = \rho V^2 D^2 \mathrm{g}\left(\frac{VD}{\nu}\right)\mathrm{h}(M) \qquad (1.42)$$

where $\mathrm{g}(VD/\nu)$ and $\mathrm{h}(M)$ are undetermined functions of the stated compound variables. Thus it can be concluded that the aerodynamic forces acting on a family of geometrically similar bodies (the similarity including the orientation to the stream), obey the law

$$\frac{F}{\rho V^2 D^2} = \text{function}\left\{\frac{VD}{\nu}; M\right\} \qquad (1.43)$$

This relationship is sometimes known as Rayleigh's equation.

The term VD/ν may also be written, from the definition of ν, as $\rho VD/\mu$, as above in the problem relating to the eddy frequency in the flow behind a circular cylinder. It is a very important parameter in fluid flows, and is called the *Reynolds number*.

Now consider any parameter representing the geometry of the flow round the bodies at any point relative to the bodies. If this parameter is expressed in a suitable non-dimensional form, it can easily be shown by dimensional analysis that this non-dimensional parameter is a function of the Reynolds number and the Mach number only. If, therefore, the values of Re (a common symbol for Reynolds number) and M are the same for a number of flows round geometrically similar bodies, it follows that all the flows are geometrically similar in all respects, differing only in geometric scale and/or speed. This is true even though some of the fluids may be gaseous and the others liquid. Flows that obey these conditions are said to be dynamically similar, and the concept of dynamic similarity is essential in wind-tunnel experiments.

It has been found, for most flows of aeronautical interest, that the effects of compressibility can be disregarded for Mach numbers less than 0.3 to 0.5, and in cases where this limit is not exceeded, Reynolds number may be used as the only criterion of dynamic similarity.

Example 1.1 An aircraft and some scale models of it are tested under various conditions, given below. Which cases are dynamically similar to the aircraft in flight, given as case (A)?

	Case (A)	Case (B)	Case (C)	Case (D)	Case (E)	Case (F)
Span (m)	15	3	3	1.5	1.5	3
Relative density	0.533	1	3	1	10	10
Temperature (°C)	−24.6	+15	+15	+15	+15	+15
Speed (TAS) (m s^{-1})	100	100	100	75	54	54

Case (A) represents the full-size aircraft at 6000 m. The other cases represent models under test in various types of wind-tunnel. Cases (C), (E) and (F), where the relative density is greater than unity, represent a special type of tunnel, the compressed-air tunnel, which may be operated at static pressures in excess of atmospheric.

From the figures given above, the Reynolds number $VD\rho/\mu$ may be calculated for each case. These are found to be

Case (A)	$Re = 5.52 \times 10^7$	Case (D)	$Re = 7.75 \times 10^6$
Case (B)	$Re = 1.84 \times 10^7$	Case (E)	$Re = 5.55 \times 10^7$
Case (C)	$Re = 5.56 \times 10^7$	Case (F)	$Re = 1.11 \times 10^8$

It is seen that the values of Re for cases (C) and (E) are very close to that for the full-size aircraft. Cases (A), (C) and (E) are therefore dynamically similar, and the flow patterns in these three cases will be geometrically similar. In addition, the ratios of the local velocity to the free stream velocity at any point on the three bodies will be the same for these three cases. Hence, from Bernoulli's equation, the pressure coefficients will similarly be the same in these three cases, and thus the forces on the bodies will be simply and directly related. Cases (B) and (D) have Reynolds numbers considerably less than (A), and are, therefore, said to represent a 'smaller aerodynamic scale'. The flows around these models, and the forces acting on them, will not be simply or directly related to the force or flow pattern on the full-size aircraft. In case (F) the value of Re is larger than that of any other case, and it has the largest aerodynamic scale of the six.

Example 1.2 An aeroplane approaches to land at a speed of $40\,\text{m s}^{-1}$ at sea level. A 1/5th scale model is tested under dynamically similar conditions in a Compressed Air Tunnel (CAT) working at 10 atmospheres pressure and $15\,°\text{C}$. It is found that the load on the tailplane is subject to impulsive fluctuations at a frequency of 20 cycles per second, owing to eddies being shed from the wing-fuselage junction. If the natural frequency of flexural vibration of the tailplane is 8.5 cycles per second, could this represent a dangerous condition?

For dynamic similarity, the Reynolds numbers must be equal. Since the temperature of the atmosphere equals that in the tunnel, $15\,°\text{C}$, the value of μ is the same in both model and full-scale cases. Thus, for similarity

$$V_\text{f} d_\text{f} \rho_\text{f} = V_\text{m} d_\text{m} \rho_\text{m}$$

In this case, then, since

$$V_\text{f} = 40\,\text{m s}^{-1}$$

$$40 \times 1 \times 1 = V_\text{m} \times \frac{1}{5} \times 10 = 2V_\text{m}$$

giving

$$V_\text{m} = 20\,\text{m s}^{-1}$$

Now Eqn (1.38) covers this case of eddy shedding, and is

$$\frac{nd}{V} = g(Re)$$

For dynamic similarity

$$\left(\frac{nd}{V}\right)_\text{f} = \left(\frac{nd}{V}\right)_\text{m}$$

Therefore

$$\frac{n_\text{f} \times 1}{40} = \frac{20 \times \frac{1}{5}}{20}$$

giving

$$n_\text{f} = 8 \text{ cycles per second}$$

This is very close to the given natural frequency of the tailplane, and there is thus a considerable danger that the eddies might excite sympathetic vibration of the tailplane, possibly leading to structural failure of that component. Thus the shedding of eddies at this frequency is very dangerous to the aircraft.

Example 1.3   An aircraft flies at a Mach number of 0.85 at 18 300 m where the pressure is 7160 N m^{-2} and the temperature is $-56.5\,°C$. A model of 1/10th scale is to be tested in a high-speed wind-tunnel. Calculate the total pressure of the tunnel stream necessary to give dynamic similarity, if the total temperature is $50\,°C$. It may be assumed that the dynamic viscosity is related to the temperature as follows:

$$\frac{\mu}{\mu_0} = \left(\frac{T}{T_0}\right)^{3/4}$$

where $T_0 = 273\,°C$ and $\mu_0 = 1.71 \times 10^{-5}\,\mathrm{kg\,m^{-1}\,s^{-1}}$

(i) *Full-scale aircraft*

$$M = 0.85, \; a = 20.05(273 - 56.5)^{1/2} = 297\,\mathrm{m\,s^{-1}}$$
$$V = 0.85 \times 297 = 252\,\mathrm{m\,s^{-1}}$$

$$\rho = \frac{p}{RT} = \frac{7160}{287.3 \times 216.5} = 0.1151\,\mathrm{kg\,m^{-3}}$$

$$\frac{\mu_0}{\mu} = \left(\frac{273}{216.5}\right)^{3/4} = 1.19$$

$$\mu = \frac{1.71}{1.19} \times 10^{-5} = 1.44 \times 10^{-5}\,\mathrm{kg\,m^{-1}\,s^{-1}}$$

Consider a dimension that, on the aircraft, has a length of 10 m. Then, basing the Reynolds number on this dimension:

$$Re_f = \frac{Vd\rho}{\mu} = \frac{252 \times 10 \times 0.1151}{1.44 \times 10^{-5}} = 20.2 \times 10^6$$

(ii) *Model*
Total temperature $T_s = 273 + 50 = 323\,\mathrm{K}$

Therefore at $M = 0.85$:

$$\frac{T_s}{T} = 1 + \frac{1}{5}(0.85)^2 = 1.1445$$
$$T = 282\,\mathrm{K}$$

Therefore

$$a = 20.05 \times (282)^{1/2} = 337\,\mathrm{m\,s^{-1}}$$
$$V = 0.85 \times 337 = 287\,\mathrm{m\,s^{-1}}$$

$$\frac{\mu}{\mu_0} = \left(\frac{282}{273}\right)^{3/4} = 1.0246$$

giving

$$\mu = 1.71 \times 1.0246 \times 10^{-5} = 1.751 \times 10^{-5} \, \text{kg} \, \text{m}^{-1} \, \text{s}^{-1}$$

For dynamic similarity the Reynolds numbers must be equal, i.e.

$$\frac{287 \times 1 \times \rho}{1.75 \times 10^{-5}} = 20.2 \times 10^{6}$$

giving

$$\rho = 1.23 \, \text{kg} \, \text{m}^{-3}$$

Thus the static pressure required in the test section is

$$p = \rho R T = 1.23 \times 287.3 \times 282 = 99\,500 \, \text{N} \, \text{m}^{-2}$$

The total pressure p_{s} is given by

$$\frac{p_{\text{s}}}{p} = \left(1 + \frac{1}{5} M^2\right)^{3.5} = (1.1445)^{3.5} = 1.605$$
$$p_{\text{s}} = 99\,500 \times 1.605 = 160\,000 \, \text{N} \, \text{m}^{-2}$$

If the total pressure available in the tunnel is less than this value, it is not possible to achieve equality of both the Mach and Reynolds numbers. Either the Mach number may be achieved at a lower value of Re or, alternatively, Re may be made equal at a lower Mach number. In such a case it is normally preferable to make the Mach number correct since, provided the Reynolds number in the tunnel is not too low, the effects of compressibility are more important than the effects of aerodynamic scale at Mach numbers of this order. Moreover, techniques are available which can alleviate the errors due to unequal aerodynamic scales.

In particular, the position at which laminar-turbulent transition (see Section 7.9) of the boundary layer occurs at full scale can be fixed on the model by roughening the model surface. This can be done by gluing on a line of carborundum powder.

1.5　Basic aerodynamics

1.5.1　Aerodynamic force and moment

Air flowing past an aeroplane, or any other body, must be diverted from its original path, and such deflections lead to changes in the speed of the air. Bernoulli's equation shows that the pressure exerted by the air on the aeroplane is altered from that of the undisturbed stream. Also the viscosity of the air leads to the existence of frictional forces tending to resist its flow. As a result of these processes, the aeroplane experiences a resultant aerodynamic force and moment. It is conventional and convenient to separate this aerodynamic force and moment into three components each, as follows.

Lift, L(–Z)

This is the component of force acting upwards, perpendicular to the direction of flight or of the undisturbed stream. The word 'upwards' is used in the same sense that the pilot's head is above his feet. Figure 1.7 illustrates the meaning in various attitudes of flight. The arrow V represents the direction of flight, the arrow L represents the lift acting upwards and the arrow W the weight of the aircraft, and shows the downward vertical. Comparison of (a) and (c) shows that this upwards is not fixed relative to the aircraft, while (a), (b), and (d) show that the meaning is not fixed relative to the earth. As a general rule, if it is remembered that the lift is always

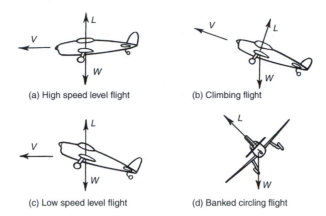

(a) High speed level flight   (b) Climbing flight

(c) Low speed level flight   (d) Banked circling flight

Fig. 1.7 The direction of the lift force

a component perpendicular to the flight direction, the exact direction in which the lift acts will be obvious, particularly after reference to Fig. 1.7. This may not apply to certain guided missiles that have no obvious top or bottom, and the exact meaning of 'up' must then be defined with care.

Drag, D(–X)

This is the component of force acting in the opposite direction to the line of flight, or in the same direction as the motion of the undisturbed stream. It is the force that resists the motion of the aircraft. There is no ambiguity regarding its direction or sense.

Cross-wind force, Y

This is the component of force mutually perpendicular to the lift and the drag, i.e. in a spanwise direction. It is reckoned positive when acting towards the starboard (right-hand to the pilot) wing-tip.

Pitching moment, M

This is the moment acting in the plane containing the lift and the drag, i.e. in the vertical plane when the aircraft is flying horizontally. It is positive when it tends to increase the incidence, or raise the nose of the aircraft upwards (using this word in the sense discussed earlier).

Rolling moment, L_R

This is the moment tending to make the aircraft roll about the flight direction, i.e. tending to depress one wing-tip and to raise the other. It is positive when it tends to depress the starboard wing-tip.

Yawing moment, N

This is the moment that tends to rotate the aircraft about the lift direction, i.e. to swing the nose to one side or the other of the flight direction. It is positive when it swings, or tends to swing, the nose to the right (starboard).

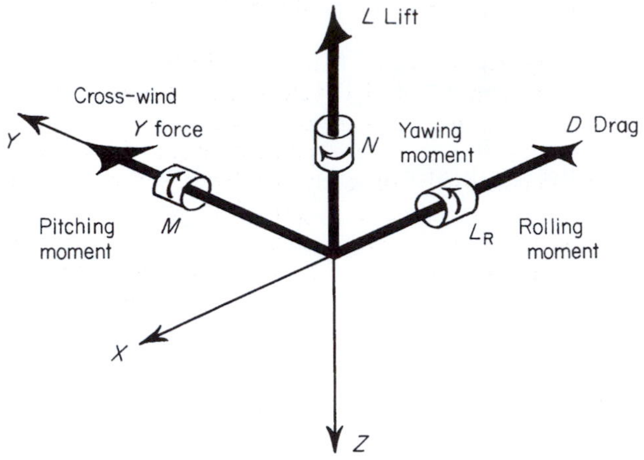

Fig. 1.8 The systems of force and moment components. The broad arrows represent forces used in elementary work; the line arrows, the system in control and stability studies. The moments are common to both systems

The relation between these components is shown in Fig. 1.8. In each case the arrow shows the direction of the positive force or moment. All three forces are mutually perpendicular, and each moment acts about the line of one of the forces.

The system of forces and moments described above is conventionally used for performance analysis and other simple problems. For aircraft stability and control studies, however, it is more convenient to use a slightly different system of forces.

1.5.2   Force and moment coefficients

The non-dimensional quantity $F/(\rho V^2 S)$ (c.f. Eqn 1.43) (where F is an aerodynamic force and S is an area) is similar to the type often developed and used in aerodynamics. It is not, however, used in precisely this form. In place of ρV^2 it is conventional for incompressible flow to use $\frac{1}{2}\rho V^2$, the dynamic pressure of the free-stream flow. The actual physical area of the body, such as the planform area of the wing, or the maximum cross-sectional area of a fuselage is usually used for S. Thus aerodynamic force coefficient is usually defined as follows:

$$C_{\mathrm{F}} = \frac{F}{\frac{1}{2}\rho V^2 S} \tag{1.44}$$

The two most important force coefficients are the lift and drag coefficients, defined by:

$$\text{lift coefficient } C_L = \text{lift}/\tfrac{1}{2}\rho V^2 S \tag{1.44a}$$

$$\text{drag coefficient } C_D = \text{drag}/\tfrac{1}{2}\rho V^2 S \tag{1.44b}$$

When the body in question is a wing the area S is almost invariably the planform area as defined in Section 1.3.1. For the drag of a body such as a fuselage, sphere or cylinder the area S is usually the projected frontal area, the maximum cross-sectional area or the (volume)$^{2/3}$. The area used for definition of the lift and drag coefficients of such a body is thus seen to be variable from case to case, and therefore needs to be stated for each case.

The impression is sometimes formed that lift and drag coefficients cannot exceed unity. This is not true; with modern developments some wings can produce lift coefficients based on their plan-area of 10 or more.

Aerodynamic moments also can be expressed in the form of non-dimensional coefficients. Since a moment is the product of a force and a length it follows that a non-dimensional form for a moment is $Q/\rho V^2 Sl$, where Q is any aerodynamic moment and l is a reference length. Here again it is conventional to replace ρV^2 by $\frac{1}{2}\rho V^2$. In the case of the pitching moment of a wing the area is the plan-area S and the length is the wing chord $\bar{c}$ or $\bar{c}_A$ (see Section 1.3.1). Then the pitching moment coefficient C_M is defined by

$$C_M = \frac{M}{\frac{1}{2}\rho V^2 S\bar{c}} \tag{1.45}$$

1.5.3 Pressure distribution on an aerofoil

The pressure on the surface of an aerofoil in flight is not uniform. Figure 1.9 shows some typical pressure distributions for a given section at various angles of incidence. It is convenient to deal with non-dimensional pressure differences with p_∞, the pressure far upstream, being used as the datum. Thus the coefficient of pressure is introduced below

$$C_p = \frac{(p - p_\infty)}{\frac{1}{2}\rho V^2}$$

Looking at the sketch for zero incidence ($\alpha = 0$) it is seen that there are small regions at the nose and tail where C_p is positive but that over most of the section C_p is negative. At the trailing edge the pressure coefficient comes close to $+1$ but does not actually reach this value. More will be said on this point later. The reduced pressure on the upper surface is tending to draw the section upwards while that on the lower

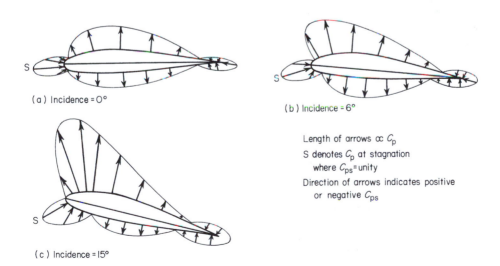

(a) Incidence = 0°

(b) Incidence = 6°

(c) Incidence = 15°

Length of arrows $\propto C_p$
S denotes C_p at stagnation
where C_{ps} = unity
Direction of arrows indicates positive
or negative C_{ps}

Fig. 1.9 Typical pressure distributions on an aerofoil section

surface has the opposite effect. With the pressure distribution as sketched, the effect on the upper surface is the larger, and there is a resultant upwards force on the section, that is the lift.

As incidence is increased from zero the following points are noted:

(i) the pressure reduction on the upper surface increases both in intensity and extent until, at large incidence, it actually encroaches on a small part of the front lower surface;
(ii) the stagnation point moves progressively further back on the lower surface, and the increased pressure on the lower surface covers a greater proportion of the surface. The pressure reduction on the lower surface is simultaneously decreased in both intensity and extent.

The large negative values of C_p reached on the upper surface at high incidences, e.g. 15 degrees, are also noteworthy. In some cases values of -6 or -7 are found. This corresponds to local flow speeds of nearly three times the speed of the undisturbed stream.

From the foregoing, the following conclusions may be drawn:

(i) at low incidence the lift is generated by the difference between the pressure reductions on the upper and lower surfaces;
(ii) at higher incidences the lift is partly due to pressure reduction on the upper surface and partly due to pressure increase on the lower surface.

At angles of incidence around $18°$ or $20°$ the pressure reduction on the upper surface suddenly collapses and what little lift remains is due principally to the pressure increase on the lower surface. A picture drawn for one small negative incidence (for this aerofoil section, about $-4°$) would show equal suction effects on the upper and lower surfaces, and the section would give no lift. At more negative incidences the lift would be negative.

The relationship between the pressure distribution and the drag of an aerofoil section is discussed later (Section 1.5.5).

1.5.4   Pitching moment

The pitching moment on a wing may be estimated experimentally by two principal methods: direct measurement on a balance, or by pressure plotting, as described in Section 1.5.6. In either case, the pitching moment coefficient is measured about some definite point on the aerofoil chord, while for some particular purpose it may be desirable to know the pitching moment coefficient about some other point on the chord. To convert from one reference point to the other is a simple application of statics.

Suppose, for example, the lift and drag are known, as also is the pitching moment M_a about a point distance a from the leading edge, and it is desired to find the pitching moment M_x about a different point, distance x behind the leading edge. The situation is then as shown in Fig. 1.10. Figure 1.10a represents the known conditions, and Fig. 1.10b the unknown conditions. These represent two alternative ways of looking at the same physical system, and must therefore give identical effects on the aerofoil.

Obviously, then, $L = L$ and $D = D$.

Taking moments in each case about the leading edge:

$$M_{\mathrm{LE}} = M_a - La \cos \alpha - Da \sin \alpha = M_x - Lx \cos \alpha - Dx \sin \alpha$$

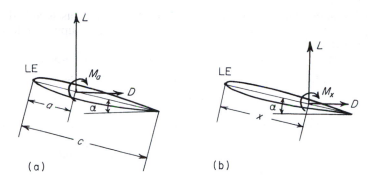

(a)   (b)

Fig. 1.10

then

$$M_x = M_a - (L \cos \alpha + D \sin \alpha)(a - x)$$

Converting to coefficient form by dividing by $\frac{1}{2}\rho V^2 Sc$ gives

$$C_{M_x} = C_{M_a} - (C_L \cos \alpha + C_D \sin \alpha)\left(\frac{a}{c} - \frac{x}{c}\right) \qquad (1.46)$$

With this equation it is easy to calculate C_{M_x}, for any value of x/c. As a particular case, if the known pitching moment coefficient is that about the leading edge, $C_{M_{\text{LE}}}$, then $a = 0$, and Eqn (1.46) becomes

$$C_{M_x} = C_{M_{\text{LE}}} + \frac{x}{c}(C_L \cos \alpha + C_D \sin \alpha) \qquad (1.47)$$

Aerodynamic centre

If the pitching moment coefficient at each point along the chord is calculated for each of several values of C_L, one very special point is found for which C_M is virtually constant, independent of the lift coefficient. This point is the aerodynamic centre. For incidences up to 10 degrees or so it is a fixed point close to, but not in general on, the chord line, between 23% and 25% of the chord behind the leading edge.

For a flat or curved plate in inviscid, incompressible flow the aerodynamic centre is theoretically exactly one quarter of the chord behind the leading edge; but thickness of the section, and viscosity of the fluid, tend to place it a few per cent further forward as indicated above, while compressibility tends to move it backwards. For a thin aerofoil of infinite aspect ratio in supersonic flow the aerodynamic centre is theoretically at 50% chord.

Knowledge of how the pitching moment coefficient about a point distance a behind the leading edge varies with C_L may be used to find the position of the aerodynamic centre behind the leading edge, and also the value of the pitching moment coefficient there, $C_{M_{\text{AC}}}$. Let the position of the aerodynamic centre be a distance x_{AC} behind the leading edge. Then, with Eqn (1.46) slightly rearranged,

$$C_{M_a} = C_{M_{\text{AC}}} - (C_L \cos \alpha + C_D \sin \alpha)\left(\frac{x_{\text{AC}}}{c} - \frac{a}{c}\right)$$

Now at moderate incidences, between say 3° and 7°:

$$C_L = O[20C_D] \quad \text{and} \quad \cos\alpha = O[10\sin\alpha]$$

where the symbol $O[\ ]$ means of the order of, i.e. C_L is of the order of 20 times C_D. Then.

$$C_L\cos\alpha = O[200\,C_D\sin\alpha]$$

and therefore $C_D\sin\alpha$ can be neglected compared with $C_L\cos\alpha$. With this approximation and the further approximation $\cos\alpha = 1$,

$$C_{M_a} = C_{M_{AC}} - C_L\left(\frac{x_{AC}}{c} - \frac{a}{c}\right) \tag{1.48}$$

Differentiating Eqn (1.48) with respect to C_L gives

$$\frac{d}{dC_L}(C_{M_a}) = \frac{d}{dC_L}(C_{M_{AC}}) - \left(\frac{x_{AC}}{c} - \frac{a}{c}\right)$$

But the aerodynamic centre is, by definition, that point about which C_M is independent of C_L, and therefore the first term on the right-hand side is identically zero, so that

$$\frac{d}{dC_L}(C_{M_a}) = 0 - \left(\frac{x_{AC}}{c} - \frac{a}{c}\right) = \frac{a}{c} - \frac{x_{AC}}{c} \tag{1.49}$$

$$\frac{x_{AC}}{c} = \frac{a}{c} - \frac{d}{dC_L}(C_{M_a}) \tag{1.50}$$

If, then, C_{M_a} is plotted against C_L, and the slope of the resulting line is measured, subtracting this value from a/c gives the aerodynamic centre position x_{AC}/c.

In addition if, in Eqn (1.48), C_L is made zero, that equation becomes

$$C_{M_a} = C_{M_{AC}} \tag{1.51}$$

i.e. the pitching moment coefficient about an axis at zero lift is equal to the constant pitching moment coefficient about the aerodynamic centre. Because of this association with zero lift, $C_{M_{AC}}$ is often denoted by C_{M_0}.

Example 1.4 For a particular aerofoil section the pitching moment coefficient about an axis 1/3 chord behind the leading edge varies with the lift coefficient in the following manner:

C_L	0.2	0.4	0.6	0.8
C_M	−0.02	0.00	+0.02	+0.04

Find the aerodynamic centre and the value of C_{M_0}.

It is seen that C_M varies linearly with C_L, the value of dC_M/dC_L being

$$\frac{0.04 - (-0.02)}{0.80 - 0.20} = +\frac{0.06}{0.60} = +0.10$$

Therefore, from Eqn (1.50), with $a/c = 1/3$

$$\frac{x_{AC}}{c} = \frac{1}{3} - 0.10 = 0.233$$

The aerodynamic centre is therefore at 23.3% chord behind the leading edge. Plotting C_M against C_L gives the value of C_{M_0}, the value of C_M when $C_L = 0$, as −0.04.

A particular case is that when the known values of C_M are those about the leading edge, namely $C_{M_{LE}}$. In this case $a = 0$ and therefore

$$\frac{x_{AC}}{c} = -\frac{d}{dC_L}(C_{M_{LE}}) \qquad (1.52)$$

Taking this equation with the statement made earlier about the normal position of the aerodynamic centre implies that, for all aerofoils at low Mach numbers:

$$\frac{d}{dC_L}(C_{M_{LE}}) \simeq -\frac{1}{4} \qquad (1.53)$$

Centre of pressure

The aerodynamic forces on an aerofoil section may be represented by a lift, a drag, and a pitching moment. At each value of the lift coefficient there will be found to be one particular point about which the pitching moment coefficient is zero, and the aerodynamic effects on the aerofoil section may be represented by the lift and the drag alone acting at that point. This special point is termed the centre of pressure.

Whereas the aerodynamic centre is a fixed point that always lies within the profile of a normal aerofoil section, the centre of pressure moves with change of lift coefficient and is not necessarily within the aerofoil profile. Figure 1.11 shows the forces on the aerofoil regarded as either

(a) lift, drag and moment acting at the aerodynamic centre; or
(b) lift and drag only acting at the centre of pressure, a fraction k_{CP} of the chord behind the leading edge.

Then, taking moments about the leading edge:

$$M_{LE} = M_{AC} - (L\cos\alpha + D\sin\alpha)x_{AC} = -(L\cos\alpha + D\sin\alpha)k_{CP}c$$

Dividing this by $\frac{1}{2}\rho V^2 Sc$, it becomes

$$C_{M_{AC}} - (C_L\cos\alpha + C_D\sin\alpha)\frac{x_{AC}}{c} = -(C_L\cos\alpha + C_D\sin\alpha)k_{CP}$$

giving

$$k_{CP} = \frac{x_{AC}}{c} - \frac{C_{M_{AC}}}{C_L\cos\alpha + C_D\sin\alpha} \qquad (1.54)$$

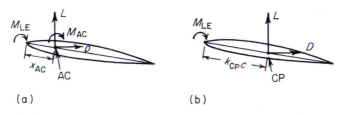

(a)   (b)

Fig. 1.11 Determination of the centre of pressure position

Again making the approximations that $\cos \alpha \simeq 1$ and $C_D \sin \alpha$ can be ignored, the Eqn (1.54), above, becomes

$$k_{CP} \simeq \frac{x_{AC}}{c} - \frac{C_{M_{AC}}}{C_L} \tag{1.55}$$

At first sight this would suggest that k_{CP} is always less than x_{AC}/c. However, $C_{M_{AC}}$ is almost invariably negative, so that in fact k_{CP} is numerically greater than x_{AC}/c and the centre of pressure is behind the aerodynamic centre.

Example 1.5   For the aerofoil section of Example 1.4, plot a curve showing the approximate variation of the position of centre of pressure with lift coefficient, for lift coefficients between zero and unity. For this case:

$$k_{CP} \simeq 0.233 - (-0.04/C_L)$$
$$\simeq 0.233 + (0.04/C_L)$$

The corresponding curve is shown as Fig. 1.12. It shows that k_{CP} tends asymptotically to x_{AC} as C_L increases, and tends to infinity behind the aerofoil as C_L tends to zero. For values of C_L less than 0.05 the centre of pressure is actually behind the aerofoil.

For a symmetrical section (zero camber) and for some special camber lines, the pitching moment coefficient about the aerodynamic centre is zero. It then follows, from Eqn (1.55), that $k_{CP} = x_{AC}/c$, i.e. the centre of pressure and the aerodynamic centre coincide, and that for moderate incidences the centre of pressure is therefore stationary at about the quarter-chord point.

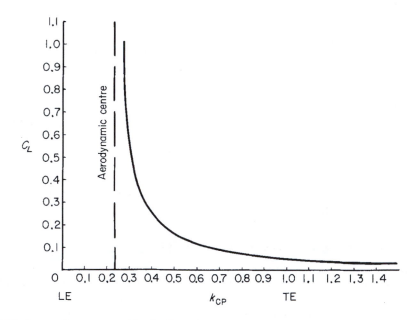

Fig. 1.12 Centre of pressure position for Example 1.5

1.5.5 Types of drag

Attempts have been made to rationalize the definitions and terminology associated with drag*. On the whole the new terms have not been widely adopted. Here we will use the widely accepted traditional terms and indicate alternatives in parentheses.

Total drag

This is formally defined as the force corresponding to the rate of decrease in momentum in the direction of the undisturbed external flow around the body, this decrease being calculated between stations at infinite distances upstream and downstream of the body. Thus it is the total force or drag in the direction of the undisturbed flow. It is also the total force resisting the motion of the body through the surrounding fluid.

There are a number of separate contributions to total drag. As a first step it may be divided into *pressure drag* and *skin-friction drag*.

Skin-friction drag (or surface-friction drag)

This is the drag that is generated by the resolved components of the traction due to the shear stresses acting on the surface of the body. This traction is due directly to viscosity and acts tangentially at all points on the surface of the body. At each point it has a component aligned with but opposing the undisturbed flow (i.e. opposite to the direction of flight). The total effect of these components, taken (i.e. integrated) over the whole exposed surface of the body, is the *skin-friction drag*. It could not exist in an invisicid flow.

Pressure drag

This is the drag that is generated by the resolved components of the forces due to pressure acting normal to the surface at all points. It may itself be considered as consisting of several distinct contributions:

(i) *Induced* drag (sometimes known as vortex drag);
(ii) *Wave* drag; and
(iii) *Form* drag (sometimes known as boundary-layer pressure drag).

Induced drag (or vortex drag)

This is discussed in more detail in Sections 1.5.7 and 5.5. For now it may be noted that induced drag depends on lift, does not depend directly on viscous effects, and can be estimated by assuming inviscid flow.

Wave drag

This is the drag associated with the formation of shock waves in high-speed flight. It is described in more detail in Chapter 6.

Form drag (or boundary-layer pressure drag)

This can be defined as the difference between the profile drag and the skin-friction drag where the former is defined as the drag due to the losses in total pressure and

* For example, the *Aeronautical Research Committee Current Paper* No. 369 which was also published in the *Journal of the Royal Aeronautical Society*, November 1958.

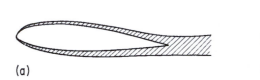

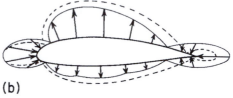

(a)   (b)

Fig. 1.13 (a) The displacement thickness of the boundary layer (hatched area) represents an effective change to the shape of the aerofoil. (Boundary-layer thickness is greatly exaggerated in this sketch.) (b) Pressure-distribution on an aerofoil section in viscous flow (dotted line) and inviscid flow (full line)

total temperature in the boundary layers. But these definitions are rather unhelpful for giving a clear idea of the physical nature and mechanisms behind form drag, so a simple explanation is attempted below.

The pressure distribution over a body in viscous flow differs from that in an ideal inviscid flow (Fig. 1.13). If the flow is inviscid, it can be shown that the flow speed at the trailing edge is zero, implying that the pressure coefficient is +1. But in a real flow (see Fig. 1.13a) the body plus the boundary-layer displacement thickness has a finite width at the trailing edge, so the flow speed does not fall to zero, and therefore the pressure coefficient is less than +1. The variation of coefficient of pressure due to real flow around an aerofoil is shown in Fig. 1.13b. This combines to generate a net drag as follows. The relatively high pressures around the nose of the aerofoil tend to push it backwards. Whereas the region of the suction pressures that follows, extending up to the point of maximum thickness, act to generate a thrust pulling the aerofoil forwards. The region of suction pressures downstream of the point of maximum thickness generates a retarding force on the aerofoil, whereas the relatively high-pressure region around the trailing edge generates a thrust. In an inviscid flow, these various contributions cancel out exactly and the net drag is zero. In a real viscous flow this exact cancellation does not occur. The pressure distribution ahead of the point of maximum thickness is little altered by real-flow effects. The drag generated by the suction pressures downstream of the point of maximum thickness is slightly reduced in a real flow. But this effect is greatly outweighed by a substantial reduction in the thrust generated by the high-pressure region around the trailing edge. Thus the exact cancellation of the pressure forces found in an inviscid flow is destroyed in a real flow, resulting in an overall rearwards force. This force is the *form* drag.

It is emphasized again that both form and skin-friction drag depend on viscosity for their existence and cannot exist in an inviscid flow.

Profile drag (or boundary-layer drag)

The profile drag is the sum of the skin-friction and form drags. See also the formal definition given at the beginning of the previous item.

Comparison of drags for various types of body

Normal flat plate (Fig. 1.14)
In the case of a flat plate set broadside to a uniform flow, the drag is entirely form drag, coming mostly from the large negative pressure coefficients over the rear face. Although viscous tractions exist, they act along the surface of the plate, and therefore have no rearwards component to produce skin-friction drag.

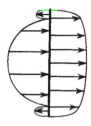

Fig. 1.14 Pressure on a normal flat plate

Parallel flat plate (Fig. 1.15)
In this case, the drag is entirely skin-friction drag. Whatever the distribution of pressure may be, it can have no rearward component, and therefore the form drag must be zero.

Circular cylinder (Fig. 1.16)
Figure 1.16 is a sketch of the distribution of pressure round a circular cylinder in inviscid flow (solid lines) (see Section 3.3.9 below) and in a viscous fluid (dotted lines). The perfect symmetry in the inviscid case shows that there is no resultant force on the cylinder. The drastic modification of the pressure distribution due to viscosity is apparent, the result being a large form drag. In this case, only some 5% of the drag is skin-friction drag, the remaining 95% being form drag, although these proportions depend on the Reynolds number.

Aerofoil or streamlined strut
The pressure distributions for this case are given in Fig. 1.13. The effect of viscosity on the pressure distribution is much less than for the circular cylinder, and the form drag is much lower as a result. The percentage of the total drag represented by skin-friction drag depends on the Reynolds number, the thickness/chord ratio, and a number of other factors, but between 40% and 80% is fairly typical.

Fig. 1.15 Viscous tractions on a tangential flat plate

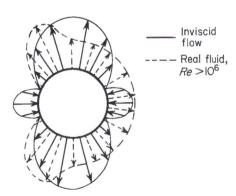

Inviscid
flow

Real fluid,
$Re > 10^6$

Fig. 1.16 Pressure on a circular cylinder with its axis normal to the stream (see also Fig. 3.23)

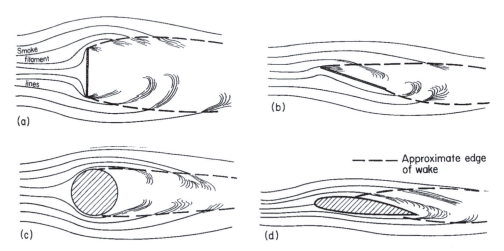

Fig. 1.17 The behaviour of smoke filaments in the flows past various bodies, showing the wakes. (a) Normal flat plate. In this case the wake oscillates up and down at several cycles per second. Half a cycle later the picture would be reversed, with the upper filaments curving back as do the lower filaments in this sketch. (b) Flat plate at fairly high incidence. (c) Circular cylinder at low Re. For pattern at higher Re, see Fig. 7.14. (d) Aerofoil section at moderate incidence and low Re

The wake

Behind any body moving in air is a wake, just as there is a wake behind a ship. Although the wake in air is not normally visible it may be felt, as when, for example, a bus passes by. The total drag of a body appears as a loss of momentum and increase of energy in this wake. The loss of momentum appears as a reduction of average flow speed, while the increase of energy is seen as violent eddying (or vorticity) in the wake. The size and intensity of the wake is therefore an indication of the profile drag of the body. Figure 1.17 gives an indication of the comparative widths of the wakes behind a few bodies.

1.5.6 Estimation of the coefficients of lift, drag and pitching moment from the pressure distribution

Let Fig. 1.18 represent an aerofoil at an angle of incidence α to a fluid flow travelling from left to right at speed V. The axes Ox and Oz are respectively aligned along and perpendicular to the chord line. The chord length is denoted by c.

Taking the aerofoil to be a wing section of constant chord and unit spanwise length, let us consider the forces acting on a small element of the upper aerofoil surface having length δs. The inward force perpendicular to the surface is given by $p_u \delta s$. This force may be resolved into components δX and δZ in the x and z directions. It can be seen that

$$\delta Z_u = -p_u \cos \varepsilon \tag{1.56}$$

and from the geometry

$$\delta s \cos \varepsilon = \delta x \tag{1.57}$$

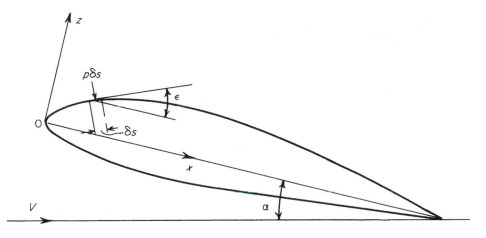

Fig. 1.18 Normal pressure force on an element of aerofoil surface

so that

$$\delta Z_u = -p_u \delta x \qquad \text{per unit span}$$

Similarly, for the lower surface

$$\delta Z_\ell = p_\ell \delta x \qquad \text{per unit span}$$

We now add these two contributions and integrate with respect to x between $x = 0$ and $x = c$ to get

$$Z = -\int_0^c p_u \mathrm{d}x + \int_0^c p_\ell \mathrm{d}x$$

But we can always subtract a constant pressure from both p_u and p_ℓ without altering the value of Z, so we can write

$$Z = -\int_0^c (p_u - p_\infty)\mathrm{d}x + \int_0^c (p_\ell - p_\infty)\mathrm{d}x \tag{1.58}$$

where p_∞ is the pressure in the free stream (we could equally well use any other constant pressure, e.g. the stagnation pressure in the free stream).

Equation (1.58) can readily be converted into coefficient form. Recalling that the aerofoil section is of unit span, the area $S = 1 \times c = c$, so we obtain

$$C_Z \equiv \frac{Z}{\frac{1}{2}\rho V^2 c} = -\frac{1}{\frac{1}{2}\rho V^2 c}\int_0^c [(p_u - p_\infty) - (p_\ell - p_\infty)]\mathrm{d}x$$

Remembering that $(1/c)\mathrm{d}x = \mathrm{d}(x/c)$ and that the definition of pressure coefficient is

$$C_\mathrm{p} = \frac{p - p_\infty}{\frac{1}{2}\rho V^2}$$

we see that

$$C_Z = -\int_0^1 (C_{p u} - C_{p \ell})\mathrm{d}(x/c) \tag{1.59a}$$

or, simply

$$C_Z = \oint_C C_p \cos \varepsilon \mathrm{d}(s/c) = \oint_C C_p \mathrm{d}(x/c), \qquad (1.59\mathrm{b})$$

where the contour integral is evaluated by following an anti-clockwise direction around the contour C of the aerofoil.

Similar arguments lead to the following relations for X.

$$\delta X_u = p_u \delta s \sin \varepsilon, \quad \delta X_\ell = p_\ell \delta s \sin \varepsilon, \quad \delta s \sin \varepsilon = \delta z,$$

giving

$$C_X = \oint_C C_p \sin \varepsilon \mathrm{d}(s/c) = \oint_C C_p \mathrm{d}(z/c) = \int_{z_{m\ell}}^{z_{mu}} \Delta C_p \mathrm{d}\left(\frac{z}{c}\right), \qquad (1.60)$$

where z_{m_u} and $z_{m\ell}$ are respectively the maximum and minimum values of z, and ΔC_p is the difference between the values of C_p acting on the fore and rear points of an aerofoil for a fixed value of z.

The pitching moment can also be calculated from the pressure distribution. For simplicity, the pitching moment about the leading edge will be calculated. The contribution due to the force δZ acting on a slice of aerofoil of length δx is given by

$$\delta M = (p_u - p_\ell)x \delta x = [(p_u - p_\infty) - (p_\ell - p_\infty)]x \delta x;$$

so, remembering that the coefficient of pitching moment is defined as

$$C_M = \frac{M}{\frac{1}{2}\rho V^2 Sc} = \frac{M}{\frac{1}{2}\rho V^2 c^2} \quad \text{in this case, as} \quad S = c,$$

the coefficient of pitching moment due to the Z force is given by

$$C_{MZ} = -\oint_C C_p \frac{x}{c} \mathrm{d}\left(\frac{x}{c}\right) = \int_0^c [C_{pu} - C_{p\ell}]\frac{x}{c}\mathrm{d}\left(\frac{x}{c}\right) \qquad (1.61)$$

Similarly, the much smaller contribution due to the X force may be obtained as

$$C_{MX} = -\oint_C C_p \sin \varepsilon \frac{z}{c} \mathrm{d}\left(\frac{s}{c}\right) = \int_{z_{m\ell}}^{z_{mu}} \Delta C_p \frac{z}{c} \mathrm{d}\left(\frac{z}{c}\right) \qquad (1.62)$$

The integrations given above are usually performed using a computer or graphically.

The force coefficients C_X and C_Z are parallel and perpendicular to the chord line, whereas the more usual coefficients C_L and C_D are defined with reference to the direction of the free-stream air flow. The conversion from one pair of coefficients to the other may be carried out with reference to Fig. 1.19, in which, C_R, the coefficient of the resultant aerodynamic force, acts at an angle γ to C_Z. C_R is both the resultant of C_X and C_Z, and of C_L and C_D; therefore from Fig. 1.19 it follows that

$$C_L = C_R \cos(\gamma + \alpha) = C_R \cos \gamma \cos \alpha - C_R \sin \gamma \sin \alpha$$

But $C_R \cos \gamma = C_Z$ and $C_R \sin \gamma = C_X$, so that

$$C_L = C_Z \cos \alpha - C_X \sin \alpha. \qquad (1.63)$$

Similarly

$$C_D = C_R \sin(\alpha + \gamma) = C_Z \sin \alpha + C_X \cos \alpha \qquad (1.64)$$

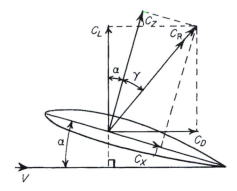

Fig. 1.19

The total pitching moment coefficient is

$$C_M = C_{M_z} + C_{M_x} \tag{1.65}$$

In Fig. 1.20 are shown the graphs necessary for the evaluation of the aerodynamic coefficients for the mid-section of a three-dimensional wing with an ellipto-Zhukovsky profile.

1.5.7 Induced drag

Section 5.5 below should also be referred to. Consider what is happening at some point y along the wing span (Fig. 1.21). Each of the trailing vortices produces a downwards component of velocity, w, at y, known as the downwash or induced velocity (see Section 5.5.1). This causes the flow over that section of the wing to be inclined slightly downwards from the direction of the undisturbed stream V (Fig. 1.22) by the angle ε, the induced angle of incidence or downwash angle. The local flow is also at a slightly different speed, q.

If the angle between the aerofoil chord line and the direction of the undisturbed stream, the geometric angle of incidence, is α, it is seen that the angle between the chord line and the actual flow at that section of the wing is equal to $\alpha - \varepsilon$, and this is called the effective incidence α_∞. It is this effective incidence that determines the lift coefficient at that section of the wing, and thus the wing is lifting less strongly than the geometric incidence would suggest. Since the circulation and therefore w and ε increase with lift coefficient, it follows that the lift of a three-dimensional wing increases less rapidly with incidence than does that for a two-dimensional wing, which has no trailing vortices.

Now the circulation round this section of the wing will have a value Γ appropriate to α_∞, and the lift force corresponding to this circulation will be $\rho q \Gamma$ per unit length, acting perpendicular to the direction of q as shown, i.e. inclined backwards from the vertical by the angle ε. This force therefore has a component perpendicular to the undisturbed stream V, that, by definition, is called the lift, and is of magnitude

$$l = \rho q \Gamma \cos \varepsilon = \rho q \Gamma \frac{V}{q} = \rho V \Gamma \text{ per unit length}$$

There is also a rearwards component of magnitude

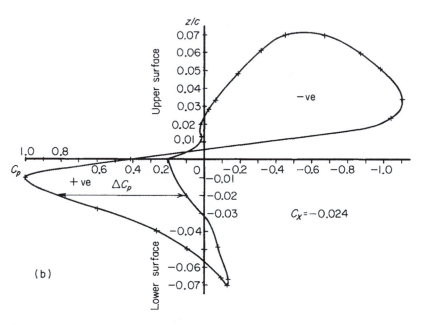

Ellipto-Zhukovsky section at the mid-section of a three-dimensional wing.
Geometric incidence = 6°
Reynolds number = 4.8×10^5

Fig. 1.20 Pressure distribution on an aerofoil surface

$$d = \rho q \Gamma \sin \varepsilon = \rho q \Gamma \frac{w}{q} = \rho w \Gamma \text{ per unit length}$$

This rearwards component must be reckoned as a drag and is, in fact, the induced drag. Thus the induced drag arises essentially from the downwards velocity induced over the wing by the wing-tip vortices.

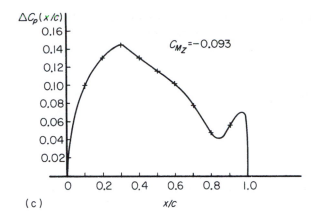

(c)

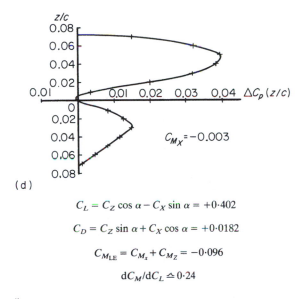

(d)

$$C_L = C_Z \cos \alpha - C_X \sin \alpha = +0 \cdot 402$$

$$C_D = C_Z \sin \alpha + C_X \cos \alpha = +0 \cdot 0182$$

$$C_{M_{LE}} = C_{M_x} + C_{M_z} = -0 \cdot 096$$

$$dC_M/dC_L \simeq 0 \cdot 24$$

Fig. 1.20 (Continued)

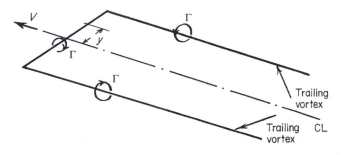

Fig. 1.21 The simplified horseshoe vortex system

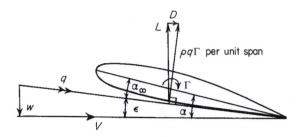

Fig. 1.22 Flow conditions and forces at a section of a three-dimensional lifting wing

The further apart the wing-tip vortices the less will be their effectiveness in producing induced incidence and drag. It is therefore to be expected that these induced quantities will depend on the wing aspect ratio, (AR). Some results obtained in Chapter 5 below are:

$$\frac{\mathrm{d}C_L}{\mathrm{d}\alpha} = a = \frac{a_\infty}{1 + a_\infty/\pi(AR)} \quad (\text{Eqn}(5.64))$$

where a_∞ is the lift curve slope for the two-dimensional wing, and the trailing vortex drag coefficient $C_{D\mathrm{v}}$ is given by

$$C_{D\mathrm{v}} = \frac{D_\mathrm{v}}{\frac{1}{2}\rho V^2 S} = \frac{C_L^2}{\pi(AR)}(1+\delta) \quad (\text{Eqn}(5.50))$$

where δ is a small positive number, constant for a given wing.

1.5.8 Lift-dependent drag

It has been seen that the induced drag coefficient is proportional to C_L^2, and may exist in an inviscid fluid. On a complete aircraft, interference at wing/fuselage, wing/engine-nacelle, and other such junctions leads to modification of the boundary layers over the isolated wing, fuselage, etc. This interference, which is actually part of the profile drag, usually varies with the lift coefficient in such a manner that it may be treated as of the form $(a + bC_L^2)$. The part of this profile drag coefficient which is represented by the term (bC_L^2) may be added to the induced drag. The sum so obtained is known as the lift-dependent drag coefficient. The lift-dependent drag is actually defined as 'the difference between the drag at a given lift coefficient and the drag at some datum lift coefficient'.

If this datum lift coefficient is taken to be zero, the total drag coefficient of a complete aeroplane may be taken, to a good approximation in most cases, as

$$C_D = C_{D_0} + kC_L^2$$

where C_{D_0} is the drag coefficient at zero lift, and kC_L^2 is the lift-dependent drag coefficient, denoted by C_{D_L}.

1.5.9 Aerofoil characteristics

Lift coefficient: incidence

This variation is illustrated in Fig. 1.23 for a two-dimensional (infinite span) wing. Considering first the full curve (a) which is for a moderately thick (13%) section of

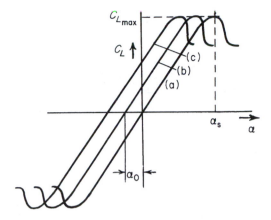

Fig. 1.23 Typical lift curves for sections of moderate thickness and various cambers

zero camber, it is seen to consist of a straight line passing through the origin, curving over at the higher values of C_L, reaching a maximum value of $C_{L_{max}}$ at an incidence of α_s, known as the stalling point. After the stalling point, the lift coefficient decreases, tending to level off at some lower value for higher incidences. The slope of the straight portion of the curve is called the two-dimensional lift-curve slope, $(\mathrm{d}C_L/\mathrm{d}\alpha)_\infty$ or a_∞. Its theoretical value for a thin section (strictly a curved or flat plate) is 2π per radian (see Section 4.4.1). For a section of finite thickness in air, a more accurate empirical value is

$$\left(\frac{\mathrm{d}C_L}{\mathrm{d}\alpha}\right)_\infty = 1.8\pi\left(1 + 0.8\frac{t}{c}\right) \tag{1.66}$$

The value of $C_{L_{max}}$ is a very important characteristic of the aerofoil since it determines the minimum speed at which an aeroplane can fly. A typical value for the type of aerofoil section mentioned is about 1.5. The corresponding value of α_s would be around 18°.

Curves (b) and (c) in Fig. 1.23 are for sections that have the same thickness distribution but that are cambered, (c) being more cambered than (b). The effect of camber is merely to reduce the incidence at which a given lift coefficient is produced, i.e. to shift the whole lift curve somewhat to the left, with negligible change in the value of the lift-curve slope, or in the shape of the curve. This shift of the curve is measured by the incidence at which the lift coefficient is zero. This is the no-lift incidence, denoted by α_0, and a typical value is $-3°$. The same reduction occurs in α_s. Thus a cambered section has the same value of $C_{L_{max}}$ as does its thickness distribution, but this occurs at a smaller incidence.

Modern, thin, sharp-nosed sections display a slightly different characteristic to the above, as shown in Fig. 1.24. In this case, the lift curve has two approximately straight portions, of different slopes. The slope of the lower portion is almost the same as that for a thicker section but, at a moderate incidence, the slope takes a different, smaller value, leading to a smaller value of $C_{L_{max}}$, typically of the order of unity. This change in the lift-curve slope is due to a change in the type of flow near the nose of the aerofoil.

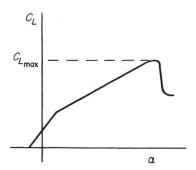

Fig. 1.24 Lift curve for a thin aerofoil section with small nose radius of curvature

Effect of aspect ratio on the $C_L : \alpha$ curve

The induced angle of incidence ε is given by

$$\varepsilon = \frac{kC_L}{\pi A}$$

where A is the aspect ratio and thus

$$\alpha_\infty = \alpha - \frac{kC_L}{\pi A}$$

Considering a number of wings of the same symmetrical section but of different aspect ratios the above expression leads to a family of C_L, α curves, as in Fig. 1.25, since the actual lift coefficient at a given section of the wing is equal to the lift coefficient for a two-dimensional wing at an incidence of α_∞.

For highly swept wings of very low aspect ratio (less than 3 or so), the lift curve slope becomes very small, leading to values of $C_{L_{max}}$ of about 1.0, occurring at stalling incidences of around 45°. This is reflected in the extreme nose-up landing attitudes of many aircraft designed with wings of this description.

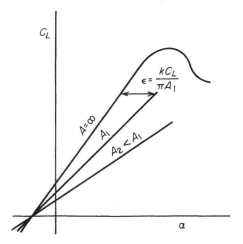

Fig. 1.25 Influence of wing aspect ratio on the lift curve

Effect of Reynolds number on the $C_L : \alpha$ curve

Reduction of Reynolds number moves the transition point of the boundary layer rearwards on the upper surface of the wing. At low values of Re this may permit a laminar boundary layer to extend into the adverse pressure gradient region of the aerofoil. As a laminar boundary layer is much less able than a turbulent boundary layer to overcome an adverse pressure gradient, the flow will separate from the surface at a lower angle of incidence. This causes a reduction of $C_{L_{max}}$. This is a problem that exists in model testing when it is always difficult to match full-scale and model Reynolds numbers. Transition can be fixed artificially on the model by roughening the model surface with carborundum powder at the calculated full-scale point.

Drag coefficient: lift coefficient

For a two-dimensional wing at low Mach numbers the drag contains no induced or wave drag, and the drag coefficient is C_{D_0}. There are two distinct forms of variation of C_D with C_L, both illustrated in Fig. 1.26.

Curve (a) represents a typical conventional aerofoil with C_{D_0} fairly constant over the working range of lift coefficient, increasing rapidly towards the two extreme values of C_L. Curve (b) represents the type of variation found for low-drag aerofoil sections. Over much of the C_L range the drag coefficient is rather larger than for the conventional type of aerofoil, but within a restricted range of lift coefficient (C_{L_1} to C_{L_2}) the profile drag coefficient is considerably less. This range of C_L is known as the favourable range for the section, and the low drag coefficient is due to the design of the aerofoil section, which permits a comparatively large extent of laminar boundary layer. It is for this reason that aerofoils of this type are also known as laminar-flow sections. The width and depth of this favourable range or, more graphically, low-drag bucket, is determined by the shape of the thickness distribution. The central value of the lift coefficient is known as the optimum or ideal lift coefficient, $C_{L_{opt}}$ or C_{L_i}. Its value is decided by the shape of the camber line, and the degree of camber, and thus the position of the favourable range may be placed where desired by suitable design of the camber line. The favourable range may be placed to cover the most common range of lift coefficient for a particular aeroplane, e.g. C_{L_2} may be slightly larger than the lift coefficient used on the climb, and C_{L_1} may be

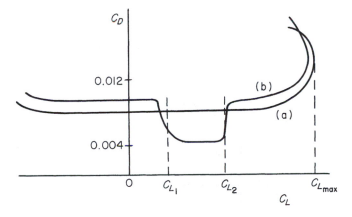

Fig. 1.26 Typical variation of sectional drag coefficient with lift coefficient

slightly less than the cruising lift coefficient. In such a case the aeroplane will have the benefit of a low value of the drag coefficient for the wing throughout most of the flight, with obvious benefits in performance and economy. Unfortunately it is not possible to have large areas of laminar flow on swept wings at high Reynolds numbers. To maintain natural laminar flow, sweep-back angles are limited to about 15°.

The effect of a finite aspect ratio is to give rise to induced drag and this drag coefficient is proportional to C_L^2, and must be added to the curves of Fig. 1.26.

Drag coefficient: (lift coefficient)^2

Since

$$C_{D_V} = \frac{C_L^2}{\pi A}(1 + \delta)$$

it follows that a curve of C_{D_V} against C_L^2 will be a straight line of slope $(1 + \delta)/\pi A$. If the curve C_{D_0} against C_L^2 from Fig. 1.26 is added to the induced drag coefficient, that is to the straight line, the result is the total drag coefficient variation with C_L^2, as shown in Fig. 1.27 for the two types of section considered in Fig. 1.26. Taking an

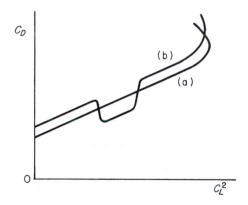

Fig. 1.27 Variation of total wing drag coefficient with (lift coefficient)^2

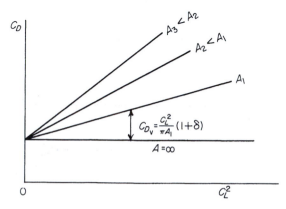

Fig. 1.28 Idealized variation of total wing drag coefficient with (lift coefficient)^2 for a family of three-dimensional wings of various aspect ratios

idealized case in which C_{D_0} is independent of lift coefficient, the $C_{D_v}:(C_L)^2$ curve for a family of wings of various aspect ratios as is shown in Fig. 1.28.

Pitching moment coefficient

In Section 1.5.4 it was shown that

$$\frac{\mathrm{d}C_M}{\mathrm{d}C_L} = \text{constant}$$

the value of the constant depending on the point of the aerofoil section about which C_M is measured. Thus a curve of C_M against C_L is theoretically as shown in Fig. 1.29.

Line (a) for which $\mathrm{d}C_M/\mathrm{d}C_L \simeq -\frac{1}{4}$ is for C_M measured about the leading edge. Line (c), for which the slope is zero, is for the case where C_M is measured about the aerodynamic centre. Line (b) would be obtained if C_M were measured about a point between the leading edge and the aerodynamic centre, while for (d) the reference point is behind the aerodynamic centre. These curves are straight only for moderate values of C_L. As the lift coefficient approaches $C_{L_{\max}}$, the C_M against C_L curve departs from the straight line. The two possibilities are sketched in Fig. 1.30.

For curve (a) the pitching moment coefficient becomes more negative near the stall, thus tending to decrease the incidence, and unstall the wing. This is known as a stable break. Curve (b), on the other hand, shows that, near the stall, the pitching moment coefficient becomes less negative. The tendency then is for the incidence to

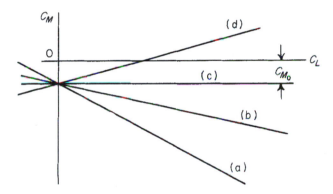

Fig. 1.29 Variation of C_M with C_L for an aerofoil section, for four different reference points

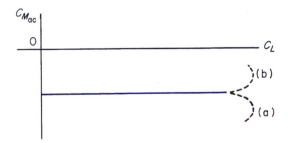

Fig. 1.30 The behaviour of the pitching moment coefficient in the region of the stalling point, showing stable and unstable breaks

increase, aggravating the stall. Such a characteristic is an unstable break. This type of characteristic is commonly found with highly swept wings, although measures can be taken to counteract this undesirable behaviour.

Exercises

1 Verify the dimensions and units given in Table 1.1.

2 The constant of gravitation G is defined by

$$F = G\frac{mM}{r^2}$$

where F is the gravitational force between two masses m and M whose centres of mass are distance r apart. Find the dimensions of G, and its units in the SI system.
(*Answer*: MT^2L^{-3}, $\text{kg s}^2\,\text{m}^{-3}$)

3 Assuming the period of oscillation of a simple pendulum to depend on the mass of the bob, the length of the pendulum and the acceleration due to gravity g, use the theory of dimensional analysis to show that the mass of the bob is not, in fact, relevant and find a suitable expression for the period of oscillation in terms of the other variables.   (*Answer*: $t = c\sqrt{l/g}$)

4 A thin flat disc of diameter D is rotated about a spindle through its centre at a speed of ω radians per second, in a fluid of density ρ and kinematic viscosity v. Show that the power P needed to rotate the disc may be expressed as:

(a)
$$P = \rho\omega^3 D^5 f\left(\frac{v}{\omega D^2}\right)$$

(b)
$$P = \frac{\rho v^3}{D} h\left(\frac{\omega D^2}{v}\right)$$

Note: for (a) solve in terms of the index of v and for (b) in terms of the index of ω.

Further, show that $\omega D^2/v$, $PD/\rho v^3$ and $P/\rho\omega^3 D^5$ are all non-dimensional quantities.   (CU)

5 Spheres of various diameters D and densities σ are allowed to fall freely under gravity through various fluids (represented by their densities ρ and kinematic viscosities v) and their terminal velocities V are measured.

Find a rational expression connecting V with the other variables, and hence suggest a suitable form of graph in which the results could be presented.

Note: there will be 5 unknown indices, and therefore 2 must remain undetermined, which will give 2 unknown functions on the right-hand side. Make the unknown indices those of σ and v.

$$\left(\textit{Answer}: V = \sqrt{Dg}\, f\left(\frac{\sigma}{\rho}\right) h\left(\frac{D}{v}\sqrt{Dg}\right), \text{ therefore plot curves of } \frac{V}{\sqrt{Dg}}\right.$$

$$\left. \text{against } \left(\frac{D}{v}\right)\sqrt{Dg} \text{ for various values of } \sigma/\rho\right)$$

6 An aeroplane weighs 60 000 N and has a wing span of 17 m. A 1/10th scale model is tested, flaps down, in a compressed-air tunnel at 15 atmospheres pressure and 15 °C at various speeds. The maximum lift on the model is measured at the various speeds, with the results as given below:

Speed (m s^{-1})	20	21	22	23	24
Maximum lift (N)	2960	3460	4000	4580	5200

Estimate the minimum flying speed of the aircraft at sea-level, i.e. the speed at which the maximum lift of the aircraft is equal to its weight. (*Answer*: 33 m s^{-1})

7 The pressure distribution over a section of a two-dimensional wing at 4° incidence may be approximated as follows: Upper surface; C_p constant at −0.8 from the leading edge to 60% chord, then increasing linearly to +0.1 at the trailing edge: Lower surface; C_p constant at −0.4 from the LE to 60% chord, then increasing linearly to +0.1 at the TE. Estimate the lift coefficient and the pitching moment coefficient about the leading edge due to lift. (*Answer*: 0.3192; −0.13)

8 The static pressure is measured at a number of points on the surface of a long circular cylinder of 150 mm diameter with its axis perpendicular to a stream of standard density at 30 m s^{-1}. The pressure points are defined by the angle θ, which is the angle subtended at the centre by the arc between the pressure point and the front stagnation point. In the table below values are given of $p - p_0$, where p is the pressure on the surface of the cylinder and p_0 is the undisturbed pressure of the free stream, for various angles θ, all pressures being in N m^{-2}. The readings are identical for the upper and lower halves of the cylinder. Estimate the form pressure drag per metre run, and the corresponding drag coefficient.

θ (degrees)	0	10	20	30	40	50	60	70	80	90	100	110	120
$p - p_0$ (N m^{-2})	+569	+502	+301	−57	−392	−597	−721	−726	−707	−660	−626	−588	−569

For values of θ between 120° and 180°, $p - p_0$ is constant at −569 N m^{-2}. (*Answer*: $C_D = 0.875$, $D = 7.25$ N m^{-1})

9 A sailplane has a wing of 18 m span and aspect ratio of 16. The fuselage is 0.6 m wide at the wing root, and the wing taper ratio is 0.3 with square-cut wing-tips. At a true air speed of 115 km h^{-1} at an altitude where the relative density is 0.7 the lift and drag are 3500 N and 145 N respectively. The wing pitching moment coefficient about the $\frac{1}{4}$-chord point is −0.03 based on the gross wing area and the aerodynamic mean chord. Calculate the lift and drag coefficients based on the gross wing area, and the pitching moment about the $\frac{1}{4}$ chord point. (*Answer*: $C_L = 0.396$, $C_D = 0.0169$, $M = -322$ N m since $\bar{c}_A = 1.245$ m)

10 Describe qualitatively the results expected from the pressure plotting of a conventional, symmetrical, low-speed, two-dimensional aerofoil. Indicate the changes expected with incidence and discuss the processes for determining the resultant forces. Are any further tests needed to complete the determination of the overall forces of lift and drag? Include in the discussion the order of magnitude expected for the various distributions and forces described. (U of L)

11 Show that for geometrically similar aerodynamic systems the non-dimensional force coefficients of lift and drag depend on Reynolds number and Mach number only. Discuss briefly the importance of this theorem in wind-tunnel testing and simple performance theory. (U of L)

2

Governing equations
of fluid mechanics

Preamble

This chapter is the first of two which set out the fundamental fluid dynamics required for the further development of aerodynamics. In it the study of air in motion starts with the physics and mathematics of one-dimensional fluid motion. Many of the physical phenomena evident in all stages of aerodynamics are most readily approached by considering the one-dimensional mode, without prejudice to the wider analysis of two- and three-dimensional motions.

The laws governing the changes in the physical properties of air are first covered and the relevant mathematics introduced. These laws are applied to the accelerating gas as it moves out of the low-speed (incompressible) regime and into the transonic and supersonic regimes where the abrupt changes in properties are manifest.

2.1 Introduction

The physical laws that govern fluid flow are deceptively simple. Paramount among them is Newton's second law of motion which states that:

Mass × Acceleration = Applied force

In fluid mechanics we prefer to use the equivalent form of

Rate of change of momentum = Applied force

Apart from the *principles of conservation of mass* and, where appropriate, *conservation of energy*, the remaining physical laws required relate solely to determining the forces involved. For a wide range of applications in aerodynamics the only forces involved are the *body forces* due to the action of gravity* (which, of course, requires the use of Newton's theory of gravity; but only in a very simple way); *pressure forces* (these are found by applying Newton's laws of motion and require no further physical laws or principles); and *viscous forces*. To determine the viscous forces we

---

* Body forces are commonly neglected in aerodynamics.

need to supplement Newton's laws of motion with a constitutive law. For pure homogeneous fluids (such as air and water) this constitutive law is provided by the Newtonian fluid model, which as the name suggests also originated with Newton. In simple terms the constitutive law for a Newtonian fluid states that:

$$\text{Viscous stress} \propto \text{Rate of strain}$$

At a fundamental level these simple physical laws are, of course, merely theoretical models. The principal theoretical assumption is that the fluid consists of continuous matter – the so-called *continuum* model. At a deeper level we are, of course, aware that the fluid is not a continuum, but is better considered as consisting of myriads of individual molecules. In most engineering applications even a tiny volume of fluid (measuring, say, $1\,\mu m^3$) contains a large number of molecules. Equivalently, a typical molecule travels on average a very short distance (known as the mean free path) before colliding with another. In typical aerodynamics applications the m.f.p. is less than 100 nm, which is very much smaller than any relevant scale characterizing quantities of engineering significance. Owing to this disparity between the m.f.p. and relevant length scales, we may expect the equations of fluid motion, based on the continuum model, to be obeyed to great precision by the fluid flows found in almost all engineering applications. This expectation is supported by experience. It also has to be admitted that the continuum model also reflects our everyday experience of the real world where air and water appear to our senses to be continuous substances. There are exceptional applications in modern engineering where the continuum model breaks down and ceases to be a good approximation. These may involve very small-scale motions, e.g. nanotechnology and Micro-Electro-Mechanical Systems (MEMS) technology,* where the relevant scales can be comparable to the m.f.p. Another example is rarefied gas dynamics (e.g. re-entry vehicles) where there are so few molecules present that the m.f.p. becomes comparable to the dimensions of the vehicle.

We first show in Section 2.2 how the principles of conservation of mass, momentum and energy can be applied to one-dimensional flows to give the governing equations of fluid motion. For this rather special case the flow variables, velocity and pressure, only vary at most with one spatial coordinate. Real fluid flows are invariably three-dimensional to a greater or lesser degree. Nevertheless, in order to understand how the conservation principles lead to equations of motion in the form of partial differential equations, it is sufficient to see how this is done for a two-dimensional flow. So this is the approach we will take in Sections 2.4–2.8. It is usually straightforward, although significantly more complicated, to extend the principles and methods to three dimensions. However, for the most part, we will be content to carry out any derivations in two dimensions and to merely quote the final result for three-dimensional flows.

2.1.1 Air flow

Consider an aeroplane in steady flight. To an observer on the ground the aeroplane is flying into air substantially at rest, assuming no wind, and any movement of the air is caused directly by the motion of the aeroplane through it. The pilot of the aeroplane, on the other hand, could consider that he is stationary, and that a stream of air is flowing past him and that the aeroplane modifies the motion of the air. In fact both

---

* Recent reviews are given by M. Gad-el-Hak (1999) The fluid mechanics of microdevices – The Freeman Scholar Lecture. *J. Fluids Engineering*, **121**, 5–33; L. Löfdahl and M. Gad-el-Hak (1999) MEMS applications in turbulence and flow control. *Prog. in Aerospace Sciences*, **35**, 101–203.

viewpoints are mathematically and physically correct. Both observers may use the same equations to study the mutual effects of the air and the aeroplane and they will both arrive at the same answers for, say, the forces exerted by the air on the aeroplane. However, the pilot will find that certain terms in the equations become, from his viewpoint, zero. He will, therefore, find that his equations are easier to solve than will the ground-based observer. Because of this it is convenient to regard most problems in aerodynamics as cases of air flowing past a body at rest, with consequent simplification of the mathematics.

Types of flow

The flow round a body may be steady or unsteady. A steady flow is one in which the flow parameters, e.g. speed, direction, pressure, may vary from point to point in the flow but at any point are constant with respect to time, i.e. measurements of the flow parameters at a given point in the flow at various times remain the same. In an unsteady flow the flow parameters at any point vary with time.

2.1.2   A comparison of steady and unsteady flow

Figure 2.1a shows a section of a stationary wing with air flowing past. The velocity of the air a long way from the wing is constant at V, as shown. The flow parameters are measured at some point fixed relative to the wing, e.g. at P(x, y). The flow perturbations produced at P by the body will be the same at all times, i.e. the flow is steady relative to a set of axes fixed in the body.

Figure 2.1b represents the same wing moving at the same speed V through air which, a long way from the body, is at rest. The flow parameters are measured at a point P$'(x', y')$ fixed relative to the stationary air. The wing thus moves past P$'$. At times t_1, when the wing is at A_1, P$'$ is a fairly large distance ahead of the wing, and the perturbations at P$'$ are small. Later, at time t_2, the wing is at A_2, directly beneath P$'$, and the perturbations are much larger. Later still, at time t_3, P$'$ is far behind the wing, which is now at A_3, and the perturbations are again small. Thus, the perturbation at P$'$ has started from a small value, increased to a maximum, and finally decreased back to a small value. The perturbation at the fixed point P$'$ is, therefore, not constant with respect to time, and so the flow, referred to axes fixed in the fluid, is not steady. Thus, changing the axes of reference from a set fixed relative to the air flow, to a different set fixed relative to the body, changes the flow from unsteady to steady. This produces the

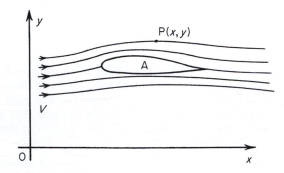

Fig. 2.1a Air moves at speed V past axes fixed relative to aerofoil

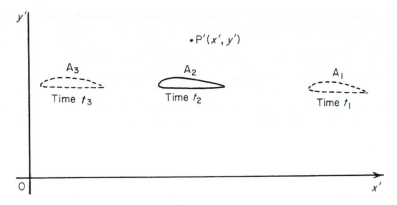

Fig. 2.1b Aerofoil moves at speed *V* through air initially at rest. Axes *Ox' Oy'* fixed relative to undisturbed air at rest

mathematical simplification mentioned earlier by eliminating time from the equations. Since the flow relative to the air flow can, by a change of axes, be made steady, it is sometimes known as 'quasi-steady'.

True unsteady flow

An example of true unsteady flow is the wake behind a bluff body, e.g. a circular cylinder (Fig. 2.2). The air is flowing from left to right, and the system of eddies or vortices behind the cylinder is moving in the same direction at a somewhat lower speed. This region of slower moving fluid is the 'wake'. Consider a point P, fixed relative to the cylinder, in the wake. Sometimes the point will be immersed in an eddy and sometimes not. Thus the flow parameters will be changing rapidly at P, and the flow there is unsteady. Moreover, it is impossible to find a set of axes relative to which the flow is steady. At a point Q well outside the wake the fluctuations are so small that they may be ignored and the flow at Q may, with little error, be regarded as steady. Thus, even though the flow in some region may be unsteady, there may be some other region where the unsteadiness is negligibly small, so that the flow there may be regarded as steady with sufficient accuracy for all practical purposes.

Three concepts that are useful in describing fluid flows are:

(i) A streamline – defined as 'an imaginary line drawn in the fluid such that there is no flow across it at any point', or alternatively as 'a line that is always in the same

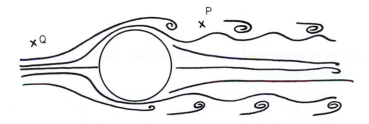

Fig. 2.2 True unsteady flow

direction as the local velocity vector'. Since this is identical to the condition at a solid boundary it follows that:

(a) any streamline may be replaced by a solid boundary without modifying the flow. (This only strictly true if viscous effects are ignored.)
(b) any solid boundary is itself a streamline of the flow around it.

(ii) A filament (or streak) line – the line taken up by successive particles of fluid passing through some given point. A fine filament of smoke injected into the flow through a nozzle traces out a filament line. The lines shown in Fig. 2.2 are examples of this.

(iii) A path line or particle path – the path traced out by any one particle of the fluid in motion.

In unsteady flow, these three are in general different, while in steady flow all three are identical. Also in steady flow it is convenient to define a *stream tube* as an imaginary bundle of adjacent streamlines.

2.2 One-dimensional flow: the basic equations

In all real flow situations the physical laws of conservation apply. These refer to the conservation respectively of mass, momentum and energy. The equation of state completes the set that needs to be solved if some or all of the parameters controlling the flow are unknown. If a real flow can be 'modelled' by a similar but simplified system then the degree of complexity in handling the resulting equations may be considerably reduced.

Historically, the lack of mathematical tools available to the engineer required that considerable simplifying assumptions should be made. The simplifications used depend on the particular problem but are not arbitrary. In fact, judgement is required to decide which parameters in a flow process may be reasonably ignored, at least to a first approximation. For example, in much of aerodynamics the gas (air) is considered to behave as an incompressible fluid (see Section 2.3.4), and an even wider assumption is that the air flow is unaffected by its viscosity. This last assumption would appear at first to be utterly inappropriate since viscosity plays an important role in the mechanism by which aerodynamic force is transmitted from the air flow to the body and vice versa. Nevertheless the science of aerodynamics progressed far on this assumption, and much of the aeronautical technology available followed from theories based on it.

Other examples will be invoked from time to time and it is salutory, and good engineering practice, to acknowledge those 'simplifying' assumptions made in order to arrive at an understanding of, or a solution to, a physical problem.

2.2.1 One-dimensional flow: the basic equations of conservation

A prime simplification of the algebra involved without any loss of physical significance may be made by examining the changes in the flow properties along a stream tube that is essentially straight or for which the cross-section changes slowly (i.e. so-called quasi-one-dimensional flow).

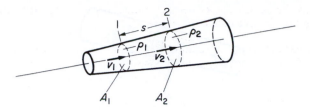

Fig. 2.3 The stream tube for conservation of mass

The conservation of mass

This law satisfies the belief that in normally perceived engineering situations matter cannot be created or destroyed. For steady flow in the stream tube shown in Fig. 2.3 let the flow properties at the stations 1 and 2 be a distance s apart, as shown. If the values for the flow velocity v and the density ρ at section 1 are the same across the tube, which is a reasonable assumption if the tube is thin, then the quantity flowing into the volume comprising the element of stream tube is:

$$\text{velocity} \times \text{area} = v_1 A_1$$

The mass flowing in through section 1 is

$$\rho_1 v_1 A_1 \tag{2.1}$$

Similarly the mass outflow at section 2, on making the same assumptions, is

$$\rho_2 v_2 A_2 \tag{2.2}$$

These two quantities (2.1) and (2.2) must be the same if the tube does not leak or gain fluid and if matter is to be conserved. Thus

$$\rho_1 v_1 A_1 = \rho_2 v_2 A_2 \tag{2.3}$$

or in a general form:

$$\rho v A = \text{constant} \tag{2.4}$$

The conservation of momentum

Conservation of momentum requires that the time rate of change of momentum in a given direction is equal to the sum of the forces acting in that direction. This is known as Newton's second law of motion and in the model used here the forces concerned are gravitational (body) forces and the surface forces.

Consider a fluid in steady flow, and take any small stream tube as in Fig. 2.4. s is the distance measured along the axis of the stream tube from some arbitrary origin. A is the cross-sectional area of the stream tube at distance s from the arbitrary origin.

p, ρ, and v represent pressure, density and flow speed respectively.

A, p, ρ, and v vary with s, i.e. with position along the stream tube, but not with time since the motion is steady.

Now consider the small element of fluid shown in Fig. 2.5, which is immersed in fluid of varying pressure. The element is the right frustrum of a cone of length δs, area A at the upstream section, area $A + \delta A$ on the downstream section. The pressure acting on one face of the element is p, and on the other face is $p + (\mathrm{d}p/\mathrm{d}s)\delta s$. Around

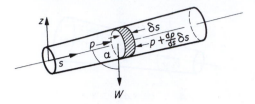

Fig. 2.4 The stream tube and element for the momentum equation

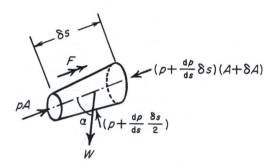

Fig. 2.5 The forces on the element

the curved surface the pressure may be taken to be the mean value $p + \frac{1}{2}(\mathrm{d}p/\mathrm{d}s)\delta s$. In addition the weight W of the fluid in the element acts vertically as shown. Shear forces on the surface due to viscosity would add another force, which is ignored here.

As a result of these pressures and the weight, there is a resultant force F acting along the axis of the cylinder where F is given by

$$F = pA - \left(p + \frac{\mathrm{d}p}{\mathrm{d}s}\delta s\right)(A + \delta A) + \left(p + \frac{\mathrm{d}p}{\mathrm{d}s}\frac{\delta s}{2}\right)\delta A - W \cos \alpha \qquad (2.5)$$

where α is the angle between the axis of the stream tube and the vertical.

From Eqn (2.5) it is seen that on neglecting quantities of small order such as $(\mathrm{d}p/\mathrm{d}s)\delta s\delta A$ and cancelling,

$$F = -\frac{\mathrm{d}p}{\mathrm{d}s}A\,\delta s - \rho g\,A\,(\delta s)\cos\alpha \qquad (2.6)$$

since the gravitational force on the fluid in the element is $\rho g\,A\,\delta s$, i.e. volume × density × g.

Now, Newton's second law of motion (force = mass × acceleration) applied to the element of Fig. 2.5, gives

$$-\rho g\,A\,\delta s\cos\alpha - \frac{\mathrm{d}p}{\mathrm{d}s}A\,\delta s = \rho A\,\delta s\frac{\mathrm{d}v}{\mathrm{d}t} \qquad (2.7)$$

where t represents time. Dividing by $A\,\delta s$ this becomes

$$-\rho g\cos\alpha - \frac{\mathrm{d}p}{\mathrm{d}s} = \rho\frac{\mathrm{d}v}{\mathrm{d}t}$$

But

$$\frac{dv}{dt} = \frac{dv}{ds}\frac{ds}{dt} = v\frac{dv}{ds}$$

and therefore

$$\rho v\frac{dv}{ds} + \frac{dp}{ds} + \rho g\cos\alpha = 0$$

or

$$v\frac{dv}{ds} + \frac{1}{\rho}\frac{dp}{ds} + g\cos\alpha = 0$$

Integrating along the stream tube, this becomes

$$\int\frac{dp}{\rho} + \int v\,dv + g\int\cos\alpha\,ds = \text{constant}$$

but since

$$\int\cos\alpha\,ds = \text{increase in vertical coordinate } z$$

and

$$\int v\,dv = \frac{1}{2}v^2$$

then

$$\int\frac{dp}{\rho} + \frac{1}{2}v^2 + gz = \text{constant} \tag{2.8}$$

This result is known as Bernoulli's equation and is discussed below.

The conservation of energy

Conservation of energy implies that changes in energy, heat transferred and work done by a system in steady operation are in balance. In seeking an equation to represent the conservation of energy in the steady flow of a fluid it is useful to consider a length of stream tube, e.g. between sections 1 and 2 (Fig. 2.6), as

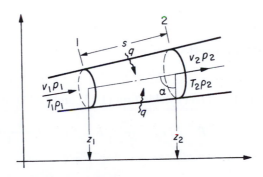

Fig. 2.6 Control volume for the energy equation

constituting the *control surface* of a 'thermodynamic system' or *control volume*. At sections 1 and 2, let the fluid properties be as shown.

Then unit mass of fluid entering the system through section will possess internal energy $c_V T_1$, kinetic energy $v_1^2/2$ and potential energy gz_1, i.e.

$$\left(c_V T_1 + \frac{v_1^2}{2} + gz_1\right) \tag{2.9a}$$

Likewise on exit from the system across section 2 unit mass will possess energy

$$\left(c_V T_2 + \frac{v_2^2}{2} + gz_2\right) \tag{2.9b}$$

Now to enter the system, unit mass possesses a volume $1/\rho_1$ which must push against the pressure p_1 and utilize energy to the value of $p_1 \times 1/\rho_1$ pressure $\times$ (specific) volume. At exit p_2/ρ_2 is utilized in a similar manner.

In the meantime, the system accepts, or rejects, heat q per unit mass. As all the quantities are flowing steadily, the energy entering plus the heat transfer must equal the energy leaving.* Thus, with a positive heat transfer it follows from conservation of energy

$$c_V T_1 + \frac{v_1^2}{2} + \frac{p_1}{\rho_1} + gz_1 + q = c_V T_2 + \frac{v_2^2}{2} + \frac{p_2}{\rho_2} + gz_2$$

However, enthalpy per unit mass of fluid is $c_V T + p/\rho = c_p T$. Thus

$$\left(c_p T_2 + \frac{v_2^2}{2} + gz_2\right) - \left(c_p T_1 + \frac{v_1^2}{2} + gz_1\right) = q$$

or in differential form

$$\frac{d}{ds}\left(c_p T + \frac{v^2}{2} + gs\cos\alpha\right) = \frac{dq}{ds} \tag{2.10}$$

For an adiabatic (no heat transfer) horizontal flow system, Eqn (2.10) becomes zero and thus

$$c_p T + \frac{v^2}{2} = \text{constant} \tag{2.11}$$

The equation of state

The equation of state for a perfect gas is

$$p/(\rho T) = R$$

Substituting for p/ρ in Eqn (1.11) yields Eqn (1.13) and (1.14), namely

$$c_p - c_V = R, \qquad c_p = \frac{\gamma}{\gamma - 1} R \qquad c_V = \frac{1}{\gamma - 1} R$$

---

* It should be noted that in a general system the fluid would also do work which should be taken into the equation, but it is disregarded here for the particular case of flow in a stream tube.

The first law of thermodynamics requires that the gain in internal energy of a mass of gas plus the work done by the mass is equal to the heat supplied, i.e. for unit mass of gas with no heat transfer

$$E + \int p\mathrm{d}\left(\frac{1}{\rho}\right) = \text{constant}$$

or

$$\mathrm{d}E + p\mathrm{d}\left(\frac{1}{\rho}\right) = 0 \tag{2.12}$$

Differentiating Eqn (1.10) for enthalpy gives

$$\mathrm{d}h = \mathrm{d}E + p\mathrm{d}\left(\frac{1}{\rho}\right) + \frac{1}{\rho}\mathrm{d}p = 0 \tag{2.13}$$

and combining Eqns (2.12) and (2.13) yields

$$\mathrm{d}h = \frac{1}{\rho}\mathrm{d}p \tag{2.14}$$

But

$$\mathrm{d}h = c_p\mathrm{d}T = \frac{c_p}{R}\mathrm{d}\left(\frac{p}{\rho}\right) = \frac{\gamma}{\gamma - 1}\left[\frac{1}{\rho}\mathrm{d}p + p\mathrm{d}\left(\frac{1}{\rho}\right)\right] \tag{2.15}$$

Therefore, from Eqns (2.14) and (2.15)

$$\frac{\mathrm{d}p}{p} + \gamma\rho\mathrm{d}\left(\frac{1}{\rho}\right) = 0$$

which on integrating gives

$$\ln p + \gamma\ln\left(\frac{1}{\rho}\right) = \text{constant}$$

or

$$p = k\rho^\gamma$$

where k is a constant. This is the isentropic relationship between pressure and density, and has been replicated for convenience from Eqn (1.24).

The momentum equation for an incompressible fluid

Provided velocity and pressure changes are small, density changes will be very small, and it is permissible to assume that the density ρ is constant throughout the flow. With this assumption, Eqn (2.8) may be integrated as

$$\int \mathrm{d}p + \frac{1}{2}\rho v^2 + \rho g z = \text{constant}$$

Performing this integration between two conditions represented by suffices 1 and 2 gives

$$(p_2 - p_1) + \frac{1}{2}\rho(v_2^2 - v_1^2) + \rho g(z_2 - z_1) = 0$$

i.e.

$$p_1 + \frac{1}{2}\rho v_1^2 + \rho g z_1 = p_2 + \frac{1}{2}\rho v_2^2 + \rho g z_2$$

In the foregoing analysis 1 and 2 were completely arbitrary choices, and therefore the same equation must apply to conditions at any other points. Thus

$$p + \frac{1}{2}\rho v^2 + \rho g z = \text{constant} \tag{2.16}$$

This is *Bernoulli's equation* for an incompressible fluid, i.e. a fluid that cannot be compressed or expanded, and for which the density is invariable. Note that Eqn (2.16) can be applied more generally to two- and three-dimensional steady flows, provided that viscous effects are neglected. In the more general case, however, it is important to note that Bernoulli's equation can only be applied along a streamline, and in certain cases the constant may vary from streamline to streamline.

2.2.2   Comments on the momentum and energy equations

Referring back to Eqn (2.8), that expresses the conservation of momentum in algebraic form,

$$\int \frac{\mathrm{d}p}{\rho} + \frac{1}{2}v^2 + gz = \text{constant}$$

the first term is the internal energy of unit mass of the air, $\frac{1}{2}v^2$ is the kinetic energy of unit mass and gz is the potential energy of unit mass. Thus, Bernoulli's equation in this form is really a statement of the principle of conservation of energy in the absence of heat exchanged and work done. As a corollary, it applies only to flows where there is no mechanism for the dissipation of energy into some form not included in the above three terms. In aerodynamics a common form of energy dissipation is that due to viscosity. Thus, strictly the equation cannot be applied in this form to a flow where the effects of viscosity are appreciable, such as that in a boundary layer.

2.3   The measurement of air speed

2.3.1   The Pitôt-static tube

Consider an instrument of the form sketched in Fig. 2.7, called a Pitôt-static tube. It consists of two concentric tubes A and B. The mouth of A is open and faces directly into the airstream, while the end of B is closed on to A, causing B to be sealed off. Some very fine holes are drilled in the wall of B, as at C, allowing B to communicate with the surrounding air. The right-hand ends of A and B are connected to opposite sides of a manometer. The instrument is placed into a stream of air, with the

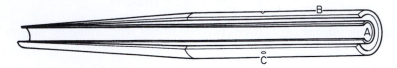

Fig. 2.7 The simple Pitôt-static tube

mouth of A pointing directly upstream, the stream being of speed v and of static pressure p. The air flowing past the holes at C will be moving at a speed very little different from v and its pressure will, therefore, be equal to p, and this pressure will be communicated to the interior of tube B through the holes C. The pressure in B is, therefore, the static pressure of the stream.

Air entering the mouth of A will, on the other hand, be brought to rest (in the ultimate analysis by the fluid in the manometer). Its pressure will therefore be equal to the total head of the stream. As a result a pressure difference exists between the air in A and that in B, and this may be measured on the manometer. Denote the pressure in A by p_A, that in B by p_B, and the difference between them by Δp. Then

$$\Delta p = p_A - p_B \tag{2.17}$$

But, by Bernoulli's equation (for incompressible flow)

$$p_A + \frac{1}{2}\rho(0)^2 = p_B + \frac{1}{2}\rho v^2$$

and therefore

$$p_A - p_B = \frac{1}{2}\rho v^2 \tag{2.18}$$

or

$$\Delta p = \frac{1}{2}\rho v^2$$

whence

$$v = \sqrt{2\Delta p/\rho} \tag{2.19}$$

The value of ρ, which is constant in incompressible flow, may be calculated from the ambient pressure and the temperature. This, together with the measured value of Δp, permits calculation of the speed v.*

The quantity $\frac{1}{2}\rho v^2$ is the *dynamic pressure* of the flow. Since $p_A = $ total pressure $= p_0$ (i.e. the pressure of the air at rest, also referred to as the stagnation pressure), and $p_B = $ static pressure $= p$, then

$$p_0 - p = \frac{1}{2}\rho v^2 \tag{2.20}$$

which may be expressed in words as

$$\text{stagnation pressure} - \text{static pressure} = \text{dynamic pressure}$$

It should be noted that this equation applies at all speeds, but the dynamic pressure is equal to $\frac{1}{2}\rho v^2$ only in incompressible flow. Note also that

$$\frac{1}{2}\rho v^2 = [\text{ML}^{-3}\text{L}^2\text{T}^{-2}] = [\text{ML}^{-1}\text{T}^{-2}]$$
$$= \text{units of pressure}$$

as is of course essential.

---

* Note that, notwithstanding the formal restriction of Bernoulli's equation to inviscid flows, the Pitôt-static tube is commonly used to determine the local velocity in wakes and boundary layers with no apparent loss of accuracy.

Defining the *stagnation pressure coefficient* as

$$C_{p_0} = \frac{p_0 - p}{\frac{1}{2}\rho v^2}$$

(2.21)

it follows immediately from Eqn (2.20) that for incompressible flow

$$C_{p_0} = 1 \text{(always)}$$

(2.22)

2.3.2   The pressure coefficient

In Chapter 1 it was seen that it is often convenient to express variables in a non-dimensional coefficient form. The coefficient of pressure is introduced in Section 1.5.3. The stagnation pressure coefficient has already been defined as

$$C_{p_0} = \frac{p_0 - p}{\frac{1}{2}\rho v^2}$$

This is a special case of the general 'pressure coefficient' defined by pressure coefficient:

$$C_{p_1} = \frac{p - p_\infty}{\frac{1}{2}\rho v^2}$$

(2.23)

where C_{p_1} = pressure coefficient
$\quad\quad p$ = static pressure at some point in the flow where the velocity is q
$\quad\quad p_\infty$ = static pressure of the undisturbed flow
$\quad\quad \rho$ = density of the undisturbed flow
$\quad\quad v$ = speed of the undisturbed flow

Now, in incompressible flow,

$$p + \frac{1}{2}\rho q^2 = p_\infty + \frac{1}{2}\rho v^2$$

Then

$$p - p_\infty = \frac{1}{2}\rho(v^2 - q^2)$$

and therefore

$$C_p = 1 - \left(\frac{q}{v}\right)^2$$

(2.24)

Then

(i) if C_p is positive $p > p_\infty$ and $q < v$
(ii) if C_p is zero $p = p_\infty$ and $q = v$
(iii) if C_p is negative $p < p_\infty$ and $q > v$

2.3.3   The air-speed indicator: indicated and equivalent air speeds

A Pitôt-static tube is commonly used to measure air speed both in the laboratory and on aircraft. There are, however, differences in the requirements for the two applications. In the laboratory, liquid manometers provide a simple and direct method for measuring pressure. These would be completely unsuitable for use on an aircraft where a pressure transducer is used that converts the pressure measurement into an electrical signal. Pressure transducers are also becoming more and more commonly used for laboratory measurements.

When the measured pressure difference is converted into air speed, the correct value for the air density should, of course, be used in Eqn (2.19). This is easy enough in the laboratory, although for accurate results the variation of density with the ambient atmospheric pressure in the laboratory should be taken into account. At one time it was more difficult to use the actual air density for flight measurements. This was because the air-speed indicator (the combination of Pitôt-static tube and transducer) would have been calibrated on the assumption that the air density took the standard sea-level International Standard Atmosphere (ISA) value. The (incorrect) value of air speed obtained from Eqn (2.19) using this standard value of pressure with a hypothetical perfect transducer is known as the equivalent air speed (EAS). A term that is still in use. The relationship between true and equivalent air speed can be derived as follows. Using the correct value of density, ρ, in Eqn (2.19) shows that the relationship between the measured pressure difference and true air speed, ν, is

$$\Delta p = \frac{1}{2}\rho\nu^2 \tag{2.25}$$

whereas if the standard value of density, $\rho_0 = 1.226 \, \text{kg/m}^3$, is used we find

$$\Delta p = \frac{1}{2}\rho_0\nu_E^2 \tag{2.26}$$

where ν_E is the equivalent air speed. But the values of Δp in Eqns (2.25) and (2.26) are the same and therefore

$$\frac{1}{2}\rho_0\nu_E^2 = \frac{1}{2}\rho\nu^2 \tag{2.27}$$

or

$$\nu_E = \nu\sqrt{\rho/\rho_0} \tag{2.28}$$

If the relative density $\sigma = \rho/\rho_0$ is introduced, Eqn (2.28) can be written as

$$\nu_E = \nu\sqrt{\sigma} \tag{2.29}$$

The term indicated air speed (IAS) is used for the measurement made with an actual (imperfect) air-speed indicator. Owing to instrument error, the IAS will normally differ from the EAS.

The following definitions may therefore be stated: IAS is the uncorrected reading shown by an actual air-speed indicator. Equivalent air speed EAS is the uncorrected reading that would be shown by a hypothetical, error-free, air-speed indicator. True air speed (TAS) is the actual speed of the aircraft relative to the air. Only when $\sigma = 1$ will true and equivalent air speeds be equal. Normally the EAS is less than the TAS.

Formerly, the aircraft navigator would have needed to calculate the TAS from the IAS. But in modern aircraft, the conversion is done electronically. The calibration of the air-speed indicator also makes an approximate correction for compressibility.

2.3.4 The incompressibility assumption

As a first step in calculating the stagnation pressure coefficient in compressible flow we use Eqn (1.6d) to rewrite the dynamic pressure as follows:

$$\frac{1}{2}\rho v^2 = \frac{1}{2}\left(\frac{\rho}{\gamma p}\right)\gamma p v^2 = \frac{1}{2}\gamma p \frac{v^2}{a^2} = \frac{1}{2}\gamma p M^2 \tag{2.30}$$

where M is Mach number.

When the ratio of the specific heats, γ, is given the value 1.4 (approximately the value for air), the stagnation pressure coefficient then becomes

$$C_{p_0} = \frac{p_0 - p}{0.7 p M^2} = \frac{1}{0.7 M^2}\left(\frac{p_0}{p} - 1\right) \tag{2.31}$$

Now

$$\frac{p_0}{p} = [1 + \frac{1}{5} M^2]^{7/2} \quad \text{(Eqn (6.16a))}$$

Expanding this by the binomial theorem gives

$$\frac{p_0}{p} = 1 + \frac{7}{2}\left(\frac{1}{5}M^2\right) + \frac{7\,5}{2\,2}\frac{1}{2!}\left(\frac{1}{5}M^2\right)^2 + \frac{7\,5\,3}{2\,2\,2}\frac{1}{3!}\left(\frac{1}{5}M^2\right)^3 + \frac{7\,5\,3\,1}{2\,2\,2\,2}\frac{1}{4!}\left(\frac{1}{5}M^2\right)^4 + \cdots$$

$$= 1 + \frac{7M^2}{10} + \frac{7M^4}{40} + \frac{7M^6}{400} + \frac{7M^8}{16\,000} + \cdots$$

Then

$$C_{p_0} = \frac{10}{7M^2}\left(\frac{p_0}{p} - 1\right)$$

$$= \frac{10}{7M^2}\left[\frac{7M^2}{10} + \frac{7M^4}{40} + \frac{7M^6}{400} + \frac{7M^8}{16\,000} + \cdots\right]$$

$$= 1 + \frac{M^2}{4} + \frac{M^4}{40} + \frac{M^6}{1600} + \cdots \tag{2.32}$$

It can be seen that this will become unity, the incompressible value, at $M = 0$. This is the practical meaning of the incompressibility assumption, i.e. that any velocity changes are small compared with the speed of sound in the fluid. The result given in Eqn (2.32) is the correct one, that applies at all Mach numbers less than unity. At supersonic speeds, shock waves may be formed in which case the physics of the flow are completely altered.

Table 2.1 shows the variation of C_{p_0} with Mach number. It is seen that the error in assuming $C_{p_0} = 1$ is only 2% at $M = 0.3$ but rises rapidly at higher Mach numbers, being slightly more than 6% at $M = 0.5$ and 27.6% at $M = 1.0$.

Table 2.1 Variation of stagnation pressure coefficient with Mach numbers less than unity

M	0	0.2	0.4	0.6	0.7	0.8	0.9	1.0
C_{p_0}	1	1.01	1.04	1.09	1.13	1.16	1.217	1.276

It is often convenient to regard the effects of compressibility as negligible if the flow speed nowhere exceeds about $100\,\mathrm{m\,s^{-1}}$. However, it must be remembered that this is an entirely arbitrary limit. Compressibility applies at all flow speeds and, therefore, ignoring it always introduces an error. It is thus necessary to consider, for each problem, whether the error can be tolerated or not.

In the following examples use will be made of the equation (1.6d) for the speed of sound that can also be written as

$$a = \sqrt{\gamma RT}$$

For air, with $\gamma = 1.4$ and $R = 287.3\,\mathrm{J\,kg^{-1}K^{-1}}$ this becomes

$$a = 20.05\sqrt{T}\,\mathrm{m\,s^{-1}} \tag{2.33}$$

where T is the temperature in K.

Example 2.1 The air-speed indicator fitted to a particular aeroplane has no instrument errors and is calibrated assuming incompressible flow in standard conditions. While flying at sea level in the ISA the indicated air speed is $950\,\mathrm{km\,h^{-1}}$. What is the true air speed?

$950\,\mathrm{km\,h^{-1}} = 264\,\mathrm{m\,s^{-1}}$ and this is the speed corresponding to the pressure difference applied to the instrument based on the stated calibration. This pressure difference can therefore be calculated by

$$p_0 - p = \Delta p = \frac{1}{2}\rho_0 v_{\mathrm{E}}^2$$

and therefore

$$p_0 - p = \frac{1}{2} \times 1.226(264)^2 = 42\,670\,\mathrm{N\,m^{-2}}$$

Now

$$\frac{p_0}{p} = \left[1 + \frac{1}{5}M^2\right]^{3.5}$$

In standard conditions $p = 101\,325\,\mathrm{N\,m^{-2}}$. Therefore

$$\frac{p_0}{p} = \frac{42\,670}{101\,325} + 1 = 1.421$$

Therefore

$$1 + \frac{1}{5}M^2 = (1.421)^{2/7} = 1.106$$

$$\frac{1}{5}M^2 = 0.106$$

$$M^2 = 0.530$$

$$M = 0.728$$

The speed of sound at standard conditions is

$$a = 20.05(288)^{\frac{1}{2}} = 340.3\,\mathrm{m\,s^{-1}}$$

Therefore, true air speed $= Ma = 0.728 \times 340.3$

$$248 \, \text{m s}^{-1} = 891 \, \text{km h}^{-1}$$

In this example, $\sigma = 1$ and therefore there is no effect due to density, i.e. the difference is due entirely to compressibility. Thus it is seen that neglecting compressibility in the calibration has led the air-speed indicator to overestimate the true air speed by $59 \, \text{km h}^{-1}$.

2.4   Two-dimensional flow

Consider flow in two dimensions only. The flow is the same as that between two planes set parallel and a little distance apart. The fluid can then flow in any direction between and parallel to the planes but not at right angles to them. This means that in the subsequent mathematics there are only two space variables, x and y in Cartesian (or rectangular) coordinates or r and θ in polar coordinates. For convenience, a unit length of the flow field is assumed in the z direction perpendicular to x and y. This simplifies the treatment of two-dimensional flow problems, but care must be taken in the matter of units.

In practice if two-dimensional flow is to be simulated experimentally, the method of constraining the flow between two close parallel plates is often used, e.g. small smoke tunnels and some high-speed tunnels.

To summarize, two-dimensional flow is fluid motion where the velocity at all points is parallel to a given plane.

We have already seen how the principles of conservation of mass and momentum can be applied to one-dimensional flows to give the continuity and momentum equations (see Section 2.2). We will now derive the governing equations for two-dimensional flow. These are obtained by applying conservation of mass and momentum to an infinitesimal rectangular control volume – see Fig. 2.8.

2.4.1   Component velocities

In general the local velocity in a flow is inclined to the reference axes Ox, Oy and it is usual to resolve the velocity vector $\vec{v}$ (magnitude q) into two components mutually at right-angles.

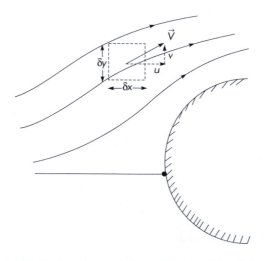

Fig. 2.8 An infinitesimal control volume in a typical two-dimensional flow field

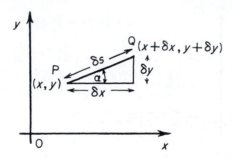

Fig. 2.9

In a Cartesian coordinate system let a particle move from point $P(x, y)$ to point $Q(x + \delta x, y + \delta y)$, a distance of δs in time δt (Fig. 2.9). Then the velocity of the particle is

$$\lim_{\delta \to 0} \frac{\delta s}{\delta t} = \frac{ds}{dt} = q$$

Going from P to Q the particle moves horizontally through δx giving the horizontal velocity $u = dx/dt$ positive to the right. Similarly going from P to Q the particle moves vertically through δy and the vertical velocity $v = dy/dt$ (upwards positive). By geometry:

$$(\delta s)^2 = (\delta x)^2 + (\delta y)^2$$

Thus

$$q^2 = u^2 + v^2$$

and the direction of q relative to the x-axis is $\alpha = \tan^{-1}(v/u)$.

In a polar coordinate system (Fig. 2.10) the particle moves distance δs from $P(r, \theta)$ to $Q(r + \delta r, \theta + \delta \theta)$ in time δt. The component velocities are:

radially (outwards positive) $q_n = \dfrac{dr}{dt}$

tangentially (anti-clockwise positive) $q_t = r\dfrac{d\theta}{dt}$

Again

$$(\delta s)^2 = (\delta r)^2 + (r\delta \theta)^2$$

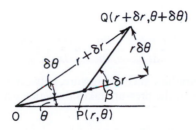

Fig. 2.10

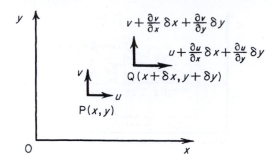

Fig. 2.11

Thus

$$q^2 = q_n^2 + q_t^2$$

and the direction of q relative to the radius vector is given by

$$\beta = \tan^{-1} \frac{q_t}{q_n}$$

Fluid acceleration

The equation of acceleration of a fluid mass is rather different from that of a vehicle, for example, and a note on fluid acceleration follows. Let a fluid particle move from P to Q in time δt in a two-dimensional flow (Fig. 2.11). At the point P(x, y) the velocity components are u and v. At the adjacent point Q($x + \delta x$, $y + \delta y$) the velocity components are $u + \delta u$ and $v + \delta v$, i.e. in general the velocity component has changed in each direction by an increment δu or δv. This incremental change is the result of a spatial displacement, and as u and v are functions of x and y the velocity components at Q are

$$u + \delta u = u + \frac{\partial u}{\partial x}\delta x + \frac{\partial u}{\partial y}\delta y \quad \text{and} \quad v + \delta v = v + \frac{\partial v}{\partial x}\delta x + \frac{\partial v}{\partial y}\delta y \tag{2.34}$$

The component of acceleration in the Ox direction is thus

$$\frac{\mathrm{d}(u + \delta u)}{\mathrm{d}t} = \frac{\partial u}{\partial t} + \frac{\partial u}{\partial x}\frac{\mathrm{d}x}{\mathrm{d}t} + \frac{\partial u}{\partial y}\frac{\mathrm{d}y}{\mathrm{d}t}$$

$$= \frac{\partial u}{\partial t} + u\frac{\partial u}{\partial x} + v\frac{\partial u}{\partial y} \tag{2.35}$$

and in the Oy direction

$$\frac{\mathrm{d}(v + \delta v)}{\mathrm{d}t} = \frac{\partial u}{\partial t} + u\frac{\partial v}{\partial x} + v\frac{\partial v}{\partial y} \tag{2.36}$$

The change in other flow variables, such as pressure, between points P and Q may be dealt with in a similar way. Thus, if the pressure takes the value p at P, at Q it takes the value

$$p + \delta p = p + \frac{\partial p}{\partial x}\delta x + \frac{\partial p}{\partial y}\delta y \tag{2.37}$$

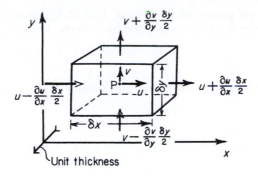

Fig. 2.12 Rectangular space of volume $\delta x \times \delta y \times 1$ at the point P (x, y) where the velocity components are u and v and the density is ρ

2.4.2   The equation of continuity or conservation of mass

Consider a typical elemental control volume like the one illustrated in Fig. 2.8. This is a small rectangular region of space of sides δx, δy and unity, centred at the point $P(x, y)$ in a fluid motion which is referred to the axes Ox, Oy. At $P(x, y)$ the local velocity components are u and v and the density ρ, where each of these three quantities is a function of x, y and t (Fig. 2.12). Dealing with the flow into the box in the Ox direction, the amount of mass flowing into the region of space per second through the left-hand vertical face is:

$$\text{mass flow per unit area} \times \text{area}$$

i.e.

$$\left(\rho u - \frac{\partial(\rho u)}{\partial x}\frac{\delta x}{2}\right)\delta y \times 1 \tag{2.38}$$

The amount of mass leaving the box per second through the right-hand vertical face is:

$$\left(\rho u + \frac{\partial(\rho u)}{\partial x}\frac{\delta x}{2}\right)\delta y \times 1 \tag{2.39}$$

The accumulation of mass per second in the box due to the horizontal flow is the difference of Eqns (2.38) and (2.39), i.e.

$$-\frac{\partial(\rho u)}{\partial x}\delta x \delta y \tag{2.40}$$

Similarly, the accumulation per second in the Oy direction is

$$-\frac{\partial(\rho v)}{\partial y}\delta x \delta y \tag{2.41}$$

so that the total accumulation per second is

$$-\left(\frac{\partial(\rho u)}{\partial x} + \frac{\partial(\rho v)}{\partial y}\right)\delta x \delta y \tag{2.42}$$

As mass cannot be destroyed or created, Eqn (2.42) must represent the rate of change of mass of the fluid in the box and can also be written as

$$\frac{\partial(\rho \times \text{volume})}{\partial t}$$

but with the elementary box having constant volume ($\delta x\, \delta y \times 1$) this becomes

$$\frac{\partial \rho}{\partial t} \delta x\, \delta y \times 1 \tag{2.43}$$

Equating (2.42) and (2.43) gives the general equation of continuity, thus:

$$\frac{\partial \rho}{\partial t} + \frac{\partial(\rho u)}{\partial x} + \frac{\partial(\rho v)}{\partial y} = 0 \tag{2.44}$$

This can be expanded to:

$$\frac{\partial \rho}{\partial t} + u\frac{\partial \rho}{\partial x} + v\frac{\partial \rho}{\partial y} + \rho\left(\frac{\partial u}{\partial x} + \frac{\partial v}{\partial y}\right) = 0 \tag{2.45}$$

and if the fluid is incompressible and the flow steady the first three terms are all zero since the density cannot change and the equation reduces for incompressible flow to

$$\frac{\partial u}{\partial x} + \frac{\partial v}{\partial y} = 0 \tag{2.46}$$

This equation is fundamental and important and it should be noted that it expresses a physical reality. For example, in the case given by Eqn (2.46)

$$\frac{\partial u}{\partial x} = -\frac{\partial v}{\partial y}$$

This reflects the fact that if the flow velocity increases in the x direction it must decrease in the y direction.

For three-dimensional flows Eqns (2.45) and (2.46) are written in the forms:

$$\frac{\partial \rho}{\partial t} + u\frac{\partial \rho}{\partial x} + v\frac{\partial \rho}{\partial y} + w\frac{\partial \rho}{\partial z} + \rho\left(\frac{\partial u}{\partial x} + \frac{\partial v}{\partial y} + \frac{\partial w}{\partial z}\right) = 0 \tag{2.47a}$$

$$\frac{\partial u}{\partial x} + \frac{\partial v}{\partial y} + \frac{\partial w}{\partial z} = 0 \tag{2.47b}$$

2.4.3 The equation of continuity in polar coordinates

A corresponding equation can be found in the polar coordinates r and θ where the velocity components are q_n and q_t radially and tangentially. By carrying out a similar development for the accumulation of fluid in a segmental elemental box of space, the equation of continuity corresponding to Eqn (2.44) above can be found as follows. Taking the element to be at $P(r, \theta)$ where the mass flow is ρq per unit length (Fig. 2.13), the accumulation per second radially is:

$$\left(\rho q_n - \frac{\partial(\rho q_n)}{\partial r}\frac{\delta r}{2}\right)\left(r - \frac{\delta r}{2}\right)\delta\theta - \left(\rho q_n + \frac{\partial(\rho q_n)}{\partial r}\frac{\delta r}{2}\right)\left(r + \frac{\delta r}{2}\right)\delta\theta$$

$$= -\rho q_n\, \delta r\, \delta\theta - \frac{\partial(\rho q_n)}{\partial r} r\, \delta r\, \delta\theta \tag{2.48}$$

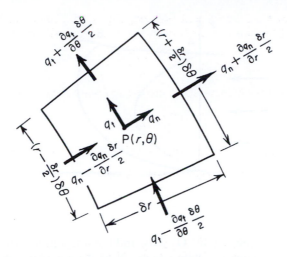

Fig. 2.13 Rectangular element at P (r, θ) in a system of polar coordinates

and accumulation per second tangentially is:

$$\left(\rho q_{\mathrm{t}} - \frac{\partial(\rho q_{\mathrm{t}})}{\partial\theta}\frac{\delta\theta}{2}\right)\delta r - \left(\rho q_{\mathrm{t}} + \frac{\partial(\rho q_{\mathrm{t}})}{\partial\theta}\frac{\delta\theta}{2}\right)\delta r = -\frac{\partial(\rho q_{\mathrm{t}})}{\partial\theta}\delta r\,\delta\theta \qquad (2.49)$$

Total accumulation per second

$$= -\left(\frac{\rho q_{\mathrm{n}}}{r} + \frac{\partial(\rho q_{\mathrm{n}})}{\partial r} + \frac{1}{r}\frac{\partial(\rho q_{\mathrm{t}})}{\partial\theta}\right)r\,\delta r\,\delta\theta \qquad (2.50)$$

and this by the previous argument equals the rate of change of mass within the region of space

$$= \frac{\partial(\rho r\delta r\delta\theta)}{\partial t} \qquad (2.51)$$

Equating (2.50) and (2.51) gives:

$$\frac{\rho q_{\mathrm{n}}}{r} + \frac{\partial\rho}{\partial t} + \frac{\partial(\rho q_{\mathrm{n}})}{\partial r} + \frac{1}{r}\frac{\partial(\rho q_{\mathrm{t}})}{\partial\theta} = 0 \qquad (2.52)$$

Hence for steady flow

$$\frac{\partial(\rho r q_{\mathrm{n}})}{\partial r} + \frac{\partial(\rho q_{\mathrm{t}})}{\partial\theta} = 0 \qquad (2.53)$$

and the incompressible equation in this form becomes:

$$\frac{q_{\mathrm{n}}}{r} + \frac{\partial q_{\mathrm{n}}}{\partial r} + \frac{1}{r}\frac{\partial q_{\mathrm{t}}}{\partial\theta} = 0 \qquad (2.54)$$

2.5 The stream function and streamline

2.5.1 The stream function ψ

Imagine being on the banks of a shallow river of a constant depth of 1 m at a position O (Fig. 2.14) with a friend directly opposite at A, 40 m away. Mathematically

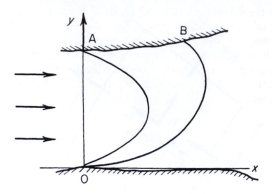

Fig. 2.14

the bank can be represented by the Ox axis, and the line joining you to your friend at A the Oy axis in the two-coordinate system. Now if the stream speed is $2\,\mathrm{m\,s^{-1}}$ the amount of water passing between you and your friend is $40 \times 1 \times 2 = 80\,\mathrm{m^3\,s^{-1}}$ and this is the amount of water flowing past any point anywhere along the river which could be measured at a weir downstream. Suppose you now throw a buoyant rope to your friend who catches the end but allows the slack to fall in the river and float into a curve as shown. The amount of water flowing under the line is still $80\,\mathrm{m^3\,s^{-1}}$ no matter what shape the rope takes, and is unaffected by the configuration of the rope.

Suppose your friend moves along to a point B somewhere downstream, still holding his end of the line but with sufficient rope paid out as he goes. The volume of water passing under the rope is still only $80\,\mathrm{m^3\,s^{-1}}$ providing he has not stepped over a tributary stream or an irrigation drain in the bank. It follows that, if no water can enter or leave the stream, the quantity flowing past the line will be the same as before and furthermore will be unaffected by the shape of the line between O and B. The amount or quantity of fluid passing such a line per second is called the *stream function* or *current function* and it is denoted by ψ.

Consider now a pair of coordinate axes set in a two-dimensional air stream that is moving generally from left to right (Fig. 2.15). The axes are arbitrary space references and in no way interrupt the fluid streaming past. Similarly the line joining O to a point P in the flow in no way interrupts the flow since it is as imaginary as the reference axes Ox and Oy. An algebraic expression can be found for the line in x and y.

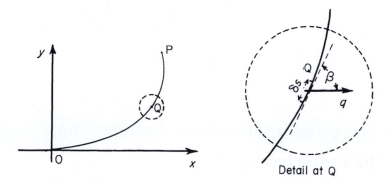

Detail at Q

Fig. 2.15

Let the flow past the line at any point Q on it be at velocity q over a small length δs of line where direction of q makes angle β to the tangent of the curve at Q. The component of the velocity q perpendicular to the element δs is $q \sin \beta$ and therefore, assuming the depth of stream flow to be unity, the amount of fluid crossing the element of line δs is $q \sin \beta \times \delta s \times 1$ per second. Adding up all such quantities crossing similar elements along the line from O to P, the total amount of flow past the line (sometimes called flux) is

$$\int_{OP} q \sin \beta \, ds$$

which is the line integral of the normal velocity component from O to P.

If this quantity of fluid flowing between O and P remains the same irrespective of the path of integration, i.e. independent of the curve of the rope then $\int_{OP} q \sin \beta \, ds$ is called the stream function of P with respect to O and

$$\psi_P = \int_{OP} q \sin \beta \, ds$$

Note: it is implicit that $\psi_0 = 0$.

Sign convention for stream functions

It is necessary here to consider a sign convention since quantities of fluid are being considered. When integrating the cross-wise component of flow along a curve, the component can go either from left to right, or vice versa, across the path of integration (Fig. 2.16). Integrating the normal flow components from O to P, the flow components are, looking in the direction of integration, either (a) from left to right or (b) from right to left. The former is considered positive flow whilst the latter is negative flow. The convention is therefore:

> Flow *across* the path of integration is *positive* if, when looking in the direction of integration, it crosses the path from left to right.

2.5.2 The streamline

From the statement above, ψ_P is the flow across the line OP. Suppose there is a point P_1 close to P which has the same value of stream function as point P (Fig. 2.17). Then the flow across any line OP_1 equals that across OP, and the amount of fluid flowing into area OPP_1O across OP equals the amount flowing out across OP_1. Therefore, no fluid crosses line PP_1 and the velocity of flow must be along, or tangential to, PP_1.

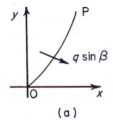

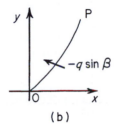

(a) (b)

Fig. 2.16

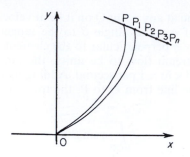

Fig. 2.17

All other points P_2, P_3, etc. which have a stream function equal in value to that of P have, by definition, the same flow across any lines joining them to O, so by the same argument the velocity of the flow in the region of P_1, P_2, P_3, etc. must be along PP_1, P_2, P_3, etc., and no fluid crosses the line PP_1, P_2, ..., P_n. Since $\psi_{P_1} = \psi_{P_2} = \psi_{P_3} = \psi_P$ = constant, the line PP_1, P_2, ... P_n, etc. is a line of constant ψ and is called a streamline. It follows further that since no flow can cross the line PP_n the velocity along the line must always be in the direction tangential to it. This leads to the two common definitions of a streamline, each of which indirectly has the other's meaning. They are:

A *streamline* is a line of constant ψ

and/or

A *streamline* is a line of fluid particles, the velocity of each particle being tangential to the line (see also Section 2.1.2).

It should be noted that the velocity can change in magnitude along a streamline but by definition the direction is always that of the tangent to the line.

2.5.3 Velocity components in terms of ψ

(a) *In Cartesian coordinates* Let point $P(x, y)$ be on the streamline AB in Fig. 2.18a of constant ψ and point $Q(x + \delta x, y + \delta y)$ be on the streamline CD of constant $\psi + \delta\psi$. Then from the definition of stream function, the amount of fluid flowing across any path between P and Q $= \delta\psi$, the change of stream function between P and Q.

The most convenient path along which to integrate in this case is PRQ, point R being given by the coordinates $(x + \delta x, y)$. Then the flow across PR $= -v\delta x$ (since the flow is from right to left and thus by convention negative), and that across RQ $= u\delta y$. Therefore, total flow across the line PRQ is

$$\delta\psi = u\delta y - v\delta x \tag{2.55}$$

Now ψ is a function of two independent variables x and y in steady motion, and thus

$$\delta\psi = \frac{\partial\psi}{\partial x}\delta x + \frac{\partial\psi}{\partial y}\delta y \tag{2.56}$$

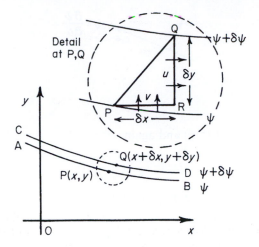

Fig. 2.18a

$\partial\psi/\partial x$ and $\partial\psi/\partial y$ being the partial derivatives with respect to x and y respectively. Then, equating terms:

$$u = \partial\psi/\partial y \qquad (2.56a)$$

and

$$v = -\partial\psi/\partial x \qquad (2.56b)$$

these being the velocity components at a point x, y in a flow given by stream function ψ.

(b) *In polar coordinates* Let the point $P(r, \theta)$ be on the streamline AB (Fig. 2.18b) of constant ψ, and point $Q(r + \delta r, \theta + \delta\theta)$ be on the streamline CD of constant $\psi + \delta\psi$. The velocity components are q_n and q_t, radially and tangentially respectively. Here the most convenient path of integration is PRQ where OP is produced to R so that PR = δr, i.e. R is given by ordinates $(r + \delta r, \theta)$. Then

$$\delta\psi = -q_t\delta r + q_n(r + \delta r)\delta\theta$$
$$= -q_t\delta r + q_n r\delta\theta + q_n\delta r\delta\theta$$

To the first order of small quantities:

$$\delta\psi = -q_t\delta r + q_n r\delta\theta \qquad (2.57)$$

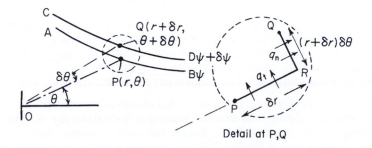

Detail at P,Q

Fig. 2.18b

Fig. 2.19

But here ψ is a function of (r, θ) and again

$$\delta\psi = \frac{\partial\psi}{\partial r}\delta r + \frac{\partial\psi}{\partial\theta}\delta\theta \tag{2.58}$$

and equating terms in Eqns (2.57) and (2.58)

$$q_{\mathrm{t}} = -\frac{\partial\psi}{\partial r} \tag{2.58a}$$

$$q_{\mathrm{n}} = \frac{1}{r}\frac{\partial\psi}{\partial\theta} \tag{2.58b}$$

these being velocity components at a point r, θ in a flow given by stream function ψ.

In general terms the velocity q in any direction s is found by differentiating the stream function ψ partially with respect to the direction n normal to q where n is taken in the anti-clockwise sense looking along q (Fig. 2.19):

$$q = \frac{\partial\psi}{\partial n}$$

2.6　The momentum equation

The momentum equation for two- or three-dimensional flow embodies the application of Newton's second law of motion (mass times acceleration = force, or rate of change of momentum = force) to an infinitesimal control volume in a fluid flow (see Fig. 2.8). It takes the form of a set of partial differential equations. Physically it states that the rate of increase in momentum within the control volume plus the net rate at which momentum flows out of the control volume equals the force acting on the fluid within the control volume.

There are two distinct classes of force that act on the fluid particles within the control volume.

(i) *Body forces.* Act on all the fluid within the control volume. Here the only body force of interest is the force of gravity or weight of the fluid.
(ii) *Surface forces.* These only act on the control surface; their effect on the fluid inside the control volume cancels out. They are always expressed in terms of stress (force per unit area). Two main types of surface force are involved namely:

(a) *Pressure force.* Pressure, p, is a stress that *always* acts perpendicular to the control surface and in the opposite direction to the unit normal (see Fig. 1.3). In other words it always tends to compress the fluid in the control volume. Although p can vary from point to point in the flow field it is invariant with direction at a particular point (in other words irrespective of the orientation of the infinitesimal control volume the pressure force on any face will be $-p\delta A$ where δA is the area of the face) – see Fig. 1.3. As is evident from Bernoulli's Eqn (2.16), the pressure depends on the flow speed.

(b) *Viscous forces*. In general the viscous force acts at an angle to any particular face of the infinitesimal control volume, so in general it will have two components in two-dimensional flow (three for three-dimensional flow) acting on each face (one due to a direct stress acting perpendicularly to the face and one shear stress (two for three-dimensional flow) acting tangentially to the face. As an example let us consider the stresses acting on two faces of a square infinitesimal control volume (Fig. 2.20). For the top face the unit normal would be **j** (unit vector in the y direction) and the unit tangential vector would be **i** (the unit vector in the x direction). In this case, then, the viscous force acting on this face and the side face would be given by

$$(\sigma_{yx}\mathbf{i} + \sigma_{yy}\mathbf{j})\delta x \times 1, \qquad (\sigma_{xx}\mathbf{i} + \sigma_{xy}\mathbf{j})\delta y \times 1$$

respectively. Note that, as in Section 2.4, we are assuming unit length in the z direction. The viscous shear stress is what is termed a second-order *tensor* – i.e. it is a quantity that is characterized by a magnitude and *two* directions (c.f. a vector or first-order tensor that is characterized by a magnitude and one direction). The stress tensor can be expressed in terms of four components (9 for three-dimensional flow) in matrix form as:

$$\begin{pmatrix} \sigma_{xx} & \sigma_{xy} \\ \sigma_{yx} & \sigma_{yy} \end{pmatrix}$$

Owing to symmetry $\sigma_{xy} = \sigma_{yx}$. Just as the components of a vector change when the coordinate system is changed, so do the components of the stress tensor. In many engineering applications the direct viscous stresses (σ_{xx}, σ_{yy}) are negligible compared with the shear stresses. The viscous stress is generated by fluid motion and cannot exist in a still fluid.

Other surface forces, e.g. surface tension, can be important in some engineering applications.

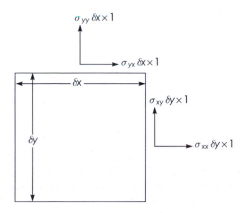

Fig. 2.20

When the momentum equation is applied to an infinitesimal control volume (c.v.), it can be written in the form:

$$\underbrace{\text{Rate of increase of momentum within the c.v.}}_{\text{(i)}}$$

$$+ \underbrace{\text{Net rate at which momentum leaves the c.v.}}_{\text{(ii)}}$$

$$= \underbrace{\text{Body force}}_{\text{(iii)}} + \underbrace{\text{pressure force}}_{\text{(iv)}} + \underbrace{\text{viscous force}}_{\text{(v)}} \qquad (2.59)$$

We will consider now the evaluation of each of terms (i) to (v) in turn for the case of two-dimensional incompressible flow.

Term (i) is dealt with in a similar way to Eqn (2.43), once it is recalled that momentum is (mass) × (velocity), so Term (i) is given by

$$\frac{\partial}{\partial t}(\rho \times \text{volume} \times \vec{v}) = \frac{\partial \rho \vec{v}}{\partial t} \delta x \delta y \times 1 = \left(\frac{\partial \rho u}{\partial t}, \frac{\partial \rho v}{\partial t}\right) \delta x \delta y \times 1 \qquad (2.60)$$

To evaluate Term (ii) we will make use of Fig. 2.21 (c.f. Fig. 2.12). Note that the rate at which momentum crosses any face of the control volume is (rate at which mass crosses the face) × velocity. So if we denote the rate at which mass crosses a face by $\dot{m}$, Term (ii) is given by

$$\dot{m}_3 \times \vec{v}_3 - \dot{m}_1 \times \vec{v}_1 + \dot{m}_4 \times \vec{v}_4 - \dot{m}_2 \times \vec{v}_2 \qquad (2.61)$$

But $\dot{m}_3$ and $\dot{m}_1$ are given by Eqns (2.38) and (2.39) respectively, and $\dot{m}_2$ and $\dot{m}_4$ by similar expressions. In a similar way it can be seen that, recalling $\vec{v} = (u, v)$

$$\vec{v}_1 = (u, v) - \left(\frac{\partial u}{\partial x}, \frac{\partial v}{\partial x}\right)\frac{\delta x}{2}, \qquad \vec{v}_3 = (u, v) + \left(\frac{\partial u}{\partial x}, \frac{\partial v}{\partial x}\right)\frac{\delta x}{2}$$

$$\vec{v}_2 = (u, v) - \left(\frac{\partial u}{\partial y}, \frac{\partial v}{\partial y}\right)\frac{\delta y}{2}, \qquad \vec{v}_4 = (u, v) + \left(\frac{\partial u}{\partial y}, \frac{\partial v}{\partial y}\right)\frac{\delta y}{2}$$

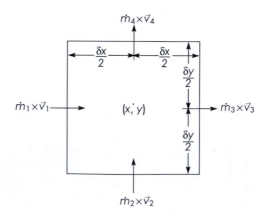

Fig. 2.21

So the x component of Eqn (2.61) becomes

$$\rho\left(u + \frac{\partial u}{\partial x}\frac{\delta x}{2}\right)\delta y \times 1\left(u + \frac{\partial u}{\partial x}\frac{\delta x}{2}\right) - \rho\left(u - \frac{\partial u}{\partial x}\frac{\delta x}{2}\right)\delta y \times 1\left(u - \frac{\partial u}{\partial x}\frac{\delta x}{2}\right)$$

$$+ \rho\left(v + \frac{\partial v}{\partial y}\frac{\delta y}{2}\right)\delta x \times 1\left(u + \frac{\partial u}{\partial y}\frac{\delta y}{2}\right) - \rho\left(v - \frac{\partial v}{\partial y}\frac{\delta y}{2}\right)\delta x \times 1\left(u - \frac{\partial u}{\partial y}\frac{\delta y}{2}\right)$$

Cancelling like terms and neglecting higher-order terms simplifies this expression to

$$\rho\left(2u\frac{\partial u}{\partial x} + v\frac{\partial u}{\partial y} + u\frac{\partial v}{\partial y}\right)\delta x\delta y \times 1$$

This can be rearranged as

$$\rho\left(u\frac{\partial u}{\partial x} + v\frac{\partial u}{\partial y} + u\underbrace{\left\{\frac{\partial u}{\partial x} + \frac{\partial v}{\partial y}\right\}}_{= 0 \text{ Eqn (2.46)}}\right)\delta x\delta y \times 1 \tag{2.62a}$$

In an exactly similar way the y component of Eqn (2.61) can be shown to be

$$\rho\left(u\frac{\partial v}{\partial x} + v\frac{\partial v}{\partial y}\right)\delta x\delta y \times 1 \tag{2.62b}$$

Term (iii) the body force, acting on the control volume, is simply given by the weight of the fluid, i.e. the mass of the fluid multiplied by the acceleration (vector) due to gravity. Thus

$$\rho\delta x\delta y \times 1 \times \vec{g} = (\rho g_x, \rho g_y)\delta x\delta y \times 1 \tag{2.63}$$

Normally, of course, gravity acts vertically downwards, so $g_x = 0$ and $g_y = -g$.

The evaluation of Term (iv), the net pressure force acting on the control volume is illustrated in Fig. 2.22. In the x direction the net pressure force is given by

$$\left(p - \frac{\partial p}{\partial x}\frac{\delta x}{2}\right)\delta y \times 1 - \left(p + \frac{\partial p}{\partial x}\frac{\delta x}{2}\right)\delta y \times 1 = -\frac{\partial p}{\partial x}\delta x\delta y \times 1 \tag{2.64a}$$

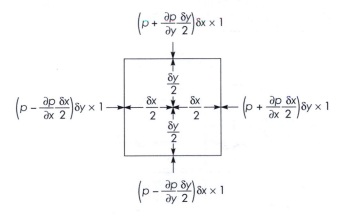

Fig. 2.22 Pressure forces acting on the infinitesimal control volume

Similarly, the y component of the net pressure force is given by

$$-\frac{\partial p}{\partial y}\delta x \delta y \times 1 \tag{2.64b}$$

The evaluation of the x component of Term (v), the net viscous force, is illustrated in Fig. 2.23. In a similar way as for Eqn (2.64a,b), we obtain the net viscous force in the x and y directions respectively as

$$\left(\frac{\partial \sigma_{xx}}{\partial x} + \frac{\partial \sigma_{xy}}{\partial y}\right)\delta x \delta y \times 1 \tag{2.65a}$$

$$\left(\frac{\partial \sigma_{yx}}{\partial x} + \frac{\partial \sigma_{yy}}{\partial y}\right)\delta x \delta y \times 1 \tag{2.65b}$$

We now substitute Eqns (2.61) to (2.65) into Eqn (2.59) and cancel the common factor $\delta x \delta y \times 1$ to obtain

$$\rho\left(\frac{\partial u}{\partial t} + u\frac{\partial u}{\partial x} + v\frac{\partial u}{\partial y}\right) = \rho g_x - \frac{\partial p}{\partial x} + \frac{\partial \sigma_{xx}}{\partial x} + \frac{\partial \sigma_{xy}}{\partial y} \tag{2.66a}$$

$$\rho\left(\frac{\partial v}{\partial t} + u\frac{\partial v}{\partial x} + v\frac{\partial v}{\partial y}\right) = \rho g_y - \frac{\partial p}{\partial y} + \frac{\partial \sigma_{yx}}{\partial x} + \frac{\partial \sigma_{yy}}{\partial y} \tag{2.66b}$$

These are the momentum equations in the form of partial differential equations.

For three dimensional flows the momentum equations can be written in the form:

$$\rho\left(\frac{\partial u}{\partial t} + u\frac{\partial u}{\partial x} + v\frac{\partial u}{\partial y} + w\frac{\partial u}{\partial z}\right) = \rho g_x - \frac{\partial p}{\partial x} + \frac{\partial \sigma_{xx}}{\partial x} + \frac{\partial \sigma_{xy}}{\partial y} + \frac{\partial \sigma_{xz}}{\partial z} \tag{2.67a}$$

$$\rho\left(\frac{\partial v}{\partial t} + u\frac{\partial v}{\partial x} + v\frac{\partial v}{\partial y} + w\frac{\partial v}{\partial z}\right) = \rho g_y - \frac{\partial p}{\partial y} + \frac{\partial \sigma_{yx}}{\partial x} + \frac{\partial \sigma_{yy}}{\partial y} + \frac{\partial \sigma_{yz}}{\partial z} \tag{2.67b}$$

$$\rho\left(\frac{\partial w}{\partial t} + u\frac{\partial w}{\partial x} + v\frac{\partial w}{\partial y} + w\frac{\partial w}{\partial z}\right) = \rho g_z - \frac{\partial p}{\partial z} + \frac{\partial \sigma_{zx}}{\partial x} + \frac{\partial \sigma_{zy}}{\partial y} + \frac{\partial \sigma_{zz}}{\partial z} \tag{2.67c}$$

where g_x, g_y, g_z are the components of the acceleration **g** due to gravity, the body force per unit volume being given by ρ**g**.

The only approximation made to derive Eqns (2.66) and (2.67) is the *continuum model*, i.e. we ignore the fact that matter consists of myriad molecules and treat it as continuous. Although we have made use of the incompressible form of the continuity

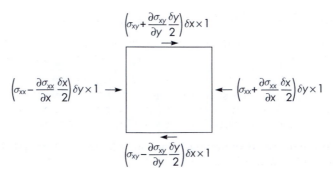

Fig. 2.23 x-component of forces due to viscous stress acting on infinitesimal control volume

Eqn (2.46) to simplify Eqn (2.58a,b), Eqns (2.62) and (2.63) apply equally well to compressible flow. In order to show this to be true, it is necessary to allow density to vary in the derivation of Term (i) and to simplify it using the compressible form of the continuity Eqn (2.45).

2.6.1 The Euler equations

For some applications in aerodynamics it can be an acceptable approximation to neglect the viscous stresses. In this case Eqns (2.66) simplify to

$$\rho\left(\frac{\partial u}{\partial t} + u\frac{\partial u}{\partial x} + v\frac{\partial u}{\partial y}\right) = \rho g_x - \frac{\partial p}{\partial x} \tag{2.68a}$$

$$\rho\left(\frac{\partial v}{\partial t} + u\frac{\partial v}{\partial x} + v\frac{\partial v}{\partial y}\right) = \rho g_y - \frac{\partial p}{\partial y} \tag{2.68b}$$

These equations are known as the *Euler* equations. In principle, Eqns (2.68a,b), together with the continuity Eqn (2.46), can be solved to give the velocity components u and v and pressure p. However, in general, this is difficult because Eqns (2.68a,b) can be regarded as the governing equations for u and v, but p does not appear explicitly in the continuity equation. Except for special cases, solution of the Euler equations can only be achieved numerically using a computer. A very special and comparatively simple case is *irrotational* flow (see Section 2.7.6). For this case the Euler equations reduce to a single simpler equation – the *Laplace* equation. This equation is much more amenable to analytical solution and this is the subject of Chapter 3.

2.7   Rates of strain, rotational flow and vorticity

As they stand, the momentum Eqns (2.66) (or 2.67), together with the continuity Eqn (2.46) (or 2.47) cannot be solved, even in principle, for the flow velocity and pressure. Before this is possible it is necessary to link the viscous stresses to the velocity field through a *constitutive* equation. Air, and all other homogeneous gases and liquids, are closely approximated by the *Newtonian* fluid model. This means that the viscous stress is proportional to the rate of strain. Below we consider the distortion experienced by an infinitesimal fluid element as it travels through the flow field. In this way we can derive the rate of strain in terms of velocity gradients. The important flow properties, vorticity and circulation will also emerge as part of this process.

2.7.1 Distortion of fluid element in flow field

Figure 2.24 shows how a fluid element is transformed as it moves through a flow field. In general the transformation comprises the following operations:

(i) *Translation* – movement from one position to another.
(ii) *Dilation/Compression* – the shape remains invariant, but volume reduces or increases. For *incompressible* flow the volume remains invariant from one position to another.
(iii) *Distortion* – change of shape keeping the volume invariant.

Distortion can be decomposed into anticlockwise *rotation* through angle $(\alpha - \beta)/2$ and a *shear* of angle $(\alpha + \beta)/2$.

The angles α and β are the shear strains.

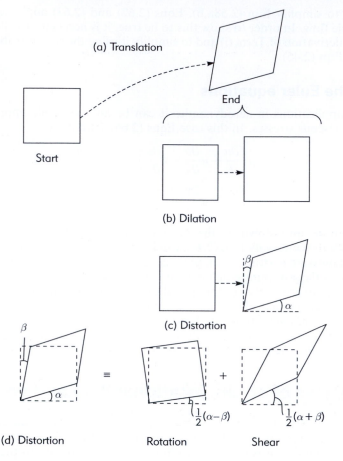

(a) Translation

End

Start

(b) Dilation

(c) Distortion

(d) Distortion   Rotation   Shear

Fig. 2.24 Transformation of a fluid element as it moves through the flow field

2.7.2   Rate of shear strain

Consider Fig. 2.25. This shows an elemental control volume $ABCD$ that initially at time $t = t_i$ is square. After an interval of time δt has elapsed $ABCD$ has moved and distorted into $A'B'C'D'$. The velocities at $t = t_i$ at A, B and C are given by

$$u_A = u - \frac{\partial u}{\partial x}\frac{\delta x}{2} - \frac{\partial u}{\partial y}\frac{\delta y}{2}, \qquad v_A = v - \frac{\partial v}{\partial x}\frac{\delta x}{2} - \frac{\partial v}{\partial y}\frac{\delta y}{2} \qquad (2.69a)$$

$$u_B = u - \frac{\partial u}{\partial x}\frac{\delta x}{2} + \frac{\partial u}{\partial y}\frac{\delta y}{2}, \qquad v_B = v - \frac{\partial v}{\partial x}\frac{\delta x}{2} + \frac{\partial v}{\partial y}\frac{\delta y}{2} \qquad (2.69b)$$

$$u_C = u + \frac{\partial u}{\partial x}\frac{\delta x}{2} - \frac{\partial u}{\partial y}\frac{\delta y}{2}, \qquad v_C = v + \frac{\partial v}{\partial x}\frac{\delta x}{2} - \frac{\partial v}{\partial y}\frac{\delta y}{2} \qquad (2.69c)$$

$$x_{A'} = u_A\delta t, \qquad y_{A'} = v_A\delta t \text{ etc.} \qquad (2.70)$$

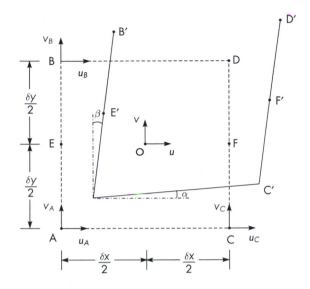

Fig. 2.25

Therefore, if we neglect the higher-order terms,

$$\alpha = \frac{y_{C'} - y_{A'}}{\delta x} = (v_C - v_A)\frac{\delta t}{\delta x} = \left\{ v + \frac{\partial v}{\partial x}\frac{\delta x}{2} - \frac{\partial v}{\partial y}\frac{\delta y}{2} - \left(v - \frac{\partial v}{\partial x}\frac{\delta x}{2} - \frac{\partial v}{\partial y}\frac{\delta y}{2} \right) \right\}\frac{\delta t}{\delta x} = \frac{\partial v}{\partial x}\delta t$$

(2.71a)

$$\beta = \frac{x_{B'} - x_{A'}}{\delta y} = (u_B - u_A)\frac{\delta t}{\delta y} = \left\{ u - \frac{\partial u}{\partial x}\frac{\delta x}{2} + \frac{\partial u}{\partial y}\frac{\delta y}{2} - \left(u - \frac{\partial u}{\partial x}\frac{\delta x}{2} - \frac{\partial u}{\partial y}\frac{\delta y}{2} \right) \right\}\frac{\delta t}{\delta y} = \frac{\partial u}{\partial y}\delta t$$

(2.71b)

The rate of shear strain in the xy plane is given by

$$\frac{d\gamma_{xy}}{dt} = \frac{d}{dt}\left(\frac{\alpha + \beta}{2} \right) = \left(\frac{\partial v}{\partial x}\delta t + \frac{\partial u}{\partial y}\delta t \right)\frac{1}{2\delta t} = \frac{1}{2}\left(\frac{\partial v}{\partial x} + \frac{\partial u}{\partial y} \right) \qquad (2.72a)$$

In much the same way, for three-dimensional flows it can be shown that there are two other components of the rate of shear strain

$$\frac{d\gamma_{xz}}{dt} = \frac{1}{2}\left(\frac{\partial w}{\partial x} + \frac{\partial u}{\partial z} \right), \qquad \frac{d\gamma_{yz}}{dt} = \frac{1}{2}\left(\frac{\partial v}{\partial z} + \frac{\partial w}{\partial y} \right), \qquad (2.72b, c)$$

2.7.3 Rate of direct strain

Following an analogous process we can also calculate the direct strains and their corresponding rates of strain, for example

$$\varepsilon_{xx} = \frac{x_{F'} - x_{E'}}{x_F - x_E} = \frac{(u_{F'} - u_{E'})\delta t}{\delta x} = \left\{ u + \frac{\partial u}{\partial x}\frac{\delta x}{2} - \left(u - \frac{\partial u}{\partial x}\frac{\delta x}{2} \right) \right\}\frac{\delta t}{\delta x} = \frac{\partial u}{\partial x}\delta t \quad (2.73)$$

The other direct strains are obtained in a similar way; thus the rates of direct strain are given by

$$\frac{d\varepsilon_{xx}}{dt} = \frac{\partial u}{\partial x}, \qquad \frac{d\varepsilon_{yy}}{dt} = \frac{\partial v}{\partial y}, \qquad \frac{d\varepsilon_{zz}}{dt} = \frac{\partial w}{\partial z} \qquad (2.74a,b,c)$$

Thus we can introduce a rate of strain tensor analogous to the stress tensor (see Section 2.6) and for which components in two-dimensional flow can be represented in matrix form as follows:

$$\begin{pmatrix} \dot{\varepsilon}_{xx} & \dot{\gamma}_{xy} \\ \dot{\gamma}_{yx} & \dot{\varepsilon}_{yy} \end{pmatrix} \qquad (2.75)$$

where (˙) is used to denote a time derivative.

2.7.4 Vorticity

The instantaneous rate of rotation of a fluid element is given by $(\dot{\alpha} - \dot{\beta})/2$ – see above. This corresponds to a fundamental property of fluid flow called the *vorticity* that, using Eqn (2.71), in two-dimensional flow is defined as

$$\zeta = \frac{d\alpha}{dt} - \frac{d\beta}{dt} = \frac{\partial v}{\partial x} - \frac{\partial u}{\partial y} \qquad (2.76)$$

In three-dimensional flow vorticity is a vector given by

$$\mathbf{\Omega} = (\xi, \eta, \zeta) = \left(\frac{\partial w}{\partial y} - \frac{\partial v}{\partial z}, \frac{\partial u}{\partial z} - \frac{\partial w}{\partial x}, \frac{\partial v}{\partial x} - \frac{\partial u}{\partial y} \right) \qquad (2.77a,b,c)$$

It can be seen that the three components of vorticity are twice the instantaneous rates of rotation of the fluid element about the three coordinate axes. Mathematically it is given by the following vector operation

$$\mathbf{\Omega} = \nabla \times \mathbf{v} \qquad (2.78)$$

Vortex lines can be defined analogously to streamlines as lines that are tangential to the vorticity vector at all points in the flow field. Similarly the concept of the *vortex tube* is analogous to that of stream tube. Physically we can think of flow structures like vortices as comprising bundles of vortex tubes. In many respects vorticity and vortex lines are even more fundamental to understanding the flow physics than are velocity and streamlines.

2.7.5 Vorticity in polar coordinates

Referring to Section 2.4.3 where polar coordinates were introduced, the corresponding definition of vorticity in polar coordinates is

$$\zeta = \frac{q_t}{r} + \frac{\partial q_t}{\partial r} - \frac{1}{r} \frac{\partial q_n}{\partial \theta} \qquad (2.79)$$

Note that consistent with its physical interpretation as rate of rotation, the units of vorticity are radians per second.

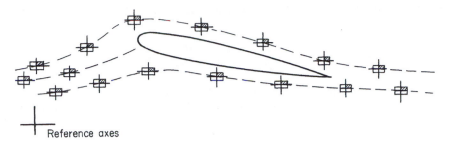

Reference axes

Fig. 2.26

2.7.6   Rotational and irrotational flow

It will be made clear in Section 2.8 that the generation of shear strain in a fluid element, as it travels through the flow field, is closely linked with the effects of viscosity. It is also plain from its definition (Eqn (2.76)) that vorticity is related to rate of shear strain. Thus, in aerodynamics, the existence of vorticity is associated with the effects of viscosity.* Accordingly, when the effects of viscosity can be neglected, the vorticity is usually equivalently zero. This means that the individual fluid elements do not rotate, or distort, as they move through the flow field. For incompressible flow, then, this corresponds to the state of pure translation that is illustrated in Fig. 2.26. Such a flow is termed *irrotational* flow. Mathematically, it is characterized by the existence of a velocity potential and is, therefore, also called *potential* flow. It is the subject of Chapter 3. The converse of irrotational flow is *rotational* flow.

2.7.7   Circulation

The total amount of vorticity passing through any plane region within a flow field is called the *circulation*, Γ. This is illustrated in Fig. 2.27 which shows a bundle of vortex tubes passing through a plane region of area A located in the flow field. The perimeter of the region is denoted by C. At a typical point P on the perimeter, the velocity vector is designated $\mathbf{q}$ or, equivalently, $\vec{q}$. At P, the infinitesimal portion of C has length δs and points in the tangential direction defined by the unit vector $\mathbf{t}$ (or $\vec{t}$). It is important to understand that the region of area A and its perimeter C have no physical existence. Like the control volumes used for the application of conservation of mass and momentum, they are purely theoretical constructs.

Mathematically, the total strength of the vortex tubes can be expressed as an integral over the area A; thus

$$\Gamma = \iint_A \mathbf{n} \cdot \boldsymbol{\Omega} \, \mathrm{d}A \tag{2.80}$$

where $\mathbf{n}$ is the unit normal to the area A. In two-dimensional flow the vorticity is in the z direction perpendicular to the two-dimensional flow field in the (x, y) plane. Thus $\mathbf{n} = \mathbf{k}$ (i.e. the unit vector in the z direction) and $\boldsymbol{\Omega} = \zeta\mathbf{k}$, so that Eqn (2.80) simplifies to

$$\Gamma = \iint_A \zeta \, \mathrm{d}A \tag{2.81}$$

---

* Vorticity can also be created by other agencies, such as the presence of spatially varying body forces in the flow field. This could correspond to the presence of particles in the flow field, for example.

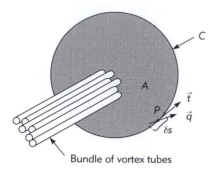

Bundle of vortex tubes

Fig. 2.27

Circulation can be regarded as a measure of the combined strength of the total number of vortex lines passing through A. It is a measure of the *vorticity flux* carried through A by these vortex lines. The relationship between circulation and vorticity is broadly similar to that between momentum and velocity or that between internal energy and temperature. Thus circulation is the property of the region A bounded by control surface C, whereas vorticity is a flow variable, like velocity, defined at a point. Strictly it makes no more sense to speak of conservation, generation, or transport of vorticity than its does to speak of conservation, generation, or transport of velocity. Logically these terms should be applied to circulation just as they are to momentum rather than velocity. But human affairs frequently defy logic and aerodynamics is no exception. We have become used to speaking in terms of conservation etc. of vorticity. It would be considered pedantic to insist on circulation in this context, even though this would be strictly correct. Our only motivation for making such fine distinctions here is to elucidate the meaning and significance of circulation. Henceforth we will adhere to the common usage of the terms vorticity and circulation.

In two-dimensional flow, in the absence of the effects of viscosity, circulation is conserved. This can be expressed mathematically as follows:

$$\frac{\partial \zeta}{\partial t} + u\frac{\partial \zeta}{\partial x} + v\frac{\partial \zeta}{\partial y} = 0 \tag{2.82}$$

In view of what was written in Section 2.7.6 about the link between vorticity and viscous effects, it may seem somewhat illogical to neglect such effects in Eqn (2.82). Nevertheless, it is often a useful approximation to use Eqn (2.82).

Circulation can also be evaluated by means of an integration around the perimeter C. This can be shown elegantly by applying *Stokes theorem* to Eqn (2.81); thus

$$\Gamma = \iint_A \mathbf{n}\cdot\mathbf{\Omega}\,\mathrm{d}A = \iint_A \mathbf{n}\cdot\nabla\times\mathbf{q}\,\mathrm{d}A = \oint_C \mathbf{q}\cdot\mathbf{t}\,\mathrm{d}s \tag{2.83}$$

This commonly serves as the definition of circulation in most aerodynamics text.

The concept of circulation is central to the theory of lift. This will become clear in Chapters 5 and 6.

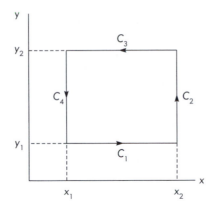

Fig. 2.28

Example 2.2 For the rectangular region of a two-dimensional flow field depicted in Fig. 2.28, starting with the definition Eqn (2.81) of circulation, show that it can also be evaluated by means of the integral around the closed circuit appearing as the last term in Eqn (2.83).

From Eqns (2.76) and (2.81) it follows that

$$\Gamma = \int_{y_1}^{y_2} \int_{x_1}^{x_2} \left(\frac{\partial v}{\partial x} - \frac{\partial u}{\partial y} \right) \mathrm{d}x \mathrm{d}y = \int_{y_1}^{y_2} \underbrace{\int_{x_1}^{x_2} \frac{\partial v}{\partial x} \mathrm{d}x}_{v(x_2, y) - v(x_1, y)} \mathrm{d}y - \int_{x_1}^{x_2} \underbrace{\int_{y_1}^{y_2} \frac{\partial u}{\partial y} \mathrm{d}y}_{u(x, y_2) - u(x, y_1)} \mathrm{d}x$$

Therefore

$$\Gamma = \int_{y_1}^{y_2} v(x_2, y) \mathrm{d}y - \int_{y_1}^{y_2} v(x_1, y) - \int_{x_1}^{x_2} u(x, y_2) \mathrm{d}x + \int_{x_1}^{x_2} u(x, y_1) \mathrm{d}x$$

$$= \int_{x_1}^{x_2} u(x, y_1) \mathrm{d}x + \int_{y_1}^{y_2} v(x_2, y) \mathrm{d}y + \int_{x_2}^{x_1} u(x, y_2) \mathrm{d}x + \int_{y_2}^{y_1} v(x_1, y) \qquad (2.84)$$

But along the lines: C_1, $\mathbf{q} = u\mathbf{i}$, $\mathbf{t} = \mathbf{i}$, $\mathrm{d}s = \mathrm{d}x$; C_2, $\mathbf{q} = v\mathbf{j}$, $\mathbf{t} = \mathbf{j}$, $\mathrm{d}s = \mathrm{d}y$; C_3, $\mathbf{q} = u\mathbf{i}$, $\mathbf{t} = -\mathbf{i}$, $\mathrm{d}s = -\mathrm{d}x$; and C_4, $\mathbf{q} = u\mathbf{j}$, $\mathbf{t} = -\mathbf{j}$, $\mathrm{d}s = -\mathrm{d}y$. It therefore follows that Eqn (2.84) is equivalent to

$$\Gamma = \oint_C \mathbf{q} \cdot \mathbf{t} \, \mathrm{d}s$$

2.8 The Navier–Stokes equations

2.8.1 Relationship between rates of strain and viscous stresses

In solid mechanics the fundamental theoretical model linking the stress and strain fields is Hooke's law that states that

$$\text{Stress} \propto \text{Strain} \qquad (2.85)$$

The equivalent in fluid mechanics is the model of the *Newtonian fluid* for which it is assumed that

$$\text{Stress} \propto \text{Rate of strain} \tag{2.86}$$

However, there is a major difference in status between the two models. At best Hooke's law is a reasonable approximation for describing small deformations of some solids, particularly structural steel. Whereas the Newtonian fluid is a very accurate model for the behaviour of almost all homogeneous fluids, in particular water and air. It does not give good results for pseudofluids formed from suspensions of particles in homogeneous fluids, e.g. blood, toothpaste, slurries. Various *Non-Newtonian fluid* models are required to describe such fluids, which are often called non-Newtonian fluids. Non-Newtonian fluids are of little interest in aerodynamics and will be considered no further here.

For two-dimensional flows, the constitutive law (2.86) can be written

$$\begin{pmatrix} \sigma_{xx} & \sigma_{xy} \\ \sigma_{yx} & \sigma_{yy} \end{pmatrix} = 2\mu \begin{pmatrix} \dot{\varepsilon}_{xx} & \dot{\gamma}_{xy} \\ \dot{\gamma}_{yx} & \dot{\varepsilon}_{yy} \end{pmatrix} \tag{2.87}$$

where (·) denotes time derivatives. The factor 2 is merely used for convenience so as to cancel out the factor 1/2 in the expression (2.72a) for the rate of shear strain. Equation (2.87) is sufficient in the case of an incompressible fluid. For a compressible fluid, however, we should also allow for the possibility of direct stress being generated by rate of change of volume or *dilation*. Thus we need to add the following to the right-hand side of (2.87)

$$\lambda \begin{pmatrix} \dot{\varepsilon}_{xx} + \dot{\varepsilon}_{yy} & 0 \\ 0 & \dot{\varepsilon}_{xx} + \dot{\varepsilon}_{yy} \end{pmatrix} \tag{2.88}$$

μ and λ are called the first and second coefficients of viscosity. More frequently μ is just termed the *dynamic viscosity* in contrast to the *kinematic viscosity* $\nu \equiv \mu/\rho$. If it is required that the actual pressure $p - \frac{1}{3}(\sigma_{xx} + \sigma_{yy}) + \sigma_{zz}$ in a viscous fluid be identical to the thermodynamic pressure p, then it is easy to show that

$$3\lambda + 2\mu = 0 \quad \text{or} \quad \lambda = -\frac{2}{3}\mu$$

This is often called *Stokes* hypothesis. In effect, it assumes that the bulk viscosity, μ', linking the average viscous direct stress to the rate of volumetric strain is zero, i.e.

$$\mu' = \lambda + \frac{2}{3}\mu \simeq 0 \tag{2.89}$$

This is still a rather controversial question. Bulk viscosity is of no importance in the great majority of engineering applications, but can be important for describing the propagation of sound waves in liquids and sometimes in gases also. Here, for the most part, we will assume incompressible flow, so that

$$\dot{\varepsilon}_{xx} + \dot{\varepsilon}_{yy} = \frac{\partial u}{\partial x} + \frac{\partial v}{\partial y} = 0$$

and Eqn (2.87) will, accordingly, be valid.

2.8.2   The derivation of the Navier–Stokes equations

Restricting our derivation to two-dimensional flow, Eqn (2.87) with (2.72a) and (2.73) gives

$$\sigma_{xx} = 2\mu \frac{\partial u}{\partial x}, \qquad \sigma_{yy} = 2\mu \frac{\partial v}{\partial y}, \qquad \sigma_{xy} = \sigma_{yx} = \mu\left(\frac{\partial u}{\partial y} + \frac{\partial v}{\partial x}\right) \tag{2.90}$$

So the right-hand side of the momentum Eqns (2.66a) becomes

$$g_x - \frac{\partial p}{\partial x} + 2\mu \frac{\partial}{\partial x}\left(\frac{\partial u}{\partial x}\right) + \mu \frac{\partial}{\partial y}\left(\frac{\partial u}{\partial y} + \frac{\partial v}{\partial x}\right)$$

$$= g_x - \frac{\partial p}{\partial x} + \mu\left(\frac{\partial^2 u}{\partial x^2} + \frac{\partial^2 u}{\partial y^2}\right) + \mu \frac{\partial}{\partial x}\underbrace{\left(\frac{\partial u}{\partial x} + \frac{\partial v}{\partial y}\right)}_{=0,\ \text{Eqn (2.46)}} \tag{2.91}$$

The right-hand side of (2.66b) can be dealt with in a similar way. Thus the momentum equations (2.66a,b) can be written in the form

$$\rho\left(\frac{\partial u}{\partial t} + u\frac{\partial u}{\partial x} + v\frac{\partial u}{\partial y}\right) = \rho g_x - \frac{\partial p}{\partial x} + \mu\left(\frac{\partial^2 u}{\partial x^2} + \frac{\partial^2 u}{\partial y^2}\right) \tag{2.92a}$$

$$\rho\left(\frac{\partial v}{\partial t} + u\frac{\partial v}{\partial x} + v\frac{\partial v}{\partial y}\right) = \rho g_y - \frac{\partial p}{\partial y} + \mu\left(\frac{\partial^2 v}{\partial x^2} + \frac{\partial^2 v}{\partial y^2}\right) \tag{2.92b}$$

This form of the momentum equations is known as the *Navier–Stokes* equations for two-dimensional flow. With the inclusion of the continuity equation

$$\frac{\partial u}{\partial x} + \frac{\partial v}{\partial y} = 0 \tag{2.93}$$

we now have three governing equations for three unknown flow variables u, v, p.

The Navier–Stokes equations for three-dimensional incompressible flows are given below:

$$\frac{\partial u}{\partial x} + \frac{\partial v}{\partial y} + \frac{\partial w}{\partial z} = 0 \tag{2.94}$$

$$\rho\left(\frac{\partial u}{\partial t} + u\frac{\partial u}{\partial x} + v\frac{\partial u}{\partial y} + w\frac{\partial u}{\partial z}\right) = \rho g_x - \frac{\partial p}{\partial x} + \mu\left(\frac{\partial^2 u}{\partial x^2} + \frac{\partial^2 u}{\partial y^2} + \frac{\partial^2 u}{\partial z^2}\right) \tag{2.95a}$$

$$\rho\left(\frac{\partial v}{\partial t} + u\frac{\partial v}{\partial x} + v\frac{\partial v}{\partial y} + w\frac{\partial v}{\partial z}\right) = \rho g_y - \frac{\partial p}{\partial y} + \mu\left(\frac{\partial^2 v}{\partial x^2} + \frac{\partial^2 v}{\partial y^2} + \frac{\partial^2 v}{\partial z^2}\right) \tag{2.95b}$$

$$\rho\left(\frac{\partial w}{\partial t} + u\frac{\partial w}{\partial x} + v\frac{\partial w}{\partial y} + w\frac{\partial w}{\partial z}\right) = \rho g_z - \frac{\partial p}{\partial z} + \mu\left(\frac{\partial^2 w}{\partial x^2} + \frac{\partial^2 w}{\partial y^2} + \frac{\partial^2 w}{\partial z^2}\right) \tag{2.95c}$$

2.9   Properties of the Navier–Stokes equations

At first sight the Navier–Stokes equations, especially the three-dimensional version, Eqns (2.95), may appear rather formidable. It is important to recall that they are nothing more than the application of Newton's second law of motion to fluid flow.

For example, the left-hand side of Eqn (2.95a) represents the total rate of change of the x component of momentum per unit volume. Indeed it is often written as:

$$\rho \frac{Du}{Dt} \qquad \text{where} \qquad \frac{D}{Dt} \equiv \frac{\partial}{\partial t} + u \frac{\partial}{\partial x} + v \frac{\partial}{\partial y} + w \frac{\partial}{\partial z} \qquad (2.96)$$

is called the *total* or *material* derivative. It represents the total rate of change with time following the fluid motion. The left-hand sides of Eqns (2.95b,c) can be written in a similar form. The three terms on the right-hand side represent the x components of body force, pressure force and viscous force respectively acting on a unit volume of fluid.

The compressible versions of the Navier–Stokes equations plus the continuity equation encompass almost the whole of aerodynamics. To be sure, applications involving combustion or rarified flow would require additional chemical and physical principles, but most of aerodynamics is contained within the Navier–Stokes equations. Why, then, do we need the rest of the book, not to mention the remaining vast, ever-growing, literature devoted to aerodynamics? Given the power of modern computers, could we not merely solve the Navier–Stokes equations numerically for any aerodynamics application of interest? The short answer is no! Moreover, there is no prospect of it ever being possible. To explain fully why this is so is rather difficult. We will, nevertheless, attempt to give a brief indication of the nature of the problem.

Let us begin by noting that the Navier–Stokes equations are a set of partial differential equations. Few analytical solutions exist that are useful in aerodynamics. (The most useful examples will be described in Section 2.10.) Accordingly, it is essential to seek approximate solutions. Nowadays, it is often possible to obtain very accurate numerical solutions by using computers. In many respects these can be regarded almost as exact solutions, although one must never forget that computer-generated solutions are subject to error. It is by no means simple to obtain such numerical solutions of the Navier–Stokes equations. There are two main sources of difficulty. First, the equations are nonlinear. The nonlinearity arises from the left-hand sides, i.e. the terms representing the rate of change of momentum – the so-called *inertial* terms. To appreciate why these terms are nonlinear, simply note that when you take a term on the right-hand side of the equations, e.g. the pressure terms, when the flow variable (e.g. pressure) is doubled the term is also doubled in magnitude. This is also true for the viscous terms. Thus these terms are proportional to the unknown flow variables, i.e. they are linear. Now consider a typical inertial term, say $u\partial u/\partial x$. This term is plainly proportional to u^2 and not u, and is therefore nonlinear. The second source of difficulty is more subtle. It involves the complex effects of viscosity.

In order to understand this second point better, it is necessary to make the Navier–Stokes equations non-dimensional. The motivation for working with non-dimensional variables and equations is that it helps to make the theory scale-invariant and accordingly more universal (see Section 1.4). In order to fix ideas, let us consider the air flowing at speed U_∞ towards a body, a circular cylinder or wing say, of length L. See Fig. 2.29. The space variables x, y, and z can be made nondimensional by dividing by L. L/U_∞ can be used as the reference time to make time non-dimensional. Thus we introduce the non-dimensional coordinates

$$X = x/L, \qquad Y = y/L, \qquad Z = z/L, \qquad \text{and} \qquad T = tU/L \qquad (2.97)$$

U_∞ can be used as the reference flow speed to make the velocity components dimensionless and ρU_∞^2 (c.f. Bernoulli equation Eqn (2.16)) used as the reference

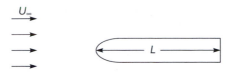

Fig. 2.29

pressure. (For incompressible flow, at least, only pressure difference is of significance and not the absolute value of the pressure.) This allows us to introduce the following non-dimensional flow variables:

$$U = u/U_\infty, \qquad V = v/U_\infty, \qquad W = w/U_\infty, \qquad \text{and} \qquad P = p/(\rho U_\infty^2) \quad (2.98)$$

If, by writing $x = XL$ etc. the non-dimensional variables given in Eqns (2.97) and (2.98) are substituted into Eqns (2.94) and (2.95) with the body-force terms omitted, we obtain the Navier–Stokes equations in the form:

$$\frac{\partial U}{\partial X} + \frac{\partial V}{\partial Y} + \frac{\partial W}{\partial Z} = 0 \tag{2.99}$$

$$\frac{DU}{DT} = -\frac{\partial P}{\partial X} + \frac{1}{Re}\left(\frac{\partial^2 U}{\partial X^2} + \frac{\partial^2 U}{\partial Y^2} + \frac{\partial^2 U}{\partial Z^2}\right) \tag{2.100a}$$

$$\frac{DV}{DT} = -\frac{\partial P}{\partial Y} + \frac{1}{Re}\left(\frac{\partial^2 V}{\partial X^2} + \frac{\partial^2 V}{\partial Y^2} + \frac{\partial^2 V}{\partial Z^2}\right) \tag{2.100b}$$

$$\frac{DW}{DT} = -\frac{\partial P}{\partial Z} + \frac{1}{Re}\left(\frac{\partial^2 W}{\partial X^2} + \frac{\partial^2 W}{\partial Y^2} + \frac{\partial^2 W}{\partial Z^2}\right) \tag{2.100c}$$

where the short-hand notation (2.96) for the material derivative has been used. A feature of Eqns (2.100) is the appearance of the dimensionless quantity known as the *Reynolds number*:

$$Re \equiv \frac{\rho U_\infty L}{\mu} \tag{2.101}$$

From the manner in which it has emerged from making the Navier–Stokes equations dimensionless, it is evident that the Reynolds number (see also Section 1.4) represents the ratio of the inertial to the viscous terms (i.e. the ratio of rate of change of momentum to the viscous force). It would be difficult to overstate the significance of Reynolds number for aerodynamics.

It should now be clear from Eqns (2.99) and (2.100) that if one were to calculate the non-dimensional flow field for a given shape – a circular cylinder, for example – the overall flow pattern obtained would depend on the Reynolds number and, in the case of unsteady flows, on the dimensionless time T. The flow around a circular cylinder is a good example for illustrating just how much the flow pattern can change over a wide range of Reynolds number. See Section 7.5 and Fig. 7.14 in particular. Incidentally, the simple dimensional analysis carried out above shows that it is not always necessary to solve equations in order to extract useful information from them.

For high-speed flows where compressibility becomes important the absolute value of pressure becomes significant. As explained in Section 2.3.4 (see also Section 1.4), this leads to the appearance of the Mach number, M (the ratio of the flow speed to the speed of sound), in the stagnation pressure coefficient. Thus, when compressibility becomes important (see Section 2.3.4), Mach number becomes a second dimension-less quantity characterizing the flow field.

The Navier–Stokes equations are deceptively simple in form, but at high Reynolds numbers the resulting flow fields can be exceedingly complex even for simple geo-metries. This is basically a consequence of the behaviour of the regions of vortical flow at high Reynolds number. Vorticity can only be created in a viscous flow and can be regarded as a marker for regions where the effects of viscosity are important in some sense.

For engineering applications of aerodynamics the Reynolds numbers are very large, values well in excess of 10^6 are commonplace. Accordingly, one would expect that to a good approximation one could drop the viscous terms on the right-hand side of the dimensionless Navier–Stokes Eqns (2.100). In general, however, this view would be mistaken and one never achieves a flow field similar to the inviscid one no matter how high the Reynolds number. The reason is that the regions of non-zero vorticity where viscous effects cannot be neglected become confined to exceedingly thin boundary layers adjacent to the body surface. As $Re \to \infty$ the boundary-layer thickness, $\delta \to 0$. If the boundary layers remained attached to the surface they would have little effect beyond giving rise to skin-friction drag. But in all real flows the boundary layers separate from the surface of the body, either because of the effects of an adverse pressure gradient or because they reach the rear of the body or its trailing edge. When these thin regions of vortical flow separate they form complex unsteady vortex-like structures in the wake. These take their most extreme form in turbulent flow which is characterized by vortical structures with a wide range of length and time scales.

As we have seen from the discussion given above, it is not necessary to solve the Navier–Stokes equations in order to obtain useful information from them. This is also illustrated by following example:

Example 2.3 *Aerodynamic modelling*
Let us suppose that we are interested carrying out tests on a model in a wind-tunnel in order to study and determine the aerodynamic forces exerted on a motor vehicle travelling at normal motorway speeds. In this case the speeds are sufficiently low to ensure that the effects of compressibility are negligible. Thus for a fixed geometry the flow field will be characterized only by Reynolds number.* In this case we can use U_∞, the speed at which the vehicle travels (the air speed in the wind-tunnel working section for the model) as the reference flow speed, and L can be the width or length of the vehicle. So the Reynolds number $Re \equiv \rho U_\infty L/\mu$. For a fixed geometry it is clear from Eqns (2.99) and (2.100) that the non-dimensional flow variables, U, V, W, and P are functions only of the dimensionless coordinates X, Y, Z, T, and the dimensionless quantity, Re. In a steady flow the aerodynamic force, being an overall characteristic of the flow field, will not depend on X, Y, Z, or T. It will, in fact, depend only on Re. Thus if we make an aerodynamic force, drag (D) say, dimensionless, by introducing a force (i.e. drag) coefficient defined as

$$C_D = \frac{D}{\frac{1}{2}\rho U_\infty^2 L^2} \qquad (2.102)$$

(see Section 1.5.2 and noting that here we have used L^2 in place of area S) it should be clear that

$$C_D = F(Re) \quad \text{i.e. a function of } Re \text{ only} \qquad (2.103)$$

---

* In fact, this statement is somewhat of an over-simplification. Technically the turbulence characteristics of the oncoming flow also influence the details of the flow field.

If we wish the model tests to produce useful information about general characteristics of the prototype's flow field, in particular estimates for its aerodynamic drag, it is necessary for the model and prototype to be *dynamically similar*, i.e. for the forces to be scale invariant. It can be seen from Eqn (2.103) that this can only be achieved provided

$$Re_m = Re_p \qquad (2.104)$$

where suffices m and p denote model and prototype respectively.

It is not usually practicable to use any other fluid but air for the model tests. For standard wind-tunnels the air properties in the wind-tunnel are not greatly different from those experienced by the prototype. Accordingly, Eqn (2.104) implies that

$$U_m = \frac{L_p}{L_m} U_p \qquad (2.105)$$

Thus, if we use a 1/5-scale model, Eqn (2.105) implies that $U_m = 5U_p$. So a prototype speed of 100 km/hr (c. 30 m/s) implies a model speed of 500 km/hr (c. 150 m/s). At such a model speed compressibility effects are no longer negligible. This illustrative example suggests that, in practice, it is rarely possible to achieve dynamic similarity in aerodynamic model tests using standard wind-tunnels. In fact, dynamic similarity can usually only be achieved in aerodynamics by using very large and expensive facilities where the dynamic similarity is achieved by compressing the air (thereby increasing its density) and using large models.

In this example we have briefly revisited the material covered in Section 1.4. The objective was to show how the dimensional analysis of the Navier–Stokes equations (effectively the exact governing equations of the flow field) could establish more rigorously the concepts introduced in Section 1.4.

2.10 Exact solutions of the Navier–Stokes equations

Few physically realizable exact solutions of the Navier–Stokes equations exist. Even fewer are of much interest in Engineering. Here we will present the two simplest solutions, namely Couette flow (simple shear flow) and plane Poiseuille flow (channel flow). These are useful for engineering applications, although not for the aerodynamics of wings and bodies. The third exact solution represents the flow in the vicinity of a stagnation point. This is important for calculating the flow around wings and bodies. It also illustrates a common and, at first sight, puzzling feature. Namely, that if the dimensionless Navier–Stokes equations can be reduced to an ordinary differential equation, this is regarded as tantamount to an exact solution. This is because the essentials of the flow field can be represented in terms of one or two curves plotted on a single graph. Also numerical solutions to ordinary differential equations can be obtained to any desired accuracy.

2.10.1 Couette flow – simple shear flow

This is the simplest exact solution. It corresponds to the flow field created between two infinite, plane, parallel surfaces; the upper one moving tangentially at speed U_T, the lower one being stationary (see Fig. 2.30). Since the flow is steady and two-dimensional, derivatives with respect to z and t are zero, and $w = 0$. The streamlines

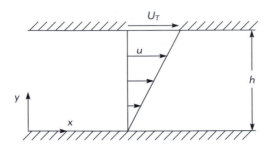

Fig. 2.30

are parallel to the x axis, so $v = 0$. Therefore Eqn (2.93) implies $\partial u/\partial x = 0$, i.e. u is a function only of y. There is no external pressure field, so Eqn (2.92a) reduces to

$$\mu \frac{\partial^2 u}{\partial y^2} = 0 \quad \text{implying} \quad u = C_1 y + C_2 \tag{2.106}$$

where C_1 and C_2 are constants of integration. $u = 0$ and U_T when $y = 0$ and h respectively, so Eqn (2.106) becomes

$$u = U_T \frac{y}{h} = \frac{\tau}{\mu} y \tag{2.107}$$

where τ is the constant viscous shear stress.

This solution approximates well the flow between two concentric cylinders with the inner one rotating at fixed speed, provided the clearance is small compared with the cylinder's radius, R. This is the basis of a viscometer – an instrument for measuring viscosity, since the torque required to rotate the cylinder at constant speed ω is proportional to τ which is given by $\mu \omega R/h$. Thus if the torque and rotational speed are measured the viscosity can be determined.

2.10.2   Plane Poiseuille flow – pressure-driven channel flow

This also corresponds to the flow between two infinite, plane, parallel surfaces (see Fig. 2.31). Unlike Couette flow, both surfaces are stationary and flow is produced by the application of pressure. Thus all the arguments used in Section 2.10.1 to simplify the Navier–Stokes equations still hold. The only difference is that the pressure term in Eqn (2.95a) is retained so that it simplifies to

$$-\frac{dp}{dx} + \mu \frac{\partial^2 u}{\partial y^2} = 0 \quad \text{implying} \quad u = \frac{1}{\mu} \frac{dp}{dx} \frac{y^2}{2} + C_1 y + C_2 \tag{2.108}$$

The no-slip condition implies that $u = 0$ at $y = 0$ and h, so Eqn (2.108) becomes

$$u = -\frac{h^2}{2\mu} \frac{dp}{dx} \frac{y}{h} \left(1 - \frac{y}{h}\right) \tag{2.109}$$

Thus the velocity profile is parabolic in shape.

The true Poiseuille flow is found in capillaries with round sections. A very similar solution can be found for this case in a similar way to Eqn (2.109) that again has

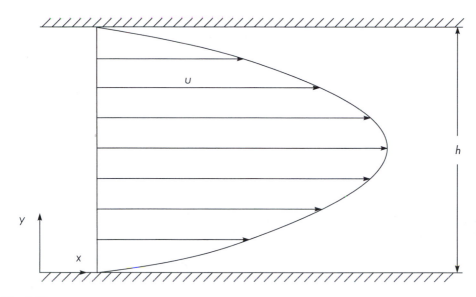

Fig. 2.31

a parabolic velocity profile. From this solution, Poiseuille's law can be derived linking the flow rate, Q, through a capillary of diameter d to the pressure gradient, namely

$$Q = -\frac{\pi d^4}{128\mu}\frac{\mathrm{d}p}{\mathrm{d}x} \tag{2.110}$$

Poiseuille was a French physician who derived his law in 1841 in the course of his studies on blood flow. His law is the basis of another type of viscometer whereby the flow rate driven through a capillary by a known pressure difference is measured. The value of viscosity can be determined from this measurement by using Eqn (2.110).

2.10.3 Hiemenz flow – two-dimensional stagnation-point flow

The simplest example of this type of flow, illustrated in Fig. 2.32, is generated by uniform flow impinging perpendicularly on an infinite plane. The flow divides equally about a stagnation point (strictly a line). The velocity field for the corresponding inviscid potential flow (see Chapter 3) is

$$u = ax \qquad v = -ay \quad \text{where } a \text{ is a const.} \tag{2.111}$$

The real viscous flow must satisfy the no-slip condition at the wall – as shown in Fig. 2.32 – but the potential flow may offer some hints on seeking the full viscous solution.

This special solution is of particular interest for aerodynamics. All two-dimensional stagnation flows behave in a similar way near the stagnation point. It can therefore be used as the starting solution for boundary-layer calculations in the case of two-dimensional bodies with rounded noses or leading edges (see Example 2.4). There is also an equivalent axisymmetric stagnation flow.

The approach used to find a solution to the two-dimensional Navier–Stokes Eqns (2.92) and (2.93) is to aim to reduce the equations to an ordinary differential equation. This is done by assuming that, when appropriately scaled, the non-dimensional

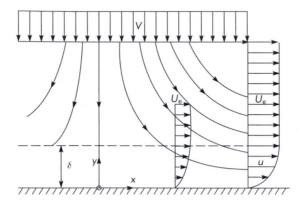

Fig. 2.32 Stagnation-zone flow field

velocity profile remains the same shape throughout the flow field. Thus the nature of the flow field suggests that the normal velocity component is independent of x, so that

$$v = -f(y) \tag{2.112}$$

where $f(y)$ is a function of y that has to be determined. Substitution of Eqn (2.112) into the continuity Eqn (2.93) gives

$$\frac{\partial u}{\partial x} = f'(y); \quad \text{integrate to get} \quad u = xf'(y) \tag{2.113}$$

where $(\ )'$ denotes differentiation with respect to y. The constant of integration in Eqn (2.113) is equivalently zero, as $u = v = 0$ at $x = 0$ (the stagnation point), and was therefore omitted.

For a potential flow the Bernoulli equation gives

$$p + \frac{1}{2}\rho(\underbrace{u^2 + v^2}_{a^2x^2 + a^2y^2}) = p_0. \tag{2.114}$$

So for the full viscous solution we will try the form:

$$p_0 - p = \frac{1}{2}\rho a^2 [x^2 + F(y)], \tag{2.115}$$

where $F(y)$ is another function of y. If the assumptions (2.112) and (2.115) are incorrect, we will fail in our objective of reducing the Navier–Stokes equations to ordinary differential equations.

Substitute Eqns (2.112), (2.113) and (2.115) into Eqn (2.92a,b) to get

$$\rho \underbrace{u\frac{\partial u}{\partial x}}_{\rho x f'^2} + \rho \underbrace{v\frac{\partial u}{\partial y}}_{-\rho x f f''} = \underbrace{-\frac{\partial p}{\partial x}}_{-\rho a^2 x} + \mu\left(\underbrace{\frac{\partial^2 u}{\partial x^2}}_{0} + \underbrace{\frac{\partial^2 u}{\partial y^2}}_{\mu x f'''}\right) \tag{2.116}$$

$$\rho \underbrace{u\frac{\partial v}{\partial x}}_{0} + \rho \underbrace{v\frac{\partial v}{\partial y}}_{-\rho f f'} = \underbrace{-\frac{\partial p}{\partial y}}_{-\rho a^2 F'/2} + \mu\left(\underbrace{\frac{\partial^2 v}{\partial x^2}}_{0} + \underbrace{\frac{\partial^2 v}{\partial y^2}}_{\mu f''}\right) \tag{2.117}$$

Simplifying these two equations gives

$$f'^2 - ff'' = a^2 + \nu f''' \tag{2.118}$$

$$ff' = \frac{1}{2}a^2 F' - \nu f'' \tag{2.119}$$

where use has been made of the definition of kinematic viscosity ($\nu = \mu/\rho$). Evidently the assumptions made above were acceptable, since we have succeeded in the aim of reducing the Navier–Stokes equations to ordinary differential equations. Also note that the second Eqn (2.119) is only required to determine the pressure field, Eqn (2.118) on its own can be solved for f, thus determining the velocity field.

The boundary conditions at the wall are straightforward, namely

$$u = v = 0 \quad \text{at} \quad y = 0 \quad \text{implying} \quad f = f' = 0 \quad \text{at} \quad y = 0 \tag{2.120}$$

As $y \to \infty$ the velocity will tend to its form in the corresponding potential flow. Thus

$$u \to ax \quad \text{as} \quad y \to \infty \quad \text{implying} \quad f' = a \quad \text{as} \quad y \to \infty \tag{2.121}$$

In its present form Eqn (2.118) contains both a and ν, so that f depends on these parameters as well as being a function of y. It is desirable to derive a universal form of Eqn (2.118), so that we only need to solve it once and for all. We attempt to achieve this by scaling the variables $f(y)$ and y, i.e. by writing

$$f(y) = \beta\phi(\eta), \quad \eta = \alpha y \tag{2.122}$$

where α and β are constants to be determined by substituting Eqn (2.122) into Eqn (2.118). Noting that

$$f' = \frac{\mathrm{d}f}{\mathrm{d}y} = \frac{\mathrm{d}\eta}{\mathrm{d}y}\beta\frac{\mathrm{d}\phi}{\mathrm{d}\eta} = \alpha\beta\phi'$$

Eqn (2.118) thereby becomes

$$\alpha^2\beta^2\phi'^2 - \alpha^2\beta^2\phi\phi'' = a^2 + \nu\alpha^3\beta\phi''' \tag{2.123}$$

Thus providing

$$\alpha^2\beta^2 = a^2 = \nu\alpha^3\beta, \quad \text{implying} \quad \alpha = \sqrt{a/\nu}, \quad \beta = \sqrt{a\nu} \tag{2.124}$$

they can be cancelled as common factors and Eqn (2.124) reduces to the universal form:

$$\phi''' + \phi\phi'' - \phi'^2 + 1 = 0 \tag{2.125}$$

with boundary conditions

$$\phi(0) = \phi'(0) = 0, \quad \phi'(\infty) = 1$$

In fact, $\phi' = u/U_e$ where $U_e = ax$ the velocity in the corresponding potential flow found when $\eta \to \infty$. It is plotted in Fig. 2.33. We can regard the point at which $\phi' = 0.99$ as marking the edge of the viscous region. This occurs at $\eta \simeq 2.4$. This viscous region can be regarded as the boundary layer in the vicinity of the stagnation point (note, though, no approximation was made to obtain the solution). Its thickness does not vary with x and is given by

$$\delta \simeq 2.4\sqrt{\nu/a} \tag{2.126}$$

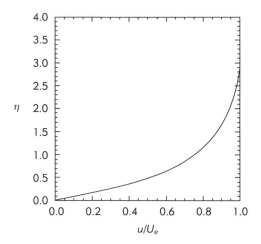

Fig. 2.33

Example 2.4 *Calculating the boundary-layer thickness in the stagnation zone at the leading edge.*
We will estimate the boundary-layer thickness in the stagnation zone of (i) a circular cylinder of 120 mm diameter in a wind-tunnel at a flow speed of 20 m/s; and (ii) the leading-edge of a Boeing 747 wing with a leading-edge radius of 150 mm at a flight speed of 250 m/s.

For a circular cylinder the potential-flow solution for the tangential velocity at the surface is given by $2U_\infty \sin\phi$ (see Eqn (3.44)). Therefore in Case (i) in the stagnation zone, $x = R\sin\phi \simeq R\phi$, so the velocity tangential to the cylinder is

$$U_e \simeq 2U_\infty\phi = 2\frac{U_\infty}{R}\underbrace{R\phi}_{x}$$

Therefore, as shown in Fig. 2.34, if we draw an analogy with the analysis in Section 2.10.3 above, $a = 2U_\infty/R = 2 \times 20/0.06 = 666.7 \text{ sec}^{-1}$. Thus from Eqn (2.126), given that for air the kinematic viscosity, $\nu \simeq 15 \times 10^{-6} \text{ m}^2/\text{s}$,

$$\delta \simeq 2.4\sqrt{\frac{\nu}{a}} = 2.4\sqrt{\frac{15 \times 10^{-6}}{666.7}} = 360 \,\mu\text{m}$$

For the aircraft wing in Case (ii) we regard the leading edge as analogous locally to a circular cylinder and follow the same procedure as for Case (i). Thus $R = 150 \text{ mm} = 0.15 \text{ m}$ and $U_\infty = 250 \text{ m/s}$, so in the stagnation zone, $a = 2U_\infty/R = 2 \times 250/0.15 = 3330 \text{ sec}^{-1}$ and

$$\delta \simeq 2.4\sqrt{\frac{\nu}{a}} = 2.4\sqrt{\frac{15 \times 10^{-6}}{3330}} = 160 \,\mu\text{m}$$

These results underline just how thin the boundary layer is! A point that will be taken up in Chapter 7.

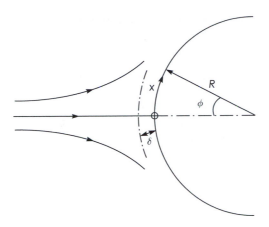

Fig. 2.34

Exercises

1 *Continuity Equation for axisymmetric flow*
(a) Consider an axisymmetric flow field expressed in terms of the cylindrical coordinate system (r, ϕ, z) where all flow variables are independent of the azimuthal angle ϕ. For example, the axial flow over a body of revolution. If the velocity components (u, w) correspond to the coordinate directions (r, z) respectively, show that the continuity equation is given by

$$\frac{\partial u}{\partial r} + \frac{u}{r} + \frac{\partial w}{\partial z} = 0$$

(b) Show that the continuity equation can be automatically satisfied by a stream-function ψ of a form such that

$$u = \frac{1}{r}\frac{\partial \psi}{\partial z}, \qquad w = -\frac{1}{r}\frac{\partial \psi}{\partial r}$$

2 *Continuity equation for two-dimensional flow in polar coordinates*
(a) Consider a two-dimensional flow field expressed in terms of the cylindrical coordinate system (r, ϕ, z) where all flow variables are independent of the azimuthal angle ϕ. For example, the flow over a circular cylinder. If the velocity components (u, v) correspond to the coordinate directions (r, ϕ) respectively, show that the continuity equation is given by

$$\frac{\partial u}{\partial r} + \frac{u}{r} + \frac{1}{r}\frac{\partial v}{\partial \phi} = 0$$

(b) Show that the continuity equation can be automatically satisfied by a stream-function ψ of a form such that

$$u = \frac{1}{r}\frac{\partial \psi}{\partial \phi}, \qquad v = -\frac{\partial \psi}{\partial r}$$

3 *Transport equation for contaminant in two-dimensional flow field*
In many engineering applications one is interested in the transport of a contaminant by the fluid flow. The contaminant could be anything from a polluting chemical to particulate matter. To derive the governing equation one needs to recognize that, provided that the contaminant is not being created within the flow field, then the mass of contaminant is conserved. The contaminant matter can be transported by two distinct physical mechanisms, namely *convection* and *molecular diffusion*. Let C be the concentration of contaminant (i.e. mass per unit volume of fluid), then the rate of transport of contamination per unit area is given by

$$-\mathcal{D}\nabla C = -\mathcal{D}\left(\mathbf{i}\frac{\partial C}{\partial x} + \mathbf{j}\frac{\partial C}{\partial y}\right)$$

where $\mathbf{i}$ and $\mathbf{j}$ are the unit vectors in the x and y directions respectively, and $\mathcal{D}$ is the diffusion coefficient (units m^2/s, the same as kinematic viscosity).

Note that diffusion transports the contaminant down the concentration gradient (i.e. the transport is from a higher to a lower concentration) hence the minus sign. It is analogous to thermal conduction.

(a) Consider an infinitesimal rectangular control volume. Assume that no contaminant is produced within the control volume and that the contaminant is sufficiently dilute to leave the fluid flow unchanged. By considering a mass balance for the control volume, show that the transport equation for a contaminant in a two-dimensional flow field is given by

$$\frac{\partial C}{\partial t} + u\frac{\partial C}{\partial x} + v\frac{\partial C}{\partial y} - \mathcal{D}\left(\frac{\partial^2 C}{\partial x^2} + \frac{\partial^2 C}{\partial y^2}\right) = 0$$

(b) Why is it necessary to assume a dilute suspension of contaminant? What form would the transport equation take if this assumption were not made? Finally, how could the equation be modified to take account of the contaminant being produced by a chemical reaction at the rate of $\dot{m}_c$ per unit volume.

4 *Euler equations for axisymmetric flow*
(a) for the flow field and coordinate system of Ex. 1 show that the Euler equations (inviscid momentum equations) take the form:

$$\rho\left(\frac{\partial u}{\partial t} + u\frac{\partial u}{\partial r} + w\frac{\partial u}{\partial z}\right) = \rho g_r - \frac{\partial p}{\partial r}$$

$$\rho\left(\frac{\partial w}{\partial t} + u\frac{\partial w}{\partial r} + w\frac{\partial w}{\partial z}\right) = \rho g_z - \frac{\partial p}{\partial z}$$

5 *The Navier–Stokes equations for two-dimensional axisymmetric flow*
(a) Show that the strain rates and vorticity for an axisymmetric viscous flow like that described in Ex. 1 are given by:

$$\dot{\varepsilon}_{rr} = \frac{\partial u}{\partial r}; \qquad \dot{\varepsilon}_{zz} = \frac{\partial w}{\partial z}; \qquad \dot{\varepsilon}_{\phi\phi} = \frac{u}{r}$$

$$\dot{\gamma}_{rz} = \frac{1}{2}\left(\frac{\partial w}{\partial r} + \frac{\partial u}{\partial z}\right); \qquad \eta = \frac{\partial w}{\partial r} - \frac{\partial u}{\partial z}$$

[Hint: Note that the azimuthal strain rate is not zero. The easiest way to determine it is to recognize that $\dot{\varepsilon}_{rr} + \dot{\varepsilon}_{\phi\phi} + \dot{\varepsilon}_{zz} = 0$ must be equivalent to the continuity equation.]

(b) Hence show that the Navier-Stokes equations for axisymmetric flow are given by

$$\rho\left(\frac{\partial u}{\partial t} + u\frac{\partial u}{\partial r} + w\frac{\partial u}{\partial z}\right) = \rho g_r - \frac{\partial p}{\partial r} + \mu\left(\frac{\partial^2 u}{\partial r^2} + \frac{1}{r}\frac{\partial u}{\partial r} - \frac{u}{r^2} + \frac{\partial^2 u}{\partial z^2}\right)$$

$$\rho\left(\frac{\partial w}{\partial t} + u\frac{\partial w}{\partial r} + w\frac{\partial w}{\partial z}\right) = \rho g_z - \frac{\partial p}{\partial z} + \mu\left(\frac{\partial^2 w}{\partial r^2} + \frac{1}{r}\frac{\partial w}{\partial r} + \frac{\partial^2 w}{\partial z^2}\right)$$

6 *Euler equations for two-dimensional flow in polar coordinates*
(a) For the two-dimensional flow described in Ex. 2 show that the Euler equations (inviscid momentum equations) take the form:

$$\rho\left(\frac{\partial u}{\partial t} + u\frac{\partial u}{\partial r} + \frac{v}{r}\frac{\partial u}{\partial \phi} - \frac{v^2}{r}\right) = \rho g_r - \frac{\partial p}{\partial r}$$

$$\rho\left(\frac{\partial v}{\partial t} + u\frac{\partial v}{\partial r} + \frac{v}{r}\frac{\partial v}{\partial \phi} + \frac{uv}{r}\right) = \rho g_\phi - \frac{1}{r}\frac{\partial p}{\partial \phi}$$

[Hints: (i) The momentum components perpendicular to and entering and leaving the side faces of the elemental control volume have small components in the radial direction that must be taken into account; likewise (ii) the pressure forces acting on these faces have small radial components.]

7 Show that the strain rates and vorticity for the flow and coordinate system of Ex. 6 are given by:

$$\dot{\varepsilon}_{rr} = \frac{\partial u}{\partial r}; \qquad \dot{\varepsilon}_{\phi\phi} = \frac{1}{r}\frac{\partial v}{\partial \phi} + \frac{u}{r}$$

$$\dot{\gamma}_{r\phi} = \frac{1}{2}\left(\frac{\partial v}{\partial r} - \frac{v}{r} + \frac{1}{r}\frac{\partial u}{\partial \phi}\right); \qquad \zeta = \frac{1}{r}\frac{\partial u}{\partial \phi} - \frac{\partial v}{\partial r} + \frac{v}{r}$$

[Hint: (i) The angle of distortion (β) of the side face must be defined relative to the line joining the origin O to the centre of the infinitesimal control volume.]

8 (a) The flow in the narrow gap (of width h) between two concentric cylinders of length L with the inner one of radius R rotating at angular speed ω can be approximated by the Couette solution to the Navier–Stokes equations. Hence show that the torque T and power P required to rotate the shaft at a rotational speed of ω rad/s are given by

$$T = \frac{2\pi\mu\omega R^3 L}{h}, \qquad P = \frac{2\pi\mu\omega^2 R^3 L}{h}$$

9 *Axisymmetric stagnation-point flow*
Carry out a similar analysis to that described in Section 2.10.3 using the axisymmetric form of the Navier–Stokes equations given in Ex. 5 for axisymmetric stagnation-point flow and show that the equivalent to Eqn (2.118) is

$$\phi''' + 2\phi\phi'' - \phi'^2 + 1 = 0$$

where ϕ' denotes differentiation with respect to the independent variable $\zeta = \sqrt{a/\nu z}$ and ϕ is defined in exactly the same way as for the two-dimensional case.

3

Potential flow

Preamble

The aim of this chapter is to describe methods for calculating the air flow around various shapes of body. The classical assumption of irrotational flow is made, meaning that the vorticity is everywhere zero. This also implies inviscid flow. Irrotational flows are potential fields. A potential function, known as the velocity potential, is introduced. It is shown how the velocity components can be determined from the velocity potential. The equations of motion for irrotational flow reduce to a single partial differential equation for velocity potential known as the Laplace equation. Classical analytical techniques are described for obtaining two-dimensional and axisymmetric solutions to the Laplace equation for aerodynamic applications. The chapter ends by showing how these classical analytical solutions can be used to develop computational methods for predicting the potential flows around the complex three-dimensional geometries typical of modern aircraft.

3.1 Introduction

The concept of irrotational flow is introduced briefly in Section 2.7.6. By definition the vorticity is everywhere zero for such flows. This does not immediately seem a very significant simplification. But it turns out that zero vorticity implies the existence of a potential field (analogous to gravitational and electric fields). In aerodynamics the main variable of the potential field is known as the *velocity potential* (it is analogous to voltage in electric fields). And another name for irrotational flow is *potential flow*. For such flows the equations of motion reduce to a single partial differential equation, the famous Laplace equation, for velocity potential. There are well-known techniques (see Sections 3.3 and 3.4) for finding analytical solutions to Laplace's equation that can be applied to aerodynamics. These analytical techniques can also be used to develop sophisticated computational methods that can calculate the potential flows around the complex three-dimensional geometries typical of modern aircraft (see Section 3.5).

In Section 2.7.6 it was explained that the existence of vorticity is associated with the effects of viscosity. It therefore follows that approximating a real flow by a potential flow is tantamount to ignoring viscous effects. Accordingly, since all real fluids are viscous, it is natural to ask whether there is any practical advantage in studying potential flows. Were we interested only in bluff bodies like circular cylinders there would indeed be little point in studying potential flow, since no matter how high the Reynolds number, the real flow around a circular cylinder never looks anything like the potential flow. (But that is not to say that there is no point in studying potential flow around a circular cylinder. In fact, the study of potential flow around a rotating cylinder led to the profound Kutta–Zhukovski theorem that links lift to circulation for all cross-sectional shapes.) But potential flow really comes into its own for slender or streamlined bodies at low angles of incidence. In such cases the boundary layer remains attached until it reaches the trailing edge or extreme rear of the body. Under these circumstances a wide low-pressure wake does not form, unlike a circular cylinder. Thus the flow more or less follows the shape of the body and the main viscous effect is the generation of skin-friction drag plus a much smaller component of form drag.

Potential flow is certainly useful for predicting the flow around fuselages and other non-lifting bodies. But what about the much more interesting case of lifting bodies like wings? Fortunately, almost all practical wings are slender bodies. Even so there is a major snag. The generation of lift implies the existence of circulation. And circulation is created by viscous effects. Happily, potential flow was rescued by an important insight known as the *Kutta condition*. It was realized that the most important effect of viscosity for lifting bodies is to make the flow leave smoothly from the trailing edge. This can be ensured within the confines of potential flow by conceptually placing one or more (potential) vortices within the contour of the wing or aerofoil and adjusting the strength so as to generate just enough circulation to satisfy the Kutta condition. The theory of lift, i.e. the modification of potential flow so that it becomes a suitable model for predicting lift-generating flows is described in Chapters 4 and 5.

3.1.1 The velocity potential

The stream function (see Section 2.5) at a point has been defined as the quantity of fluid moving across some convenient imaginary line in the flow pattern, and lines of constant stream function (amount of flow or flux) may be plotted to give a picture of the flow pattern (see Section 2.5). Another mathematical definition, giving a different pattern of curves, can be obtained for the same flow system. In this case an expression giving the amount of flow *along* the convenient imaginary line is found.

In a general two-dimensional fluid flow, consider any (imaginary) line OP joining the origin of a pair of axes to the point $P(x, y)$. Again, the axes and this line do not impede the flow, and are used only to form a reference datum. At a point Q on the line let the local velocity q meet the line OP in β (Fig. 3.1). Then the component of velocity parallel to δs is $q \cos \beta$. The amount of fluid flowing along δs is $q \cos \beta \, \delta s$. The total amount of fluid flowing along the line towards P is the sum of all such amounts and is given mathematically as the integral $\int q \cos \beta \, \mathrm{d}s$. This function is called the *velocity potential* of P with respect to O and is denoted by ϕ.

Now OQP can be any line between O and P and a necessary condition for $\int q \cos \beta \, \mathrm{d}s$ to be the velocity potential ϕ is that the value of ϕ is unique for the point P, irrespective of the path of integration. Then:

$$\text{Velocity potential } \phi = \int_{OP} q \cos \beta \, \mathrm{d}s \qquad (3.1)$$

If this were not the case, and integrating the tangential flow component from O to P via A (Fig. 3.2) did not produce the same magnitude of ϕ as integrating from O to P

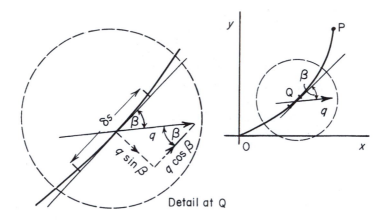

Detail at Q

Fig. 3.1

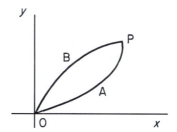

Fig. 3.2

via some other path such as B, there would be some flow components circulating in the circuit OAPBO. This in turn would imply that the fluid within the circuit possessed vorticity. The existence of a velocity potential must therefore imply zero vorticity in the flow, or in other words, a flow without circulation (see Section 2.7.7), i.e. an *irrotational* flow. Such flows are also called *potential* flows.

Sign convention for velocity potential

The tangential flow along a curve is the product of the local velocity component and the elementary length of the curve. Now, if the velocity component is in the direction of integration, it is considered a *positive* increment of the velocity potential.

3.1.2 The equipotential

Consider a point P having a velocity potential ϕ (ϕ is the integral of the flow component along OP) and let another point P_1 close to P have the same velocity potential ϕ. This then means that the integral of flow along OP_1 equals the integral of flow along OP (Fig. 3.3). But by definition OPP_1 is another path of integration from O to P_1. Therefore

$$\phi = \int_{OP} q \cos \beta \, ds = \int_{OP_1} q \cos \beta \, ds = \int_{OPP_1} q \cos \beta \, ds,$$

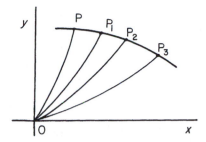

Fig. 3.3

but since the integral along OP equals that along OP_1 there can be no flow along the remaining portions of the path of the third integral, that is along PP_1. Similarly for other points such as P_2, P_3, having the same velocity potential, there can be no flow along the line joining P_1 to P_2.

The line joining P, P_1, P_2, P_3 is a line joining points having the same velocity potential and is called an *equipotential* or a line of constant velocity potential, i.e. a line of constant ϕ. The significant characteristic of an equipotential is that there is no flow along such a line. Notice the correspondence between an equipotential and a streamline that is a line across which there is no flow.

The flow in the region of points P and P_1 should be investigated more closely. From the above there can be no flow along the line PP_1, but there is fluid flowing in this region so it must be flowing in such a way that there is no component of velocity in the direction PP_1. So the flow can only be at right-angles to PP_1, that is the flow in the region PP_1 must be normal to PP_1. Now the streamline in this region, the line to which the flow is tangential, must also be at right-angles to PP_1 which is itself the local equipotential.

This relation applies at all points in a homogeneous continuous fluid and can be stated thus: streamlines and equipotentials meet orthogonally, i.e. always at right-angles. It follows from this statement that for a given streamline pattern there is a unique equipotential pattern for which the equipotentials are everywhere normal to the streamlines.

3.1.3   Velocity components in terms of ϕ

(a) *In Cartesian coordinates*   Let a point P(x, y) be on an equipotential ϕ and a neighbouring point Q($x + \delta x$, $y + \delta y$) be on the equipotential $\phi + \delta\phi$ (Fig. 3.4). Then by definition the increase in velocity potential from P to Q is the line integral of the *tangential* velocity component along any path between P and Q. Taking PRQ as the most convenient path where the local velocity components are u and v:

$$\delta\phi = u\delta x + v\delta y$$

but

$$\delta\phi = \frac{\partial\phi}{\partial x}\delta x + \frac{\partial\phi}{\partial y}\delta y$$

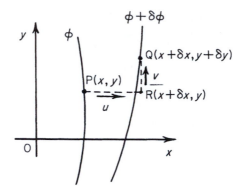

Fig. 3.4

Thus, equating terms

$$u = \frac{\partial \phi}{\partial x}$$

and

$$v = \frac{\partial \phi}{\partial y}$$

(3.2)

(b) *In polar coordinates* Let a point $P(r, \theta)$ be on an equipotential ϕ and a neighbouring point $Q(r + \delta r, \theta + \delta\theta)$ be on an equipotential $\phi + \delta\phi$ (Fig. 3.5). By definition the increase $\delta\phi$ is the line integral of the *tangential* component of velocity along any path. For convenience choose PRQ where point R is $(r + \delta r, \theta)$. Then integrating along PR and RQ where the velocities are q_n and q_t respectively, and are both in the direction of integration:

$$\delta\phi = q_n\delta r + q_t(r + \delta r)\delta\theta$$
$$= q_n\delta r + q_t r\delta\theta \text{ to the first order of small quantities.}$$

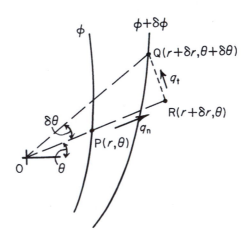

Fig. 3.5

But, since ϕ is a function of two independent variables;

$$\delta\phi = \frac{\partial\phi}{\partial r}\,\delta r + \frac{\partial\phi}{\partial\theta}\,\delta\theta$$

and
$$\left.\begin{array}{c} q_n = \dfrac{\partial\phi}{\partial r} \\[3mm] q_t = \dfrac{1}{r}\dfrac{\partial\phi}{\partial\theta} \end{array}\right\} \qquad (3.3)$$

Again, in general, the velocity q in any direction s is found by differentiating the velocity potential ϕ partially with respect to the direction s of q:

$$q = \frac{\partial\phi}{\partial s}$$

3.2 Laplace's equation

As a focus of the new ideas met so far that are to be used in this chapter, the main fundamentals are summarized, using Cartesian coordinates for convenience, as follows:

(1) The equation of continuity in two dimensions (incompressible flow)

$$\frac{\partial u}{\partial x} + \frac{\partial v}{\partial y} = 0 \qquad (i)$$

(2) The equation of vorticity

$$\frac{\partial v}{\partial x} - \frac{\partial u}{\partial y} = \zeta \qquad (ii)$$

(3) The stream function (incompressible flow) ψ describes a continuous flow in two dimensions where the velocity at any point is given by

$$u = \frac{\partial\psi}{\partial y} \qquad v = -\frac{\partial\psi}{\partial x} \qquad (iii)$$

(4) The velocity potential ϕ describes an irrotational flow in two dimensions where the velocity at any point is given by

$$u = \frac{\partial\phi}{\partial x} \qquad v = \frac{\partial\phi}{\partial y} \qquad (iv)$$

Substituting (iii) in (i) gives the identity

$$\frac{\partial^2\psi}{\partial x\partial y} - \frac{\partial^2\psi}{\partial x\partial y} = 0$$

which demonstrates the validity of (iii), while substituting (iv) in (ii) gives the identity

$$\frac{\partial^2\phi}{\partial x\partial y} - \frac{\partial^2\phi}{\partial x\partial y} = 0$$

demonstrating the validity of (iv), i.e. a flow described by a unique velocity potential must be irrotational.

Alternatively substituting (iii) in (ii) and (iv) in (i) the criteria for irrotational continuous flow are that

$$\frac{\partial^2 \phi}{\partial x^2} + \frac{\partial^2 \phi}{\partial y^2} = 0 = \frac{\partial^2 \psi}{\partial x^2} + \frac{\partial^2 \psi}{\partial y^2} \tag{3.4}$$

also written as $\nabla^2 \phi = \nabla^2 \psi = 0$, where the operator *nabla* squared

$$\nabla^2 = \frac{\partial^2}{\partial x^2} + \frac{\partial^2}{\partial y^2}$$

Eqn (3.4) is Laplace's equation.

3.3 Standard flows in terms of ψ and ϕ

There are three basic two-dimensional flow fields, from combinations of which all other steady flow conditions may be modelled. These are the *uniform parallel flow, source (sink)* and *point vortex*.

The three flows, the source (sink), vortex and uniform stream, form standard flow states, from combinations of which a number of other useful flows may be derived.

3.3.1 Two-dimensional flow from a source (or towards a sink)

A source (sink) of strength $m(-m)$ is a point at which fluid is appearing (or disappearing) at a uniform rate of $m(-m)\,\text{m}^2\,\text{s}^{-1}$. Consider the analogy of a small hole in a large flat plate through which fluid is welling (the source). If there is no obstruction and the plate is perfectly flat and level, the fluid puddle will get larger and larger all the while remaining circular in shape. The path that any particle of fluid will trace out as it emerges from the hole and travels outwards is a purely radial one, since it cannot go sideways, because its fellow particles are also moving outwards.

Also its velocity must get less as it goes outwards. Fluid issues from the hole at a rate of $m\,\text{m}^2\,\text{s}^{-1}$. The velocity of flow over a circular boundary of 1 m radius is $m/2\pi\,\text{m}\,\text{s}^{-1}$. Over a circular boundary of 2 m radius it is $m/(2\pi \times 2)$, i.e. half as much, and over a circle of diameter $2r$ the velocity is $m/2\pi r\,\text{m}\,\text{s}^{-1}$. Therefore the velocity of flow is inversely proportional to the distance of the particle from the source.

All the above applies to a sink except that fluid is being drained away through the hole and is moving towards the sink radially, increasing in speed as the sink is approached. Hence the particles all move radially, and the streamlines must be radial lines with their origin at the source (or sink).

To find the stream function ψ of a source

Place the source for convenience at the origin of a system of axes, to which the point P has ordinates (x, y) and (r, θ) (Fig. 3.6). Putting the line along the x-axis as $\psi = 0$

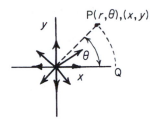

Fig. 3.6

(a datum) and taking the most convenient contour for integration as OQP where QP is an arc of a circle of radius r,

$$\psi = \text{flow across OQ} + \text{flow across QP}$$

$$= \text{velocity across OQ} \times \text{OQ} + \text{velocity across QP} \times \text{QP}$$

$$= 0 + \frac{m}{2\pi r} \times r\theta$$

Therefore

$$\psi = m\theta/2\pi$$

or putting $\theta = \tan^{-1}(y/x)$

$$\psi = \frac{m}{2\pi} \tan^{-1} \frac{y}{x} \tag{3.5}$$

There is a limitation to the size of θ here. θ can have values only between 0 and 2π. For $\psi = m\theta/2\pi$ where θ is greater than 2π would mean that ψ, i.e. the amount of fluid flowing, was greater than $m\,\text{m}^2\,\text{s}^{-1}$, which is impossible since m is the capacity of the source and integrating a circuit round and round a source will not increase its strength. Therefore $0 \leq \theta \leq 2\pi$.

For a sink

$$\psi = -(m/2\pi)\theta$$

To find the velocity potential ϕ of a source

The velocity everywhere in the field is radial, i.e. the velocity at any point $P(r, \theta)$ is given by $q = \sqrt{q_n^2 + q_t^2}$ and $q = q_n$ here, since $q_t = 0$. Integrating round OQP where Q is point $(r, 0)$

$$\phi = \int_{OQ} q \cos\beta \, ds + \int_{QP} q \cos\beta \, ds$$

$$= \int_{OQ} q_n dr + \int_{QP} q_t r \, \delta\theta = \int_{OQ} q_n \, dr + 0$$

But

$$q_n = \frac{m}{2\pi r}$$

Therefore

$$\phi = \int_{r_0}^{r} \frac{m}{2\pi r} dr = \frac{m}{2\pi} \ln \frac{r}{r_0} \tag{3.6}$$

where r_0 is the radius of the equipotential $\phi = 0$.

Alternatively, since the velocity q is always radial ($q = q_n$) it must be some function of r only and the tangential component is zero. Now

$$q_n = \frac{m}{2\pi r} = \frac{\partial \phi}{\partial r}$$

Therefore

$$\phi = \int_{r_0}^{r} \frac{m}{2\pi r} \, dr = \frac{m}{2\pi} \ln \frac{r}{r_0} \tag{3.7}$$

In Cartesian coordinates with $\phi = 0$ on the curve $r_0 = 1$

$$\phi = \frac{m}{4\pi} \ln(x^2 + y^2) \tag{3.8}$$

The equipotential pattern is given by $\phi = \text{constant}$. From Eqn (3.7)

$$\phi = \frac{m}{2\pi} \ln r - C \quad \text{where} \quad C = \frac{m}{2\pi} \ln r_0$$

$$r = e^{2\pi(\phi + C)/m} \tag{3.9}$$

and

$$r^2 = e^{4\pi(\phi + C)/m}$$

which is the equation of a circle of centre at the origin and radius $e^{2\pi(\phi + C)/m}$ when ϕ is constant. Thus equipotentials for a source (or sink) are concentric circles and satisfy the requirement of meeting the streamlines orthogonally.

3.3.2 Line (point) vortex

This flow is that associated with a straight line vortex. A line vortex can best be described as a string of rotating particles. A chain of fluid particles are spinning on their common axis and carrying around with them a swirl of fluid particles which flow around in circles. A cross-section of such a string of particles and its associated flow shows a spinning *point* outside of which is streamline flow in concentric circles (Fig. 3.7).

Vortices are common in nature, the difference between a real vortex as opposed to a theoretical line (potential) vortex is that the former has a core of fluid which is rotating as a solid, although the associated swirl outside is similar to the flow outside the point vortex. The streamlines associated with a line vortex are circular and therefore the particle velocity at any point must be tangential only.

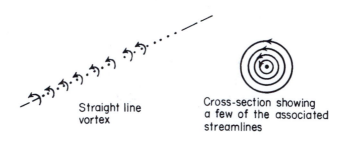

Straight line
vortex

Cross-section showing
a few of the associated
streamlines

Fig. 3.7

Consider a vortex located at the origin of a polar system of coordinates. But the flow is irrotational, so the vorticity everywhere is zero. Recalling that the streamlines are concentric circles, centred on the origin, so that $q_\theta = 0$, it therefore follows from Eqn (2.79), that

$$\zeta = \frac{q_t}{r} + \frac{dq_t}{dr} = 0, \quad \text{i.e.} \quad \frac{1}{r}\frac{d}{dr}(rq_t) = 0$$

So $d(rq_t)/dr = 0$ and integration gives

$$rq_t = C$$

where C is a constant. Now, recall Eqn (2.83) which is one of the two equivalent definitions of circulation, namely

$$\Gamma = \oint \vec{q} \cdot \vec{t}\, ds$$

In the present example, $\vec{q} \cdot \vec{t} = q_t$ and $ds = rd\theta$, so

$$\Gamma = 2\pi r q_t = 2\pi C.$$

Thus $C = \Gamma/(2\pi)$ and

$$q_t = -\frac{d\psi}{dr} = \frac{\Gamma}{2\pi r}$$

and

$$\psi = \int -\frac{\Gamma}{2\pi r}\, dr$$

Integrating along the most convenient boundary from radius r_0 to $P(r, \theta)$ which in this case is any radial line (Fig. 3.8):

$$\psi = -\int_{r_0}^{r} \frac{\Gamma}{2\pi r}\, dr \quad (r_0 = \text{radius of streamline}, \psi = 0)$$

$$\psi = -\left[\frac{\Gamma}{2\pi}\ln r\right]_{r_0}^{r} = -\frac{\Gamma}{2\pi}\ln\frac{r}{r_0} \tag{3.10}$$

Circulation is a measure of how fast the flow circulates the origin. (It is introduced and defined in Section 2.7.7.) Here the circulation is denoted by Γ and, by convention, is positive when anti-clockwise.

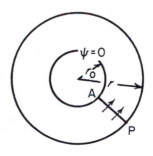

Fig. 3.8

Since the flow due to a line vortex gives streamlines that are concentric circles, the equipotentials, shown to be always normal to the streamlines, must be radial lines emanating from the vortex, and since

$$q_n = 0, \quad \phi \text{ is a function of } \theta, \text{ and}$$

$$q_t = \frac{1}{r}\frac{d\phi}{d\theta} = \frac{\Gamma}{2\pi r}$$

Therefore

$$d\phi = \frac{\Gamma}{2\pi}d\theta$$

and on integrating

$$\phi = \frac{\Gamma}{2\pi}\theta + \text{constant}$$

By defining $\phi = 0$ when $\theta = 0$:

$$\phi = \frac{\Gamma}{2\pi}\theta \qquad (3.11)$$

Compare this with the stream function for a source, i.e.

$$\psi = \frac{m\theta}{2\pi} \quad (\text{Eqn}(3.5))$$

Also compare the stream function for a vortex with the function for a source. Then consider two orthogonal sets of curves: one set is the set of radial lines emanating from a point and the other set is the set of circles centred on the same point. Then, if the point represents a source, the radial lines are the streamlines and the circles are the equipotentials. But if the point is regarded as representing a vortex, the roles of the two sets of curves are interchanged. This is an example of a general rule: consider the streamlines and equipotentials of a two-dimensional, continuous, irrotational flow. Then the streamlines and equipotentials correspond respectively to the equipotentials and streamlines of another flow, also two-dimensional, continuous and irrotational.

Since, for one of these flows, the streamlines and equipotentials are orthogonal, and since its equipotentials are the streamlines of the other flow, it follows that the streamlines of one flow are orthogonal to the streamlines of the other flow. The same is therefore true of the velocity vectors at any (and every) point in the two flows. If this principle is applied to the source–sink pair of Section 3.3.6, the result is the flow due to a pair of parallel line vortices of opposite senses. For such a vortex pair, therefore the streamlines are the circles sketched in Fig. 3.17, while the equipotentials are the circles sketched in Fig. 3.16.

3.3.3 Uniform flow

Flow of constant velocity parallel to Ox axis from left to right

Consider flow streaming past the coordinate axes Ox, Oy at velocity U parallel to Ox (Fig. 3.9). By definition the stream function ψ at a point P(x, y) in the flow is given by the amount of fluid crossing any line between O and P. For convenience the contour

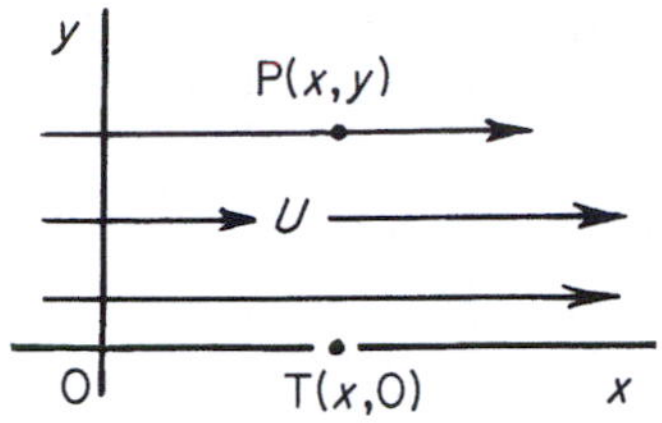

Fig. 3.9

OTP is taken where T is on the Ox axis x along from O, i.e. point T is given by $(x, 0)$. Then

$$\psi = \text{flow across line OTP}$$

$$= \text{flow across line OT plus flow across line TP}$$

$$= 0 + U \times \text{length TP}$$

$$= 0 + Uy$$

Therefore

$$\psi = Uy \qquad (3.12)$$

The streamlines (lines of constant ψ) are given by drawing the curves

$$\psi = \text{constant} = Uy$$

Now the velocity is constant, therefore

$$y = \frac{\psi}{U} = \text{constant on streamlines}$$

The lines $\psi = \text{constant}$ are all straight lines parallel to Ox.

By definition the velocity potential at a point P(x, y) in the flow is given by the line integral of the *tangential* velocity component along any curve from O to P. For convenience take OTP where T has ordinates $(x, 0)$. Then

$$\phi = \text{flow along contour OTP}$$

$$= \text{flow along OT } + \text{ flow along TP}$$

$$= Ux + 0$$

Therefore

$$\phi = Ux \qquad (3.13)$$

The lines of constant ϕ, the equipotentials, are given by $Ux = \text{constant}$, and since the velocity is constant the equipotentials must be lines of constant x, or lines parallel to Oy that are everywhere normal to the streamlines.

Flow of constant velocity parallel to Oy axis

Consider flow streaming past the Ox, Oy axes at velocity V parallel to Oy (Fig. 3.10). Again by definition the stream function ψ at a point P(x, y) in the flow is given by the

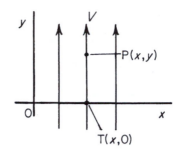

Fig. 3.10

amount of fluid crossing any curve between O and P. For convenience take OTP where T is given by $(x, 0)$. Then

$$\psi = \text{flow across OT} + \text{flow across TP}$$
$$= -Vx + 0$$

Note here that when going from O towards T the flow appears from the right and disappears to the left and therefore is of negative sign, i.e.

$$\psi = -Vx \qquad (3.14)$$

The streamlines being lines of constant ψ are given by $x = -\psi/V$ and are parallel to Oy axis.

Again consider flow streaming past the Ox, Oy axes with velocity V parallel to the Oy axis (Fig. 3.10). Again, taking the most convenient boundary as OTP where T is given by $(x, 0)$

$$\phi = \text{flow along OT} + \text{flow along TP}$$
$$= 0 + Vy$$

Therefore

$$\phi = Vy \qquad (3.15)$$

The lines of constant velocity potential, ϕ (equipotentials), are given by $Vy = $ constant, which means, since V is constant, lines of constant y, are lines parallel to Ox axis.

Flow of constant velocity in any direction

Consider the flow streaming past the x, y axes at some velocity Q making angle θ with the Ox axis (Fig. 3.11). The velocity Q can be resolved into two components U and V parallel to the Ox and Oy axes respectively where $Q^2 = U^2 + V^2$ and $\tan \theta = V/U$.

Again the stream function ψ at a point P in the flow is a measure of the amount of fluid flowing past any line joining OP. Let the most convenient contour be OTP, T being given by $(x, 0)$. Therefore

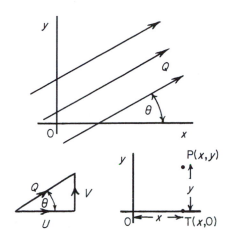

Fig. 3.11

$\psi =$ flow across OT (going right to left, therefore negative in sign)

   + flow across TP

   $=-$ component of Q parallel to Oy times x

   + component of Q parallel to Ox times y

$\psi = -Vx + Uy$ (3.16)

Lines of constant ψ or streamlines are the curves

$$-Vx + Uy = \text{constant}$$

assigning a different value of ψ for every streamline. Then in the equation V and U are constant velocities and the equation is that of a series of straight lines depending on the value of constant ψ.

Here the velocity potential at P is a measure of the flow along any curve joining P to O. Taking OTP as the line of integration [T(x, O)]:

$$\phi = \text{flow along OT} + \text{flow along TP}$$
$$= Ux + Vy$$
$$\phi = Ux + Vy \qquad (3.17)$$

Example 3.1   Interpret the flow given by the stream function (units: $\mathrm{m^2\,s^{-1}}$)

$$\psi = 6x + 12y$$

The constant velocity in the horizontal direction $= \dfrac{\partial \psi}{\partial y} = +12\,\mathrm{m\,s^{-1}}$

The constant velocity in the vertical direction $= -\dfrac{\partial \psi}{\partial x} = -6\,\mathrm{m\,s^{-1}}$

Therefore the flow equation represents uniform flow inclined to the Ox axis by angle θ where $\tan\theta = -6/12$, i.e. inclined downward.

   The speed of flow is given by

$$Q = \sqrt{6^2 + 12^2} = \sqrt{180}\,\mathrm{m\,s^{-1}}$$

3.3.4   Solid boundaries and image systems

The fact that the flow is always along a streamline and not through it has an important fundamental consequence. This is that a streamline of an *inviscid* flow can be replaced by a solid boundary of the same shape without affecting the remainder of the flow pattern. If, as often is the case, a streamline forms a closed curve that separates the flow pattern into two separate streams, one inside and one outside, then a solid body can replace the closed curve and the flow made outside without altering the shape of the flow (Fig. 3.12a). To represent the flow in the region of a contour or body it is only necessary to replace the contour by a similarly shaped streamline. The following sections contain examples of simple flows which provide continuous streamlines in the shapes of circles and aerofoils, and these emerge as consequences of the flow combinations chosen.

When arbitrary contours and their adjacent flows have to be replaced by identical flows containing similarly shaped streamlines, image systems have to be placed within the contour that are the reflections of the external flow system in the solid streamline.

Figure 3.12b shows the simple case of a source A placed a short distance from an infinite plane wall. The effect of the solid boundary on the flow from the source is exactly represented by considering the effect of the image source A′ reflected in the wall. The source pair has a long straight streamline, i.e. the vertical axis of symmetry, that separates the flows from the two sources and that may be replaced by a solid boundary without affecting the flow.

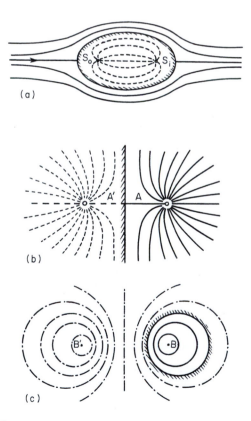

Fig. 3.12 Image systems

Figure 3.12c shows the flow in the cross-section of a vortex lying parallel to the axis of a circular duct. The circular duct wall can be replaced by the corresponding streamline in the vortex-pair system given by the original vortex B and its image B'. It can easily be shown that B' is a distance r^2/s from the centre of the duct on the diameter produced passing through B, where r is the radius of the duct and s is the distance of the vortex axis from the centre of the duct.

More complicated contours require more complicated image systems and these are left until discussion of the cases in which they arise. It will be seen that Fig. 3.12a, which is the flow of Section 3.3.7, has an internal image system, the source being the image of a source at $-\infty$ and the sink being the image of a sink at $+\infty$. This external source and sink combination produces the undisturbed uniform stream as has been noted above.

3.3.5   A source in a uniform horizontal stream

Let a source of strength m be situated at the origin with a uniform stream of $-U$ moving from right to left (Fig. 3.13).

Then

$$\psi = \frac{m\theta}{2\pi} - Uy \tag{3.18}$$

which is a combination of two previous equations. Eqn (3.18) can be rewritten

$$\psi = \frac{m}{2\pi}\tan^{-1}\frac{y}{x} - Uy \tag{3.19}$$

to make the variables the same in each term.

Combining the velocity potentials:

$$\phi = \frac{m}{2\pi}\ln\frac{r}{r_0} - Ux$$

or

$$\phi = \frac{m}{4\pi}\ln\left(\frac{x^2}{r_0^2} + \frac{y^2}{r_0^2}\right) - Ux \tag{3.20}$$

or in polar coordinates

$$\phi = \frac{m}{2\pi}\ln\frac{r}{r_0} - Ur\cos\theta \tag{3.21}$$

These equations give, for constant values of ϕ, the equipotential lines everywhere normal to the streamlines.

Streamline patterns can be found by substituting constant values for ψ and plotting Eqn (3.18) or (3.19) or alternatively by adding algebraically the stream functions due to the two cases involved. The second method is easier here.

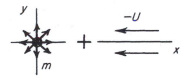

Fig. 3.13

Method (see Fig. 3.14)

(1) Plot the streamlines due to a source at the origin taking the strength of the source equal to $20\,\mathrm{m^2\,s^{-1}}$ (say). The streamlines are $\pi/10$ apart. It is necessary to take positive values of y only since the pattern is symmetrical about the Ox axis.

(2) *Superimpose on the plot horizontal lines to a scale so that* $\psi = -Uy = -1$, -2, -3, etc., are lines about 1 unit apart on the paper. Where the lines intersect, add the values of ψ at the lines of intersection. Connect up all points of constant ψ (streamlines) by smooth lines.

The resulting flow pattern shows that the streamlines can be separated into two distinct groups: (a) the fluid from the source moves from the source to infinity without mingling with the uniform stream, being constrained within the streamline $\psi = 0$; (b) the uniform stream is split along the Ox axis, the two resulting streams being deflected in their path towards infinity by $\psi = 0$.

It is possible to replace any streamline by a solid boundary without interfering with the flow in any way. If $\psi = 0$ is replaced by a solid boundary the effects of the source are truly cut off from the horizontal flow and it can be seen that here is a mathematical expression that represents the flow round a curved fairing (say) in a uniform flow. The same expression can be used for an approximation to the behaviour of a wind sweeping in off a plain or the sea and up over a cliff. The upward components of velocity of such an airflow are used in soaring.

The vertical velocity component at any point in the flow is given by $-\partial\psi/\partial x$. Now

$$\psi = \frac{m}{2\pi}\tan^{-1}\left(\frac{y}{x}\right) - Uy \quad (\text{Eqn}(3.19))$$

$$-\frac{\partial\psi}{\partial x} = -\frac{m}{2\pi}\frac{\partial\tan^{-1}(y/x)}{\partial(y/x)}\frac{\partial(y/x)}{\partial x}$$

$$= -\frac{m}{2\pi}\frac{1}{1+(y/x)^2}\frac{-y}{x^2}$$

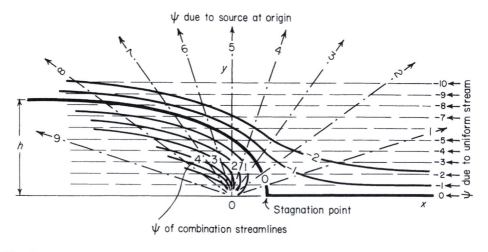

ψ due to source at origin

ψ of combination streamlines

Fig. 3.14

or

$$v = \frac{m}{2\pi} \frac{y}{x^2 + y^2}$$

and this is upwards.

This expression also shows, by comparing it, in the rearranged form $x^2 + y^2 - (m/2\pi v)y = 0$, with the general equation of a circle $(x^2 + y^2 + 2gx + 2hy + f = 0)$, that lines of constant vertical velocity are circles with centres $(0, m/4\pi v)$ and radii $m/4\pi v$.

The ultimate thickness, $2h$ (or height of cliff h) of the shape given by $\psi = 0$ for this combination is found by putting $y = h$ and $\theta = \pi$ in the general expression, i.e. substituting the appropriate data in Eqn (3.18):

$$\psi = \frac{m\pi}{2\pi} - Uh = 0$$

Therefore

$$h = m/2U \tag{3.22}$$

Note that when $\theta = \pi/2$, $y = h/2$.

The position of the stagnation point

By finding the stagnation point, the distance of the foot of the cliff, or the front of the fairing, from the source can be found. A stagnation point is given by $u = 0$, $v = 0$, i.e.

$$u = \frac{\partial \psi}{\partial y} = 0 = \frac{m}{2\pi} \frac{x}{x^2 + y^2} - U \tag{3.23}$$

$$v = -\frac{\partial \psi}{\partial x} = 0 = \frac{m}{2\pi} \frac{y}{x^2 + y^2} \tag{3.24}$$

From Eqn (3.24) $v = 0$ when $y = 0$, and substituting in Eqn (3.23) when $y = 0$ and $x = x_0$:

$$u = 0 = \frac{m}{2\pi} \frac{1}{x_0} - U$$

when

$$x_0 = m/2\pi U \tag{3.25}$$

The local velocity

The local velocity $q = \sqrt{u^2 + v^2}$.

$$u = \frac{\partial \psi}{\partial y} \quad \text{and} \quad \psi = \frac{m}{2\pi} \tan^{-1} \frac{y}{x} - Uy$$

Therefore

$$u = \frac{m}{2\pi} \frac{1/x}{1 + (y/x)^2} - U$$

giving

$$u = \frac{m}{2\pi}\frac{x}{x^2 + y^2} - U$$

and from $v = -\partial\psi/\partial x$

$$v = \frac{m}{2\pi}\frac{y}{x^2 + y^2}$$

from which the local velocity can be obtained from $q = \sqrt{u^2 + v^2}$ and the direction given by $\tan^{-1}(v/u)$ in any particular case.

3.3.6 Source–sink pair

This is a combination of a source and sink of equal (but opposite) strengths situated a distance $2c$ apart. Let $\pm m$ be the strengths of a source and sink situated at points A $(c, 0)$ and B $(-c, 0)$, that is at a distance of c m on either side of the origin (Fig. 3.15). The stream function at a point P(x, y), (r, θ) due to the combination is

$$\psi = \frac{m\theta_1}{2\pi} - \frac{m\theta_2}{2\pi} = \frac{m}{2\pi}(\theta_1 - \theta_2)$$

$$\psi = \frac{m}{2\pi}\beta \tag{3.26}$$

Transposing the equation to Cartesian coordinates:

$$\tan\theta_1 = \frac{y}{x - c}, \qquad \tan\theta_2 = \frac{y}{x + c}$$

$$\tan(\theta_1 - \theta_2) = \frac{\tan\theta_1 - \tan\theta_2}{1 + \tan\theta_1 \tan\theta_2} = \frac{\frac{y}{x-c} - \frac{y}{x+c}}{1 + \frac{y^2}{x^2 - c^2}}$$

Therefore

$$\beta = \theta_1 - \theta_2 = \tan^{-1}\frac{2cy}{x^2 + y^2 - c^2} \tag{3.27}$$

and substituting in Eqn (3.26):

$$\psi = \frac{m}{2\pi}\tan^{-1}\frac{2cy}{x^2 + y^2 - c^2} \tag{3.28}$$

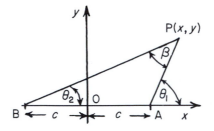

Fig. 3.15

To find the shape of the streamlines associated with this combination it is necessary to investigate Eqn (3.28). Rearranging:

$$\tan\left(\frac{2\pi}{m}\psi\right) = \frac{2cy}{x^2 + y^2 - c^2}$$

or

$$x^2 + y^2 - c^2 = \frac{2cy}{\tan\left(\frac{2\pi\psi}{m}\right)}$$

or

$$x^2 + y^2 - 2c\cot\frac{2\pi\psi}{m}y - c^2 = 0$$

which is the equation of a circle of radius $c\sqrt{\cot^2(2\pi\psi/m) + 1}$, and centre $c\cot(2\pi\psi/m)$.

Therefore streamlines for this combination consist of a series of circles with centres on the Oy axis and intersecting in the source and sink, the flow being from the source to the sink (Fig. 3.16).

Consider the velocity potential at any point $P(r, \theta)(x, y)$.*

$$\phi = \frac{m}{2\pi}\ln\frac{r_1}{r_0} - \frac{m}{2\pi}\ln\frac{r_2}{r_0} = \frac{m}{2\pi}\ln\frac{r_1}{r_2} \tag{3.29}$$

$$r_1^2 = (x - c)^2 + y^2 = x^2 + y^2 + c^2 - 2xc$$
$$r_2^2 = (x + c)^2 + y^2 = x^2 + y^2 + c^2 + 2xc$$

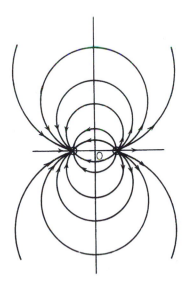

Fig. 3.16 Streamlines due to a source and sink pair

---

* Note that here r_0 is the radius of the equipotential $\phi = 0$ for the isolated source and the isolated sink, but not for the combination.

Therefore

$$\phi = \frac{m}{4\pi} \ln \frac{x^2 + y^2 + c^2 - 2xc}{x^2 + y^2 + c^2 + 2xc}$$

Rearranging

$$e^{4\pi\phi/m} = \frac{x^2 + y^2 + c^2 - 2xc}{x^2 + y^2 + c^2 + 2xc} = \lambda \text{(say)}$$

Then

$$(x^2 + y^2 + c^2 + 2xc)\lambda = x^2 + y^2 + c^2 - 2xc$$
$$(x^2 + y^2 + c^2)[\lambda - 1] + 2xc(\lambda + 1) = 0$$

$$x^2 + y^2 + 2xc\left(\frac{\lambda + 1}{\lambda - 1}\right) + c^2 = 0 \qquad (3.30)$$

which is the equation of a circle of centre

$$x = -c\left(\frac{\lambda + 1}{\lambda - 1}\right), \qquad y = 0$$

i.e.

$$x = -c\frac{e^{(4\pi\phi/m)} + 1}{e^{(4\pi\phi/m)} - 1} = -c \coth \frac{2\pi\phi}{m}, \qquad y = 0$$

and radius

$$c\sqrt{\left(\frac{\lambda + 1}{\lambda - 1}\right)^2 - 1} = 2c\frac{\sqrt{\lambda}}{\lambda - 1} = 2c\frac{e^{2\pi\phi/m}}{e^{(4\pi\phi/m)} - 1}$$

$$= 2c \operatorname{cosech} \frac{2\pi\phi}{m}$$

Therefore, the equipotentials due to a source and sink combination are sets of eccentric non-intersecting circles with their centres on the Ox axis (Fig. 3.17). This pattern is exactly the same as the streamline pattern due to point vortices of opposite sign separated by a distance $2c$.

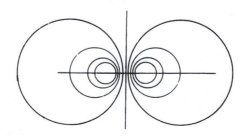

Fig. 3.17 Equipotential lines due to a source and sink pair

3.3.7 A source set upstream of an equal sink in a uniform stream

The stream function due to this combination is:

$$\psi = \frac{m}{2\pi} \tan^{-1} \frac{2cy}{x^2 + y^2 - c^2} - Uy \qquad (3.31)$$

Here the first term represents a source and sink combination set with the source to the right of the sink. For the source to be upstream of the sink the uniform stream must be from right to left, i.e. negative. If the source is placed downstream of the sink an entirely different stream pattern is obtained.

The velocity potential at any point in the flow due to this combination is given by:

$$\phi = \frac{m}{2\pi} \ln \frac{r_1}{r_2} - Ur \sin \theta \qquad (3.32)$$

or

$$\phi = \frac{m}{4\pi} \ln \frac{x^2 + y^2 + c^2 - 2xc}{x^2 + y^2 + c^2 + 2xc} - Ux \qquad (3.33)$$

The streamline $\psi = 0$ gives a closed oval curve (not an ellipse), that is symmetrical about the Ox and Oy axes. Flow of stream function ψ greater than $\psi = 0$ shows the flow round such an oval set at zero incidence in a uniform stream. Streamlines can be obtained by plotting or by superposition of the separate standard flows (Fig. 3.18). The streamline $\psi = 0$ again separates the flow into two distinct regions. The first is wholly contained within the closed oval and consists of the flow out of the source and into the sink. The second is that of the approaching uniform stream which flows around the oval curve and returns to its uniformity again. Again replacing $\psi = 0$ by a solid boundary, or indeed a solid body whose shape is given by $\psi = 0$, does not influence the flow pattern in any way.

Thus the stream function ψ of Eqn (3.31) can be used to represent the flow around a long cylinder of oval section set with its major axis parallel to a steady stream. To find the stream function representing a flow round such an oval cylinder it must be possible to obtain m and c (the strengths of the source and sink and distance apart) in terms of the size of the body and the speed of the incident stream.

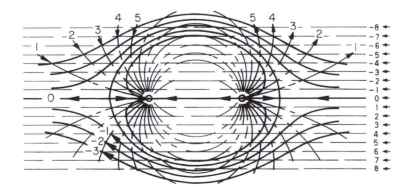

Fig. 3.18

Suppose there is an oval of breadth $2b_0$ and thickness $2t_0$ set in a uniform flow of U. The problem is to find m and c in the stream function, Eqn (3.31), which will then represent the flow round the oval.

(a) The oval must conform to Eqn (3.31):

$$\psi = 0 = \frac{m}{2\pi}\tan^{-1}\frac{2cy}{x^2 + y^2 - c^2} - Uy$$

(b) On streamline $\psi = 0$ maximum thickness t_0 occurs at $x = 0$, $y = t_0$. Therefore, substituting in the above equation:

$$0 = \frac{m}{2\pi}\tan^{-1}\frac{2ct_0}{t_0^2 - c^2} - Ut_0$$

and rearranging

$$\tan\frac{2\pi Ut_0}{m} = \frac{2t_0 c}{t_0^2 - c^2} \tag{3.34}$$

(c) A stagnation point (point where the local velocity is zero) is situated at the 'nose' of the oval, i.e. at the point $y = 0$, $x = b_0$, i.e.:

$$u = 0 = \frac{\partial \psi}{\partial y} = \frac{\partial}{\partial y}\left(\frac{m}{2\pi}\tan^{-1}\frac{2cy}{x^2 + y^2 - c^2} - Uy\right)$$

$$\frac{\partial \psi}{\partial y} = \frac{m}{2\pi}\frac{1}{1 + \left(\frac{2cy}{x^2+y^2-c^2}\right)^2}\frac{(x^2 + y^2 - c^2)2c - 2y\,2cy}{(x^2 + y^2 - c^2)^2} - U$$

and putting $y = 0$ and $x = b_0$ with $\partial \psi/\partial y = 0$:

$$0 = \frac{m}{2\pi}\frac{(b_0^2 - c^2)2c}{(b_0^2 - c^2)^2} - U = \frac{m}{2\pi}\frac{2c}{b_0^2 - c^2} - U$$

Therefore

$$m = \pi U\frac{b_0^2 - c^2}{c} \tag{3.35}$$

The simultaneous solution of Eqns (3.34) and (3.35) will furnish values of m and c to satisfy any given set of conditions. Alternatively (a), (b) and (c) above can be used to find the thickness and length of the oval formed by the streamline $\psi = 0$. This form of the problem is more often set in examinations than the preceding one.

3.3.8 Doublet

A doublet is a source and sink combination, as described above, but with the separation infinitely small. A doublet is considered to be at a point, and the definition of the strength of a doublet contains the measure of separation. The strength (μ) of a doublet is the product of the infinitely small distance of separation, and the strength of source and sink. The doublet axis is the line from the sink to the source in that sense.

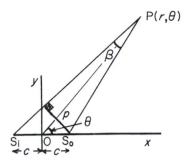

Fig. 3.19

The streamlines due to a source and sink combination are circles each intersecting in the source and sink. As the source and sink approach, the points of intersection also approach until in the limit, when separated by an infinitesimal distance, the circles are all touching (intersecting) at one point – the doublet. This can be shown as follows. For the source and sink:

$$\psi = (m/2\pi)\beta \quad \text{from Eqn (3.26)}$$

By constructing the perpendicular of length p from the source to the line joining the sink and P it can be seen that as the source and sink approach (Fig. 3.19),

$$p \to 2c\sin\theta \quad \text{and also} \quad p \to r\beta$$

Therefore in the limit

$$2c\sin\theta = r\beta \quad \text{or} \quad \beta = \frac{2c\sin\theta}{r}$$

$$\psi = \frac{m}{2\pi}\frac{2c}{r}\sin\theta$$

and putting $\mu = 2\,\text{cm} = $ strength of the doublet:

$$\psi = \frac{\mu}{2\pi r}\sin\theta \tag{3.36}$$

On converting to Cartesian coordinates where

$$r = \sqrt{x^2 + y^2}, \qquad \sin\theta = \frac{y}{\sqrt{x^2 + y^2}}, \qquad \psi = \frac{\mu}{2\pi}\frac{y}{x^2 + y^2} \tag{3.37}$$

and rearranging gives

$$(x^2 + y^2) - \frac{\mu}{2\pi\psi}y = 0$$

which, when ψ is a constant, is the equation of a circle.

Therefore, lines of constant ψ are circles of radius $\mu/(4\pi\psi)$ and centres $(0, \mu/(4\pi\psi))$ (Fig. 3.20), i.e. circles, with centres lying on the Oy axis, passing through the origin as deduced above.

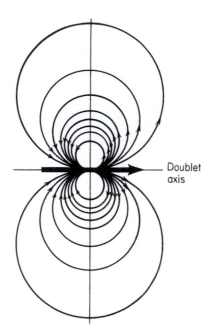

Fig. 3.20 Streamlines due to a doublet

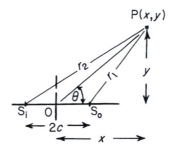

Fig. 3.21

Consider again a source and sink set a very small distance, $2c$, apart (Fig. 3.21). Then*

$$\phi = \frac{m}{2\pi} \ln \frac{r_1}{r_0} - \frac{m}{2\pi} \ln \frac{r_2}{r_0}$$

where $\pm m$ is the strength of the source and sink respectively. Then

$$\phi = \frac{m}{2\pi} \ln \frac{r_1}{r_2} = \frac{m}{4\pi} \ln \frac{r_1^2}{r_2^2}$$

* Here r_0 is the radius of the equipotential $\phi = 0$ for the isolated source and the isolated sink, but not for the combination.

Now

$$r_1^2 = x^2 + y^2 - 2xc + c^2$$

and

$$r_2^2 = x^2 + y^2 + 2xc + c^2$$

Therefore

$$\phi = \frac{m}{4\pi} \ln \frac{x^2 + y^2 - 2xc + c^2}{x^2 + y^2 + 2xc + c^2}$$

and dividing out

$$\phi = \frac{m}{4\pi} \ln \left(1 - \frac{4xc}{x^2 + y^2 + c^2 + 2xc} \right)$$

On expanding,

$$\ln(1 - t) = -t - \frac{t^2}{2} - \frac{t^3}{3} - \frac{t^4}{4} \cdots$$

Therefore:

$$\phi = \frac{m}{4\pi} \left[-\frac{4xc}{x^2 + y^2 + c^2 + 2xc} - \frac{16x^2c^2}{2(x^2 + y^2 + c^2 + 2xc)^2} - \cdots \right]$$

Since c is very small c^2 can be neglected. Therefore, ignoring c^2 and higher powers of c

$$\phi = -\frac{m}{4\pi} \frac{4xc}{x^2 + y^2 + 2xc}$$

and as $c \to 0$, and $2mc = \mu$ (which is the strength of the doublet) a limiting value of ϕ is given by

$$\phi = -\frac{\mu}{2\pi} \frac{x}{x^2 + y^2} \quad \text{but} \quad \frac{x}{\sqrt{x^2 + y^2}} = \cos\theta$$

Therefore

$$\phi = -\frac{\mu}{2\pi r} \cos\theta \qquad (3.38)$$

3.3.9 Flow around a circular cylinder given by a doublet in a uniform horizontal flow

The stream function due to this combination is:

$$\psi = \frac{\mu}{2\pi r} \sin\theta - Uy \qquad (3.39)$$

It should be noted that the terms in the stream functions must be opposite in sign to obtain the useful results below. Here again the source must be upstream of the sink in the flow system. Equation (3.39) converted to rectangular coordinates gives:

$$\psi = \frac{\mu}{2\pi} \frac{y}{x^2 + y^2} - Uy \tag{3.40}$$

and for the streamline $\psi = 0$

$$y\left(\frac{\mu}{2\pi(x^2 + y^2)} - U\right) = 0$$

i.e.

$$y = 0 \quad \text{or} \quad x^2 + y^2 = \frac{\mu}{2\pi U}$$

This shows the streamline $\psi = 0$ to consist of the Ox axis together with a circle, centre O, of radius $\sqrt{\mu/(2\pi U)} = a$ (say).

Alternatively by converting Eqn (3.39) to polar coordinates:

$$\psi = \frac{\mu}{2\pi r} \sin\theta - Ur\sin\theta$$

Therefore

$$\psi = \sin\theta\left(\frac{\mu}{2\pi r} - Ur\right) = 0 \quad \text{for} \quad \psi = 0$$

giving

$$\sin\theta = 0 \quad \text{so} \quad \theta = 0 \quad \text{or} \quad \pm\pi$$

or

$$\frac{\mu}{2\pi r} - Ur = 0 \quad \text{giving} \quad r = \sqrt{\frac{\mu}{2\pi U}} = a$$

the two solutions as before.

The streamline $\psi = 0$ thus consists of a circle and a straight line on a diameter produced (Fig. 3.22). Again in this case the streamline $\psi = 0$ separates the flow into two distinct patterns: that outside the circle coming from the undisturbed flow a long

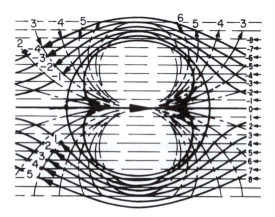

Fig. 3.22 Streamlines due to a doublet in a uniform stream

way upstream, to flow around the circle and again to revert to uniform flow down-stream. That inside the circle is from the doublet. This is confined within the circle and does not mingle with the horizontal stream at all. This inside flow pattern is usually neglected. This combination is consequently a mathematical device for giving expression to the ideal two-dimensional flow around a circular cylinder.

The velocity potential due to this combination is that corresponding to a uniform stream flowing parallel to the Ox axis, superimposed on that of a doublet at the origin. Putting $x = r \cos \theta$:

$$\phi = -Ur \cos \theta + \frac{\mu}{2\pi r} \cos \theta$$

$$\phi = -U \cos \theta \left(r + \frac{a^2}{r} \right) \tag{3.41}$$

where $a = \sqrt{\mu/(2\pi U)}$ is the radius of the streamline $\psi = 0$.

The streamlines can be obtained directly by plotting using the superposition method outlined in previous cases. Rewriting Eqn (3.39) in polar coordinates

$$\psi = \frac{\mu}{2\pi r} \sin \theta - Ur \sin \theta$$

and rearranging, this becomes

$$\psi = U \sin \theta \left(\frac{\mu}{2\pi r U} - r \right)$$

and with $\mu/(2\pi U) = a^2$ a constant (a = radius of the circle $\psi = 0$)

$$\psi = U \sin \theta \left(\frac{a^2}{r} - r \right) \tag{3.42}$$

Differentiating this partially with respect to r and θ in turn will give expressions for the velocity everywhere, i.e.:

$$\left. \begin{aligned} q_n &= \frac{1}{r} \frac{\partial \psi}{\partial \theta} = U \cos \theta \left(\frac{a^2}{r^2} - 1 \right) \\ q_t &= -\frac{\partial \psi}{\partial r} = U \sin \theta \left(\frac{a^2}{r^2} + 1 \right) \end{aligned} \right\} \tag{3.43}$$

Putting $r = a$ (the cylinder radius) in Eqns (3.43) gives:

(i) $q_n = U \cos \theta [1 - 1] = 0$ which is expected since the velocity must be parallel to the surface everywhere, and

(ii) $q_t = U \sin \theta [1 + 1] = 2U \sin \theta$.

Therefore the velocity on the surface is $2U \sin \theta$ and it is important to note that the velocity at the surface is independent of the radius of the cylinder.

The pressure distribution around a cylinder

If a long circular cylinder is set in a uniform flow the motion around it will, ideally, be given by the expression (3.42) above, and the velocity anywhere on the surface by the formula

$$q = 2U \sin \theta \tag{3.44}$$

By the use of Bernoulli's equation, the pressure p acting on the surface of the cylinder where the velocity is q can be found. If p_0 is the static pressure of the free stream where the velocity is U then by Bernoulli's equation:

$$p_0 + \frac{1}{2}\rho U^2 = p + \frac{1}{2}\rho q^2$$

$$= p + \frac{1}{2}\rho(2U\sin\theta)^2$$

Therefore

$$p - p_0 = \frac{1}{2}\rho U^2[1 - 4\sin^2\theta] \tag{3.45}$$

Plotting this expression gives a curve as shown on Fig. 3.23. Important points to note are:

(1) At the stagnation points (0° and 180°) the pressure difference $(p - p_0)$ is positive and equal to $\frac{1}{2}\rho U^2$.
(2) At 30° and 150° where $\sin\theta = \frac{1}{2}$, $(p - p_0)$ is zero, and at these points the local velocity is the same as that of the free stream.
(3) Between 30° and 150° C_p is negative, showing that p is less than p_0.
(4) The pressure distribution is symmetrical about the vertical axis and therefore there is no drag force. Comparison of this ideal pressure distribution with that obtained by experiment shows that the actual pressure distribution is similar to the theoretical value up to about 70° but departs radically from it thereafter. Furthermore, it can be seen that the pressure coefficient over the rear portion of the cylinder remains negative. This destroys the symmetry about the vertical axis and produces a force in the direction of the flow (see Section 1.5.5).

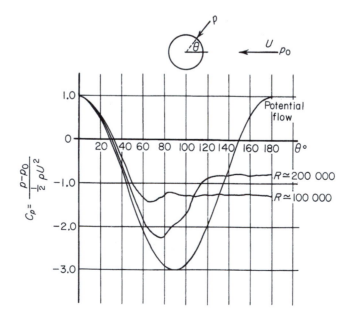

Fig. 3.23

3.3.10 A spinning cylinder in a uniform flow

This is given by the stream function due to a doublet, in a uniform horizontal flow, with a line vortex superimposed at the origin. By adding these cases

$$\psi = \frac{\mu}{2\pi r}\sin\theta - Uy - \frac{\Gamma}{2\pi}\ln\frac{r}{r_0}$$

Converting to homogeneous coordinates

$$\psi = Ur\sin\theta\left(\frac{\mu}{2\pi r^2 U} - 1\right) - \frac{\Gamma}{2\pi}\ln\frac{r}{r_0}$$

but from the previous case $\sqrt{\mu/(2\pi U)} = a$, the radius of the cylinder.

Also since the cylinder periphery marks the inner limit of the vortex flow, $r_0 = a$; therefore the stream function becomes:

$$\psi = Ur\sin\theta\left(\frac{a^2}{r^2} - 1\right) - \frac{\Gamma}{2\pi}\ln\frac{r}{a} \qquad (3.46)$$

and differentiating partially with respect to r and θ the velocity components of the flow anywhere on or outside the cylinder become, respectively:

$$\left.\begin{array}{l} q_t = -\dfrac{\partial\psi}{\partial r} = U\sin\theta\left(\dfrac{a^2}{r} + 1\right) + \dfrac{\Gamma}{2\pi r} \\[3mm] q_n = \dfrac{1}{r}\dfrac{\partial\psi}{\partial\theta} = U\cos\theta\left(\dfrac{a^2}{r^2} - 1\right) \end{array}\right\} \qquad (3.47)$$

and

$$q = \sqrt{q_n^2 + q_t^2}$$

On the surface of the spinning cylinder $r = a$. Therefore,

$$q_n = 0$$

$$q_t = 2U\sin\theta + \frac{\Gamma}{2\pi a} \qquad (3.48)$$

Therefore

$$q = q_t = 2U\sin\theta + \frac{\Gamma}{2\pi a}$$

and applying Bernoulli's equation between a point a long way upstream and a point on the cylinder where the static pressure is p:

$$p_0 + \frac{1}{2}\rho U^2 = p + \frac{1}{2}\rho q^2$$

$$= p + \frac{1}{2}\rho\left(2U\sin\theta + \frac{\Gamma}{2\pi a}\right)^2$$

Therefore

$$p - p_0 = \frac{1}{2}\rho U^2\left[1 - \left(2\sin\theta + \frac{\Gamma}{2\pi Ua}\right)^2\right] \qquad (3.49)$$

This equation differs from that of the non-spinning cylinder in a uniform stream of the previous section by the addition of the term $(\Gamma/(2\pi Ua)) = B$ (a constant), in the squared bracket. This has the effect of altering the symmetry of the pressure distribution about a horizontal axis. This is indicated by considering the extreme top and bottom of the cylinder and denoting the pressures there by p_T and p_B respectively. At the top $p = p_T$ when $\theta = \pi/2$ and $\sin \theta = 1$. Then Eqn (3.49) becomes

$$p_T - p_0 = \frac{1}{2}\rho U^2(1 - [2 + B]^2)$$
$$= -\frac{1}{2}\rho U^2(3 + 4B + B^2) \tag{3.50}$$

At the bottom $p = p_B$ when $\theta = -\pi/2$ and $\sin \theta = -1$:

$$p_B - p_0 = -\frac{1}{2}\rho U^2(3 - 4B + B^2) \tag{3.51}$$

Clearly (3.50) does not equal (3.51) which shows that a pressure difference exists between the top and bottom of the cylinder equal in magnitude to

$$p_T - p_B = 8B\left(-\frac{1}{2}\rho U^2\right) = -\frac{2}{\pi a}\rho U\Gamma$$

which suggests that if the pressure distribution is integrated round the cylinder then a resultant force would be found normal to the direction of motion.

The normal force on a spinning circular cylinder in a uniform stream

Consider a surface element of cylinder of unit span and radius a (Fig. 3.24). The area of the element $= a\,\delta\theta \times 1$, the static pressure acting on element $= p$, resultant force $= (p - p_0)a\,\delta\theta$, vertical component $= (p - p_0)a\,\delta\theta \sin \theta$.

Substituting for $(p - p_0)$ from Eqn (3.49) and retaining the notation $B = \Gamma/2\pi Ua$, the vertical component of force acting on the element $= \frac{1}{2}\rho U^2[1 - (2\sin\theta + B)^2]a\,\delta\theta \sin \theta$. The total vertical force per unit span by integration is (l positive upwards):

$$l = \int_0^{2\pi} -\frac{1}{2}\rho U^2 a[1 - (2\sin\theta + B)^2]\sin\theta\,d\theta$$

which becomes

$$l = -\frac{1}{2}\rho U^2 a \int_0^{2\pi}[\sin\theta(1 - B^2) - 4B\sin^2\theta - 4\sin^3\theta]d\theta$$

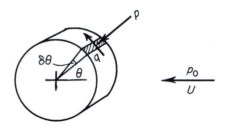

Fig. 3.24 The pressure and velocity on the surface of unit length of a cylinder of radius a

On integrating from 0 to 2π the first and third terms vanish leaving

$$\int_0^{2\pi} 4B \sin^2 \theta \, d\theta = 4B\pi$$

Therefore

$$l = \frac{1}{2} \rho U^2 a \, 4B\pi$$

Replacing B by $\Gamma/2\pi Ua$ and cancelling gives the equation for the lift force per unit span

$$l = \rho U \Gamma \tag{3.52}$$

The lift per unit span in N is equal to the product of density ρ, the linear velocity U, and the circulation Γ.

This expression is the algebraic form of the Kutta–Zhukovsky theorem, and is valid for any system that produces a circulation superimposed on a linear velocity (see Section 4.1.3). The spinning cylinder is used here as it lends itself to stream function theory as well as being of interest later.

It is important to note that the diameter of the cylinder has no influence on the final expression, so if a line vortex of strength Γ moved with velocity U in a uniform flow of density ρ, the same sideways force $l = \rho U \Gamma$ per unit length of vortex would be found. This sideways force commonly associated with a spinning object moving through the air has been recognized and used in ball games since ancient times. It is usually referred to as the *Magnus effect* after the scholar and philosopher Magnus.

The flow pattern around a spinning cylinder

The flow pattern around the spinning cylinder is also altered as the strength of the circulation increases. In Fig. 3.25 when $\Gamma = 0$ the flow pattern is that associated with the previous non-spinning case with front and rear stagnation points S_1 and S_2

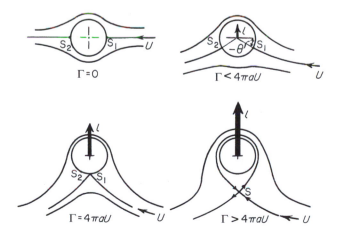

Fig. 3.25

respectively, occurring on the horizontal axis. As Γ is increased positively a small amount the stagnation points move down below the horizontal axis.

Since from the equation for the velocity anywhere on the surface

$$q_t = 2U \sin\theta + \frac{\Gamma}{2\pi a} = 0 \text{ at the stagnation points}$$

$$\theta = \arcsin(-\Gamma/4\pi a U)$$

which is negative. As Γ is further increased a limiting condition occurs when $\theta = -\pi/2$, i.e. $\Gamma = 4\pi a U$, the stagnation points merge at the bottom of the cylinder. When Γ is greater than $4\pi a U$ the stagnation point (S) leaves the cylinder. The cylinder continues to rotate within the closed loop of the stagnation streamline, carrying round with it a region of fluid confined within the loop.

3.3.11　Bernoulli's equation for rotational flow

Consider fluid moving in a circular path. Higher pressure must be exerted from the outside, towards the centre of rotation, in order to provide the centripetal force. That is, some outside pressure force must be available to prevent the particles moving in a straight line. This suggests that the pressure is growing in magnitude as the radius increases, and a corollary is that the velocity of flow must fall as the distance from the centre increases.

With a segmental element at $P(r, \theta)$ where the velocity is q_t only and the pressure p, the pressures on the sides will be shown as in Fig. 3.26 and the resultant pressure thrust inwards is

$$\left(p + \frac{\partial p}{\partial r}\frac{\partial r}{2}\right)\left(r + \frac{\delta r}{2}\right)\delta\theta - \left(p - \frac{\partial p}{\partial r}\frac{\partial r}{2}\right)\left(r - \frac{\delta r}{2}\right)\delta\theta - p\,\delta r\,\delta\theta$$

which reduces to

$$\frac{\partial p}{\partial r}r\,\delta r\,\delta\theta \tag{3.53}$$

This must provide the centripetal force = mass × centripetal acceleration

$$= \rho r\,\delta r\,\delta\theta\,q_t^2/r \tag{3.54}$$

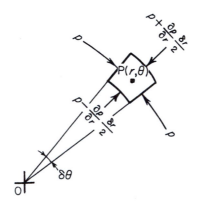

Fig. 3.26

Equating (3.53) and (3.54):

$$\frac{\partial p}{\partial r} = \frac{\rho q_t^2}{r} \qquad (3.55)$$

The rate of change of total pressure H is

$$\frac{\partial H}{\partial r} = \frac{\partial(p + \frac{1}{2}\rho q_t^2)}{\partial r} = \frac{\partial p}{\partial r} + \rho q_t \frac{\partial q_t}{\partial r}$$

and substituting for Eqn (3.55):

$$\frac{\partial H}{\partial r} = \rho \frac{q_t^2}{r} + \rho q_t \frac{\partial q_t}{\partial r} = \rho q_t \left(\frac{q_t}{r} + \frac{\partial q_t}{\partial r}\right)$$

Now for this system $(1/r)(\partial q_n/\partial \theta)$ is zero since the streamlines are circular and therefore the vorticity is $(q_t/r) + (\partial q_t/\partial r)$ from Eqn (2.79), giving

$$\frac{\partial H}{\partial r} = \rho q_t \zeta \qquad (3.56)$$

3.4 Axisymmetric flows (inviscid and incompressible flows)

Consider now axisymmetric potential flows, i.e. the flows around bodies such as cones aligned to the flow and spheres. In order to analyse, and for that matter to define, axisymmetric flows it is necessary to introduce cylindrical and spherical coordinate systems. Unlike the Cartesian coordinate system these coordinate systems can exploit the underlying symmetry of the flows.

3.4.1 Cylindrical coordinate system

The cylindrical coordinate system is illustrated in Fig. 3.27. The three coordinate surfaces are the planes $z = $ constant and $\theta = $ constant and the surface of the cylinder having radius r. In contrast, for the Cartesian system all three coordinate surfaces are

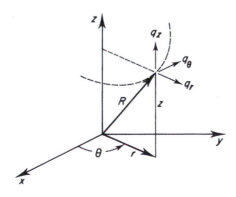

Fig. 3.27 Cylindrical coordinates

planes. As a consequence for the Cartesian system the directions (x, y, z) of the velocity components, say, are fixed throughout the flow field. For the cylindrical coordinate system, though, only one of the directions (z) is fixed throughout the flow field; the other two (r and θ) vary throughout the flow field depending on the value of the angular coordinate θ. In this respect there is a certain similarity to the polar coordinates introduced earlier in the chapter. The velocity component q_r is always locally perpendicular to the cylindrical coordinate surface and q_θ is always tangential to that surface. Once this elementary fact is properly understood cylindrical coordinates become as easy to use as the Cartesian system.

In a similar way as the relationships between velocity potential and velocity components are derived for polar coordinates (see Section 3.1.3 above), the following relationships are obtained for cylindrical coordinates

$$q_r = \frac{\partial \phi}{\partial r}, \qquad q_\theta = \frac{1}{r}\frac{\partial \phi}{\partial \theta}, \qquad q_z = \frac{\partial \phi}{\partial z} \tag{3.57}$$

An axisymmetric flow is defined as one for which the flow variables, i.e. velocity and pressure, do not vary with the angular coordinate θ. This would be so, for example, for a body of revolution about the z axis with the oncoming flow directed along the z axis. For such an axisymmetric flow a stream function can be defined. The continuity equation for axisymmetric flow in cylindrical coordinates can be derived in a similar manner as it is for two-dimensional flow in polar coordinates (see Section 2.4.3); it takes the form

$$\frac{1}{r}\frac{\partial r q_r}{\partial r} + \frac{\partial q_z}{\partial z} = 0 \tag{3.58}$$

The relationship between stream function and velocity component must be such as to satisfy Eqn (3.58); hence it can be seen that

$$q_r = -\frac{1}{r}\frac{\partial \psi}{\partial z}, \qquad q_z = \frac{1}{r}\frac{\partial \psi}{\partial r} \tag{3.59}$$

3.4.2 Spherical coordinates

For analysing certain two-dimensional flows, for example the flow over a circular cylinder with and without circulation, it is convenient to work with polar coordinates. The axisymmetric equivalents of polar coordinates are spherical coordinates, for example those used for analysing the flow around spheres. Spherical coordinates are illustrated in Fig. 3.28. In this case none of the coordinate surfaces are plane and the directions of all three velocity components vary over the flow field, depending on the values of the angular coordinates θ and φ. In this case the relationships between the velocity components and potential are given by

$$q_R = \frac{\partial \phi}{\partial R}, \qquad q_\theta = \frac{1}{R \sin \varphi}\frac{\partial \phi}{\partial \theta}, \qquad q_\varphi = \frac{1}{R}\frac{\partial \phi}{\partial \varphi} \tag{3.60}$$

For axisymmetric flows the variables are independent of θ and in this case the continuity equation takes the form

$$\frac{1}{R^2}\frac{\partial (R^2 q_R)}{\partial R} + \frac{1}{R \sin \varphi}\frac{\partial (\sin \varphi q_\varphi)}{\partial \varphi} = 0 \tag{3.61}$$

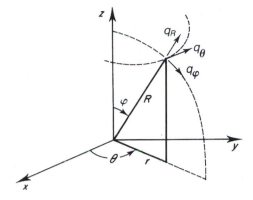

Fig. 3.28 Spherical coordinates

Again the relationship between the stream function and the velocity components must be such as to satisfy the continuity Eqn (3.61); hence

$$q_R = \frac{1}{R^2 \sin\varphi} \frac{\partial\psi}{\partial\varphi} \qquad q_\varphi = -\frac{1}{R \sin\varphi} \frac{\partial\psi}{\partial R} \tag{3.62}$$

3.4.3 Axisymmetric flow from a point source (or towards a point sink)

The point source and sink are similar in concept to the line source and sink discussed in Section 3.3. A close physical analogy can be found if one imagines the flow into or out of a very (strictly infinitely) thin round pipe – as depicted in Fig. 3.29. As suggested in this figure the streamlines would be purely radial in direction.

Let us suppose that the flow rate out of the point source is given by Q. Q is usually referred to as the *strength* of the point source. Now since the flow is purely radial away from the source the total flow rate across the surface of any sphere having its centre at the source will also be Q. (Note that this sphere is purely notional and does not represent a solid body or in any way hinder the flow.) Thus the radial velocity component at any radius R is related to Q as follows

$$4\pi R^2 q_R = Q$$

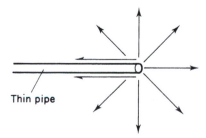

Thin pipe

Fig. 3.29

It therefore follows from Eqn (3.60) that

$$q_R = \frac{\partial \phi}{\partial R} = \frac{Q}{4\pi R^2}$$

Integration then gives the expression for the velocity potential of a point source as

$$\phi = -\frac{Q}{4\pi R} \tag{3.63}$$

In a similar fashion an expression for stream function can be derived using Eqn (3.62) giving

$$\psi = -\frac{Q}{4\pi} \cos \varphi \tag{3.64}$$

3.4.4 Point source and sink in a uniform axisymmetric flow

Placing a point source and/or sink in a uniform horizontal stream of $-U$ leads to very similar results as found in Section 3.3.5 for the two-dimensional case with line sources and sinks.

First the velocity potential and stream function for uniform flow, $-U$, in the z direction must be expressed in spherical coordinates. The velocity components q_R and q_φ are related to $-U$ as follows

$$q_R = -U \cos \varphi \text{ and } q_\varphi = U \sin \varphi$$

Using Eqn (3.60) followed by integration then gives

$$\frac{\partial \phi}{\partial R} = -U \cos \varphi \rightarrow \phi = -U R \cos \varphi + f(\varphi)$$

$$\frac{\partial \phi}{\partial \varphi} = U R \sin \varphi \rightarrow \phi = -U R \cos \varphi + g(R)$$

$f(\varphi)$ and $g(R)$ are arbitrary functions that take the place of constants of integration when partial integration is carried out. Plainly in order for the two expressions for ϕ derived above to be in agreement $f(\varphi) = g(R) = 0$. The required expression for the velocity potential is thereby given as

$$\phi = -U R \cos \varphi \tag{3.65}$$

Similarly using Eqn (3.62) followed by integration gives

$$\frac{\partial \psi}{\partial \varphi} = -U R^2 \cos \varphi \sin \varphi = -\frac{U R^2}{2} \sin 2\varphi \rightarrow \psi = \frac{U R^2}{4} \cos 2\varphi + f(R)$$

$$\frac{\partial \psi}{\partial R} = -U R \sin^2 \varphi \rightarrow \psi = -\frac{U R^2}{2} \sin^2 \varphi + g(\varphi)$$

Recognizing that $\cos 2\varphi = 1 - 2\sin^2 \varphi$ it can be seen that the two expressions given above for ψ will agree if the arbitrary functions of integration take the values $f(R) = -U R^2/4$ and $g(\varphi) = 0$. The required expression for the stream function is thereby given as

$$\psi = -\frac{U R^2}{2} \sin^2 \varphi \tag{3.66}$$

Using Eqns (3.63) and (3.65) and Eqns (3.64) and (3.66) it can be seen that for a point source at the origin placed in a uniform flow $-U$ along the z axis

$$\phi = -U\,R\cos\varphi - \frac{Q}{4\pi R} \tag{3.67a}$$

$$\psi = -\frac{1}{2}U\,R^2\sin^2\varphi - \frac{Q}{4\pi}\cos\varphi \tag{3.67b}$$

The flow field represented by Eqns (3.67) corresponds to the potential flow around a semi-finite body of revolution – very much like its two-dimensional counterpart described in Section 3.3.5. In a similar way to the procedure described in Section 3.3.5 it can be shown that the stagnation point occurs at the point $(-a, 0)$ where

$$a = \sqrt{\frac{Q}{4\pi U}} \tag{3.68}$$

and that the streamlines passing through this stagnation point define a body of revolution given by

$$R^2 = 2a^2(1 + \cos\varphi)/\sin^2\varphi \tag{3.69}$$

The derivation of Eqns (3.68) and (3.69) are left as an exercise (see Ex. 19) for the reader.

In a similar fashion to the two-dimensional case described in Section 3.3.6 a point source placed on the z axis at $z = -a$ combined with an equal-strength point sink also placed on the z axis at $z = a$ (see Fig. 3.30) below gives the following velocity potential and stream function at the point P.

$$\phi = \frac{Q}{4\pi[(R\cos\varphi + a)^2 + R^2\sin^2\varphi]^{1/2}} - \frac{Q}{4\pi[(R\cos\varphi - a)^2 + R^2\sin^2\varphi]^{1/2}} \tag{3.70}$$

$$\psi = \frac{Q}{4\pi}(\cos\varphi_1 - \cos\varphi_2) \tag{3.71}$$

where

$$\cos\varphi_1 = \frac{R\cos\varphi + a}{[(R\cos\varphi + a)^2 + R^2\sin^2\varphi]^{1/2}}$$

$$\cos\varphi_2 = \frac{R\cos\varphi - a}{[(R\cos\varphi - a)^2 + R^2\sin^2\varphi]^{1/2}}$$

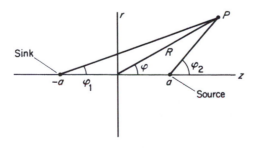

Fig. 3.30

If this source–sink pair is placed in a uniform stream $-U$ in the z direction it generates the flow around a body of revolution known as a Rankine body. The shape is very similar to the two-dimensional Rankine oval shown in Fig. 3.18 and described in Section 3.3.7.

3.4.5 The point doublet and the potential flow around a sphere

A point doublet is produced when the source–sink pair in Fig. 3.30 become infinitely close together. This is closely analogous to line doublet described in Section 3.3.8. Mathematically the expressions for the velocity potential and stream function for a point doublet can be derived from Eqns (3.70) and (3.71) respectively by allowing $a \to 0$ keeping $\mu = 2Qa$ fixed. The latter quantity is known as the strength of the doublet.

If a is very small a^2 may be neglected compared to $2Ra \cos\varphi$ in Eqn (3.70) then it can be written as

$$
\phi = \frac{Q}{4\pi}\left[\frac{1}{\{R^2\cos^2\varphi + R^2\sin^2\varphi + 2aR\cos\varphi\}^{1/2}} \right.
$$

$$
\left. - \frac{1}{\{R^2\cos^2\varphi + R^2\sin^2\varphi - 2aR\cos\varphi\}^{1/2}} \right]
$$

$$
= \frac{Q}{4\pi R}\left[\frac{1}{\{1 + 2(a/R)\cos\varphi\}^{1/2}} - \frac{1}{\{1 - 2(a/R)\cos\varphi\}^{1/2}} \right] \tag{3.72}
$$

On expanding

$$
\frac{1}{\sqrt{1 \pm x}} = 1 \mp \frac{1}{2}x + \cdots
$$

Therefore as $a \to 0$ Eqn (3.72) reduces to

$$
\phi = \frac{Q}{4\pi R}\left(1 - \frac{a}{R}\cos\varphi - 1 - \frac{a}{R}\cos\varphi \right)
$$

$$
= -\frac{Qa}{2\pi R^2}\cos\varphi = -\frac{\mu}{4\pi R^2}\cos\varphi \tag{3.73}
$$

In a similar way write

$$
\cos\varphi_{1,2} = \frac{R\cos\varphi \pm a}{R}\left(1 \mp \frac{a}{R}\cos\varphi \right)
$$

$$
= \cos\varphi \pm \frac{a}{R} \mp \frac{a}{R}\cos^2\varphi
$$

Thus as $a \to 0$ Eqn (3.71) reduces to

$$
\psi = \frac{Qa}{2\pi R}(1 - \cos^2\varphi) = \frac{\mu}{4\pi R}\sin^2\varphi \tag{3.74}
$$

The streamline patterns corresponding to the point doublet are similar to those depicted in Fig. 3.20. It is apparent from this streamline pattern and from the form of Eqn (3.74) that, unlike the point source, the flow field for the doublet is not omnidirectional. On the contrary the flow field is strongly directional. Moreover, the case analysed above is something of a special case in that the source–sink pair lies on the z axis. In fact the *axis* of the doublet can be in any direction in three-dimensional space.

For two-dimensional flow it was shown in Section 3.3.9 that the line doublet placed in a uniform stream produces the potential flow around a circular cylinder. Similarly it will be shown below that a point doublet placed in a uniform stream corresponds to the potential flow around a sphere.

From Eqns (3.65) and (3.73) the velocity potential for a point doublet in a uniform stream, with both the uniform stream and doublet axis aligned in the negative z direction, is given by

$$\phi = -U\,R\cos\varphi - \frac{\mu}{4\pi R^2}\cos\varphi \qquad (3.75)$$

From Eqn (3.60) the velocity components are given by

$$q_R = \frac{\partial\phi}{\partial R} = -\left(U - \frac{\mu}{2\pi R^3}\right)\cos\varphi \qquad (3.76)$$

$$q_\varphi = \frac{1}{R}\frac{\partial\phi}{\partial\varphi} = \left(U + \frac{\mu}{4\pi R^3}\right)\sin\varphi \qquad (3.77)$$

The stagnation points are defined by $q_R = q_\varphi = 0$. Let the coordinates of the stagnation points be denoted by (R_s, φ_s). Then from Eqn (3.77) it can be seen that either

$$R_s^3 = -\frac{\mu}{4\pi U} \quad\text{or}\quad \sin\varphi_s = 0$$

The first of these two equations cannot be satisfied as it implies that R_s is not a positive number. Accordingly, the second of the two equations must hold implying that

$$\varphi_s = 0 \quad\text{and}\quad \pi \qquad (3.78a)$$

It now follows from Eqn (3.76) that

$$R_s = \left(\frac{\mu}{2\pi U}\right)^{1/3} \qquad (3.78b)$$

Thus there are two stagnation points on the z axis at equal distances from the origin.

From Eqns (3.66) and (3.74) the stream function for a point doublet in a uniform flow is given by

$$\psi = -\frac{U\,R^2}{2}\sin^2\varphi + \frac{\mu}{4\pi R}\sin^2\varphi \qquad (3.79)$$

It follows from substituting Eqns (3.78b) in Eqn (3.79) that at the stagnation points $\psi = 0$. So the streamlines passing through the stagnation points are described by

$$\psi = -\left(\frac{U\,R^2}{2} - \frac{\mu}{4\pi R}\right)\sin^2\varphi = 0 \qquad (3.80)$$

Equation (3.79) shows that when $\varphi \neq 0$ or π the radius R of the stream-surface, containing the streamlines that pass through the stagnation points, remains fixed equal to R_s. R can take any value when $\varphi = 0$ or π. Thus these streamlines define the surface of a sphere of radius R_s. This is very similar to the two-dimensional case of the flow over a circular cylinder described in Section 3.3.9.

From Eqns (3.77) and (3.78b) it follows that the velocity on the surface of the sphere is given by

$$q = \frac{3}{2} U \sin \varphi$$

So that using the Bernoulli equation gives that

$$p_0 + \frac{1}{2}\rho U^2 = p + \frac{1}{2}\rho q^2$$
$$= p + \frac{1}{2}\rho \left(\frac{3}{2} U \sin \varphi\right)^2$$

Therefore the pressure variation over the sphere's surface is given by

$$p - p_0 = \frac{1}{2} U^2 (1 - \frac{9}{4}\sin^2 \varphi) \tag{3.81}$$

Again this result is quite similar to that for the circular cylinder described in Section 3.3.9 and depicted in Fig. 3.23.

3.4.6 Flow around slender bodies

In the foregoing part of this section it has been shown that the flow around a class of bodies of revolution can be modelled by the use of a source and sink of equal strength. Accordingly, it would be natural to speculate whether the flow around more general body shapes could be obtained by using several sources and sinks or a distribution of them along the z axis. It is indeed possible to do this as first shown by Fuhrmann.[*] Two examples similar to those presented by him are shown in Fig. 3.31. Although Fuhrmann's method could model the flow around realistic-looking bodies it suffered an important defect from the design point of view. One could calculate the body of revolution corresponding to a specified distribution of sources and sinks, but a designer would wish to be able to solve the inverse problem of how to choose the variation of source strength in order to obtain the flow around a given shape. This more practical approach became possible after Munk[†] introduced his slender-body theory for calculating the forces on airship hulls. A brief description of this approach is given below.

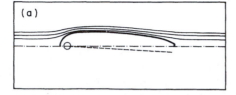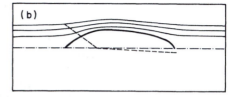

Fig. 3.31 Two examples of flow around bodies of revolution generated by (a) a point source plus a linear distribution of source strength; and (b) two linear distributions of source strength. The source distributions are denoted by broken lines

---

[*] Fuhrmann, G. (1911), Drag and pressure measurements on balloon models (in German), *Z. Flugtech.*, **11**, 165.
[†] Munk, M.M. (1924), The Aerodynamic Forces on Airship Hulls, NACA Report 184.

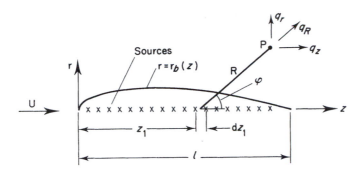

Fig. 3.32 Flow over a slender body of revolution modelled by source distribution

For Munk's slender-body theory it is assumed that the radius of the body is very much smaller than its total length. The flow is modelled by a distribution of sources and sinks placed on the z axis as depicted in Fig. 3.32. In many respects this theory is analogous to the theory for calculating the two-dimensional flow around symmetric wing sections – the so-called thickness problem (see Section 4.9).

For an element of source distribution located at $z = z_1$ the velocity induced at point P (r, z) is

$$q_R = \frac{\sigma(z_1)}{4\pi R^2}\,dz_1 \tag{3.82}$$

where $\sigma(z_1)$ is the source strength per unit length and $\sigma(z_1)dz_1$ takes the place of Q in Eqn (3.63). Thus to obtain the velocity components in the r and z directions at P due to all the sources we resolve the velocity given by Eqn (3.82) in the two coordinate directions and integrate along the length of the body. Thus

$$
\begin{aligned}
q_r &= \int_0^l q_R \sin\varphi \; dz_1 \\
&= \frac{1}{4\pi} \int_0^l \sigma(z_1) \frac{r}{[(z - z_1)^2 + r^2]^{3/2}}\,dz_1
\end{aligned}
\tag{3.83}
$$

$$
\begin{aligned}
q_z &= \int_0^l q_R \cos\varphi \; dz_1 \\
&= \frac{1}{4\pi} \int_0^l \sigma(z_1) \frac{z - z_1}{[(z - z_1)^2 + r^2]^{3/2}}\,dz_1
\end{aligned}
\tag{3.84}
$$

The source strength can be related to the body geometry by the following physical argument. Consider the elemental length of the body as shown in Fig. 3.33. If the body radius r_b is very small compared to the length, l, then the limit $r \to 0$ can be considered. For this limit the flow from the sources may be considered purely radial so that the flow across the body surface of the element is entirely due to the sources within the element itself. Accordingly

$$2\pi r q_r dz_1 = \sigma(z_1)dz_1 \text{ at } r = r_b \text{ provided } r_b \to 0$$

But the effects of the oncoming flow must also be considered as well as the sources. The net perpendicular velocity on the body surface due to both the oncoming flow and the sources must be zero. Provided that the slope of the body contour is very

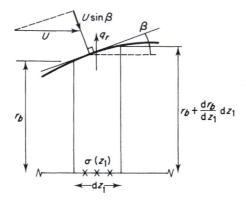

Fig. 3.33

small (i.e. $dr_b/dz \ll 1$) then the perpendicular and radial velocity components may be considered the same. Thus the requirement that the net normal velocity be zero becomes (see Fig. 3.33)

$$\underbrace{q_r}_{\text{Sources}} = \underbrace{U \sin \beta = U \frac{dr_b}{dz_1}}_{\text{Oncoming flow}}$$

So that the source strength per unit length and body shape are related as follows

$$\sigma(z) = U \frac{dS}{dz} \tag{3.85}$$

where S is the frontal area of a cross-section and is given by $S = \pi r_b^2$.

In the limit as $r \to 0$ Eqn (3.84) simplifies to

$$q_z = \frac{1}{4\pi} \int_0^l \frac{\sigma(z_1)}{(z - z_1)^2} dz_1 \tag{3.86}$$

Thus once the variation of source strength per unit length has been determined according to Eqn (3.85) the axial velocity can be obtained by evaluating Eqn (3.86) and hence the pressure evaluated from the Bernoulli equation.

It can be seen from the derivation of Eqn (3.86) that both r_b and dr_b/dz must be very small. Plainly the latter requirement would be violated in the vicinity of $z = 0$ if the body had a rounded nose. This is a major drawback of the method.

The slender-body theory was extended by Munk* to the case of a body at an angle of incidence or yaw. This case is treated as a superposition of two distinct flows as shown in Fig. 3.34. One of these is the slender body at zero angle of incidence as discussed above. The other is the slender body in a crossflow. For such a slender body the flow around a particular cross-section is closely analogous to that around a circular cylinder (see Section 3.3.9). Accordingly this flow can be modelled by a distribution of point doublets with axes aligned in the direction

* Munk, M.M. (1934), Fluid Mechanics, Part VI, Section Q, in *Aerodynamic Theory*, volume 1 (ed. W. Durand), Springer, Berlin; Dover, New York.

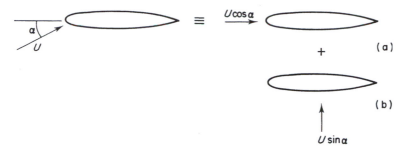

Fig. 3.34 Flow at angle of yaw around a body of revolution as the superposition of two flows

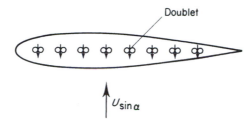

Fig. 3.35 Cross-flow over slender body of revolution modelled as distribution of doublets

of the cross-flow, as depicted in Fig. 3.35. Slender-body theory will not be taken further here. The reader is referred to Thwaites and Karamcheti for further details.*

3.5 Computational (panel) methods

In Section 3.3.7, it was shown how the two-dimensional potential flow around an oval-shaped contour, the Rankine oval, could be generated by the superposition of a source and sink on the x axis and a uniform flow. An analogous three-dimensional flow can also be generated around a Rankine body – see Section 3.4.4 above – by using a point source and sink. Thus it can be demonstrated that the potential flow around certain bodies can be modelled by placing sources and sinks in the interior of the body. However, it is only possible to deal with particular cases in this way. It is possible to model the potential flow around slender bodies or thin aerofoils of any shape by a distribution of sources lying along the x axis in the interior of the body. This slender-body theory is discussed in Section 3.4 and the analogous thin-wing theory is described in Section 4.3. However, calculations based on this theory are only approximate unless the body is infinitely thin and the slope of the body contour is very small. Even in this case the theory breaks down if the nose or leading edge is rounded because there the slope of the contour is infinite. The panel methods described here model the potential flow around a body by distributing sources over the body surface. In this way the potential flow around a body of any shape can be

*see Bibliography.

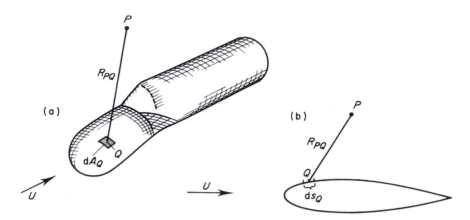

Fig. 3.36

calculated to a very high degree of precision. The method was developed by Hess and Smith* at Douglas Aircraft Company.

If a body is placed in a uniform flow of speed U, in exactly the same way as for the Rankine oval of Section 3.3.7, or the Rankine body of Section 3.4.4, the velocity potential for the uniform flow may be superimposed on that for the disturbed flow around the body to obtain a total velocity potential of the form

$$\Phi = Ux + \phi \tag{3.87}$$

where ϕ denotes the so-called disturbance potential: i.e. the departure from free-stream conditions. It can be shown that the disturbance potential flow around a body of any given shape can be modelled by a distribution of sources over the body surface (Fig. 3.36). Let the source strength per unit arc of contour (or per area in the three-dimensional case) be σ_Q. In the two-dimensional case $\sigma_Q \, ds_Q$ would replace $m/2\pi$ in Eqn (3.7) and constant C can be set equal to zero without loss of generality. Thus the velocity potential at P due to sources on an element ds_Q of arc of contour centred at point Q is given by

$$\phi_{PQ} = \sigma_Q \ln R_{PQ} ds_Q \tag{3.88a}$$

where R_{PQ} is the distance from P to Q. For the three-dimensional body $\sigma_Q dA_Q$ would replace $-Q/(4\pi)$ in Eqn (3.63) and the velocity potential due to the sources on an element, dA_Q, of surface area at point Q is given by

$$\phi_{PQ} = \frac{\sigma_Q}{R_{PQ}} dA_Q \tag{3.88b}$$

The velocity potential due to all the sources on the body surface is obtained by integrating (3.88b) over the body surface. Thus following Eqn (3.87) the total velocity potential at P can be written as

$$\phi_P = Ux + \oint \sigma_Q \ln R_{PQ} ds_Q \text{ for the two-dimensional case, (a)}$$

$$\phi_P = Ux + \iint \frac{\sigma_Q}{R_{PQ}} dA_Q \text{ for the three-dimensional case, (b)} \tag{3.89}$$

---

*J.L. Hess and A.M.O. Smith 'Calculation of Potential Flow about Arbitrary Bodies' *Prog. in Aero. Sci.*, **8** (1967).

where the integrals are to be understood as being carried out over the contour (or surface) of the body. Until the advent of modern computers the result (3.89) was of relatively little practical use. Owing to the power of modern computers, however, it has become the basis of a computational technique that is now commonplace in aerodynamic design.

In order to use Eqn (3.89) for numerical modelling it is first necessary to 'discretize' the surface, i.e. break it down into a finite but quite possibly large number of separate parts. This is achieved by representing the surface of the body by a collection of quadrilateral 'panels' – hence the name – see Fig. 3.37a. In the case of a two-dimensional shape the surface is represented by a series of straight line segments – see Fig. 3.37b. For simplicity of presentation concentrate on the two-dimensional case. Analogous procedures can be followed for the three-dimensional body.

The use of panel methods to calculate the potential flow around a body may be best understood by way of a concrete example. To this end the two-dimensional flow around a symmetric aerofoil is selected for illustrative purposes. See Fig. 3.37b.

The first step is to number all the end points or *nodes* of the panels from 1 to N as indicated in Fig. 3.37b. The individual panels are assigned the same number as the node located to the left when facing in the outward direction from the panel. The mid-points of each panel are chosen as *collocation* points. It will emerge below that the boundary condition of zero flow perpendicular to the surface is applied at these points. Also define for each panel the unit normal and tangential vectors, $\hat{n}_i$ and $\hat{t}_i$ respectively. Consider panels i and j in Fig. 3.37b. The sources distributed over panel j induce a velocity, which is denoted by the vector $\vec{v}_{ij}$, at the collocation point of panel i. The components of $\vec{v}_{ij}$ perpendicular and tangential to the surface at the collocation

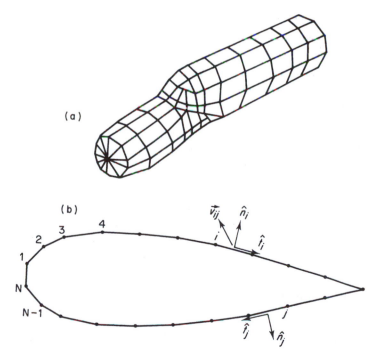

Fig. 3.37 Discretization of (a) three-dimensional body surface into panels; and (b) aerofoil contour into straight line segments

point i are given by the scalar (or dot) products $\vec{v}_{ij} \cdot \hat{n}_i$ and $\vec{v}_{ij} \cdot \hat{t}_i$ respectively. Both of these quantities are proportional to the strength of the sources on panel j and therefore they can be written in the forms

$$\vec{v}_{ij} \cdot \hat{n}_i = \sigma_j N_{ij} \quad \text{and} \quad \vec{v}_{ij} \cdot \hat{t}_i = \sigma_j T_{ij} \tag{3.90}$$

N_{ij} and T_{ij} are the perpendicular and tangential velocities induced at the collocation point of panel i by sources of unit strength distributed over panel j; they are known as the *normal* and *tangential influence coefficients*.

The actual velocity perpendicular to the surface at collocation point i is the sum of the perpendicular velocities induced by each of the N panels plus the contribution due to the free stream. It is given by

$$v_{n_i} = \sum_{j=1}^{N} \sigma_j N_{ij} + \vec{U} \cdot \hat{n}_i \tag{3.91}$$

In a similar fashion the tangential velocity at collocation point i is given by

$$v_{s_i} = \sum_{j=1}^{N} \sigma_j T_{ij} + \vec{U} \cdot \hat{t}_i \tag{3.92}$$

If the surface represented by the panels is to correspond to a solid surface then the actual perpendicular velocity at each collocation point must be zero. This condition may be expressed mathematically as $v_{n_i} = 0$ so that Eqn (3.91) becomes

$$\sum_{j=1}^{N} \sigma_j N_{ij} = -\vec{U} \cdot \hat{n}_i (i = 1, 2, \ldots, N) \tag{3.93}$$

Equation (3.93) is a system of linear algebraic equations for the N unknown source strengths, $\sigma_i(i = 1, 2, \ldots, N)$. It takes the form of a matrix equation

$$\mathbf{N}\boldsymbol{\sigma} = \mathbf{b} \tag{3.94}$$

where $\mathbf{N}$ is an $N \times N$ matrix composed of the elements N_{ij}, $\boldsymbol{\sigma}$ is a column matrix composed of the N elements σ_i, and $\mathbf{b}$ is a column matrix composed of the N elements $-\vec{U} \cdot \hat{n}_i$. Assuming for the moment that the perpendicular influence coefficients N_{ij} have been calculated and that the elements of the right-hand column matrix $\mathbf{b}$ have also been calculated, then Eqn (3.94) may, in principle at least, be solved for the source strengths comprising the elements of the column matrix $\boldsymbol{\sigma}$. Systems of linear equations like (3.94) can be readily solved numerically using standard methods. For the results presented here the LU decomposition was used to solve for the source strengths. This method is described by Press *et al.** who also give listings for the necessary computational routines.

Once the influence coefficients N_{ij} have been calculated the source strengths can be determined by solving the system of Eqn (3.93) by some standard numerical technique. If the tangential influence coefficients T_{ij} have also been calculated then, once the source strengths have been determined, the tangential velocities may be obtained from Eqn (3.92). The Bernoulli equation can then be used to calculate the pressure acting at collocation point i, in particular the coefficient of pressure is given by Eqn (2.24) as:

$$C_{p_i} = 1 - \left(\frac{v_{s_i}}{U}\right)^2 \tag{3.95}$$

---

* W.H. Press *et al.* (1992) *Numerical Recipes. The Art of Scientific Computing.* 2nd ed. Cambridge University Press.

The calculation of the influence coefficient is a central and essential part of the panel method, and this is the question now addressed. As a first step consider the calculation of the velocity induced at a point P by sources of unit strength distributed over a panel centred at point Q.

In terms of a coordinate system (x_Q, y_Q) measured relative to the panel (Fig. 3.38), the disturbance potential is given by integrating Eqn (3.88) over the panel. Mathematically this is expressed as follows

$$\phi_{PQ} = \int_{-\Delta s/2}^{\Delta s/2} \ln \sqrt{(x_Q - \xi)^2 + y_Q^2}\, d\xi \tag{3.96}$$

The corresponding velocity components at P in the x_Q and y_Q directions can be readily obtained from Eqn (3.96) as

$$v_{x_Q} = \frac{\partial \phi_{PQ}}{\partial x_Q} = \int_{-\Delta s/2}^{\Delta s/2} \frac{x_Q - \xi}{(x_Q - \xi)^2 + y_Q^2}\, d\xi$$

$$= -\frac{1}{2} \ln \left[\frac{(x_Q + \Delta s/2)^2 + y_Q^2}{(x_Q - \Delta s/2)^2 + y_Q^2} \right] \tag{3.97}$$

$$v_{y_Q} = \frac{\partial \phi_{PQ}}{\partial y_Q} = \int_{-\Delta s/2}^{\Delta s/2} \frac{y_Q}{(x_Q - \xi)^2 + y_Q^2}\, d\xi$$

$$= -\left[\tan^{-1}\left(\frac{x_Q + \Delta s/2}{y_Q} \right) - \tan^{-1}\left(\frac{x_Q - \Delta s/2}{y_Q} \right) \right] \tag{3.98}$$

Armed with these results for the velocity components induced at point P due to the sources on a panel centred at point Q return now to the problem of calculating the influence coefficients. Suppose that points P and Q are chosen to be the collocation points i and j respectively. Equations (3.97) and (3.98) give the velocity components in a coordinate system relative to panel j, whereas what are required are the velocity components perpendicular and tangential to panel i. In vector form the velocity at collocation point i is given by

$$\vec{v}_{PQ} = v_{x_Q}\hat{t}_j + v_{y_Q}\hat{n}_j$$

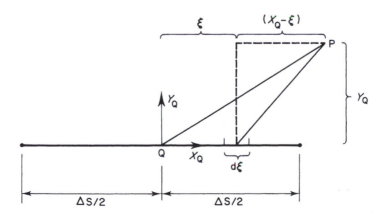

Fig. 3.38

Therefore to obtain the components of this velocity vector perpendicular and tangential to panel i take the scalar product of the velocity vector with $\hat{n}_i$ and $\hat{t}_i$ respectively to obtain

$$N_{ij} = \vec{v}_{PQ} \cdot \hat{n}_i = v_{x_Q} \hat{n}_i \cdot \hat{t}_j + v_{y_Q} \hat{n}_i \cdot \hat{n}_j \tag{3.99a}$$

$$T_{ij} = \vec{v}_{PQ} \cdot \hat{t}_i = v_{x_Q} \hat{t}_i \cdot \hat{t}_j + v_{y_Q} \hat{t}_i \cdot \hat{n}_j \tag{3.99b}$$

A computational routine in FORTRAN 77

In order to see how the calculation of the influence coefficients works in practice, a computational routine written in standard FORTRAN 77 is given below, with a description of each step.

```
SUBROUTINE INFLU(XC, YC, AN, AT, NHAT, THAT, N, NM)
    On exit XC and YC are column matrices of length N containing the co-ordinates of
    the collocation points; AN and AT are the N*N influence coefficient matrices; and
    NHAT and THAT are the N*2 matrices containing the co-ordinates of the unit normal
    and tangent vectors, the first and second columns contain the x and y co-ordinates
    respectively. N is the number of panels and NM is the maximum number of panels.

    PARAMETER(NMAX=200,PI=3.141592654)
    REAL NHAT,NTIJ,NNIJ
    DIMENSION XC(NM),YC(NM),AN(NM,NM),AT(NM,NM)
    DIMENSION XP(NMAX),YP(NMAX),NHAT(NM,2),
   &            THAT(NM,2), S(NMAX)
    OPEN(7,FILE='POINTS.DAT',STATUS='OLD')
    DO 10 I=1,N                    Reading in co-ordinates of panel
10    READ(7,*) XP(I), YP(I)       end-points.
    CLOSE(7)
    DO 20 J=1,N
     IF (J.EQ.1) THEN
       XPL=XP(N)
       YPL=YP(N)
               ELSE
       XPL=XP(J-1)
       YPL=YP(J-1)
     ENDIF
     XC(J)=0.5*(XP(J)+XPL)         Calculating co-ordinates of
     YC(J)=0.5*(YP(J)+YPL)         collocation points.
     S(J)=SQRT((XP(J)-XPL)**2+(YP(J)-YPL)**2)   Calculating panel length.
     THAT(J,1)=(XP(J)-XPL)/S(J)   Calculating x co-ordinate of unit tangent vector.
     THAT(J,2)=(YP(J)-YPL)/S(J)   Calculating y co-ordinate of unit tangent vector.
     NHAT(J,1)=-THAT(J,2)          Calculating x co-ordinate of unit normal vector.
     NHAT(J,2)=  THAT(J,1)         Calculating y co-ordinate of unit normal vector.
20  CONTINUE

    Calculation of the influence coefficients.

    DO 30 I=1,N
     DO 40 J=1,N
       IF (I.EQ.J) THEN
         AN(I,J)=PI      Case of i=j.
         AT(I,J)=0.0
               ELSE
       DX=XC(I)-XC(J)   Calculating x and y components of line
       DY=YC(I)-YC(J)   joining collocation point i and j
```

```
        XQ=DX*THAT(J,1)+DY*THAT(J,2)   Converting to co-ordinate system
        YQ=DX*NHAT(J,1)+DY*NHAT(J,2)   based on panel j.
        VX=0.5*(LOG((XQ+0.5*S(J))**2+YQ*YQ)   Using Eqn. (3.97)
   &         -LOG((XQ-0.5*S(J))**2+YQ*YQ))
        VY=ATAN((XQ+0.5*S(J))/YQ) -            Using Eqn. (3.98)
   &            ATAN((XQ-0.5*S(J))/YQ)
   Begin calculation of various scalar products of unit vectors used in Eqn. (3.99)
        NTIJ=0.0
        NNIJ=0.0
        TTIJ=0.0
        TNIJ=0.0
        DO 50 K=1,2
          NTIJ=NHAT(I,K)*THAT(J,K)+NTIJ
          NNIJ=NHAT(I,K)*NHAT(J,K)+NNIJ
          TTIJ=THAT(I,K)*THAT(J,K)+TTIJ
          TNIJ=THAT(I,K)*NHAT(J,K)+TNIJ
50      CONTINUE
        End calculation of scalar products.
        AN(I,J)=VX*NTIJ+VY*NNIJ  Using Eqn. (3.99a)
        AN(I,J)=VX*TTIJ+VY*TNIJ  Using Eqn. (3.99b)
      ENDIF
40      CONTINUE
30  CONTINUE
    RETURN
    END
```

The routine, step by step, performs the following.

1 Discretizes the surface by assigning numbers from 1 to N to points on the surface of the aerofoil as suggested in Fig. 3.37. The x and y coordinates of these points are entered into a file named POINTS.DAT. The subroutine starts with reading these coordinates $XP(I)$, $YP(I)$, say x'_i, y'_i, from this file for $I = 1$ to N.
For each panel from $J = 1$ to N:
2 The collocation points are calculated by taking an average of the coordinates at either end of the panel in question.
3 The length $S(J)$, i.e. Δs_j, of each panel is calculated.
4 The x and y components of the unit tangent vectors for each panel are calculated as follows:

$$t_{j_x} = \frac{x'_j - x'_{j-1}}{\Delta s_j}, \quad t_{j_y} = \frac{y'_j - y'_{j-1}}{\Delta s_j}$$

5 The unit normal vectors are then calculated from $n_{j_x} = -t_{j_y}$ and $n_{j_y} = t_{j_x}$. The main task of the routine, that of calculating the influence coefficients, now begins.
For each possible combination of panels, i.e. I and $J = 1$ to N.
6 First the special case is dealt with when $i = j$, i.e. the velocity induced by the sources on the panel itself at its collocation point. From Eqn (3.93, 3.97, 3.98) it is seen that

$$v_{PQx} = \ln(1) = 0 \quad \text{when} \quad x_Q = y_Q = 0 \qquad (3.100a)$$

$$v_{PQy} = \tan^{-1}(\infty) - \tan^{-1}(-\infty) = \pi \quad \text{when} \quad x_Q = y_Q = 0 \qquad (3.100b)$$

When $i \neq j$ the influence coefficients have to be calculated from Eqns (3.97, 3.98, 3.99).

7 The components DX and DY of R_{PQ} are calculated in terms of the x and y coordinates.

8 The components of R_{PQ} in terms of the coordinate system based on panel j are then calculated as

$$X_Q = \vec{R}_{PQ} \cdot \hat{t}_j \text{ and } Y_Q = \vec{R}_{PQ} \cdot \hat{n}_j$$

9 VX and VY (i.e. v_{x_Q} and v_{y_Q}) are evaluated using Eqns (3.97) and (3.98).

10 $\hat{n}_i \cdot \hat{t}_j$, $\hat{n}_i \cdot \hat{n}_j$, $\hat{t}_i \cdot \hat{t}_j$, and $\hat{t}_i \cdot \hat{n}_j$ are evaluated.

11 Finally the influence coefficients are evaluated from Eqn (3.99).

The routine presented above is primarily intended for educational purposes and has not been optimized to economize on computing time. Nevertheless, using a computer program based on the above routine and LU decomposition, accurate computations of the pressure distribution around two-dimensional aerofoils can be obtained in a few seconds with a modern personal computer. An example of such a calculation for an NACA 0024 aerofoil is presented in Fig. 3.39. In this case 29 panels were used for the complete aerofoil consisting of upper and lower surfaces.

The extension of the panel method to the case of lifting bodies, i.e. wings, is described in Sections 4.10 and 5.8. When the methods described there are used it is possible to compute the flow around the entire aircraft. Such computations are carried out routinely during aerodynamic design and have replaced wind-tunnel testing to a considerable extent. However, calculation of the potential flow around complex three-dimensional bodies is very demanding in terms of computational time and memory. In most cases around 70 to 80 per cent of the computing time is consumed in calculating the influence coefficients. Accordingly considerable effort has been devoted to developing routines for carrying out these calculations efficiently.

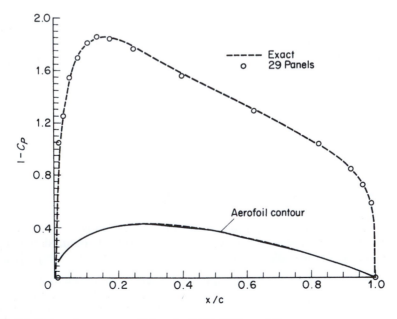

Fig. 3.39 Calculation of pressure coefficient for NACA 0024 aerofoil

What are the advantages of the panel method compared to other numerical methods such as finite differences and finite elements? Both of the latter are field methods that require that the whole of the flow field be discretized. The panel method, on the other hand, only requires the discretization of the body surface – the boundary of the flow field. The dimensions of the solution are thereby reduced by one compared to the field method. Thus for the aerofoil calculation presented above the panel method required N node points along the aerofoil contour, whereas a field method would require $N \times N$ points throughout the flow field. However, this advantage is more apparent than real, since for the panel method the $N \times N$ influence coefficients need to be calculated. The real advantages of panel methods lie elsewhere. First, like finite-element methods, but unlike finite difference methods, the panel method can readily accommodate complex geometries. In fact, an alternative and perhaps more appropriate term to *panel method* is *boundary-element method*. This name makes the connection with finite elements more clear. A second advantage compared to any field method is the ease with which panel methods can deal with an infinite flow field; note that the aerofoil in Fig. 3.39 is placed in an airflow of infinite extent, as is usual. Thirdly, as can readily be seen from the example in Fig. 3.39, accurate results can be obtained by means of a relatively coarse discretization, i.e. using a small number of panels. Lastly, and arguably the most important advantage from the viewpoint of aerodynamic design, is the ease with which modifications of the design can be incorporated with a panel method. For example, suppose the effects of under-wing stores, such as additional fuel tanks or missiles, were being investigated. If an additional store were to be added it would not be necessary to repeat the entire calculation with a panel method. It would be necessary only to calculate the additional influence coefficients involving the new under-wing store. This facility of panel methods allows the effects of modifications to be investigated rapidly during aerodynamic design.

Exercises

1 Define vorticity in a fluid and obtain an expression for vorticity at a point with polar coordinates (r, θ), the motion being assumed two-dimensional. From the definition of a line vortex as irrotational flow in concentric circles determine the variation of velocity with radius, hence obtain the stream function (ψ), and the velocity potential (ϕ), for a line vortex.   (U of L)

2 A sink of strength $120\,\text{m}^2\text{s}^{-1}$ is situated $2\,\text{m}$ downstream from a source of equal strength in an irrotational uniform stream of $30\,\text{m s}^{-1}$. Find the fineness ratio of the oval formed by the streamline $\psi = 0$.   (*Answer*: 1.51)(CU)

3 A sink of strength $20\,\text{m}^2\,\text{s}^{-1}$ is situated $3\,\text{m}$ upstream of a source of $40\,\text{m}^2\,\text{s}^{-1}$, in a uniform irrotational stream. It is found that at the point $2.5\,\text{m}$ equidistant from both source and sink, the local velocity is normal to the line joining the source and sink. Find the velocity at this point and the velocity of the undisturbed stream.
     (*Answer*: $1.02\,\text{m s}^{-1}$, $2.29\,\text{m s}^{-1}$)(CU)

4 A line source of strength m and a sink of strength $2m$ are separated a distance c. Show that the field of flow consists in part of closed curves. Locate any stagnation points and sketch the field of flow.   (U of L)

5 Derive the expression giving the stream function for irrotational flow of an incompressible fluid past a circular cylinder of infinite span. Hence determine the position of generators on the cylinder at which the pressure is equal to that of the undisturbed stream.   (*Answer*: $\pm 30°$, $\pm 150°$)(U of L)

6 Determine the stream function for a two-dimensional source of strength m. Sketch the resultant field of flow due to three such sources, each of strength m, located at the vertices of an equilateral triangle.   (U of L)

7 Derive the irrotational flow formula

$$p - p_0 = \frac{1}{2}\rho U^2 (1 - 4\sin^2 \theta)$$

giving the intensity of normal pressure p on the surface of a long, circular cylinder set at right-angles to a stream of velocity U. The undisturbed static pressure in the fluid is p_0 and θ is the angular distance round from the stagnation point. Describe briefly an experiment to test the accuracy of the above formula and comment on the results obtained.   (U of L)

8 A long right circular cylinder of diameter a m is set horizontally in a steady stream of velocity U m s^{-1} and caused to rotate at ω rad s^{-1}. Obtain an expression in terms of ω and U for the ratio of the pressure difference between the top and the bottom of the cylinder to the dynamic pressure of the stream. Describe briefly the behaviour of the stagnation lines of such a system as ω is increased from zero, keeping U constant.

$$\left(Answer: \frac{8a\omega}{U} \right) (\text{CU})$$

9 A line source is immersed in a uniform stream. Show that the resultant flow, if irrotational, may represent the flow past a two-dimensional fairing. If a maximum thickness of the fairing is 0.15 m and the undisturbed velocity of the stream 6.0 m s^{-1}, determine the strength and location of the source. Obtain also an expression for the pressure at any point on the surface of the fairing, taking the pressure at infinity as datum.   (*Answer*: 0.9 m^2 s^{-1}, 0.0237 m)(U of L)

10 A long right circular cylinder of radius a m is held with its axis normal to an irrotational inviscid stream of U. Obtain an expression for the drag force acting on unit length of the cylinder due to the pressures exerted on the front half only.

$$\left(Answer: -\frac{1}{3}\rho U^2 a \right) (\text{CU})$$

11 Show that a velocity potential exists in a two-dimensional steady irrotational incompressible fluid motion. The stream function of a two-dimensional motion of an incompressible fluid is given by

$$\psi = \frac{a}{2}x^2 + bxy - \frac{c}{2}y^2$$

where a, b and c are arbitrary constants. Show that, if the flow is irrotational, the lines of constant pressure never coincide with either the streamlines or the equipotential lines. Is this possible for rotational motion?   (U of L)

12 State the stream function and velocity potential for each of the motions induced by a source, vortex and doublet in a two-dimensional incompressible fluid. Show that a doublet may be regarded, either as

(i) the limiting case of a source and sink, or
(ii) the limiting case of equal and opposite vortices, indicating clearly the direction of the resultant doublet.   (U of L)

13 Define (a) the stream function, (b) irrotational flow and (c) the velocity potential for two-dimensional motion of an incompressible fluid, indicating the conditions under which they exist. Determine the stream function for a point source of strength σ at the origin. Hence, or otherwise, show that for the flow due to any number of sources at points on a circle, the circle is a streamline provided that the algebraic sum of the strengths of the sources is zero. (U of L)

14 A line vortex of strength Γ is mechanically fixed at the point $(l, 0)$ referred to a system of rectangular axes in an inviscid incompressible fluid at rest at infinity bounded by a plane wall coincident with the y-axis. Find the velocity in the fluid at the point $(0, y)$ and determine the force that acts on the wall (per unit depth) if the pressure on the other side of the wall is the same as at infinity. Bearing in mind that this must be equal and opposite to the force acting on unit length of the vortex show that your result is consistent with the Kutta–Zhukovsky theorem. (U of L)

15 Write down the velocity potential for the two-dimensional flow about a circular cylinder with a circulation Γ in an otherwise uniform stream of velocity U. Hence show that the lift on unit span of the cylinder is $\rho U \Gamma$. Produce a brief but plausible argument that the same result should hold for the lift on a cylinder of arbitrary shape, basing your argument on consideration of the flow at large distances from the cylinder. (U of L)

16 Define the terms velocity potential, circulation, and vorticity as used in two-dimensional fluid mechanics, and show how they are related. The velocity distribution in the laminar boundary layer of a wide flat plate is given by

$$u = u_0 \left[\frac{3}{2} \frac{y}{\delta} - \frac{1}{2} \left(\frac{y}{\delta} \right)^3 \right]$$

where u_0 is the velocity at the edge of the boundary layer where y equals δ. Find the vorticity on the surface of the plate.

$$\left(Answer: -\frac{3}{2} \frac{u_0}{\delta} \right) \text{ (U of L)}$$

17 A two-dimensional fluid motion is represented by a point vortex of strength Γ set at unit distance from an infinite straight boundary. Draw the streamlines and plot the velocity distribution on the boundary when $\Gamma = \pi$. (U of L)

18 The velocity components of a two-dimensional inviscid incompressible flow are given by

$$u = 2y - \frac{y}{(x^2 + y^2)^{1/2}}, \qquad v = -2x - \frac{x}{(x^2 + y^2)^{1/2}}$$

Find the stream function, and the vorticity, and sketch the streamlines.

$$\left(Answer: \psi = x^2 + y^2 + (x^2 + y^2)^{1/2}, \zeta = -\left[4 + \frac{1}{(x^2 + y^2)^{1/2}} \right] \right) \text{(U of L)}$$

19 (a) Given that the velocity potential for a point source takes the form

$$\phi = -\frac{Q}{4\pi R}$$

where in axisymmetric cylindrical coordinates $(r, z) R = \sqrt{z^2 + r^2}$, show that when a uniform stream, U, is superimposed on a point source located at the origin, there is a stagnation point located on the z-axis upstream of the origin at distance

$$a = \sqrt{\frac{Q}{4\pi U}}$$

(b) Given that in axisymmetric spherical coordinates (R, φ) the stream function for the point source takes the form

$$\psi = -\frac{Q}{4\pi R}$$

show that the streamlines passing through the stagnation point found in (a) define a body of revolution given by

$$R^2 = \frac{2a^2(1 + \cos\varphi)}{\sin^2\varphi}$$

Make a rough sketch of this body.

4

Two-dimensional wing theory

Preamble

Here the basic fluid mechanics outlined previously is applied to the analysis of the flow about a lifting wing section. It is explained that potential flow theories of themselves offer little further scope for this problem unless modified to simulate certain effects of real flows. The result is a powerful but elementary aerofoil theory capable of wide exploitation. This is derived in the general form and applied to a number of discrete aeronautical situations, including the flapped aerofoil and the jet flap. The 'reverse' problem is also presented: to determine the rudimentary aerofoil shape that produces certain aerodynamic performance requirements. This theory is essentially relevant to thin aerofoils but thickness parameters are added to enhance the practical applications of the method. Classical mathematical solutions are referred to, also the solutions offered towards the end of the chapter that employ computational panel methods.

4.1 Introduction

By the end of the nineteenth century the theory of ideal, or potential, flow (see Chapter 3) was extremely well-developed. The motion of an inviscid fluid was a well-defined mathematical problem. It satisfied a relatively simple linear partial differential equation, the Laplace equation (see Section 3.2), with well-defined boundary conditions. Owing to this state of affairs many distinguished mathematicians were able to develop a wide variety of analytical methods for predicting such flows. Their work was and is very useful for many practical problems, for example the flow around airships, ship hydrodynamics and water waves. But for the most important practical applications in aerodynamics potential flow theory was almost a complete failure.

Potential flow theory predicted the flow field absolutely exactly for an inviscid fluid, that is for *infinite Reynolds number*. In two important respects, however, it did not correspond to the flow field of real fluid, no matter how large the Reynolds number. Firstly, real flows have a tendency to separate from the surface of the body. This is especially pronounced when the bodies are bluff like a circular cylinder, and in such cases the real flow bears no resemblance to the corresponding potential flow. Secondly, steady potential flow around a body can produce no force irrespective of the shape. This result is usually known as *d'Alembert's paradox* after the French mathematician who first discovered it in 1744. Thus there is no prospect of using potential flow theory in its pure form to estimate the lift or drag of wings and thereby to develop aerodynamic design methods.

Flow separation and d'Alembert's paradox both result from the subtle effects of viscosity on flows at high Reynolds number. The necessary understanding and knowledge of viscous effects came largely from work done during the first two decades of the twentieth century. It took several more decades, however, before this knowledge was fully exploited in aerodynamic design. The great German aeronautical engineer Prandtl and his research team at the University of Göttingen deserve most of the credit both for explaining these paradoxes and showing how potential flow theory can be modified to yield useful predictions of the flow around wings and thus of their aerodynamic characteristics. His boundary-layer theory explained why flow separation occurs and showed how skin-friction drag could be calculated. This theory and its later developments are described in Chapter 7 below. He also showed how a theoretical model based on vortices could be developed for the flow field of a wing having large aspect ratio. This theory is described in Chapter 5. There it is shown how a knowledge of the aerodynamic characteristics, principally the lift coefficient, of a wing of infinite span – an aerofoil – can be adapted to give estimates of the aerodynamic characteristics of a wing of finite span. This work firmly established the relevance of studying the two-dimensional flow around aerofoils that is the subject of the present chapter.

4.1.1 The Kutta condition

How can potential flow be adapted to provide a reasonable theoretical model for the flow around an aerofoil that generates lift? The answer lies in drawing an analogy between the flow around an aerofoil and that around a spinning cylinder (see Section 3.3.10). For the latter it can be shown that when a point vortex is superimposed with a doublet on a uniform flow, a lifting flow is generated. It was explained in Section 3.3.9 that the doublet and uniform flow alone constitutes a non-circulatory irrotational flow with zero vorticity everywhere. In contrast, when the vortex is present the vorticity is zero everywhere except at the origin. Thus, although the flow is still irrotational everywhere save at the origin, the net effect is that the circulation is non-zero. The generation of lift is always associated with circulation. In fact, it can be shown (see Eqn 3.52) that for the spinning cylinder the lift is directly proportional to the circulation. It will be shown below that this important result can also be extended to aerofoils. The other point to note from Fig. 3.25 is that as the vortex strength, and therefore circulation, rise both the fore and aft stagnation points move downwards along the surface of the cylinder.

Now suppose that in some way it is possible to use vortices to generate circulation, and thereby lift, for the flow around an aerofoil. The result is shown schematically in Fig. 4.1. Figure 4.1a shows the pure non-circulatory potential flow around an aerofoil at an angle of incidence. If a small amount of circulation is added the fore and aft stagnation points, S_F and S_A, move as shown in Fig. 4.1b. In this case the rear stagnation point remains on the upper surface. On the other hand, if the circulation is relatively large the rear stagnation point moves to the lower surface, as shown in Fig. 4.1c. For all three of these cases the flow has to pass around the trailing edge. For an inviscid flow this implies that the flow speed becomes infinite at the trailing edge. This is evidently impossible in a real viscous fluid because viscous effects ensure that such flows cannot be sustained in nature. In fact, the only position for the rear stagnation point that is sustainable in a real flow is at the trailing edge, as illustrated in Fig. 4.1d. Only with the rear stagnation point at the trailing edge does

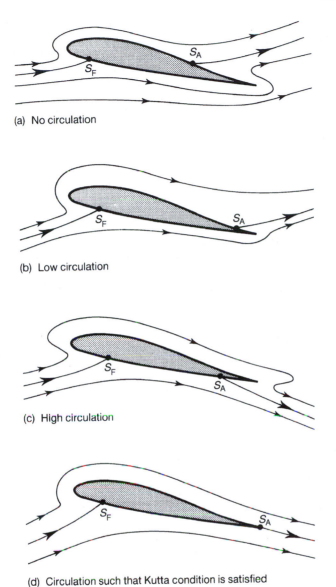

(a) No circulation

(b) Low circulation

(c) High circulation

(d) Circulation such that Kutta condition is satisfied

Fig. 4.1 Effect of circulation on the flow around an aerofoil at an angle of incidence

the flow leave the upper and lower surfaces smoothly at the trailing edge. This is the essence of the *Kutta condition* first introduced by the German mathematician Kutta.[*]

Imposing the Kutta condition gives a unique way of choosing the circulation for an aerofoil, and thereby determining the lift. This is extremely important because otherwise there would be an infinite number of different lifting flows, each corresponding to a different value of circulation, just as in the case of the spinning cylinder

---

[*] W. Kutta (1902) 'Lift forces in flowing fluids' (in German), *Ill. Aeronaut. Mitt.*, **6**, 133.

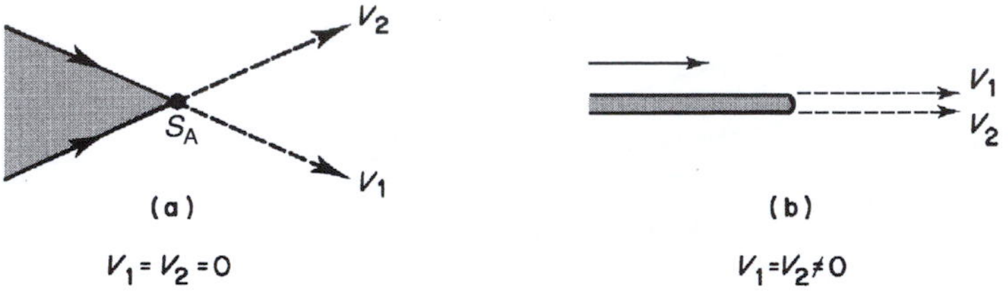

(a)

$V_1 = V_2 = 0$

(b)

$V_1 = V_2 \neq 0$

Fig. 4.2

for which the lift generated depends on the rate of spin. In summary, the Kutta condition can be expressed as follows.

- For a given aerofoil at a given angle of attack the value of the circulation must take the unique value which ensures that the flow leaves the trailing edge smoothly.
- For practical aerofoils with trailing edges that subtend a finite angle – see Fig. 4.2a – this condition implies that the rear stagnation point is located at the trailing edge.

All real aerofoils are like Fig. 4.2a, of course, but (as in Section 4.2) for theoretical reasons it is frequently desirable to consider infinitely thin aerofoils, Fig. 4.2b. In this case and for the more general case of a cusped trailing edge the trailing edge need not be a stagnation point for the flow to leave the trailing edge smoothly.

- If the angle subtended by the trailing edge is zero then the velocities leaving the upper and lower surfaces at the trailing edge are finite and equal in magnitude and direction.

4.1.2 Circulation and vorticity

From the discussion above it is evident that *circulation* and *vorticity*, introduced in Section 2.7, are key concepts in understanding the generation of lift. These concepts are now explored further, and the precise relationship between the lift force and circulation is derived.

Consider an imaginary open curve AB drawn in a purely potential flow as in Fig. 4.3a. The difference in the velocity potential ϕ evaluated at A and B is given by the line integral of the tangential velocity component of flow along the curve, i.e. if the flow velocity across AB at the point P is q, inclined at angle α to the local tangent, then

$$\phi_A - \phi_B = \int_{AB} q \cos \alpha \, ds \tag{4.1}$$

which could also be written in the form

$$\phi_A - \phi_B = \int_{AB} (u \, dx + v \, dy)$$

Equation (4.1) could be regarded as an alternative definition of velocity potential.

Consider next a closed curve or circuit in a circulatory flow (Fig. 4.3b) (remember that the circuit is imaginary and does not influence the flow in any way, i.e. it is not a boundary). The *circulation* is defined in Eqn (2.83) as the line integral taken around the circuit and is denoted by Γ, i.e.

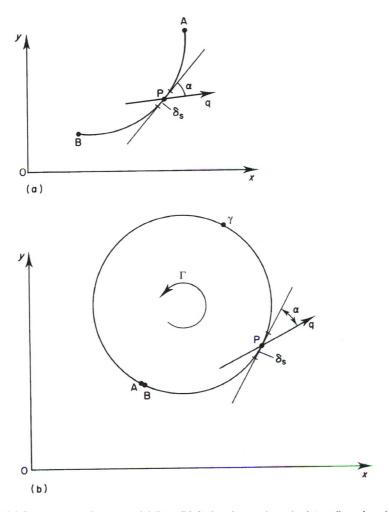

Fig. 4.3 (a) An open curve in a potential flow. (b) A closed curve in a circulatory flow; A and B coincide

$$\Gamma = \oint q\cos\alpha\, ds \qquad \text{or} \qquad \Gamma = \oint (u\,dx + v\,dy) \tag{4.2}$$

It is evident from Eqns (4.1) and (4.2) that in a purely potential flow, for which ϕ_A must equal ϕ_B when the two points coincide, the circulation must be zero.

Circulation implies a component of *rotation* of flow in the system. This is not to say that there are circular streamlines, or that elements of fluid are actually moving around some closed loop although this is a possible flow system. Circulation in a flow means that the flow system could be resolved into a uniform irrotational portion and a circulating portion. Figure 4.4 shows an idealized concept. The implication is that if circulation is present in a fluid motion, then vorticity must be present, even though it may be confined to a restricted space, e.g. as in the core of a point vortex. Alternatively, as in the case of the circular cylinder with circulation, the vorticity at the centre of the cylinder may actually be excluded from the region of flow considered, namely that outside the cylinder.

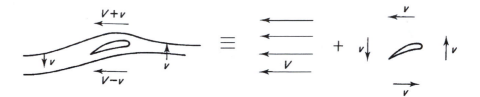

Fig. 4.4

Consider this by the reverse argument. Look again at Fig. 4.3b. By definition the velocity potential of C relative to A (ϕ_{CA}) must be equal to the velocity potential of C relative to B (ϕ_{CB}) in a potential flow. The integration continued around ACB gives

$$\Gamma = \phi_{CA} \pm \phi_{CB} = 0$$

This is for a potential flow only. Thus, if Γ is finite the definition of the velocity potential breaks down and the curve ACB must contain a region of rotational flow. If the flow is not potential then Eqn (ii) in Section 3.2 must give a non-zero value for vorticity.

An alternative equation for Γ is found by considering the circuit of integration to consist of a large number of rectangular elements of side $\delta x\, \delta y$ (e.g. see Section 2.7.7 and Example 2.2). Applying the integral $\Gamma = \int (u\, dx + v\, dy)$ round $abcd$, say, which is the element at $P(x, y)$ where the velocity is u and v, gives (Fig. 4.5).

$$\Delta\Gamma = \left(v + \frac{\partial v}{\partial x}\frac{\delta x}{2}\right)\delta y - \left(u + \frac{\partial u}{\partial y}\frac{\delta y}{2}\right)\delta x - \left(v - \frac{\partial v}{\partial x}\frac{\delta x}{2}\right)\delta y$$
$$+ \left(u - \frac{\partial u}{\partial y}\frac{\delta y}{2}\right)\delta x$$
$$\Delta\Gamma = \left(\frac{\partial v}{\partial x} - \frac{\partial u}{\partial y}\right)\delta x\, \delta y$$

The sum of the circulations of all the areas is clearly the circulation of the circuit as a whole because, as the $\Delta\Gamma$ of each element is added to the $\Delta\Gamma$ of the neighbouring element, the contributions of the common sides disappear.

Applying this argument from element to neighbouring element throughout the area, the only sides contributing to the circulation when the $\Delta\Gamma$s of all areas are summed together are those sides which actually form the circuit itself. This means that for the circuit as a whole

$$\Gamma = \underbrace{\iint\left(\frac{\partial v}{\partial x} - \frac{\partial u}{\partial y}\right)dx\, dy}_{\text{over the area}} = \underbrace{\oint(u\, dx + v\, dy)}_{\text{round the circuit}}$$

and

$$\frac{\partial v}{\partial x} - \frac{\partial u}{\partial y} = \zeta$$

This shows explicitly that the circulation is given by the integral of the vorticity contained in the region enclosed by the circuit.

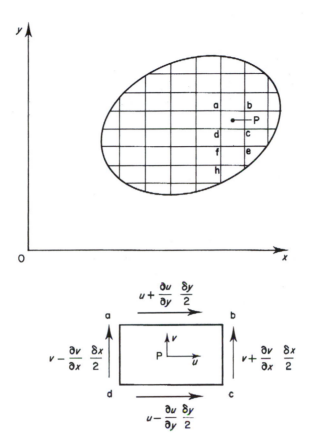

Fig. 4.5

If the strength of the circulation Γ remains constant whilst the circuit shrinks to encompass an ever smaller area, i.e. until it shrinks to an area the size of a rectangular element, then:

$$\Gamma = \zeta \times \delta x \, \delta y = \zeta \times \text{area of element}$$

Therefore,

$$\text{vorticity} = \lim_{\text{area} \to 0} \frac{\Gamma}{\text{area of circuit}} \tag{4.3}$$

Here the (potential) line vortex introduced in Section 3.3.2 will be re-visited and the definition (4.2) of circulation will now be applied to two particular circuits around a point (Fig. 4.6). One of these is a circle, of radius r_1, centred at the centre of the vortex. The second circuit is ABCD, composed of two circular arcs of radii r_1 and r_2 and two radial lines subtending the angle β at the centre of the vortex. For the concentric circuit, the velocity is constant at the value

$$q = \frac{C}{r_1}$$

where C is the constant value of qr.

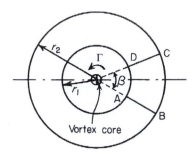

Fig. 4.6 Two circuits in the flow around a point vortex

Since the flow is, by the definition of a vortex, along the circle, α is everywhere zero and therefore $\cos\alpha = 1$. Then, from Eqn (4.2)

$$\Gamma = \oint \frac{C}{r_1}\,ds$$

Now suppose an angle θ to be measured in the anti-clockwise sense from some arbitrary axis, such as OAB. Then

$$ds = r_1 d\theta$$

whence

$$\Gamma = \int_0^{2\pi} \frac{C}{r_1} r_1 d\theta = 2\pi C \tag{4.4}$$

Since C is a constant, it follows that Γ is also a constant, independent of the radius. It can be shown that, provided the circuit encloses the centre of the vortex, the circulation round it is equal to Γ, whatever the shape of the circuit. The circulation Γ round a circuit enclosing the centre of a vortex is called the strength of the vortex. The dimensions of circulation and vortex strength are, from Eqn (4.2), velocity times length, i.e. $L^2 T^{-1}$, the units being $m^2\ s^{-1}$. Now $\Gamma = 2\pi C$, and C was defined as equal to qr; hence

$$\Gamma = 2\pi qr$$

and

$$q = \frac{\Gamma}{2\pi r} \tag{4.5}$$

Taking now the second circuit ABCD, the contribution towards the circulation from each part of the circuit is calculated as follows:

(i) *Radial line* **AB** Since the flow around a vortex is in concentric circles, the velocity vector is everywhere perpendicular to the radial line, i.e. $\alpha = 90°$, $\cos\alpha = 0$. Thus the tangential velocity component is zero along AB, and there is therefore no contribution to the circulation.
(ii) *Circular arc* **BC** Here $\alpha = 0$, $\cos\alpha = 1$. Therefore

$$\delta\Gamma = \int_{BC} q\cos\alpha\,ds = \int_0^\beta qr_2 d\theta$$

But, by Eqn (4.5),

$$q = \frac{\Gamma}{2\pi r_2}$$

$$\delta\Gamma = \int_0^\beta \frac{\Gamma}{2\pi r_2} r_2 \, d\theta = \frac{\beta\Gamma}{2\pi}$$

(iii) *Radial line* CD   As for AB, there is no contribution to the circulation from this part of the circuit.

(iv) *Circular arc* DA Here the path of integration is from D to A, while the direction of velocity is from A to D. Therefore $\alpha = 180°$, $\cos\alpha = -1$. Then

$$\delta\Gamma = \int_0^\beta \frac{\Gamma}{2\pi r_1} (-1) r_1 d\theta = -\frac{\beta\Gamma}{2\pi}$$

Therefore the total circulation round the complete circuit ABCD is

$$\Gamma = 0 + \frac{\beta\Gamma}{2\pi} + 0 - \frac{\beta\Gamma}{2\pi} = 0 \tag{4.6}$$

Thus the total circulation round this circuit, that does not enclose the core of the vortex, is zero. Now any circuit can be split into infinitely short circular arcs joined by infinitely short radial lines. Applying the above process to such a circuit would lead to the result that the circulation round a circuit of any shape that does not enclose the core of a vortex is zero. This is in accordance with the notion that potential flow is irrotational (see Section 3.1).

4.1.3   Circulation and lift (Kutta–Zhukovsky theorem)

In Eqn (3.52) it was shown that the lift l per unit span and the circulation Γ of a spinning circular cylinder are simply related by

$$l = \rho V \Gamma$$

where ρ is the fluid density and V is the speed of the flow approaching the cylinder. In fact, as demonstrated independently by Kutta* and Zhukovsky[†], the Russian physicist, at the beginning of the twentieth century, this result applies equally well to a cylinder of any shape and, in particular, applies to aerofoils. This powerful and useful result is accordingly usually known as the *Kutta–Zhukovsky Theorem*. Its validity is demonstrated below.

The lift on any aerofoil moving relative to a bulk of fluid can be derived by direct analysis. Consider the aerofoil in Fig. 4.7 generating a circulation of Γ when in a stream of velocity V, density ρ, and static pressure p_0. The lift produced by the aerofoil must be sustained by any boundary (imaginary or real) surrounding the aerofoil.

For a circuit of radius r, that is very large compared to the aerofoil, the lift of the aerofoil upwards must be equal to the sum of the pressure force on the whole periphery of the circuit and the reaction to the rate of change of downward momentum of the air through the periphery. At this distance the effects of the aerofoil thickness distribution may be ignored, and the aerofoil represented only by the circulation it generates.

---

* see footnote on page 161.

[†] N. Zhukovsky 'On the shape of the lifting surfaces of kites' (in German), *Z. Flugtech. Motorluftschiffahrt*, **1**, 281 (1910) and **3**, 81 (1912).

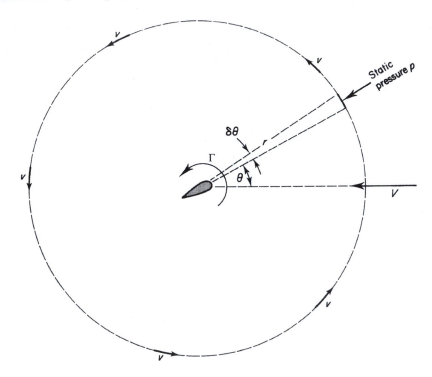

Fig. 4.7

The vertical static pressure force or buoyancy l_b on the circular boundary is the sum of the vertical pressure components acting on elements of the periphery. At the element subtending $\delta\theta$ at the centre of the aerofoil the static pressure is p and the local velocity is the resultant of V and the velocity v induced by the circulation. By Bernoulli's equation

$$p_0 + \frac{1}{2}\rho V^2 = p + \frac{1}{2}\rho[V^2 + v^2 + 2Vv\sin\theta]$$

giving

$$p = p_0 - \rho Vv\sin\theta$$

if v^2 may be neglected compared with V^2, which is permissible since r is large.

The vertical component of pressure force on this element is

$$-pr\sin\theta\,\delta\theta$$

and, on substituting for p and integrating, the contribution to lift due to the force acting on the boundary is

$$l_b = -\int_0^{2\pi}(p_0 - \rho Vv\sin\theta)r\sin\theta\,d\theta \qquad (4.7)$$
$$= +\rho Vvr\pi$$

with p_0 and r constant.

The mass flow through the elemental area of the boundary is given by $\rho V r \cos\theta \, \delta\theta$. This mass flow has a vertical velocity increase of $v \cos\theta$, and therefore the rate of change of downward momentum through the element is $-\rho V v r \cos^2\theta \, \delta\theta$; therefore by integrating round the boundary, the inertial contribution to the lift, l_i, is

$$l_i = +\int_0^{2\pi} \rho V v r \cos^2\theta \, d\theta$$
$$= \rho V v r \pi \tag{4.8}$$

Thus the total lift is:

$$l = 2\rho V v r \pi \tag{4.9}$$

From Eqn (4.5):

$$v = \frac{\Gamma}{2\pi r}$$

giving, finally, for the lift per unit span, l:

$$l = \rho V \Gamma \tag{4.10}$$

This expression can be obtained without consideration of the behaviour of air in a boundary circuit, by integrating pressures on the surface of the aerofoil directly.

It can be shown that this lift force is theoretically independent of the shape of the aerofoil section, the main effect of which is to produce a pitching moment in potential flow, plus a drag in the practical case of motion in a real viscous fluid.

4.2 The development of aerofoil theory

The first successful aerofoil theory was developed by Zhukovsky.* This was based on a very elegant mathematical concept – the conformal transformation – that exploits the theory of complex variables. Any two-dimensional potential flow can be represented by an analytical function of a complex variable. The basic idea behind Zhukovsky's theory is to take a circle in the complex $\zeta = (\xi + i\eta)$ plane (noting that here ζ does not denote vorticity) and map (or transform) it into an aerofoil-shaped contour. This is illustrated in Fig. 4.8.

A potential flow can be represented by a complex potential defined by $\Phi = \phi + i\psi$ where, as previously, ϕ and ψ are the velocity potential and stream function respectively. The same Zhukovsky mapping (or transformation), expressed mathematically as

$$z = \zeta + \frac{C^2}{\zeta}$$

(where C is a parameter), would then map the complex potential flow around the circle in the ζ-plane to the corresponding flow around the aerofoil in the z-plane. This makes it possible to use the results for the cylinder with circulation (see Section 3.3.10) to calculate the flow around an aerofoil. The magnitude of the circulation is chosen so as to satisfy the Kutta condition in the z-plane.

From a practical point of view Zhukovsky's theory suffered an important drawback. It only applied to a particular family of aerofoil shapes. Moreover, all the

* see footnote on page 161.

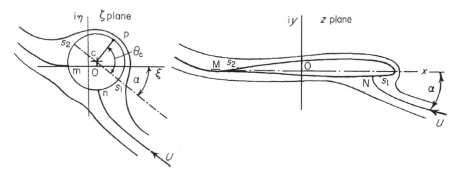

Fig. 4.8 Zhukovsky transformation, of the flow around a circular cylinder with circulation, to that around an aerofoil generating lift

members of this family of shapes have a cusped trailing edge whereas the aerofoils used in practical aerodynamics have trailing edges with finite angles. Kármán and Trefftz[*] later devised a more general conformal transformation that gave a family of aerofoils with trailing edges of finite angle. Aerofoil theory based on conformal transformation became a practical tool for aerodynamic design in 1931 when the American engineer Theodorsen[†] developed a method for aerofoils of arbitrary shape. The method has continued to be developed well into the second half of the twentieth century. Advanced versions of the method exploited modern computing techniques like the Fast Fourier Transform.[**]

If aerodynamic design were to involve only two-dimensional flows at low speeds, design methods based on conformal transformation would be a good choice. However, the technique cannot be extended to three-dimensional or high-speed flows. For this reason it is no longer widely used in aerodynamic design. Methods based on conformal transformation are not discussed further here. Instead two approaches, namely *thin aerofoil theory* and computational *boundary element (or panel) methods*, which can be extended to three-dimensional flows will be described.

The Zhukovsky theory was of little or no direct use in practical aerofoil design. Nevertheless it introduced some features that are basic to any aerofoil theory. Firstly, the overall lift is proportional to the circulation generated, and secondly, the magnitude of the circulation must be such as to keep the velocity finite at the trailing edge, in accordance with the Kutta condition.

It is not necessary to suppose the vorticity that gives rise to the circulation be due to a single vortex. Instead the vorticity can be distributed throughout the region enclosed by the aerofoil profile or even on the aerofoil surface. But the magnitude of circulation generated by all this vorticity must still be such as to satisfy the Kutta condition. A simple version of this concept is to concentrate the vortex distribution on the camber line as suggested by Fig. 4.9. In this form, it becomes the basis of the classic thin aerofoil theory developed by Munk[‡] and Glauert.[§]

Glauert's version of the theory was based on a sort of reverse Zhukovsky transformation that exploited the not unreasonable assumption that practical aerofoils are

[*] *Z. Flugtech. Motorluftschiffahrt*, **9**, 111 (1918).

[†] NACA Report, No. 411 (1931).

[**] N.D. Halsey (1979) Potential flow analysis of multi-element airfoils using conformal mapping, *AIAA J.*, **12**, 1281.

[‡] NACA Report, No. 142 (1922).

[§] Aeronautical Research Council, *Reports and Memoranda* No. 910 (1924).

Fig. 4.9

thin. He was thereby able to determine the aerofoil shape required for specified aerofoil characteristics. This made the theory a practical tool for aerodynamic design. However, as remarked above, the use of conformal transformation is restricted to two dimensions. Fortunately, it is not necessary to use Glauert's approach to obtain his final results. In Section 4.3, later developments are followed using a method that does not depend on conformal transformation in any way and, accordingly, in principle at least, can be extended to three dimensions.

Thin aerofoil theory and its applications are described in Sections 4.3 to 4.9. As the name suggests the method is restricted to thin aerofoils with small camber at small angles of attack. This is not a major drawback since most practical wings are fairly thin. A modern computational method that is not restricted to thin aerofoils is described in Section 4.10. This is based on the extension of the panel method of Section 3.5 to lifting flows. It was developed in the late 1950s and early 1960s by Hess and Smith at Douglas Aircraft Company.

4.3   The general thin aerofoil theory

For the development of this theory it is assumed that the maximum aerofoil thickness is small compared to the chord length. It is also assumed that the camber-line shape only deviates slightly from the chord line. A corollary of the second assumption is that the theory should be restricted to low angles of incidence.

Consider a typical cambered aerofoil as shown in Fig. 4.10. The upper and lower curves of the aerofoil profile are denoted by y_u and y_l respectively. Let the velocities in the x and y directions be denoted by u and v and write them in the form:

$$u = U\cos\alpha + u', \qquad v = U\sin\alpha + v'$$

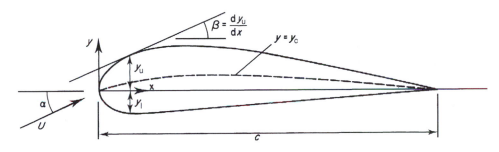

Fig. 4.10

u' and v' represent the departure of the local velocity from the undisturbed free stream, and are commonly known as the *disturbance* or *perturbation* velocities. In fact, thin-aerofoil theory is an example of a small perturbation theory.

The velocity component perpendicular to the aerofoil profile is zero. This constitutes the boundary condition for the potential flow and can be expressed mathematically as:

$$-u \sin \beta + v \cos \beta = 0 \quad \text{at} \quad y = y_u \quad \text{and} \quad y_l$$

Dividing both sides by $\cos \beta$, this boundary condition can be rewritten as

$$-(U \cos \alpha + u') \frac{dy}{dx} + U \sin \alpha + v' = 0 \quad \text{at} \quad y = y_u \quad \text{and} \quad y_l \tag{4.11}$$

By making the thin-aerofoil assumptions mentioned above, Eqn (4.11) may be simplified. Mathematically, these assumptions can be written in the form

$$y_u \text{ and } y_l \ll c; \ \alpha, \frac{dy_u}{dx} \text{ and } \frac{dy_l}{dx} \ll 1$$

Note that the additional assumption is made that the slope of the aerofoil profile is small. These thin-aerofoil assumptions imply that the disturbance velocities are small compared to the undisturbed free-steam speed, i.e.

$$u' \quad \text{and} \quad v' \ll U$$

Given the above assumptions Eqn (4.11) can be simplified by replacing $\cos \alpha$ and $\sin \alpha$ by 1 and α respectively. Furthermore, products of small quantities can be neglected, thereby allowing the term $u'dy/dx$ to be discarded so that Eqn (4.11) becomes

$$v' = U \frac{dy_u}{dx} - U\alpha \quad \text{and} \quad v' = U \frac{dy_l}{dx} - U\alpha \tag{4.12}$$

One further simplification can be made by recognizing that if y_u and $y_l \ll c$ then to a sufficiently good approximation the boundary conditions Eqn (4.12) can be applied at $y = 0$ rather than at $y = y_u$ or y_l.

Since potential flow with Eqn (4.12) as a boundary condition is a linear system, the flow around a cambered aerofoil at incidence can be regarded as the superposition of two separate flows, one circulatory and the other non-circulatory. This is illustrated in Fig. 4.11. The circulatory flow is that around an infinitely thin cambered plate and the non-circulatory flow is that around a symmetric aerofoil at zero incidence. This superposition can be demonstrated formally as follows. Let

$$y_u = y_c + y_t \quad \text{and} \quad y_l = y_c - y_t$$

$y = y_c(x)$ is the function describing the camber line and $y = y_t = (y_u - y_l)/2$ is known as the thickness function. Now Eqn (4.12) can be rewritten in the form

$$v' = U \underbrace{\frac{dy_c}{dx} - U\alpha}_{\text{Circulatory}} \pm \underbrace{U \frac{dy_t}{dx}}_{\text{Non-circulatory}}$$

where the plus sign applies for the upper surface and the minus sign for the lower surface.

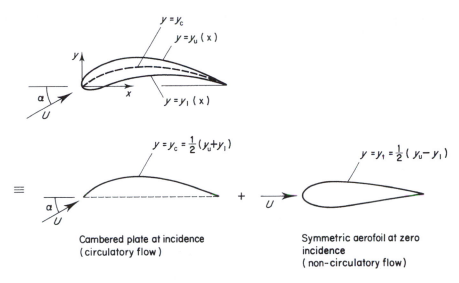

Fig. 4.11 Cambered thin aerofoil at incidence as superposition of a circulatory and non-circulatory flow

Thus the non-circulatory flow is given by the solution of potential flow subject to the boundary condition $v' = \pm U\, dy_t/dx$ which is applied at $y = 0$ for $0 \le x \le c$. The solution of this problem is discussed in Section 4.9. The lifting characteristics of the aerofoil are determined solely by the circulatory flow. Consequently, it is the solution of this problem that is of primary importance. Turn now to the formulation and solution of the mathematical problem for the circulatory flow.

It may be seen from Sections 4.1 and 4.2 that vortices can be used to represent lifting flow. In the present case, the lifting flow generated by an infinitely thin cambered plate at incidence is represented by a string of line vortices, each of infinitesimal strength, along the camber line as shown in Fig. 4.12. Thus the camber line is replaced by a line of variable vorticity so that the total circulation about the chord is the sum of the vortex elements. This can be written as

$$\Gamma = \int_0^c k\,\mathrm{d}s \qquad (4.13)$$

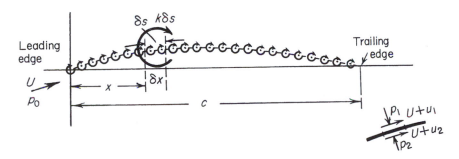

Fig. 4.12 Insert shows velocity and pressure above and below δs

where k is the distribution of vorticity over the element of camber line δs and circulation is taken as positive in the clockwise direction. The problem now becomes one of determining the function $k(x)$ such that the boundary condition

$$v' = U\frac{dy_c}{dx} - U\alpha \quad \text{at} \quad y = 0, \quad 0 \leq x \leq 1 \tag{4.14}$$

is satisfied as well as the Kutta condition (see Section 4.1.1).

There should be no difficulty in accepting this idealized concept. A lifting wing may be replaced by, and produces forces and disturbances identical to, a vortex system, and Chapter 5 presents the classical theory of finite wings in which the idea of a bound vortex system is fully exploited. A wing replaced by a sheet of spanwise vortex elements (Fig. 5.21), say, will have a section that is essentially that of the replaced camber line above.

The leading edge is taken as the origin of a pair of coordinate axes x and y; Ox along the chord, and Oy normal to it. The basic assumptions of the theory permit the variation of vorticity along the camber line to be assumed the same as the variation along the Ox axis, i.e. δs differs negligibly from δx, so that Eqn (4.13) becomes

$$\Gamma = \int_0^c k\,dx \tag{4.15}$$

Hence from Eqn (4.10) for unit span of this section the lift is given by

$$l = \rho U\Gamma = \rho U\int_0^c k\,dx \tag{4.16}$$

Alternatively Eqn (4.16) could be written with $\rho U k = p$:

$$l = \int_0^c \rho U k\,dx = \int_0^c p\,dx \tag{4.17}$$

Now considering unit spanwise length, p has the dimensions of force per unit area or pressure and the moment of these chordwise pressure forces about the leading edge or origin of the system is simply

$$M_{\text{LE}} = -\int_0^c px\,dx = -\rho U\int_0^c kx\,dx \tag{4.18}$$

Note that pitching 'nose up' is positive.

The thin wing section has thus been replaced for analytical purposes by a line discontinuity in the flow in the form of a vorticity distribution. This gives rise to an overall circulation, as does the aerofoil, and produces a chordwise pressure variation.

For the aerofoil in a flow of undisturbed velocity U and pressure p_0, the insert to Fig. 4.12 shows the static pressures p_1 and p_2 above and below the element δs where the local velocities are $U + u_1$ and $U + u_2$, respectively. The overall pressure difference p is $p_2 - p_1$. By Bernoulli:

$$p_1 + \frac{1}{2}\rho(U + u_1)^2 = p_0 + \frac{1}{2}\rho U^2$$

$$p_2 + \frac{1}{2}\rho(U + u_2)^2 = p_0 + \frac{1}{2}\rho U^2$$

and subtracting

$$p_2 - p_1 = \frac{1}{2}\rho U^2 \left[2\left(\frac{u_1}{U} - \frac{u_2}{U}\right) + \left(\frac{u_1}{U}\right)^2 - \left(\frac{u_2}{U}\right)^2 \right]$$

and with the aerofoil thin and at small incidence the perturbation velocity ratios u_1/U and u_2/U will be so small compared with unity that $(u_1/U)^2$ and $(u_2/U)^2$ are neglected compared with u_1/U and u_2/U, respectively. Then

$$p = p_2 - p_1 = \rho U(u_1 - u_2) \tag{4.19}$$

The equivalent vorticity distribution indicates that the circulation due to element δs is $k\,\delta x$ (δx because the camber line deviates only slightly from the Ox axis). Evaluating the circulation around δs and taking clockwise as positive in this case, by taking the algebraic sum of the flow of fluid along the top and bottom of δs, gives

$$k\,\delta x = +(U + u_1)\delta x - (U + u_2)\delta x = (u_1 - u_2)\delta x \tag{4.20}$$

Comparing (4.19) and (4.20) shows that $p = \rho U k$ as introduced in Eqn (4.17).

For a trailing edge angle of zero the Kutta condition (see Section 4.1.1) requires $u_1 = u_2$ at the trailing edge. It follows from Eqn (4.20) that the Kutta condition is satisfied if

$$k = 0 \quad \text{at} \quad x = c \tag{4.21}$$

The induced velocity v in Eqn (4.14) can be expressed in terms of k, by considering the effect of the elementary circulation $k\,\delta x$ at x, a distance $x - x_1$ from the point considered (Fig. 4.13). Circulation $k\,\delta x$ induces a velocity at the point x_1 equal to

$$\frac{1}{2\pi} \frac{k\,\delta x}{x - x_1}$$

from Eqn (4.5).

The effect of all such elements of circulation along the chord is the induced velocity v' where

$$v' = \frac{1}{2\pi}\int_0^c \frac{k\,\mathrm{d}x}{x - x_1}$$

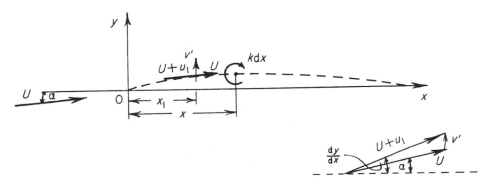

Fig. 4.13 Velocities at x_1 from 0: $U + u_1$, resultant tangential to camber lines; v', induced by chordwise variation in circulation; U, free stream velocity inclined at angle α to Ox

and introducing this in Eqn (4.14) gives

$$U\left[\frac{dy_c}{dx} - \alpha\right] = \frac{1}{2\pi} \int_0^c \frac{k \, dx}{x - x_1} \tag{4.22}$$

The solution for $k \, dx$ that satisfies Eqn (4.22) for a given shape of camber line (defining dy_c/dx) and incidence can be introduced in Eqns (4.17) and (4.18) to obtain the lift and moment for the aerofoil shape. The characteristics C_L and $C_{M_{LE}}$ follow directly and hence k_{CP}, the centre of pressure coefficient, and the angle for zero lift.

4.4 The solution of the general equation

In the general case Eqn (4.22) must be solved directly to determine the function $k(x)$ that corresponds to a specified camber-line shape. Alternatively, the inverse design problem may be solved whereby the pressure distribution or, equivalently, the tangential velocity variation along the upper and lower surfaces of the aerofoil is given. The corresponding $k(x)$ may then be simply found from Eqns (4.19) and (4.20). The problem then becomes one of finding the requisite camber line shape from Eqn (4.22). The present approach is to work up to the general case through the simple case of the flat plate at incidence, and then to consider some practical applications of the general case. To this end the integral in Eqn (4.22) will be considered and expressions for some useful definite integrals given.

In order to use certain trigonometric relationships it is convenient to change variables from x to θ, through $x = (c/2)(1 - \cos\theta)$, and x_1 to θ_1, then the limits change as follows:

$$\theta \sim 0 \to \pi \quad \text{as} \quad x \sim 0 \to c, \text{ and}$$

$$dx = \frac{c}{2} \sin\theta \, d\theta$$

So

$$\int_0^c \frac{k \, dx}{x - x_1} = -\int_0^\pi \frac{k \sin\theta \, d\theta}{(\cos\theta - \cos\theta_1)} \tag{4.23}$$

Also the Kutta condition (4.21) becomes

$$k = 0 \quad \text{at} \quad \theta = \pi \tag{4.24}$$

The expressions found by evaluating two useful definite integrals are given below

$$\int_0^\pi \frac{\cos n\theta}{(\cos\theta - \cos\theta_1)} \, d\theta = \pi \frac{\sin n\theta_1}{\sin\theta_1} : n = 0, 1, 2, \ldots \tag{4.25}$$

$$\int_0^\pi \frac{\sin n\theta \sin\theta}{(\cos\theta - \cos\theta_1)} \, d\theta = -\pi \cos n\theta_1 : n = 0, 1, 2, \ldots \tag{4.26}$$

The derivations of these results are given in Appendix 3. However, it is not necessary to be familiar with this derivation in order to use Eqns (4.25) and (4.26) in applications of the thin-aerofoil theory.

4.4.1 The thin symmetrical flat plate aerofoil

In this simple case the camber line is straight along Ox, and $dy_c/dx = 0$. Using Eqn (4.23) the general equation (4.22) becomes

$$U_\alpha = \frac{1}{2\pi} \int_0^\pi \frac{k \sin \theta}{(\cos \theta - \cos \theta_1)} \qquad (4.27)$$

What value should k take on the right-hand side of Eqn (4.27) to give a left-hand side which does not vary with x or, equivalently, θ? To answer this question consider the result (4.25) with $n = 1$. From this it can be seen that

$$\int_0^\pi \frac{\cos \theta \, d\theta}{(\cos \theta - \cos \theta_1)} = \pi$$

Comparing this result with Eqn (4.27) it can be seen that if $k = k_1 = 2U\alpha \cos \theta / \sin \theta$ it will satisfy Eqn (4.27). The only problem is that far from satisfying the Kutta condition (4.24) this solution goes to infinity at the trailing edge. To overcome this problem it is necessary to recognize that if there exists a function k_2 such that

$$\int_0^\pi \frac{k_2 \sin \theta \, d\theta}{(\cos \theta - \cos \theta_1)} = 0 \qquad (4.28)$$

then $k = k_1 + k_2$ will also satisfy Eqn (4.27).

Consider Eqn (4.25) with $n = 0$ so that

$$\int_0^\pi \frac{1}{(\cos \theta - \cos \theta_1)} \, d\theta = 0$$

Comparing this result to Eqn (4.28) shows that the solution is

$$k_2 = \frac{C}{\sin \theta}$$

where C is an arbitrary constant.

Thus the complete (or general) solution for the flat plate is given by

$$k = k_1 + k_2 = \frac{2U\alpha \cos \theta + C}{\sin \theta}$$

The Kutta condition (4.24) will be satisfied if $C = 2U\alpha$ giving a final solution of

$$k = 2U\alpha \frac{(1 + \cos \theta)}{\sin \theta} \qquad (4.29)$$

Aerodynamic coefficients for a flat plate

The expression for k can now be put in the appropriate equations for lift and moment by using the pressure:

$$p = \rho U k = 2\rho U^2 \alpha \frac{1 + \cos \theta}{\sin \theta} \qquad (4.30)$$

The lift per unit span

$$l = \rho U \int_0^\pi 2U\alpha \left(\frac{1+\cos\theta}{\sin\theta}\right) \frac{c}{2} \sin\theta \, d\theta$$

$$= \alpha\rho U^2 c \int_0^\pi (1+\cos\theta) \, d\theta = \pi\alpha\rho U^2 c$$

It therefore follows that for unit span

$$C_L = \frac{l}{\left(\frac{1}{2}\rho U_c^2\right)} = 2\pi\alpha \tag{4.31}$$

The moment about the leading edge per unit span

$$M_{LE} = -\int_0^c px \, dx$$

Changing the sign

$$-M_{LE} = 2\rho U^2 \alpha \int_0^\pi \frac{(1+\cos\theta)}{\sin\theta} \frac{c}{2}(1-\cos\theta)\frac{c}{2}\sin\theta \, d\theta$$

$$= \frac{1}{2}\rho U^2 \alpha c^2 \int_0^\pi (1-\cos^2\theta) \, d\theta$$

Therefore for unit span

$$-C_{M_{LE}} \equiv -\frac{M_{LE}}{\frac{1}{2}\rho U^2 c^2} = \alpha \int_0^\pi \left(\frac{1}{2} - \frac{\cos 2\theta}{2}\right) d\theta$$

$$= \alpha\frac{\pi}{2} \tag{4.32}$$

Comparing Eqns (4.31) and (4.32) shows that

$$C_{M_{LE}} = -\frac{C_L}{4} \tag{4.33}$$

The centre of pressure coefficient k_{CP} is given for small angles of incidence approximately by

$$k_{CP} = \frac{-C_{M_{LE}}}{C_L} = \frac{1}{4} \tag{4.34}$$

and this shows a fixed centre of pressure coincident with the aerodynamic centre as is necessarily true for any symmetrical section.

4.4.2   The general thin aerofoil section

In general, the camber line can be any function of x (or θ) provided that $y_c = 0$ at $x = 0$ and c (i.e. at $\theta = 0$ and π). When trigonometric functions are involved a convenient way to express an arbitrary function is to use a Fourier series. Accordingly, the slope of the camber line appearing in Eqn (4.22) can be expressed in terms of a Fourier cosine series

$$\frac{dy_c}{dx} = A_0 + \sum_{n=1}^{\infty} A_n \cos n\theta \tag{4.35}$$

Sine terms are not used here because practical camber lines must go to zero at the leading and trailing edges. Thus y_c is an odd function which implies that its derivative is an even function.

Equation (4.22) now becomes

$$U(\alpha - A_0) - U\sum_{n=1}^{\infty} A_n \cos n\theta = \frac{1}{2\pi}\int_0^{\pi} \frac{k\sin\theta\,\mathrm{d}\theta}{(\cos\theta - \cos\theta_1)} \tag{4.36}$$

The solution for k as a function of θ can be considered as comprising three parts so that $k = k_1 + k_2 + k_3$ where

$$\frac{1}{2\pi}\int_0^{\pi} k_1(\theta)\frac{\sin\theta}{(\cos\theta - \cos\theta_1)}\,\mathrm{d}\theta = U(\alpha - A_0) \tag{4.37}$$

$$\frac{1}{2\pi}\int_0^{\pi} k_2(\theta)\frac{\sin\theta}{(\cos\theta - \cos\theta_1)}\,\mathrm{d}\theta = 0 \tag{4.38}$$

$$\frac{1}{2\pi}\int_0^{\pi} k_3(\theta)\frac{\sin\theta}{(\cos\theta - \cos\theta_1)}\,\mathrm{d}\theta = -U\sum_{n=1}^{\infty} A_n \cos n\theta \tag{4.39}$$

The solutions for k_1 and k_2 are identical to those given in Section 4.4.1 except that $U(\alpha - A_0)$ replaces $U\alpha$ in the case of k_1. Thus it is only necessary to solve Eqn (4.39) for k_3. By comparing Eqn (4.26) with Eqn (4.39) it can be seen that the solution to Eqn (4.39) is given by

$$k_3(\theta) = 2U\sum_{n=1}^{\infty} A_n \sin n\theta$$

Thus the complete solution is given by

$$k(\theta) = k_1 + k_2 + k_3 = 2U(\alpha - A_0)\frac{\cos\theta}{\sin\theta} + \frac{C}{\sin\theta} + 2U\sum_{n=1}^{\infty} A_n \sin n\theta$$

The constant C has to be chosen so as to satisfy the Kutta condition (4.24) which gives $C = 2U(\alpha - A_0)$. Thus the final solution is

$$k(\theta) = 2U\left[(\alpha - A_0)\frac{\cos\theta + 1}{\sin\theta} + \sum_{n=1}^{\infty} A_n \sin n\theta\right] \tag{4.40}$$

To obtain the coefficients A_0 and A_n in terms of the camberline slope, the usual procedures for Fourier series are followed. On integrating both sides of Eqn (4.35) with respect to θ, the second term on the right-hand side vanishes leaving

$$\int_0^{\pi} \frac{\mathrm{d}y_c}{\mathrm{d}x}\,\mathrm{d}\theta = \int_0^{\pi} A_0\,\mathrm{d}\theta = A_0\pi$$

Therefore

$$A_0 = \frac{1}{\pi} \int_0^\pi \frac{dy_c}{dx} d\theta \qquad (4.41)$$

Multiplying both sides of Eqn (4.35) by $\cos m\theta$, where m is an integer, and integrating with respect to θ

$$\int_0^\pi \frac{dy_c}{dx} \cos m\theta \, d\theta = \int_0^\pi \cos m\theta \, d\theta + \int_0^\pi \Sigma A_n \cos n\theta \cos m\theta \, d\theta$$

$$\int_0^\pi A_n \cos n\theta \cos m\theta \, d\theta = 0 \quad \text{except when} \quad n = m$$

Then the first term on the right-hand side vanishes, and also the second term, except for $n = m$, i.e.

$$\int_0^\pi \frac{dy_c}{dx} \cos n\theta \, d\theta = \int_0^\pi A_n \cos^2 n\theta \, d\theta = \frac{\pi}{2} A_n$$

whence

$$A_n = \frac{2}{\pi} \int_0^\pi \frac{dy_c}{dx} \cos n\theta \, d\theta \qquad (4.42)$$

Lift and moment coefficients for a general thin aerofoil

From Eqn (4.7)

$$l = \int_0^c \rho U k \, dx = \int_0^\pi \rho U \frac{c}{2} k \sin \theta \, d\theta$$

$$= 2\rho U^2 \frac{c}{2} \int_0^\pi \left[(\alpha - A_0)(1 + \cos \theta) + \sum_1^\infty A_n \sin n\theta \sin \theta \right] d\theta$$

$$= 2\rho U^2 \frac{c}{2} \left[\pi(\alpha - A_0) + \frac{\pi}{2} A_1 \right] = C_L \frac{1}{2} \rho U^2 c$$

Since

$$\int_0^\pi \sin n\theta \, d\theta = 0 \quad \text{when} \quad n \neq 1, \text{giving}$$

$$C_L = (C_{L_0}) + \frac{dC_L}{d\alpha} \alpha = \pi(A_1 - 2A_0) + 2\pi\alpha \qquad (4.43)$$

The first term on the right-hand side of Eqn (4.43) is the coefficient of lift at zero incidence. It contains the effects of camber and is zero for a symmetrical aerofoil. It is also worth noting that, according to general thin aerofoil theory, the lift curve slope takes the same value 2π for all aerofoils.

$$-M_{\text{LE}} = \rho U \int_0^c kx \, dx = -C_{M_{\text{LE}}} \frac{1}{2} \rho U^2 c^2$$

With the usual substitution

$$-C_{M_{LE}} = \frac{2\rho U^2 (c/2)^2}{\frac{1}{2}\rho U^2 c^2}$$

$$\times \int_0^\pi \left[(\alpha - A_0) \frac{(1 + \cos\theta)}{\sin\theta} + \sum_1^\infty A_n \sin n\theta \right] \sin\theta (1 - \cos\theta) \, d\theta$$

$$= \int_0^\pi (\alpha - A_0)(1 - \cos^2\theta) d\theta + \int_0^\pi \sum_1^\infty A_n \sin n\theta \sin\theta \, d\theta$$

$$- \int_0^\pi \sum_1^\infty A_n \sin n\theta \cos\theta \sin\theta \, d\theta$$

$$= \frac{\pi}{2}(\alpha - A_0) + \frac{\pi}{2} A_1 - \frac{\pi}{4} A_2$$

since

$$\int_0^\pi \sin n\theta \sin m\theta \, d\theta = 0 \quad \text{when} \quad n \neq m, \text{or}$$

$$C_{M_{LE}} = -\frac{\pi}{2} \left[(\alpha - A_0) + A_1 - \frac{A_2}{2} \right] \tag{4.44}$$

In terms of the lift coefficient, $C_{M_{LE}}$ becomes

$$C_{M_{LE}} = -\frac{C_L}{4} \left[1 + \frac{A_1 - A_2}{C_L/\pi} \right]$$

Then the centre of pressure coefficient is

$$k_{CP} = -\frac{C_{M_{LE}}}{C_L} = \frac{1}{4} + \frac{\pi}{4 C_L} (A_1 - A_2) \tag{4.45}$$

and again the centre of pressure moves as the lift or incidence is changed. Now, from Section 1.5.4,

$$k_{CP} = -\frac{C_{M_{1/4}}}{C_L} + \frac{1}{4} \tag{4.46}$$

and comparing Eqns (4.44) and (4.45) gives

$$-C_{M_{1/4}} = \frac{\pi}{4} (A_1 - A_2) \tag{4.47}$$

This shows that, theoretically, the pitching moment about the quarter chord point for a thin aerofoil is constant, depending on the camber parameters only, and the quarter chord point is therefore the aerodynamic centre.

It is apparent from this analysis that no matter what the camber-line shape, only the first three terms of the cosine series describing the camber-line slope have any influence on the usual aerodynamic characteristics. This is indeed the case, but the terms corresponding to $n > 2$ contribute to the pressure distribution over the chord.

Owing to the quality of the basic approximations used in the theory it is found that the theoretical chordwise pressure distribution p does not agree closely with experimental data, especially near the leading edge and near stagnation points where the small perturbation theory, for example, breaks down. Any local inaccuracies tend to vanish in the overall integration processes, however, and the aerofoil coefficients are found to be reliable theoretical predictions.

4.5 The flapped aerofoil

Thin aerofoil theory lends itself very readily to aerofoils with variable camber such as flapped aerofoils. The distribution of circulation along the camber line for the general aerofoil has been found to consist of the sum of a component due to a flat plate at incidence and a component due to the camber-line shape. It is sufficient for the assumptions in the theory to consider the influence of a flap deflection as an addition to the two components above. Figure 4.14 shows how the three contributions can be combined. In fact the deflection of the flap about a hinge in the camber line effectively alters the camber so that the contribution due to flap deflection is the effect of an additional camber-line shape.

The problem is thus reduced to the general case of finding a distribution to fit a camber line made up of the chord of the aerofoil and the flap chord deflected through η (see Fig. 4.15). The thin aerofoil theory does not require that the leading and/or trailing edges be on the x axis, only that the surface slope is small and the displacement from the x axis is small.

With the camber defined as hc the slope of the part AB of the aerofoil is zero, and that of the flap – h/F. To find the coefficients of k for the flap camber, substitute these values of slope in Eqns (4.41) and (4.42) but with the limits of integration confined to the parts of the aerofoil over which the slopes occur. Thus

$$A_0 = \left\{ \frac{1}{\pi} \int_0^\pi 0 \, d\theta + \frac{1}{\pi} \int_\phi^\pi -\frac{h}{F} d\theta \right\} \tag{4.48}$$

where ϕ is the value of θ at the hinge, i.e.

$$(1 - F)c = \frac{c}{2}(1 - \cos \phi)$$

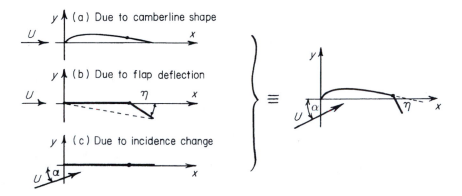

Fig. 4.14 Subdivision of lift contributions to total lift of cambered flapped aerofoil

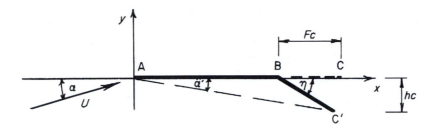

Fig. 4.15

whence $\cos \phi = 2F - 1$. Evaluating the integral

$$A_0 = -\left(1 - \frac{\phi}{\pi}\right)\frac{h}{F}$$

i.e. since all angles are small $h/F = \tan \eta \simeq \eta$, so

$$A_0 = -\left(1 - \frac{\phi}{\pi}\right)\eta \qquad (4.49)$$

Similarly from Eqn (4.42)

$$A_n = \frac{2}{\pi}\left\{\int_0^{\phi} 0 \cos n\theta \, d\theta + \int_{\phi}^{\pi} -\frac{h}{F}\cos n\theta \, d\theta\right\}$$

$$= \frac{2\sin n\phi}{n\pi}\eta \qquad (4.50)$$

Thus

$$A_1 = \frac{2\sin\phi}{\pi}\eta \qquad \text{and} \qquad A_2 = \frac{\sin 2\phi}{\pi}\eta$$

The distribution of chordwise circulation due to flap deflection becomes

$$k = 2U\alpha\frac{1+\cos\theta}{\sin\theta} + 2U\left[\left(1 - \frac{\phi}{\pi}\right)\frac{1+\cos\theta}{\sin\theta} + \sum_1^{\infty}\frac{2\sin n\phi}{n\pi}\sin n\theta\right]\eta \qquad (4.51)$$

and this for a constant incidence α is a linear function of η, as is the lift coefficient, e.g. from Eqn (4.43)

$$C_L = 2\pi\alpha + 2\pi\eta\left(1 - \frac{\phi}{\pi}\right) + 2\eta\sin\phi$$

giving

$$C_L = 2\pi\alpha + 2(\pi - \phi + \sin\phi)\eta \qquad (4.52)$$

Likewise the moment coefficient $C_{M_{LE}}$ from Eqn (4.44) is

$$-C_{M_{LE}} = \frac{\pi}{2}\alpha + \frac{\pi}{2}\left[\eta\left(1 - \frac{\phi}{\pi}\right) + \frac{2\sin\phi}{\pi}\eta - \frac{\sin 2\phi}{2\pi}\eta\right]$$

$$C_{M_{LE}} = -\frac{\pi}{2}\alpha - \frac{1}{2}[\pi - \phi + \sin\phi(2 - \cos\phi)]\eta \qquad (4.53)$$

Note that a positive flap deflection, i.e. a downwards deflection, decreases the moment coefficient, tending to pitch the main aerofoil nose down and vice versa.

4.5.1   The hinge moment coefficient

A flapped-aerofoil characteristic that is of great importance in stability and control calculations, is the aerodynamic moment about the hinge line, shown as H in Fig. 4.16.

Taking moments of elementary pressures p, acting on the flap about the hinge,

$$H = -\int_{hinge}^{trailing\ edge} px'\,\mathrm{d}x$$

where $p = \rho Uk$ and $x' = x - (1 - F)c$. Putting

$$x' = \frac{c}{2}(1 - \cos\theta) - \frac{c}{2}(1 - \cos\phi) = \frac{c}{2}(\cos\phi - \cos\theta)$$

and k from Eqn (4.51):

$$H = -\int_{\phi}^{\pi} 2\rho U^2 \left[\left\{\alpha + \eta\left(1 - \frac{\phi}{\pi}\right)\right\} \frac{(1 + \cos\theta)}{\sin\theta}\right.$$
$$\left. + \eta\sum_{1}^{\infty} \frac{2\sin n\phi}{n\pi}\sin n\theta\right] \frac{c}{2}(\cos\phi - \cos\theta)\frac{c}{2}\sin\theta\,\mathrm{d}\theta$$

Substituting $H = C_H \frac{1}{2}\rho U^2 (Fc)^2$ and cancelling

$$-C_H F^2 = \alpha \int_{\phi}^{\pi} (1 + \cos\theta)(\cos\phi - \cos\theta)\mathrm{d}\theta$$
$$+ \eta\left\{\left(1 - \frac{\theta}{\pi}\right)\cos\phi I_1 - \left(1 - \frac{\phi}{\pi}\right)I_2\right.$$
$$\left. + \sum_{1}^{\infty}\frac{2\sin n\phi}{n\pi}\cos\phi I_3 + \sum_{1}^{\infty}\frac{2\sin n\phi}{n\pi}I_4\right\} \tag{4.54}$$

where

$$I_1 = \int_{\phi}^{\pi}(1 + \cos\theta)\,\mathrm{d}\theta = \pi - \phi - \sin\phi$$

$$I_2 = \int_{\phi}^{\pi}(1 + \cos\theta)\cos\theta\,\mathrm{d}\theta = \left[\frac{\pi - \phi}{2}\sin\phi - \frac{\sin 2\phi}{4}\right]$$

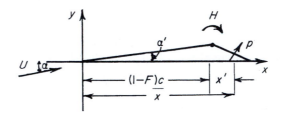

Fig. 4.16

$$I_3 = \int_\phi^\pi \sin n\theta \sin\theta \, d\theta = \frac{1}{2}\left[\frac{\sin(n+1)\phi}{n+1} - \frac{\sin(n-1)\phi}{n-1}\right]$$

$$I_4 = \int_\phi^\pi \sin n\theta \sin\theta \cos\theta \, d\theta = \frac{1}{2}\left[\frac{\sin(n+2)\phi}{n+2} - \frac{\sin(n-2)\phi}{n-2}\right]$$

In the usual notation $C_H = b_1\alpha + b_2\eta$, where

$$b_1 = \frac{\partial C_H}{\partial\alpha} \quad \text{and} \quad b_2 = \frac{\partial C_H}{\partial\eta}$$

From Eqn (4.54):

$$b_1 = -\frac{1}{F^2}\int_\phi^\pi (1+\cos\theta)(\cos\phi - \cos\theta)d\theta$$

giving

$$b_1 = -\frac{1}{4F^2}\{2(\pi-\phi)(2\cos\phi - 1) + 4\sin\phi - \sin 2\phi\} \tag{4.55}$$

Similarly from Eqn (4.54)

$$b_2 = \frac{\partial C_H}{\partial\eta} = \frac{1}{F^2} \times \text{coefficient of } \eta \text{ in Eqn (4.54)}$$

This somewhat unwieldy expression reduces to*

$$b_2 = -\frac{1}{4\pi F^2}\{(1-\cos 2\phi) - 2(\pi-\phi)^2(1-2\cos\phi) + 4(\pi-\phi)\sin\phi\} \tag{4.56}$$

The parameter $a_1 = \partial C_L/\partial\alpha$ is 2π and $a_2 = \partial C_L/\partial\eta$ from Eqn (4.52) becomes

$$a_2 = 2(\pi - \phi + \sin\phi) \tag{4.57}$$

Thus thin aerofoil theory provides an estimate of all the parameters of a flapped aerofoil.

Note that aspect-ratio corrections have not been included in this analysis which is essentially two-dimensional. Following the conclusions of the finite wing theory in Chapter 5, the parameters a_1, a_2, b_1 and b_2 may be suitably corrected for end effects. In practice, however, they are always determined from computational studies and wind-tunnel tests and confirmed by flight tests.

4.6 The jet flap

Considering the jet flap (see also Section 8.4.2) as a high-velocity sheet of air issuing from the trailing edge of an aerofoil at some downward angle τ to the chord line of the aerofoil, an analysis can be made by replacing the jet stream as well as the aerofoil by a vortex distribution.[†]

---

* See *R and M*, No. 1095, for the complete analysis.
[†] D.A. Spence, The lift coefficient of a thin, jet flapped wing, *Proc. Roy. Soc. A.*, No. 1212, Dec. 1956. D.A. Spence, The lift of a thin aerofoil with jet augmented flap, *Aeronautical Quarterly*, Aug. 1958.

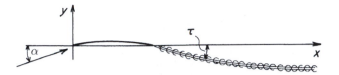

Fig. 4.17

The flap contributes to the lift on two accounts. Firstly, the downward deflection of the jet efflux produces a lifting component of reaction and secondly, the jet affects the pressure distribution on the aerofoil in a similar manner to that obtained by an addition to the circulation round the aerofoil.

The jet is shown to be equivalent to a band of spanwise vortex filaments which for small deflection angles τ can be assumed to lie along the Ox axis (Fig. 4.17). In the analysis, which is not proceeded with here, both components of lift are considered in order to arrive at the expression for C_L:

$$C_L = 4\pi A_0 \tau + 2\pi(1 + 2B_0)\alpha \tag{4.58}$$

where A_0 and B_0 are the initial coefficients in the Fourier series associated with the deflection of the jet and the incidence of the aerofoil respectively and which can be obtained in terms of the momentum (coefficient) of the jet.

It is interesting to notice in the experimental work on jet flaps at National Gas Turbine Establishment, Pyestock, good agreement was obtained with the theoretical C_L even at large values of τ.

4.7 The normal force and pitching moment derivatives due to pitching*

4.7.1 $(Z_q)(M_q)$ wing contributions

Thin-aerofoil theory can be used as a convenient basis for the estimation of these important derivatives. Although the use of these derivatives is beyond the general scope of this volume, no text on thin-aerofoil theory is complete without some reference to this common use of the theory.

When an aeroplane is rotating with pitch velocity q about an axis through the centre of gravity (CG) normal to the plane of symmetry on the chord line produced (see Fig. 4.18), the aerofoil's effective incidence is changing with time as also, as a consequence, are the aerodynamic forces and moments.

The rates of change of these forces and moments with respect to the pitching velocity q are two of the aerodynamic quasi-static derivatives that are in general commonly abbreviated to derivatives. Here the rate of change of normal force on the aircraft, i.e. resultant force in the normal or Z direction, with respect to pitching velocity is, in the conventional notation, $\partial Z/\partial q$. This is symbolized by Z_q. Similarly the rate of change of M with respect to q is $\partial M/\partial q = M_q$.

In common with other aerodynamic forces and moments these are reduced to non-dimensional or coefficient form by dividing through in this case by $\rho V l_t$ and $\rho V l_t^2$ respectively, where l_t is the tail plane moment arm, to give the non-dimensional

---

* It is suggested that this section be omitted from general study until the reader is familiar with these derivatives and their use.

Fig. 4.18

normal force derivative due to pitching z_q, and the non-dimensional pitching moment derivative due to pitching m_q.

   The contributions to these two, due to the mainplanes, can be considered by replacing the wing by the equivalent thin aerofoil. In Fig. 4.19, the centre of rotation (CG) is a distance hc behind the leading edge where c is the chord. At some point x from the leading edge of the aerofoil the velocity induced by the rotation of the aerofoil about the CG is $v' = -q(hc - x)$. Owing to the vorticity replacing the camber line a velocity v is induced. The incident flow velocity is V inclined at α to the chord line, and from the condition that the local velocity at x must be tangential to the aerofoil (camber line) (see Section 4.3) Eqn (4.14) becomes for this case

$$V\left(\frac{\mathrm{d}y}{\mathrm{d}x} - \alpha\right) = v - v'$$

or

$$\frac{\mathrm{d}y}{\mathrm{d}x} - \alpha = \frac{v}{V} - \frac{q}{V}(hc - x) \tag{4.59}$$

and with the substitution $x = \dfrac{c}{2}(1 - \cos\theta)$

$$\frac{\mathrm{d}y}{\mathrm{d}x} - \alpha = \frac{v}{V} - \frac{qc}{V}\left(h - \frac{1}{2} + \frac{\cos\theta}{2}\right)$$

From the general case in steady straight flight, Eqn (4.35), gives

$$\frac{\mathrm{d}y}{\mathrm{d}x} - \alpha = A_0 - \alpha + \Sigma A_n \cos n\theta \tag{4.60}$$

but in the pitching case the loading distribution would be altered to some general form given by, say,

$$\frac{v}{V} = B_0 + \Sigma B_n \, \cos n\theta \tag{4.61}$$

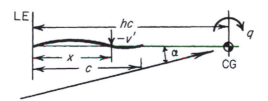

Fig. 4.19

where the coefficients are changed because of the relative flow changes, while the camber-line shape remains constant, i.e. the form of the function remains the same but the coefficients change. Thus in the pitching case

$$\frac{dy}{dx} - \alpha = B_0 + \Sigma B_n \cos n\theta - \frac{qc}{V}\left(h - \frac{1}{2} + \frac{\cos\theta}{2}\right) \tag{4.62}$$

Equations (4.60) and (4.62) give:

$$B_0 = A_0 - \alpha - \frac{qc}{V}\left(\frac{1}{2} - h\right), \qquad B_1 = A_1 + \frac{qc}{2V} \qquad \text{and} \qquad B_n = A_n$$

In analogy to the derivation of Eqn (4.40), the vorticity distribution here can be written

$$k = 2V\left[-B_0\left(\frac{1 + \cos\theta}{\sin\theta}\right) + \Sigma B_n \sin n\theta\right]$$

and following similar steps for those of the derivation of Eqn (4.43), this leads to

$$C_L = 2\pi(\alpha - B_0) + \pi B_1 = 2\pi\left[\alpha - A_0 + \frac{A_1}{2} + \left(\frac{3}{4} - h\right)\frac{qc}{V}\right] \tag{4.63}$$

It should be remembered that this is for a two-dimensional wing. However, the effect of the curvature of the trailing vortex sheet is negligible in three dimensions, so it remains to replace the ideal $\partial C_L/\partial\alpha = 2\pi$ by a reasonable value, a, that accounts for the aspect ratio change (see Chapter 5). The lift coefficient of a pitching rectangular wing then becomes

$$C_L = a\left[\alpha - A_0 + \frac{A_1}{2} + \left(\frac{3}{4} - h\right)\frac{qc}{V}\right] \tag{4.64}$$

Similarly the pitching-moment coefficient about the leading edge is found from Eqn (4.44):

$$C_{M_{LE}} = \frac{\pi}{4}(B_2 - B_1) - \frac{C_L}{4}$$

$$= \frac{\pi}{4}(A_2 - A_1) - \frac{\pi qC}{8V} - \frac{1}{4}C_L \tag{4.65}$$

which for a rectangular wing, on substituting for C_L, becomes

$$C_{M_{LE}} = \frac{\pi}{4}(A_2 - A_1) - \frac{\pi}{8}\frac{c}{V}q - \frac{a}{4}\left\{\alpha - A_0 + \frac{A_1}{2} + \left(\frac{3}{4} - h\right)\frac{qc}{V}\right\} \tag{4.66}$$

The moment coefficient of importance in the derivative is that about the CG and this is found from

$$C_{M_{CG}} = C_{M_{LE}} + hC_L \tag{4.67}$$

and substituting appropriate values

$$C_{M_{CG}} = \frac{\pi}{4}(A_2 - A_1) - \frac{2\pi}{16}\frac{qc}{V} + \left(h - \frac{1}{4}\right)a\left[\alpha - A_0 + \frac{A_1}{2} + \left(\frac{3}{4} - h\right)\frac{qc}{V}\right]$$

which can be rearranged in terms of a function of coefficients A_n plus a term involving q, thus:

$$C_{M_{CG}} = f(A_n) - \left[\frac{a}{4}(1-2h)^2 + \frac{2\pi - a}{16}\right]\frac{qc}{V} \tag{4.68}$$

The contribution of the wings to Z_q or z_q thus becomes

$$Z_q = \frac{\partial Z}{\partial q} = -\frac{\partial L}{\partial q} = -\frac{\partial C_L}{\partial q}\frac{1}{2}\rho V^2 S$$

$$= -\frac{1}{2}\rho V^2 Sa\left(\frac{3}{4}-h\right)\frac{c}{V}$$

by differentiating Eqn (4.64) with respect to q.

Therefore for a rectangular wing, defining z_q by $Z_q/(\rho VSl_t)$,

$$z_q = \frac{-a}{2}\left(\frac{3}{4}-h\right)\frac{c}{l_t} \tag{4.69}$$

For other than rectangular wings an approximate expression can be obtained by using the strip theory, e.g.

$$Z_q = -\rho V\int_{-s}^{s}\frac{a}{2}\left(\frac{3}{4}-h\right)c^2\,\mathrm{d}y$$

giving

$$z_q = \frac{-1}{Sl_t}\int_{-s}^{s}\frac{a}{2}\left(\frac{3}{4}-h\right)c^2\,\mathrm{d}y \tag{4.70}$$

In a similar fashion the contribution to M_q and m_q can be found by differentiating the expression for M_{CG}, with respect to q, i.e.

$$M_q = \frac{\partial M_{CG}}{\partial q} = \frac{\partial C_{M_{CG}}}{\partial q}\frac{1}{2}\rho V^2 Sc$$

$$= -\frac{1}{2}\rho V^2 Sc\left\{\frac{a}{4}(1-2h)^2 + \frac{2\pi - a}{16}\right\}\frac{c}{V} \quad \text{from Eqn (4.68)}$$

$$= -\left[\frac{a}{8}(1-2h)^2 + \frac{2\pi - a}{32}\right]VSc^2 \tag{4.71}$$

giving for a rectangular wing

$$m_q = \frac{M_q}{\rho VSl_t^2} = -\left[\frac{a}{8}(1-2h)^2 + \frac{2\pi - a}{32}\right]\frac{c^2}{l_t^2} \tag{4.72}$$

For other than rectangular wings the contribution becomes, by strip theory:

$$M_q = -\rho V\int_{-s}^{s}\left(\frac{a}{8}(1-2h)^2 + \frac{2\pi - a}{32}\right)c^3\,\mathrm{d}y \tag{4.73}$$

and

$$m_q = -\frac{1}{Sl_t^2}\int_{-s}^{s}\left(\frac{a}{8}(1-2h)^2 + \frac{2\pi - a}{32}\right)c^3\,\mathrm{d}y \tag{4.74}$$

For the theoretical estimation of z_q and m_q, of the complete aircraft, the contributions of the tailplane must be added. These are given here for completeness.

$$
\left.
\begin{aligned}
z_{q\text{tail}} &= -\frac{1}{2}\frac{S'}{S}\left(\frac{\partial C_L'}{\partial \alpha'} + C_0'\right) \\
m_{q\text{tail}} &= -\frac{1}{2}\frac{S'}{S}\left(\frac{\partial C_L'}{\partial \alpha'} + C_0'\right)
\end{aligned}
\right\}
\tag{4.75}
$$

where the terms with dashes refer to tailplane data.

4.8   Particular camber lines

It has been shown that quite general camber lines may be used in the theory satisfactorily and reasonable predictions of the aerofoil characteristics obtained. The reverse problem may be of more interest to the aerofoil designer who wishes to obtain the camber-line shape to produce certain desirable characteristics. The general design problem is more comprehensive than this simple statement suggests and the theory so far dealt with is capable of considerable extension involving the introduction of thickness functions to give shape to the camber line. This is outlined in Section 4.9.

4.8.1   Cubic camber lines

Starting with a desirable aerodynamic characteristic the simpler problem will be considered here. Numerous authorities* have taken a cubic equation as the general shape and evaluated the coefficients required to give the aerofoil the characteristic of a fixed centre of pressure. The resulting camber line has the reflex trailing edge which is the well-known feature of this characteristic.

Example 4.1   Find the cubic camber line that will provide zero pitching moment about the quarter chord point for a given camber.

The general equation for a cubic can be written as $y = a'x(x + b')(x + d')$ with the origin at the leading edge. For convenience the new variables $x_1 = x/c$ and $y_1 = y/\delta$ can be introduced. δ is the camber. The conditions to be satisfied are that:

  (i) $y = 0$ when $x = 0$, i.e. $y_1 = x_1 = 0$ at leading edge
 (ii) $y = 0$ when $x = c$, i.e. $y_1 = 0$ when $x_1 = 1$
(iii) $\mathrm{d}y/\mathrm{d}x = 0$ and $y = \delta$, i.e. $\mathrm{d}y_1/\mathrm{d}x_1 = 0$ when $y_1 = 1$ (when $x_1 = x_0$)
(iv) $C_{M_{1/4}} = 0$, i.e. $A_1 - A_2 = 0$

Rewriting the cubic in the dimensionless variables x_1 and y_1

$$
y_1 = ax_1(x_1 + b)(x_1 + d)
\tag{4.76}
$$

this satisfies condition (i).

   To satisfy condition (ii), $(x_1 + d) = 0$ when $x_1 = 1$, therefore $d = -1$, giving

$$
y_1 = ax_1(x_1 + b)(x_1 - 1)
\tag{4.77}
$$

or multiplying out

$$
y_1 = ax_1^3 + a(b-1)x_1^2 - abx_1
\tag{4.78}
$$

* H. Glauert, *Aerofoil and Airscrew Theory*; N.A.V. Piercy, *Aerodynamics*; etc.

Differentiating Eqn (4.78) to satisfy (iii)

$$\frac{dy_1}{dx_1} = 3ax_1^2 + 2a(b-1)x_1 - ab = 0 \quad \text{when} \quad y_1 = 1 \tag{4.79}$$

and if x_0 corresponds to the value of x_1 when $y_1 = 1$, i.e. at the point of maximum displacement from the chord the two simultaneous equations are

$$\left. \begin{array}{l} 1 = ax_0^3 + a(b-1)x_0^2 - abx_0 \\ 0 = 3ax_0^2 + 2a(b-1)x_0 - ab \end{array} \right\} \tag{4.80}$$

To satisfy (iv) above, A_1 and A_2 must be found. dy_1/dx_1 can be converted to expressions suitable for comparison with Eqn (4.35) by writing

$$x = \frac{c}{2}(1 - \cos\theta) \quad \text{or} \quad x_1 = \frac{1}{2}(1 - \cos\theta)$$

$$\frac{dy_1}{dx_1} = \frac{3}{4}a(1 - 2\cos\theta + \cos^2\theta) + a(b-1) - a(b-1)\cos\theta - ab$$

$$= \left(\frac{3}{4}a + ab - a - ab\right) - \left(\frac{3}{2}a + ab - a\right)\cos\theta + \frac{3}{4}a\cos^2\theta$$

$$\frac{dy_c}{dx} = \frac{\delta}{c}\frac{dy_1}{dx_1} = \frac{\delta}{c}\left[\frac{3}{4}a\cos^2\theta - \left(\frac{a}{2} + ab\right)\cos\theta - \frac{a}{4}\right] \tag{4.81}$$

Comparing Eqn (4.81) and (4.35) gives

$$A_0 = -\frac{a\delta}{4c} + \frac{3}{8}a\frac{\delta}{c} = \frac{9}{8}\frac{\delta}{c}$$

$$A_1 = -\left(\frac{a}{2} + ab\right)\frac{\delta}{c}$$

$$A_2 = \frac{3a}{8}\frac{\delta}{c}$$

Thus to satisfy (iv) above, $A_1 = A_2$, i.e.

$$-\left(\frac{a}{2} + ab\right)\frac{\delta}{c} = a\frac{3}{8}\frac{\delta}{c} \quad \text{giving} \quad b = -\frac{7}{8} \tag{4.82}$$

The quadratic in Eqn (4.80) gives for x_0 on cancelling a,

$$x_0 = \frac{-2(b-1) \pm \sqrt{2^2(b-1)^2 + 4 \times 3b}}{6} = \frac{(1-b) \pm \sqrt{b^2 + b + 1}}{3}$$

From Eqn (4.82), $b = -\frac{7}{8}$ gives

$$x_0 = \frac{22.55}{24} \quad \text{or} \quad \frac{7.45}{24}$$

i.e. taking the smaller value since the larger only gives the point of reflexure near the trailing edge:

$$y = \delta \quad \text{when} \quad x = 0.31 \times \text{chord}$$

Substituting $x_0 = 0.31$ in the cubic of Eqn (4.80) gives

$$a = \frac{1}{0.121} = 8.28$$

The camber-line equation then is

$$
\left.
\begin{aligned}
y &= 8.28\delta\, x_1 \left(x_1 - \frac{7}{8} \right)(x_1 - 1) \\
y &= 8.28\delta \left(x_1^3 - \frac{15}{8}x_1^2 + \frac{7}{8}x_1 \right)
\end{aligned}
\right\}
\tag{4.83}
$$

This cubic camber-line shape is shown plotted on Fig. 4.20 and the ordinates given on the inset table.

Lift coefficient   The lift coefficient is given from Eqn (4.43) by

$$
C_L = 2\pi \left(\alpha - A_0 + \frac{A_1}{2} \right)
$$

So with the values of A_0 and A_1 given above

$$
C_L = 2\pi \left[\alpha - \frac{a\delta}{8\,c} - \frac{1}{2}\left(\frac{a}{2} + ab \right)\frac{\delta}{c} \right]
$$

Substituting for $a = 8.28$ and $b = -\dfrac{7}{8}$

$$
C_L = 2\pi \left(\alpha + 0.518\frac{\delta}{c} \right)
$$

giving a no-lift angle

$$
\alpha_0 = -0.518\frac{\delta}{c} \text{ radians}
$$

or with $\beta =$ the percentage camber $= 100\delta/c$

$$
\alpha_0 = -0.3\beta \text{ degrees}
$$

The load distribution   From Eqn (4.40)

$$
k = 2U \left\{ \left(\alpha - \frac{1.04\delta}{c} \right)\frac{1+\cos\theta}{\sin\theta} + \frac{3.12\delta}{c}\sin\theta + \frac{3.12\delta}{c}\sin 2\theta \right\}
$$

for the first three terms. This has been evaluated for the incidence $\alpha° = 29.6(\delta/c)$ and the result shown plotted and tabulated in Fig. 4.20.

It should be noted that the leading-edge value has been omitted, since it is infinite according to this theory. This is due to the term

$$
(\alpha - A_0)\frac{1+\cos\theta}{\sin\theta}
$$

becoming infinite at $\theta = 0$. When

$$
\alpha = \frac{a}{8}\left(\frac{\delta}{c} \right) = 1.04\frac{\delta}{c}
$$

$(\alpha - A_0)$ becomes zero so $(\alpha - A_0)\dfrac{1+\cos\theta}{\sin\theta}$

becomes zero. Then the intensity of circulation at the leading edge is zero and the stream flows smoothly on to the camber line at the leading edge, the leading edge being a stagnation point.

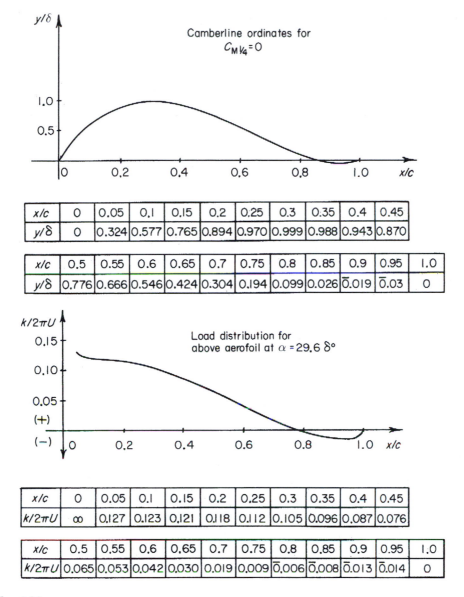

Camberline ordinates for $C_{M\frac{1}{4}} = 0$

x/c	0	0.05	0.1	0.15	0.2	0.25	0.3	0.35	0.4	0.45
y/δ	0	0.324	0.577	0.765	0.894	0.970	0.999	0.988	0.943	0.870

x/c	0.5	0.55	0.6	0.65	0.7	0.75	0.8	0.85	0.9	0.95	1.0
y/δ	0.776	0.666	0.546	0.424	0.304	0.194	0.099	0.026	$\overline{0}$.019	$\overline{0}$.03	0

Load distribution for above aerofoil at $\alpha = 29.6\,\delta°$

x/c	0	0.05	0.1	0.15	0.2	0.25	0.3	0.35	0.4	0.45
k/2πU	∞	0.127	0.123	0.121	0.118	0.112	0.105	0.096	0.087	0.076

x/c	0.5	0.55	0.6	0.65	0.7	0.75	0.8	0.85	0.9	0.95	1.0
k/2πU	0.065	0.053	0.042	0.030	0.019	0.009	$\overline{0}$.006	$\overline{0}$.008	$\overline{0}$.013	$\overline{0}$.014	0

Fig. 4.20

This is the so-called Theodorsen condition, and the appropriate C_L is the ideal, optimum, or design lift coefficient, $C_{L\text{opt}}$.

4.8.2 The NACA four-digit wing sections

According to Abbott and von Doenhoff when the NACA four-digit wing sections were first derived in 1932, it was found that the thickness distributions of efficient wing sections such as the Göttingen 398 and the Clark Y were nearly the same when the maximum thicknesses were set equal to the same value. The thickness distribution for the NACA four-digit sections was selected to correspond closely to those for these earlier wing sections and is given by the following equation:

$$y_t = \pm 5ct[0.2969\sqrt{\xi} - 0.1260\xi - 0.3516\xi^2 + 0.2843\xi^3 - 0.1015\xi^4] \qquad (4.84)$$

where t is the maximum thickness expressed as a fraction of the chord and $\xi = x/c$. The leading-edge radius is

$$r_t = 1.1019ct^2 \qquad (4.85)$$

It will be noted from Eqns (4.84) and (4.85) that the ordinate at any point is directly proportional to the thickness ratio and that the leading-edge radius varies as the square of the thickness ratio.

In order to study systematically the effect of variation in the amount of camber and the shape of the camber line, the shapes of the camber lines were expressed analytically as two parabolic arcs tangent at the position of the maximum camber-line ordinate. The equations used to define the camber line are:

$$y_c = \frac{mc}{p^2}(2p\xi - \xi^2) \quad \xi \le p$$

$$y_c = \frac{mc}{(1-p)^2}[(1 - 2p) + 2p\xi - \xi^2] \quad \xi \ge p \qquad (4.86)$$

where m is the maximum value of y_c expressed as a fraction of the chord c, and p is the value of x/c corresponding to this maximum.

The numbering system for the NACA four-digit wing sections is based on the section geometry. The first integer equals $100m$, the second equals $10p$, and the final two taken together equal $100t$. Thus the NACA 4412 wing section has 4 per cent camber at $x = 0.4c$ from the leading edge and is 12 per cent thick.

To determine the lifting characteristics using thin-aerofoil theory the camber-line slope has to be expressed as a Fourier series. Differentiating Eqn (4.86) with respect to x gives

$$\frac{dy_c}{dx} = \frac{d(y_c/c)}{d\xi} = \frac{2m}{p^2}(p - \xi) \quad \xi \le p$$

$$\frac{dy_c}{dx} = \frac{d(y_c/c)}{d\xi} = \frac{2m}{(1-p)^2}(p - \xi) \quad \xi \ge p$$

Changing variables from ξ to θ where $\xi = (1 - \cos\theta)/2$ gives

$$\frac{dy_c}{dx} = \frac{m}{p^2}(2p - 1 + \cos\theta) \quad \theta \le \theta_p$$

$$\frac{dy_c}{dx} = \frac{m}{(1-p)^2}(2p - 1 + \cos\theta) \quad \theta \ge \theta_p \qquad (4.87)$$

where θ_p is the value of θ corresponding to $x = pc$.

Substituting Eqn (4.87) into Eqn (4.41) gives

$$A_0 = \frac{1}{\pi} \int_0^\pi \frac{dy_c}{dx} d\theta$$

$$= \frac{1}{\pi} \int_0^{\theta_p} \frac{m}{p^2} (2p - 1 + \cos\theta) d\theta + \frac{1}{\pi} \int_{\theta_p}^\pi \frac{m}{(1-p)^2} (2p - 1 + \cos\theta) \, d\theta$$

$$= \frac{m}{\pi p^2} [(2p - 1)\theta_p + \sin\theta_p] + \frac{m}{\pi(1-p)^2} [(2p - 1)(\pi - \theta_p) - \sin\theta_p] \qquad (4.88)$$

Similarily from Eqn (4.42)

$$A_1 = \frac{2}{\pi} \int_0^\pi \frac{dy_c}{dx} \cos\theta \, d\theta$$

$$= \frac{2m}{\pi p^2} \int_0^{\theta_p} [(2p-1)\cos\theta + \cos^2\theta] d\theta + \frac{2m}{\pi(1-p)^2} \int_{\theta_p}^\pi [(2p-1)\cos\theta + \cos^2\theta] d\theta$$

$$= \frac{2m}{\pi p^2} \left[(2p-1)\sin\theta_p + \frac{1}{4}\sin 2\theta_p + \frac{\theta_p}{2} \right]$$

$$- \frac{2m}{\pi(1-p)^2} \left[(2p-1)\sin\theta_p + \frac{1}{4}\sin 2\theta_p - \frac{1}{2}(\pi - \theta_p) \right] \qquad (4.89)$$

$$A_2 = \frac{2}{\pi} \int_0^\pi \frac{dy_c}{dx} \cos^2\theta \, d\theta$$

$$= \frac{2m}{\pi p^2} \int_0^{\theta_p} [(2p-1)\cos^2\theta + \cos^3\theta] d\theta + \frac{2m}{\pi(1-p)^2} \int_{\theta_p}^\pi [(2p-1)\cos^2\theta + \cos^3\theta] d\theta$$

$$= \frac{2m}{\pi p^2} \left[(2p-1)\left(\frac{1}{4}\sin 2\theta_p + \frac{\theta_p}{2}\right) + \sin\theta_p - \frac{1}{3}\sin^3\theta_p \right]$$

$$- \frac{2m}{\pi(1-p)^2} \left[(2p-1)\left(\frac{1}{4}\sin 2\theta_p - \frac{\pi - \theta_p}{2}\right) + \sin\theta_p - \frac{1}{3}\sin^3\theta_p \right] \qquad (4.90)$$

Example 4.2 The NACA 4412 wing section

For a NACA 4412 wing section $m = 0.04$ and $p = 0.4$ so that

$$\theta_p = \cos^{-1}(1 - 2 \times 0.4) = 78.46° = 1.3694 \, \text{rad}$$

making these substitutions into Eqns (4.88) to (4.90) gives

$$A_0 = 0.0090, \qquad A_1 = 0.163 \qquad \text{and} \qquad A_2 = 0.0228$$

Thus Eqns (4.43) and (4.47) give

$$C_L = \pi(A_1 - 2A_0) + 2\pi\alpha = \pi(0.163 - 2 \times 0.009) + 2\pi\alpha = 0.456 + 6.2832\alpha \qquad (4.91)$$

$$C_{M_{1/4}} = -\frac{\pi}{4}(A_1 - A_2) = -\frac{\pi}{4}(0.163 - 0.0228) = -0.110 \qquad (4.92)$$

In Section 4.10 (Fig. 4.26), the predictions of thin-aerofoil theory, as embodied in Eqns (4.91) and (4.92), are compared with accurate numerical solutions and experimental data. It can be seen that the predictions of thin-aerofoil theory are in satisfactory agreement with the accurate numerical results, especially bearing in mind the considerable discrepancy between the latter and the experimental data.

4.9 Thickness problem for thin-aerofoil theory

Before extending the theory to take account of the thickness of aerofoil sections, it is useful to review the parts of the method. Briefly, in thin-aerofoil theory, above, the two-dimensional thin wing is replaced by the vortex sheet which occupies the camber surface or, to the first approximation, the chordal plane. Vortex filaments comprising the sheet extend to infinity in both directions normal to the plane, and all velocities are confined to the xy plane. In such a situation, as shown in Fig. 4.12, the sheet supports a pressure difference producing a normal (upward) increment of force of $(p_1 - p_2)\delta s$ per unit spanwise length. Suffices 1 and 2 refer to under and upper sides of the sheet respectively. But from Bernoulli's equation:

$$p_1 - p_2 = \frac{1}{2}\rho(u_2^2 - u_1^2) = \rho(u_2 - u_1)\frac{u_2 + u_1}{2} \tag{4.93}$$

Writing $(u_2 + u_1)/2 \simeq U$ the free-stream velocity, and $u_2 - u_1 = k$, the local loading on the wing becomes

$$(p_1 - p_2)\delta s = \rho U k\, \delta s \tag{4.94}$$

The lift may then be obtained by integrating the normal component and similarly the pitching moment. It remains now to relate the local vorticity to the thin shape of the aerofoil and this is done by introducing the solid boundary condition of zero velocity normal to the surface. For the vortex sheet to simulate the aerofoil completely, the velocity component induced locally by the distributed vorticity must be sufficient to make the resultant velocity be tangential to the surface. In other words, the component of the free-stream velocity that is normal to the surface at a point on the aerofoil must be completely nullified by the normal-velocity component induced by the distributed vorticity. This condition, which is satisfied completely by replacing the surface line by a streamline, results in an integral equation that relates the strength of the vortex distribution to the shape of the aerofoil.

So far in this review no assumptions or approximations have been made, but thin-aerofoil theory utilizes, in addition to the thin assumption of zero thickness and small camber, the following assumptions:

(a) That the magnitude of total velocity at any point on the aerofoil is that of the local chordwise velocity $\equiv U + u'$.
(b) That chordwise perturbation velocities u' are small in relation to the chordwise component of the free stream U.
(c) That the vertical perturbation velocity v anywhere on the aerofoil may be taken as that (locally) at the chord.

Making use of these restrictions gives

$$v = \int_0^c \frac{k}{2\pi}\frac{\mathrm{d}x}{x - x_1}$$

and thus Eqn (4.42) is obtained:

$$U\left[\frac{dy_c}{dx} - \alpha\right] = \int_0^c \frac{k}{2\pi} \frac{dx}{x - x_1} \quad \text{(Eqn (4.42))}$$

This last integral equation relates the chordwise loading, i.e. the vorticity, to the shape and incidence of the thin aerofoil and by the insertion of a suitable series expression for k in the integral is capable of solution for both the direct and indirect aerofoil problems. The aerofoil is reduced to what is in essence a thin lifting sheet, infinitely long in span, and is replaced by a distribution of singularities that satisfies the same conditions at the boundaries of the aerofoil system, i.e. at the surface and at infinity. Further, the theory is a linearized theory that permits, for example, the velocity at a point in the vicinity of the aerofoil to be taken to be the sum of the velocity components due to the various characteristics of the system, each treated separately. As shown in Section 4.3, these linearization assumptions permit an extension to the theory by allowing a perturbation velocity contribution due to thickness to be added to the other effects.

4.9.1 The thickness problem for thin aerofoils

A symmetrical closed contour of small thickness-chord ratio may be obtained from a distribution of sources, and sinks, confined to the chord and immersed in a uniform undisturbed stream parallel to the chord. The typical model is shown in Fig. 4.21 where $\sigma(x)$ is the chordwise source distribution. It will be recalled that a system of discrete sources and sinks in a stream may result in a closed streamline.

Consider the influence of the sources in the element δx_1 of chord, x_1 from the origin. The strength of these sources is

$$\delta m = \sigma(x_1)\delta x_1$$

Since the elements of upper and lower surface are impermeable, the strength of the sources between x_1 and $x_1 + \delta x_1$ are found from continuity as:

$$\delta m = \text{outflow across boundary}\left(y_t \pm \frac{dy_t}{dx_1}\delta x_1\right) - \text{inflow across} \pm y_t$$

$$= 2\left[\left(U + u' - \frac{\partial u'}{\partial x_1}\delta x_1\right)\left(y_t + \frac{dy_t}{dx_1}\delta x_1\right) - (U + u')y_t\right] \tag{4.95}$$

Neglecting second-order quantities,

$$\delta m = 2U\frac{dy_t}{dx_1}\delta x_1 \tag{4.96}$$

The velocity potential at a general point P for a source of this strength is given by (see Eqn (3.6))

$$\delta\phi = \frac{\delta m}{2\pi}\ln r$$

$$= \frac{U}{\pi}\frac{dy_t}{dx_1}\delta x_1 \ln r \tag{4.97}$$

where $r = \sqrt{(x - x_1)^2 + y^2}$. The velocity potential for the complete distribution of sources lying between 0 and c on the x axis becomes

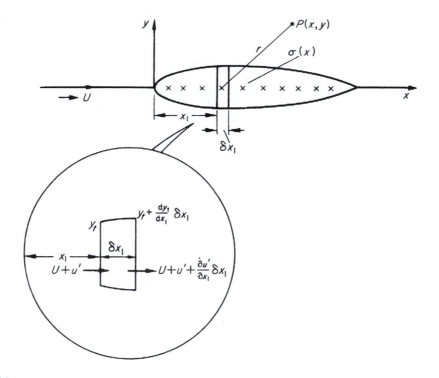

Fig. 4.21

$$\phi = \frac{U}{\pi}\int_0^c \frac{dy_t}{dx_1}\ln r\,dx_1 \tag{4.98}$$

and adding the free stream gives

$$\phi = Ux + \frac{U}{\pi}\int_0^c \frac{dy_t}{dx_1}\ln r\,dx_1 \tag{4.99}$$

Differentiating to find the velocity components

$$u = \frac{\partial\phi}{\partial x} = U + \frac{U}{\pi}\int_0^c \frac{dy_t}{dx_1}\frac{(x-x_1)}{(x-x_1)^2+y^2}\,dx_1 \tag{4.100}$$

$$v = \frac{\partial\phi}{\partial y} = \frac{U}{\pi}\int_0^c \frac{dy_t}{dx_1}\frac{y}{(x-x_1)^2+y^2}\,dx_1 \tag{4.101}$$

To obtain the tangential velocity at the surface of the aerofoil the limit as $y \to 0$ is taken for Eqn (4.100) so that

$$u = U + u' = U + \frac{U}{\pi}\int_0^c \frac{dy_t}{dx_1}\frac{1}{x-x_1}\,dx_1 \tag{4.102}$$

The coefficient of pressure is then given by

$$C_p = -2\frac{u'}{U} = -\frac{2}{\pi}\int_0^c \frac{dy_t}{dx_1}\frac{1}{x-x_1}\,dx_1 \tag{4.103}$$

The theory in the form given above is of limited usefulness for practical aerofoil sections because most of these have rounded leading edges. At a rounded leading edge dy_t/dx_1 becomes infinite thereby violating the assumptions made to develop the thin-aerofoil theory. In fact from Example 4.3 given below it will be seen that the theory even breaks down when dy_t/dx_1 is finite at the leading and trailing edges. There are various refinements of the theory that partially overcome this problem* and others that permit its extension to moderately thick aerofoils.[†]

Example 4.3   Find the pressure distribution on the bi-convex aerofoil

$$\frac{y_t}{c} = \frac{t}{2c}\left[1 - \left(\frac{2x}{c}\right)^2\right]$$

(with origin at mid-chord) set at zero incidence in an otherwise undisturbed stream. For the given aerofoil

$$\frac{y_t}{c} = \frac{t}{2c}\left[1 - \left(\frac{2x_1}{c}\right)^2\right]$$

and

$$\frac{dy_t}{dx_1} = -4t\frac{x_1}{c^2}$$

From above:

$$u' = \frac{u}{\pi}\int_{-c/2}^{c/2} -4\frac{t}{c^2}\frac{x_1}{x - x_1}\,dx_1$$

or

$$C_p = \frac{2}{\pi}\frac{4t}{c^2}\int_{-c/2}^{c/2}\frac{x_1}{x - x_1}\,dx_1$$

$$= \frac{-8}{\pi}\frac{t}{c^2}\left[x\ln(x - x_1) + x_1\right]_{-c/2}^{c/2}$$

Thus

$$C_p = -\frac{8}{\pi}\frac{t}{c}\left[1 + \frac{x}{c}\ln\frac{2x - c}{2x + c}\right]$$

At the mid-chord point:

$$x = 0 \qquad C_p = \frac{-8t}{\pi c}$$

At the leading and trailing edges, $x = \pm c$, $C_p \to -\infty$. The latter result shows that the approximations involved in the linearization do not permit the method to be applied for local effects in the region of stagnation points, even when the slope of the thickness shape is finite.

---

* Lighthill, M.J. (1951) 'A new approach to thin aerofoil theory', *Aero. Quart.*, **3**, 193.
[†] J. Weber (1953) Aeronautical Research Council, *Reports & Memoranda* No. 2918.

4.10 Computational (panel) methods for two-dimensional lifting flows

The extension of the computational method, described in Section 3.5, to two-dimensional lifting flows is described in this section. The basic panel method was developed by Hess and Smith at Douglas Aircraft Co. in the late 1950s and early 1960s. The method appears to have been first extended to lifting flows by Rubbert* at Boeing. The two-dimensional version of the method can be applied to aerofoil sections of any thickness or camber. In essence, in order to generate the circulation necessary for the production of lift, vorticity in some form must be introduced into the modelling of the flow.

It is assumed in the present section that the reader is familiar with the panel method for non-lifting bodies as described in Section 3.5. In a similar way to the computational method in the non-lifting case, the aerofoil section must be modelled by panels in the form of straight-line segments – see Section 3.5 (Fig. 3.37). The required vorticity can either be distributed over internal panels, as suggested by Fig. 4.22a, or on the panels that model the aerofoil contour itself, as shown in Fig. 4.22b.

The central problem of extending the panel method to lifting flows is how to satisfy the Kutta condition (see Section 4.1.1). It is not possible with a computational scheme to satisfy the Kutta condition directly, instead the aim is to satisfy some of the implied conditions namely:

(a) The streamline leaves the trailing edge with a direction along the bisector of the trailing-edge angle.
(b) As the trailing edge is approached the magnitudes of the velocities on the upper and lower surfaces approach the same limiting value.

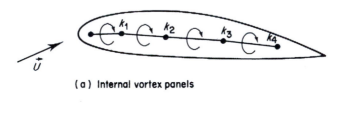

(a) Internal vortex panels

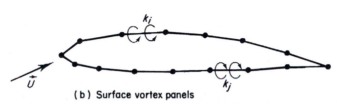

(b) Surface vortex panels

Fig. 4.22 Vortex panels: (a) internal; (b) surface

---

*P.E. Rubbert (1964) *Theoretical Characteristics of Arbitrary Wings by a Nonplanar Vortex Lattice Method* D6-9244, The Boeing Co.

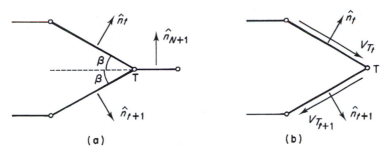

Fig. 4.23 Two methods of implementing the Kutta condition at the trailing edge *T*

(c) In the practical case of an aerofoil with a finite trailing-edge angle the trailing edge must be a stagnation point so the common limiting value of (b) must be zero.

(d) The source strength per unit length must be zero at the trailing edge.

Computational schemes either use conditions (a) or (b). It is not generally possible to satisfy (c) and (d) as well because, as will be shown below, this leads to an over-specification of the problem. The methods of satisfying (a) and (b) are illustrated in Fig. 4.23. For condition (a) an additional panel must be introduced oriented along the bisector of the trailing-edge angle. The value of the circulation is then fixed by requiring the normal velocity to be zero at the collocation point of the additional $(N + 1)$th panel. For condition (b) the magnitudes of the tangential velocity vectors at the collocation points of the two panels, that define the trailing edge, are required to be equal. Hess* has shown that the use of condition (b) gives more accurate results than (a), other things being equal. The use of surface, rather than interior, vorticity panels is also preferable from the viewpoint of computational accuracy.

There are two main ways that surface vorticity panels can be used. One method[†] is to use vorticity panels alone. In this case each of the N panels carries a vorticity distribution of uniform strength per unit length, $\gamma_i (i = 1, 2, \ldots, N)$. In general, the vortex strength will vary from panel to panel. Let $i = t$ for the panel on the upper surface at the trailing edge so that $i = t + 1$ for the panel on the lower surface at the trailing edge. Condition (b) above is equivalent to requiring that

$$\gamma_t = -\gamma_{t+1} \tag{4.104}$$

The normal velocity component at the collocation point of each panel must be zero, as it is for the non-lifting case. This gives N conditions to be satisfied for each of the N panels. So when account is also taken of condition Eqn (4.104) there are $N + 1$ conditions to be satisfied in total. Unfortunately, there are only N unknown vortex strengths. Accordingly, it is not possible to satisfy all $N + 1$ conditions. In order to proceed further, therefore, it is necessary to ignore the requirement that the normal velocity should be zero for one of the panels. This is rather unsatisfactory since it is not at all clear which panel would be the best choice.

---

* J.L. Hess (1972) *Calculation of Potential Flow about Arbitrary Three-Dimensional Lifting Bodies* Douglas Aircraft Co. Rep. MDC J5679/01.

[†] A full description is given in J.D. Anderson (1985) *Fundamentals of Aerodynamics* McGraw-Hill.

An alternative and more satisfactory method is to distribute both sources and vortices of uniform strength per unit length over each panel. In this case, though, the vortex strength is the same for all panels, i.e.

$$\gamma_i = \gamma(i = 1, 2, \ldots, N) \tag{4.105}$$

Thus there are now $N + 1$ unknown quantities, namely the N source strengths and the uniform vortex strength per unit length, γ, to match the $N + 1$ conditions. With this approach it is perfectly feasible to use internal vortex panels instead of surface ones. However these internal panels must carry vortices that are either of uniform strength or of predetermined variable strength, providing the variation is characterized by a single unknown parameter. Generally, however, the use of surface vortex panels leads to better results. Also Condition (a) can be used in place of (b). Again, however, the use of Condition (b) generally gives more accurate results.

A practical panel method for lifting flows around aerofoils is described in some detail below. This method uses Condition (b) and is based on a combination of surface vortex panels of uniform strength and source panels. First, however, it is necessary to show how the normal and tangential influence coefficients may be evaluated for vortex panels. It turns out that the procedure is very similar to that for source panels.

The velocity at point P due to vortices on an element of length $\delta\xi$ in Fig. 4.24 is given by

$$\delta V_\theta = \frac{\gamma}{R} d\xi \tag{4.106}$$

where $\gamma d\xi$ replaces $\Gamma/(2\pi)$ used in Section 3.3.2. δV_θ is oriented at angle θ as shown.

Therefore, the velocity components in the panel-based coordinate directions, i.e. in the x_Q and y_Q directions, are given by

$$\delta V_{x_Q} = \delta V_\theta \sin\theta = \frac{\gamma y_Q}{(x_Q - \xi)^2 + y_Q^2} \delta\xi \tag{4.107}$$

$$\delta V_{y_Q} = -\delta V_\theta \cos\theta = -\frac{\gamma(x_Q - \xi)}{(x_Q - \xi)^2 + y_Q^2} \delta\xi \tag{4.108}$$

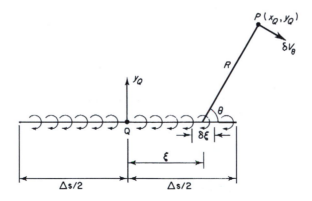

Fig. 4.24

To obtain the corresponding velocity components at P due to all the vortices on the panel, integration along the length of the panel is carried out to give

$$V_{x_Q} = \gamma \int_{-\Delta s/2}^{\Delta s/2} \frac{y_Q}{(x_Q - \xi)^2 + y_Q^2} \, d\xi$$

$$= \gamma \left[\tan^{-1} \left(\frac{x_Q + \Delta s/2}{y_Q} \right) - \tan^{-1} \left(\frac{x_Q - \Delta s/2}{y_Q} \right) \right] \quad (4.109)$$

$$V_{y_Q} = -\gamma \int_{-\Delta s/2}^{\Delta s/2} \frac{x_Q - \xi}{(x_Q - \xi)^2 + y_Q^2} \, d\xi$$

$$= -\frac{\gamma}{2} \ln \left[\frac{(x_Q + \Delta s/2)^2 + y_Q^2}{(x_Q - \Delta s/2)^2 + y_Q^2} \right] \quad (4.110)$$

Following the basic method described in Section 3.5 normal and tangential influence coefficients, N'_{ij} and T'_{ij} are introduced, the primes are used to distinguish these coefficients from those introduced in Section 3.5 for the source panels. N'_{ij} and T'_{ij} represent the normal and tangential velocity components at collocation point i due to vortices of unit strength per unit length distributed on panel j. Let $\hat{t}_i$ and $\hat{n}_i (i = 1, 2, \ldots, N)$ denote the unit tangent and normal vectors for each of the panels, and let the point P correspond to collocation point i, then in vector form the velocity at collocation point i is given by

$$\vec{V}_{PQ} = V_{x_Q} \hat{t}_j + V_{y_Q} \hat{n}_j$$

To obtain the components of this velocity vector perpendicular and tangential to panel i take the scalar product of the velocity vector with $\hat{n}_i$ and $\hat{t}_i$ respectively. If furthermore γ is set equal to 1 in Eqns (4.109) and (4.110) the following expressions are obtained for the influence coefficients

$$N'_{ij} = \vec{V}_{PQ} \cdot \hat{n}_i = V_{x_Q} \hat{n}_i \cdot \hat{t}_j + V_{y_Q} \hat{n}_i \cdot \hat{n}_j \quad (4.111a)$$

$$T'_{ij} = \vec{V}_{PQ} \cdot \hat{t}_i = V_{x_Q} \hat{t}_i \cdot \hat{t}_j + V_{y_Q} \hat{t}_i \cdot \hat{n}_j \quad (4.111b)$$

Making a comparison between Eqns (4.109) and (4.111) for the vortices and the corresponding expressions (3.97) and (3.99) for the source panels shows that

$$[V_{x_Q}]_{\text{vortices}} = [V_{y_Q}]_{\text{sources}} \quad \text{and} \quad [V_{y_Q}]_{\text{vortices}} = -[V_{x_Q}]_{\text{sources}} \quad (4.112)$$

With the results given above it is now possible to describe how the basic panel method of Section 3.5 may be extended to lifting aerofoils. Each of the N panels now carries a source distribution of strength σ_i per unit length and a vortex distribution of strength γ per unit length. Thus there are now $N + 1$ unknown quantities. The $N \times N$ influence coefficient matrices N_{ij} and T_{ij} corresponding to the sources must now be expanded to $N \times (N + 1)$ matrices. The $(N + 1)$th column now contains the velocities induced at the collocation points by vortices of unit strength per unit length on all the panels. Thus $N_{i, N+1}$ represents the normal velocity at the ith collocation point induced by the vortices over all the panels and similarly for $T_{i, N+1}$. Thus using Eqns (4.111)

$$N_{i, N+1} = \sum_{j=1}^{N} N'_{i,j} \quad \text{and} \quad T_{i, N+1} = \sum_{j=1}^{N} T'_{i,j} \quad (4.113)$$

In a similar fashion as for the non-lifting case described in Section 3.5 the total normal velocity at each collocation point, due to the net effect of all the sources, the vortices and the oncoming flow, is required to be zero. This requirement can be written in the form:

$$\underbrace{\sum_{j=1}^{N} \sigma_j N_{ij}}_{\text{Sources}} + \underbrace{\gamma N_{i,N+1}}_{\text{Vortices}} + \underbrace{\vec{U} \cdot \hat{n}_i}_{\text{Oncoming flow}} = 0 \quad (i = 1, 2, \ldots, N) \quad (4.114)$$

These N equations are supplemented by imposing Condition (b). The simplest way to do this is to equate the magnitudes of the tangential velocities at the collocation point of the two panels defining the trailing edge (see Fig. 4.23b). Remembering that the unit tangent vectors $\hat{t}_t$ and $\hat{t}_{t+1}$ are in opposite directions Condition (b) can be expressed mathematically as

$$\sum_{j=1}^{N} \sigma_j T_{t,j} + \gamma T_{t,N+1} + \vec{U} \cdot \hat{t}_t = -\left(\sum_{j=1}^{N} \sigma_j T_{t+1,j} + \gamma T_{t+1,N+1} + \vec{U} \cdot \hat{t}_{t+1} \right) \quad (4.115)$$

Equations (4.114) and (4.115) combine to form a matrix equation that can be written as

$$\mathbf{Ma} = \mathbf{b} \quad (4.116)$$

where $\mathbf{M}$ is an $(N + 1) \times (N + 1)$ matrix and $\mathbf{a}$ and $\mathbf{b}$ are $(N + 1)$ column vectors. The elements of the matrix and vectors are as follows:

$$M_{i,j} = N_{i,j} \quad i = 1, 2, \ldots, N \quad j = 1, 2, \ldots, N + 1$$
$$M_{N+1,j} = T_{t,j} + T_{t+1,j} \quad j = 1, 2, \ldots, N + 1$$
$$a_i = \sigma_i \quad i = 1, 2, \ldots, N \quad \text{and} \quad a_{N+1} = \gamma$$
$$b_i = -\vec{U} \cdot \hat{n}_i \quad i = 1, 2, \ldots, N$$
$$b_{N+1} = -\vec{U} \cdot (\hat{t}_t + \hat{t}_{t+1})$$

Systems of linear equations like (4.116) can be readily solved numerically for the unknowns a_i using standard methods (see Section 3.5). Also it is now possible to see why the Condition (c), requiring that the tangential velocities on the upper and lower surfaces both tend to zero at the trailing edge, cannot be satisfied in this sort of numerical scheme. Condition (c) could be imposed approximately by requiring, say, that the tangential velocities on panels t and $t + 1$ are both zero. Referring to Eqn (4.115) this approximate condition can be expressed mathematically as

$$\sum_{j=1}^{N} \sigma_j T_{t,j} + \gamma T_{t,N+1} + \vec{U} \cdot \hat{t}_t = 0$$

$$\sum_{j=1}^{N} \sigma_j T_{t+1,j} + \gamma T_{t+1,N+1} + \vec{U} \cdot \hat{t}_{t+1} = 0$$

Equation (4.115) is now replaced by the above two equations so that $\mathbf{M}$ in Eqn (4.116) is now a $(N + 2) \times (N + 1)$ matrix. The problem is now overdetermined, i.e. there is one more equation than the number of unknowns, and Eqn (4.116) can no longer be solved for the vector $\mathbf{a}$, i.e. for the source and vortex strengths.

The calculation of the influence coefficients is at the heart of a panel method. In Section 3.5 a computational routine in FORTRAN 77 is given for computing the influence coefficients for the non-lifting case. It is shown below how this routine can be extended to include the calculation of the influence coefficients due to the vortices required for a lifting flow.

Two modifications to SUBROUTINE INFLU in Section 3.5 are required to extend it to the lifting case.

(1) The first two execution statements i.e.

```
      DO 10 I = 1,N
10    READ(7,*) XP(I),YP(I)
```

should be replaced by

```
      NP1 = N + 1
      DO 10 I = 1,N
        AN(I,NP1) = 0.0
        AT(I,NP1) = PI
10    READ(7,*)XP(I),YP(I)
```

The additional lines initialize the values of the influence coefficients, $N_{i,N+1}$ and $T_{i,N+1}$ in preparation for their calculation later in the program. Note that the initial value of $T_{i,N+1}$ is set at π because in Eqn (4.113)

$$T'_{N+1,N-1} = N_{i,i} = \pi$$

that is the tangential velocity induced on a panel by vortices of unit strength per unit length distributed over the same panel is, from Eqn (4.112), the same as the normal velocity induced by sources of unit strength per unit length distributed over the panel. This was shown to take the value π in Eqn (3.100b).

(2) It remains to insert the two lines of instruction that calculate the additional influence coefficients according to Eqn (4.113). This is accomplished by inserting two additional lines below the last two execution statements in the routine, as shown

```
      AN(I, J) = VX * NTIJ + VY * NNIJ              Existing line
      AT(I, J) = VX * TTIJ + VY * TNIJ              Existing line
      AN(I,NP1) = AN(1,NP1) + VY * NTIJ - VX * NNIJ  New line
      AT(I,NP1) = AT(I,NP1) + VY * TTIJ - VX * TNIJ  New line
```

As with the original routine presented in Section 3.5 this modified routine is primarily intended for educational purposes. Nevertheless, as is shown by the example computation for a NACA 4412 aerofoil presented below, a computer program based on this routine and LU decomposition gives accurate results for the pressure distribution and coefficients of lift and pitching moment. The computation times required are typically a few seconds using a modern personal computer.

The NACA 4412 wing section has been chosen to illustrate the use of the panel method. The corresponding aerofoil profile is shown inset in Fig. 4.25. As can be seen it is a moderately thick aerofoil with moderate camber. The variation of the pressure coefficient around a NACA 4412 wing section at an angle of attack of 8 degrees is presented in Fig. 4.25. Experimental data are compared with the computed

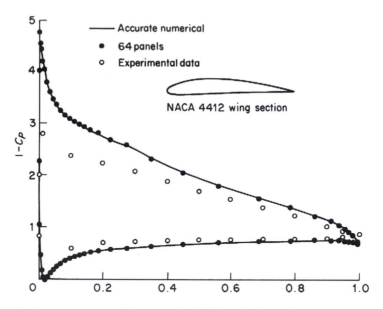

Fig. 4.25 Variation of pressure coefficient around a NACA 4412 wing section at an angle of attack of 8°

results for 64 panels and 160 panels. The latter can be regarded as exact and are plotted as the solid line in the figure. It can be seen that the agreement between the two sets of computed data is very satisfactory. The agreement between the experimental and computed data is not good, especially for the upper surface. This is undoubtedly a result of fairly strong viscous effects at this relatively high angle of attack. The discrepancy between the computed and experimental pressure coefficients is particularly marked on the upper surface near the leading edge. In this region, according to the computed results based on inviscid theory, there is a very strong favourable pressure gradient followed by a strong adverse one. This scenario is very likely to give rise to local boundary-layer separation (see Section 7.4.1 below) near the leading edge leading to greatly reduced peak suction pressures near the leading edge.

The computed and experimental lift and pitching-moment coefficients, C_L and $C_{M_{1/4}}$ are plotted as functions of the angle of attack in Fig. 4.26. Again there is good agreement between the two sets of computed results. For the reasons explained above the agreement between the computed and experimental lift coefficients is not all that satisfactory, especially at the higher angles of attack. Also shown in Fig. 4.25 are the predictions of thin-aerofoil theory – see Eqns (4.91) and (4.92). In view of the relatively poor agreement between theory and experiment evidenced in Fig. 4.26 it might be thought that there is little to choose between thin-aerofoil theory and computations using the panel method. For predictions of C_L and $C_{M_{1/4}}$ this is probably a reasonable conclusion, although for aerofoils that are thicker or more cambered than the NACA 4412, the thin-aerofoil theory would perform much less well. The great advantage of the panel method, however, is that it provides accurate results for the pressure distribution according to inviscid theory. Accordingly, a panel method can be used in conjunction with a method for computing the viscous (boundary-layer) effects and ultimately produce a corrected pressure distribution that is much closer to the experimental one (see Section 7.11).

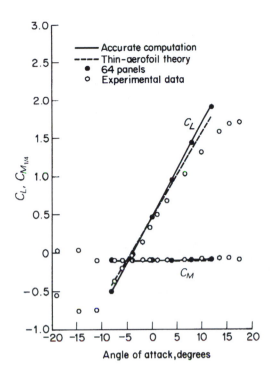

Fig. 4.26 Variation of lift and moment coefficients with angle of attack for NACA 4412 aerofoil

Exercises

1 A thin two-dimensional aerofoil of chord c is operating at its ideal lift coefficient C_{Li}. Assume that the loading (i.e. the pressure difference across the aerofoil) varies linearly with its maximum value at the leading edge. Show that

$$\frac{dy_c}{dx} - \alpha = \frac{C_{Li}}{2\pi}\left[(1-\xi)\ln\left(\frac{1-\xi}{\xi}\right) - 1\right]$$

where y_c defines the camber line, α is the angle of incidence, and $\xi = x/c$.

[Hint: Do not attempt to make the transformation $x = (c/2)(1 - \cos\theta)$, instead write the singular integral as follows:

$$\int_0^1 \frac{1}{\xi - \xi_1}d\xi = \lim_{\varepsilon \to 0}\left\{\int_0^{\xi_1-\varepsilon} \frac{1}{\xi - \xi_1}d\xi + \int_{\xi_1+\varepsilon}^1 \frac{1}{\xi - \xi_1}d\xi\right\}]$$

Then, using this result, show that the angle of incidence and the camber-line shape are given by

$$\alpha = \frac{C_L}{4\pi}; \qquad \frac{y_c}{c} = \frac{C_L}{4\pi}\left\{-(1-\xi)^2\ln(1-\xi) + \xi(\xi - 2)\ln\xi\right\}$$

[Hint: Write $-1 = C - 1 - C$ where $1 + C = 2\pi\alpha/C_L$ and the constant C is determined by requiring that $y_c = 0$ at $\xi = 0$ and $\xi = 1$.]

2 A thin aerofoil has a camber line defined by the relation $y_c = kc\xi(\xi - 1)(\xi - 2)$. Show that if the maximum camber is 2% of chord then $k = 0.052$. Determine the coefficients of lift and pitching moment, i.e. C_L and $C_{M_{1/4}}$, at 3° incidence.

(*Answer*: 0.535, −0.046)

3 Use thin-aerofoil theory to estimate the coefficient of lift at zero incidence and the pitching-moment coefficient $C_{M_{1/4}}$ for a NACA 8210 wing section.

(*Answer*: 0.789, −0.172)

4 Use thin-aerofoil theory to select a NACA four-digit wing section with a coefficient of lift at zero incidence approximately equal to unity. The maximum camber must be located at 40% chord and the thickness ratio is to be 0.10. Estimate the required maximum camber as a percentage of chord to the nearest whole number. [Hint: Use a spreadsheet program to solve by trial and error.]

(*Answer*: NACA 9410)

5 Use thin-aerofoil theory to select a NACA four-digit wing section with a coefficient of lift at zero incidence approximately equal to unity and pitching-moment coefficient $C_{M_{1/4}} = -0.25$. The thickness ratio is to be 0.10. Estimate the required maximum camber as a percentage of chord to the nearest whole number and its position to the nearest tenth of a chord. The C_L value must be within 1% of the required value and $C_{M_{1/4}}$ within 3%. [Hint: Use a spreadsheet program to solve by trial and error.]

(*Answer*: NACA 7610, but NACA 9410 and NACA 8510 are also close.)

6 A thin two-dimensional flat-plate aerofoil is fitted with a trailing-edge flap of chord $100e$ per cent of the aerofoil chord. Show that the flap effectiveness,

$$\frac{a_2}{a_1} \equiv \frac{\frac{\partial C_L}{\partial \eta}}{\frac{\partial C_L}{\partial \alpha}}$$

where α is the angle of incidence and η is the flap angle, is approximately $4\sqrt{e}/\pi$ for flaps of small chord.

7 A thin aerofoil has a circular-arc camber line with a maximum camber of 0.025 chord. Determine the theoretical pitching-moment coefficient $C_{M_{1/4}}$ and indicate methods by which this could be reduced without changing maximum camber.

The camber line may be approximated by the expression

$$y_c = kc\left[\frac{1}{4} - \left(\frac{x'}{c}\right)^2\right]$$

where $x' = x - 0.5c$.

(*Answer*: -0.025π)

8 The camber line of a circular-arc aerofoil is given by

$$\frac{y_c}{c} = 4h\frac{x}{c}\left(1 - \frac{x}{c}\right)$$

Derive an expression for the load distribution (pressure difference across the aerofoil) at incidence α. Show that the zero-lift angle $\alpha_0 = -2h$, and sketch the load distribution at this incidence. Compare the lift curve of this aerofoil with that of a flat plate.

9 A flat-plate aerofoil is aligned along the x-axis with the origin at the leading edge and trailing edge at $x = c$. The plate is at an angle of incidence α to a free stream of air speed U. A vortex of strength Γ_v is located at (x_v, y_v). Show that the distribution, $k(x)$, of vorticity along the aerofoil from $x = 0$ to $x = c$ satisfies the integral equation

$$\frac{1}{2\pi}\int_0^c \frac{k(x)}{x - x_1}dx = -U\alpha - \frac{\Gamma_v(x_v - x_1)}{(x_v - x_1)^2 + y_v^2}$$

where $x = x_1$ is a particular location on the chord of the aerofoil. If $x_v = c/2$ and $y_v = h \gg x_v$ show that the additional increment of lift produced by the vortex (which could represent a nearby aerofoil) is given approximately by

$$\frac{4\Gamma}{3\pi h^2}.$$

5

Finite wing theory

Preamble

Whatever the operating requirements of an aeroplane may be in terms of speed; endurance, pay-load and so on, a critical stage in its eventual operation is in the low-speed flight regime, and this must be accommodated in the overall design process. The fact that low-speed flight was the classic flight regime has meant that over the years a vast array of empirical data has been accumulated from flight and other tests, and a range of theories and hypotheses set up to explain and extend these observations. Some theories have survived to provide successful working processes for wing design that are capable of further exploitation by computational methods.

In this chapter such a classic theory is developed to the stage of initiating the preliminary low-speed aerodynamic design of straight, swept and delta wings. Theoretical fluid mechanics of vortex systems are employed, to model the loading properties of lifting wings in terms of their geometric and attitudinal characteristics and of the behaviour of the associated flow processes.

The basis on which historical solutions to the finite wing problem were arrived at are explained in detail and the work refined and extended to take advantage of more modern computing techniques.

A great step forward in aeronautics came with the vortex theory of a lifting aerofoil due to Lanchester[*] and the subsequent development of this work by Prandtl.[†] Previously, all aerofoil data had to be obtained from experimental work and fitted to other aspect ratios, planforms, etc., by empirical formulae based on past experience with other aerofoils.

Among other uses the Lanchester–Prandtl theory showed how knowledge of two-dimensional aerofoil data could be used to predict the aerodynamic characteristics of (three-dimensional) wings. It is this derivation of the aerodynamic characteristics of wings that is the concern of this chapter. The aerofoil data can either be obtained empirically from wind-tunnel tests or by means of the theory described in Chapter 4. Provided the aspect ratio is fairly large and the assumptions of thin-aerofoil theory are met (see Section 4.3 above), the theory can be applied to wing planforms and sections of any shape.

[*] see Bibliography.
[†] Prandtl, L. (1918), Tragflügeltheorie, *Nachr. Ges. Wiss.*, Göttingen, 107 and 451.

5.1 The vortex system

Lanchester's contribution was essentially the replacement of the lifting wing by a theoretical model consisting of a system of vortices that imparted to the surrounding air a motion similar to the actual flow, and that sustained a force equivalent to the lift known to be created. The vortex system can be divided into three main parts: the starting vortex; the trailing vortex system; and the bound vortex system. Each of these may be treated separately but it should be remembered that they are all component parts of one whole.

5.1.1 The starting vortex

When a wing is accelerated from rest the circulation round it, and therefore the lift, is not produced instantaneously. Instead, at the instant of starting the streamlines over the rear part of the wing section are as shown in Fig. 5.1, with a stagnation point occurring on the rear upper surface. At the sharp trailing edge the air is required to change direction suddenly while still moving at high speed. This high speed calls for extremely high local accelerations producing very large viscous forces and the air is unable to turn round the trailing edge to the stagnation point. Instead it leaves the surface and produces a vortex just above the trailing edge. The stagnation point moves towards the trailing edge, the circulation round the wing, and therefore its lift, increasing progressively as the stagnation point moves back. When the stagnation point reaches the trailing edge the air is no longer required to flow round the trailing edge. Instead it decelerates gradually along the aerofoil surface, comes to rest at the trailing edge, and then accelerates from rest in a different direction (Fig. 5.2). The vortex is left behind at the point reached by the wing when the stagnation point

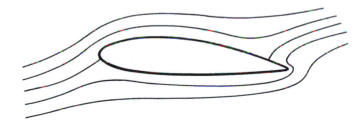

Fig. 5.1 Streamlines of the flow around an aerofoil with zero circulation, stagnation point on the rear upper surface

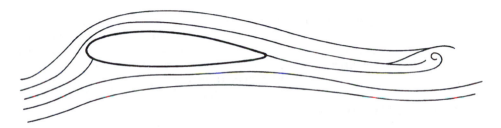

Fig. 5.2 Streamlines of the flow around an aerofoil with full circulation, stagnation point at the trailing edge. The initial eddy is left way behind

reached the trailing edge. Its reaction, the circulation round the wing, has become stabilized at the value necessary to place the stagnation point at the trailing edge (see Section 4.1.1).* The vortex that has been left behind is equal in strength and opposite in sense to the circulation round the wing and is called the starting vortex or initial eddy.

5.1.2   The trailing vortex system

The pressure on the upper surface of a lifting wing is lower than that of the surrounding atmosphere, while the pressure on the lower surface is greater than that on the upper surface, and may be greater than that of the surrounding atmosphere. Thus, over the upper surface, air will tend to flow inwards towards the root from the tips, being replaced by air that was originally outboard of the tips. Similarly, on the undersurface air will either tend to flow inwards to a lesser extent, or may tend to flow outwards. Where these two streams combine at the trailing edge, the difference in spanwise velocity will cause the air to roll up into a number of small streamwise vortices, distributed along the whole span. These small vortices roll up into two large vortices just inboard of the wing-tips. This is illustrated in Fig. 5.3. The strength of

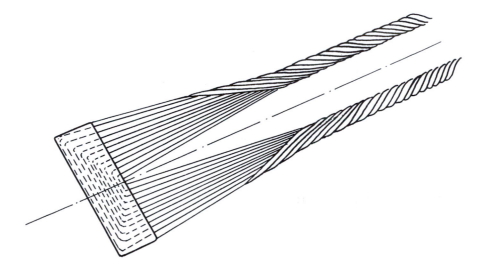

Fig. 5.3 The horseshoe vortex

*There is no fully convincing physical explanation for the production of the starting vortex and the generation of the circulation around the aerofoil. Various incomplete explanations will be found in the references quoted in the bibliography. The most usual explanation is based on the large viscous forces associated with the high velocities round the trailing edge, from which it is inferred that circulation cannot be generated, and aerodynamic lift produced, in an inviscid fluid. It may be, however, that local flow acceleration is equally important and that this is sufficiently high to account for the failure of the flow to follow round the sharp trailing edge, without invoking viscosity. Certainly it is now known, from the work of T. Weis-Fogh [Quick estimates of flight fitness in hovering animals, including novel mechanisms for lift production, *J. Expl. Biol.*, **59**, 169–230, 1973] and M.J. Lighthill [On the Weis-Fogh mechanism of lift generation, *J. Fluid Mech.*, **60**, 1–17, 1973] on the hovering flight of the small wasp *Encarsia formosa*, that it is possible to generate circulation and lift in the complete absence of viscosity.

In practical aeronautics, fluid is not inviscid and the complete explanation of this phenomenon must take account of viscosity and the consequent growth of the boundary layer as well as high local velocities as the motion is generated.

each of these two vortices will equal the strength of the vortex replacing the wing itself.

The existence of the trailing and starting vortices may easily be verified visually. When a fast aeroplane pulls out of a dive in humid air the reduction of pressure and temperature at the centres of the trailing vortices is often sufficient to cause some of the water vapour to condense into droplets, which are seen as a thin streamer for a short distance behind each wing-tip (see frontispiece).

To see the starting vortex all that is needed is a tub of water and a small piece of board, or even a hand. If the board is placed upright into the water cutting the surface and then suddenly moved through the water at a moderate incidence, an eddy will be seen to leave the rear, and move forwards and away from the 'wing'. This is the starting vortex, and its movement is induced by the circulation round the plate.

5.1.3   The bound vortex system

Both the starting vortex and the trailing system of vortices are physical entities that can be explored and seen if conditions are right. The bound vortex system, on the other hand, is a hypothetical arrangement of vortices that replace the real physical wing in every way except that of thickness, in the theoretical treatments given in this chapter. This is the essence of finite wing theory. It is largely concerned with developing the equivalent bound vortex system that simulates accurately, at least a little distance away, all the properties, effects, disturbances, force systems, etc., due to the real wing.

Consider a wing in steady flight. What effect has it on the surrounding air, and how will changes in basic wing parameters such as span, planform, aerodynamic or geometric twist, etc., alter these disturbances? The replacement bound vortex system must create the same disturbances, and this mathematical model must be sufficiently flexible to allow for the effects of the changed parameters. A real wing produces a trailing vortex system. The hypothetical bound vortex must do the same. A consequence of the tendency to equalize the pressures acting on the top and bottom surfaces of an aerofoil is for the lift force per unit span to fall off towards the tips. The bound vortex system must produce the same grading of lift along the span.

For complete equivalence, the bound vortex system should consist of a large number of spanwise vortex elements of differing spanwise lengths all turned backwards at each end to form a pair of the vortex elements in the trailing system. The varying spanwise lengths accommodate the grading of the lift towards the wing-tips, the ends turned back produce the trailing system and the two physical attributes of a real wing are thus simulated.

For partial equivalence the wing can be considered to be replaced by a single bound vortex of strength equal to the mid-span circulation. This, bent back at each end, forms the trailing vortex pair. This concept is adequate for providing good estimations of wing effects at distances greater than about two chord lengths from the centre of pressure.

5.1.4   The horseshoe vortex

The total vortex system associated with a wing, plus its replacement bound vortex system, forms a complete vortex ring that satisfies all physical laws (see Section 5.2.1). The starting vortex, however, is soon left behind and the trailing pair stretches effectively to infinity as steady flight proceeds. For practical purposes the system consists of the bound vortices and the trailing vortex on either side close to the wing. This three-sided vortex has been called the *horseshoe vortex* (Fig. 5.3).

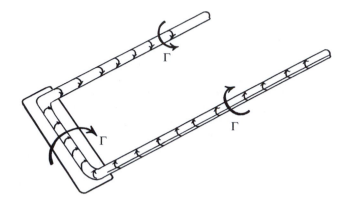

Fig. 5.4 The simplified horseshoe vortex

Study of the completely equivalent vortex system is largely confined to investigating wing effects in close proximity to the wing. For estimation of distant phenomena the system is simplified to a single bound vortex and trailing pair, known as the *simplified horseshoe vortex* (Fig. 5.4). This is dealt with in Section 5.3, before the more involved and complete theoretical treatments of wing aerodynamics.

5.2 Laws of vortex motion

The theoretical modelling of the flow around wings was discussed in the previous section. There the use of an equivalent vortex system to model the lifting effects of a wing was described. In order to use this theoretical model to obtain quantitative predictions of the aerodynamic characteristics of a wing it is necessary first to study the laws of vortex motion. These laws also act as a guide for understanding how modern computationally based wing theories may be developed.

In the analysis of the point vortex (Chapter 3) it was considered to be a string of rotating particles surrounded by fluid at large moving irrotationally under the influence of the rotating particles. Further, the flow investigation was confined to a plane section normal to the length or axis of the vortex. A more general definition is that a vortex is a flow system in which a finite area in a normal section plane contains vorticity. Figure 5.5 shows the section area S of a vortex so called because S possesses vorticity. The axis of the vortex (or of the vorticity, or spin) is clearly always normal

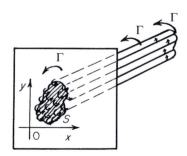

Fig. 5.5 The vorticity of a section of vortex tube

to the two-dimensional flow plane considered previously and the influence of the so-called line vortex is the influence, in a section plane, of an infinitely long, straight-line vortex of vanishingly small area.

In general, the vortex axis will be a curve in space and area S will have finite size. It is convenient to assume that S is made up of several elemental areas or, alternatively, that the vortex consists of a bundle of elemental vortex lines or filaments. Such a bundle is often called a *vortex tube* (c.f. a *stream tube* which is a bundle of streamlines), being a tube bounded by vortex filaments.

Since the vortex axis is a curve winding about within the fluid, capable of flexure and motion as a whole, the estimation of its influence on the fluid at large is somewhat complex and beyond the present intentions. All the vortices of significance to the present theory are fixed relative to some axes in the system or free to move in a very controlled fashion and can be assumed to be linear. Nonetheless, the vortices will not all be of infinite length and therefore some three-dimensional or end influence must be accounted for.

Vortices conform to certain laws of motion. A rigorous treatment of these is precluded from a text of this standard but may be acquired with additional study of the basic references.*

5.2.1 Helmholtz's theorems

The four fundamental theorems of vortex motion in an inviscid flow are named after their author, Helmholtz. The first theorem has been discussed in part in Sections 2.7 and 4.1, and refers to a fluid particle in general motion possessing all or some of the following: linear velocity, vorticity, and distortion. The second theorem demonstrates the constancy of strength of a vortex along its length. This is sometimes referred to as the equation of vortex continuity. It is not difficult to prove that the strength of a vortex cannot grow or diminish along its axis or length. The strength of a vortex is the magnitude of the circulation around it and this is equal to the product of the vorticity ζ and area S. Thus

$$\Gamma = \zeta S$$

It follows from the second theorem that ζS is constant along the vortex tube (or filament), so that if the section area diminishes, the vorticity increases and vice versa. Since infinite vorticity is unacceptable the cross-sectional area S cannot diminish to zero.

In other words a vortex line cannot end in the fluid. In practice the vortex line must form a closed loop, or originate (or terminate) in a discontinuity in the fluid such as a solid body or a surface of separation. A refinement of this is that a vortex tube cannot change in strength between two sections unless vortex filaments of equivalent strength join or leave the vortex tube (Fig. 5.6). This is of great importance in the vortex theory of lift.

The third and fourth theorems demonstrate respectively that a vortex tube consists of the same particles of fluid, i.e. there is no fluid interchange between tube and surrounding fluid, and the strength of a vortex remains constant as the vortex moves through the fluid.

The theorem of most consequence to the present chapter is theorem two, although the third and fourth are tacitly accepted as the development proceeds.

* Saffman, P.G. 1992 *Vortex Dynamics*, Cambridge University Press.

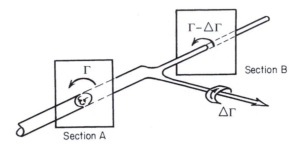

Fig. 5.6

5.2.2 The Biot–Savart law

The original application of this law was in electromagnetism, where it relates the intensity of the magnetic field in the vicinity of a conductor carrying an electric current to the magnitude of the current. In the present application velocity and vortex strength (circulation) are analogous to the magnetic field strength and electric current respectively, and a vortex filament replaces the electrical conductor. Thus the Biot–Savart law can also be interpreted as the relationship between the velocity *induced* by a vortex tube and the strength (circulation) of the vortex tube. Only the fluid motion aspects will be further pursued here, except to remark that the term *induced velocity*, used to describe the velocity generated at a distance by the vortex tube, was borrowed from electromagnetism.

Allow a vortex tube of strength Γ, consisting of an infinite number of vortex filaments, to terminate in some point P. The total strength of the vortex filaments will be spread over the surface of a spherical boundary of radius R (Fig. 5.7) as the filaments diverge from the point P in all directions. The vorticity in the spherical surface will thus have the total strength Γ.

Owing to symmetry the velocity of flow in the surface of the sphere will be tangential to the circular line of intersection of the sphere with a plane normal to the axis of the vortex. Moreover, the direction will be in the sense of the circulation about the vortex. Figure 5.8 shows such a circle ABC of radius r subtending a conical angle of 2θ at P. If the velocity on the sphere at R, θ from P is v, then the circulation round the circuit ABC is Γ' where

$$\Gamma' = 2\pi R \sin \theta v \tag{5.1}$$

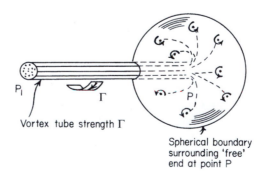

Vortex tube strength Γ

Spherical boundary
surrounding 'free'
end at point P

Fig. 5.7

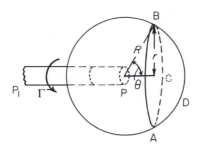

Fig. 5.8

Putting r = radius of circuit = $R \sin \theta$, Eqn (5.1) becomes

$$\Gamma' = 2\pi r v \tag{5.2}$$

Now the circulation round the circuit is equal to the strength of the vorticity in the contained area. This is on the cap ABCD of the sphere. Since the distribution of the vorticity is constant over the surface

$$\Gamma' = \frac{\text{surface area of cap}}{\text{surface area of sphere}} \Gamma = \frac{2\pi R^2 (1 - \cos \theta)}{4\pi R^2} \Gamma$$

$$= \frac{\Gamma}{2}(1 - \cos \theta) \tag{5.3}$$

Equating (5.2) and (5.3) gives

$$v = \frac{\Gamma}{4\pi r}(1 - \cos \theta) \tag{5.4}$$

Now let the length, $P_1 P$, of the vortex decrease until it is very short (Fig. 5.9). The circle ABC is now influenced by the opposite end P_1. Working through Eqns (5.1), (5.2) and (5.3) shows that the induced velocity due to P_1 is now

$$v_1 = \frac{-\Gamma}{4\pi r}(1 - \cos \theta_1) \tag{5.5}$$

since $r = R_1 \sin \theta_1$ and the sign of the vorticity is reversed on the sphere of radius R_1 as the vortex elements are now entering the sphere to congregate on P_1.

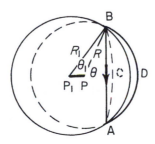

Fig. 5.9

The net velocity in the circuit ABC is the sum of Eqns (5.4) and (5.5):

$$v - v_1 = \frac{\Gamma}{4\pi r}[1 - \cos\theta - (1 - \cos\theta_1)]$$

$$= \frac{\Gamma}{4\pi r}(\cos\theta_1 - \cos\theta)$$

As P_1 approaches P

$$\cos\theta_1 \rightarrow \cos(\theta - \delta\theta) = \cos\theta + \sin\theta\,\delta\theta$$

and

$$v - v_1 \rightarrow \delta v$$

giving

$$\delta v = \frac{\Gamma}{4\pi r}\sin\theta\,\delta\theta \qquad (5.6)$$

This is the induced velocity at a point in the field of an elementary length δs of vortex of strength Γ that subtends an angle $\delta\theta$ at P located by the coordinates R, θ from the element. Since $r = R\sin\theta$ and $R\,\delta\theta = \delta s\sin\theta$ it is more usefully quoted as:

$$\delta v = \frac{\Gamma}{4\pi R^2}\sin\theta\,\delta s \qquad (5.7)$$

Special cases of the Biot–Savart law

Equation (5.6) needs further treatment before it yields working equations. This treatment, of integration, varies with the length and shape of the finite vortex being studied. The vortices of immediate interest are all assumed to be straight lines, so no shape complexity arises. They will vary only in their overall length.

A linear vortex of finite length AB   Figure 5.10 shows a length AB of vortex with an adjacent point P located by the angular displacements α and β from A and B respectively. Point P has, further, coordinates r and θ with respect to any elemental length δs of the length AB that may be defined as a distance s from the foot of the perpendicular h. From Eqn (5.7) the velocity at P induced by the elemental length δs is

$$\delta v = \frac{\Gamma}{4\pi r^2}\sin\theta\,\delta s \qquad (5.8)$$

in the sense shown, i.e. normal to the plane APB.

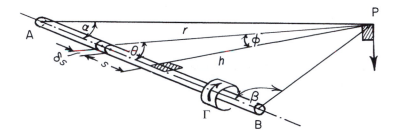

Fig. 5.10

To find the velocity at P due to the length AB the sum of induced velocities due to all such elements is required. Before integrating, however, all the variables must be quoted in terms of a single variable. A convenient variable is ϕ (see Fig. 5.10) and the limits of the integration are

$$\phi_A = -\left(\frac{\pi}{2} - \alpha\right) \quad \text{to} \quad \phi_B = +\left(\frac{\pi}{2} - \beta\right)$$

since ϕ passes through zero when integrating from A to B.

$$\sin\theta = \cos\phi, \quad r^2 = h^2 \sec^2\phi$$
$$ds = d(h\tan\phi) = h\sec^2\phi\, d\phi$$

The integration of Eqn (5.8) is thus

$$v = \int_{-(\pi/2-\alpha)}^{+(\pi/2-\beta)} \frac{\Gamma}{4\pi h}\cos\phi\, d\phi = \frac{\Gamma}{4\pi h}\left[\sin\left(\frac{\pi}{2}-\beta\right) + \sin\left(\frac{\pi}{2}-\alpha\right)\right]$$
$$= \frac{\Gamma}{4\pi h}(\cos\alpha + \cos\beta) \tag{5.9}$$

This result is of the utmost importance in what follows and is so often required that it is best committed to memory. All the values for induced velocity now to be used in this chapter are derived from this Eqn (5.9), that is limited to a straight line vortex of length AB.

The influence of a semi-infinite vortex (Fig. 5.11a)   If one end of the vortex stretches to infinity, e.g. end B, then $\beta = 0$ and $\cos\beta = 1$, so that Eqn (5.9) becomes

$$v = \frac{\Gamma}{4\pi h}(\cos\alpha + 1) \tag{5.10}$$

When the point P is opposite the end of the vortex (Fig. 5.11b), so that $\alpha = \pi/2$, $\cos\alpha = 0$, Eqn (5.9) becomes

$$v = \frac{\Gamma}{4\pi h} \tag{5.11}$$

The influence of an infinite vortex (Fig. 5.11c)   When $\alpha = \beta = 0$, Eqn (5.9) gives

$$v = \frac{\Gamma}{2\pi h} \tag{5.12}$$

and this will be recognized as the familiar expression for velocity due to the line vortex of Section 3.3.2. Note that this is twice the velocity induced by a semi-infinite vortex, a result that can be seen intuitively.

In nature, a vortex is a core of fluid rotating as though it were solid, and around which air flows in concentric circles. The vorticity associated with the vortex is confined to its core, so although an element of outside air is flowing in circles the element itself does not rotate. This is not easy to visualize, but a good analogy is with a car on a fairground big wheel. Although the car circulates round the axis of the wheel, the car does not rotate about its own axis. The top of the car is always at the top and the passengers are never upside down. The elements of air in the flow outside a vortex core behave in a very similar way.

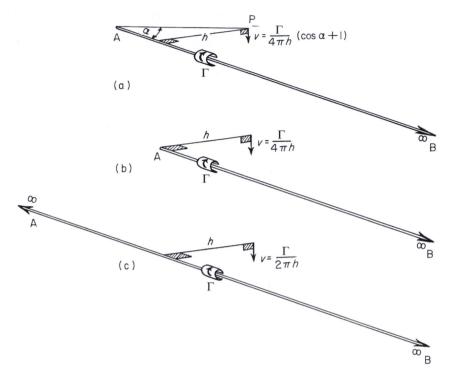

$$v = \frac{\Gamma}{4\pi h} (\cos \alpha + 1)$$

(a)

$$v = \frac{\Gamma}{4\pi h}$$

(b)

$$v = \frac{\Gamma}{2\pi h}$$

(c)

Fig. 5.11

5.2.3 Variation of velocity in vortex flow

To confirm how the velocity outside a vortex core varies with distance from the centre consider an element in a thin shell of air (Fig. 5.12). Here, flow conditions depend only on the distance from the centre and are constant all round the vortex at any given radius. The small element, which subtends the angle $\delta\theta$ at the centre, is

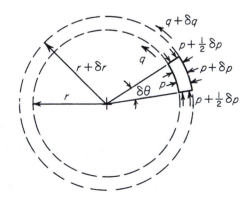

Fig. 5.12 Motion of an element outside a vortex core

circulating round the centre in steady motion under the influence of the force due to the radial pressure gradient.

Considering unit axial length, the inwards force due to the pressures is:

$$(p + \delta p)(r + \delta r)\delta\theta - pr\,\delta\theta - 2(p + \tfrac{1}{2}\delta p)\delta r\tfrac{1}{2}\delta\theta$$

which reduces to $\delta p(r - \tfrac{1}{2}\delta r)\delta\theta$. Ignoring $\tfrac{1}{2}\delta r$ in comparison with r, this becomes $r\,\delta p\,\delta\theta$. The volume of unit length of the element is $r\,\delta r\,\delta\theta$ and therefore its mass is $\rho r\,\delta r\,\delta\theta$. Its centripetal acceleration is (velocity)^2/radius, and the force required to produce this acceleration is:

$$\text{mass}\,\frac{(\text{velocity})^2}{\text{radius}} = \rho r\,\delta r\,\delta\theta\frac{q^2}{r} = \rho q^2\,\delta r\,\delta\theta$$

Equating this to the force produced by the pressure gradient leads to

$$r\,\delta p = \rho q^2\,\delta r \quad \text{since} \quad \delta\theta \neq 0 \tag{5.13}$$

Now, since the flow outside the vortex core is assumed to be inviscid, Bernoulli's equation for incompressible flow can be used to give, in this case,

$$p + \frac{1}{2}\rho q^2 = (p + \delta p) + \frac{1}{2}\rho(q + \delta q)^2$$

Expanding the term in $q + \delta q$, ignoring terms such as $(\delta q)^2$ as small, and cancelling, leads to:

$$\delta p + \rho q\,\delta q = 0$$

i.e.

$$\delta p = -\rho q\,\delta q \tag{5.14}$$

Substituting this value for δp in Eqn (5.13) gives

$$\rho q^2\,\delta r + \rho q r\,\delta q = 0$$

which when divided by ρq becomes

$$q\,\delta r + r\,\delta q = 0$$

But the left-hand side of this equation is $\delta(qr)$. Thus

$$\delta(qr) = 0$$

$$qr = \text{constant} \tag{5.15}$$

This shows that, in the inviscid flow round a vortex core, the velocity is inversely proportional to the radius (see also Section 3.3.2).

When the core is small, or assumed concentrated on a line axis, it is apparent from Eqn (5.15) that when r is small q can be very large. However, within the core the air behaves as though it were a solid cylinder and rotates at a uniform angular velocity. Figure 5.13 shows the variation of velocity with radius for a typical vortex.

The solid line represents the idealized case, but in reality the boundary is not so distinct, and the velocity peak is rounded off, after the style of the dotted lines.

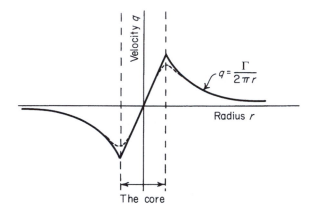

Fig. 5.13 Velocity distribution in a real vortex with a core

5.3 The simplified horseshoe vortex

A simplified system may replace the complete vortex system of a wing when considering the influence of the lifting system on distant points in the flow. Many such problems do exist and simple solutions, although not all exact, can be readily obtained using the suggested simplification. This necessitates replacing the wing by a single bound spanwise vortex of constant strength that is turned through 90° at each end to form the trailing vortices that extend effectively to infinity behind the wing. The general vortex system and its simplified equivalent must have two things in common:

(i) each must provide the same total lift
(ii) each must have the same value of circulation about the trailing vortices and hence the same circulation at mid-span.

These equalities provide for the complete definition of the simplified system.
 The spanwise distributions created for the general vortex system and its simplified equivalent are shown in Fig. 5.14. Both have the same mid-span circulation Γ_0 that is now constant along part of the span of the simplified equivalent case. For equivalence in area under the curve, which is proportional to the total lift, the span length of the single vortex must be less than that of the wing.

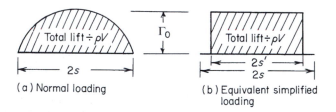

(a) Normal loading   (b) Equivalent simplified loading

Fig. 5.14

Thus

$$\Gamma_0 2s' = \text{area under general distribution} = \frac{\text{lift}}{\rho V}$$

Hence

$$\frac{s'}{s} = \frac{\text{total lift}}{2s\rho V\Gamma_0} \tag{5.16}$$

$2s'$ is the distance between the trailing vortex core centres. From Eqn (5.47a) (see page 246) it follows that

$$L = \rho V^2 s^2 2\pi A_1$$

and substituting also

$$\Gamma_0 = 4sV\Sigma A_n \sin n\frac{\pi}{2}$$

$$\frac{s'}{s} = \frac{\rho V^2 s^2 2\pi A_1}{2\rho V^2 4s^2 \Sigma A_n \sin n\frac{\pi}{2}}$$

$$= \frac{\pi}{4} \frac{A_1}{[A_1 - A_3 + A_5 - A_7 \ldots]}$$

For the general case then:

$$\frac{s}{s'} = \frac{4}{\pi} \left[1 - \frac{A_3}{A_1} + \frac{A_5}{A_1} - \frac{A_7}{A_1} \cdots \right] \tag{5.17}$$

For the simpler elliptic distribution (see Section 5.5.3 below):

$$A_3 = A_5 = A_7 = 0$$

$$s' = \left(\frac{\pi}{4}\right)s \tag{5.18}$$

In the absence of other information it is usual to assume that the separation of the trailing vortices is given by the elliptic case.

5.3.1 Formation flying effects

Aircraft flying in close proximity experience mutual interference effects and good estimates of these influences are obtained by replacing each aircraft in the formation by its equivalent simplified horseshoe vortex.

Consider the problem shown in Fig. 5.15 where three identical aircraft are flying in a vee formation at a forward speed V in the same horizontal plane. The total mutual interference is the sum of (i) that of the followers on the leader (1), (ii) that of the leader and follower (2) on (3), and (iii) that of leader and follower (3) on (2). (ii) and (iii) are identical.

(i) The leader is flying in a flow regime that has additional vertical flow components induced by the following vortices. Upward components appear from the bound vortices a_2c_2, a_3c_3, trailing vortices c_2d_2, a_3b_3 and downward

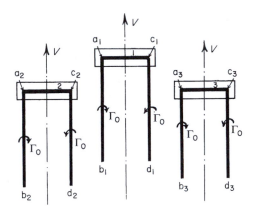

Fig. 5.15

components from the trailing vortices a_2b_2 and c_3d_3. The net result is an upwash on the leader.

(ii) These wings have additional influences to their own trails due to the leader and the other follower. Bound vortex a_1c_1 and trailing vortices a_1b_1, a_2b_2 produce downwashes. Again the net influence is an upwash.

From these simple considerations it appears that each aircraft is flying in a regime in which upward components are induced by the presence of the others. The upwash components reduce the downward velocities induced by the aircraft's own trail and hence its trailing vortex drag. Because of the reduction in drag, less power is required to maintain the forward velocity and the well-known operational fact emerges that each aircraft of a formation has a better performance than when flying singly. In most problems it is usual to assume that the wings have an elliptic distribution, and that the influence calculated for mid-span position is typical of the whole wing span. Also any curvature of the trails is neglected and the special forms of the Biot–Savart law (Section 5.2.2) are used unreservedly.

5.3.2 Influence of the downwash on the tailplane

On most aircraft the tailplane is between the trailing vortices springing from the mainplanes ahead and the flow around it is considerably influenced by these trails. Forces on aerofoils are proportional to the square of the velocity and the angle of incidence. Small velocity changes, therefore, have negligible effect unless they alter the incidence of the aerofoil, when they then have a significant effect on the force on the aerofoil.

Tailplanes work at incidences that are altered appreciably by the tilting of the relative wind due to the large downward induced velocity components. Each particular aircraft configuration will have its own geometry. The solution of a particular problem will be given here to show the method.

Example 5.1 Let the tailplane of an aeroplane be at distance x behind the wing centre of pressure and in the plane of the vortex trail (Fig. 5.16).

Assuming elliptic distribution, the semi-span of the bound vortex is given by Eqn (5.18) as

$$s' = \left(\frac{\pi}{4}\right)s$$

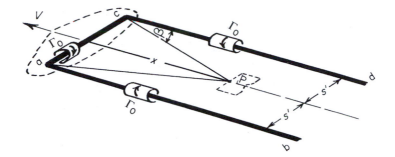

Fig. 5.16

The downwash at the mid-span point P of the tailplane caused by the wing is the sum of that caused by the bound vortex ac and that of each of the trailing vortices ab and cd. Using the special form of Biot–Savart equations (Section 5.2.2) the downwash at P:

$$w_P \downarrow = \frac{\Gamma_0}{4\pi x} 2\sin\beta + \frac{2\Gamma_0}{4\pi s'}(1+\cos\beta)$$
$$= \frac{\Gamma_0}{2\pi}\left(\frac{\sin\beta}{x} + \frac{1+\cos\beta}{s'}\right)$$

From the sketch $x = s'\cot\beta$ and $s' = (\pi/4)s$

$$w_P \downarrow = \frac{\Gamma_0}{2\pi}\left(\frac{\sin\beta}{s'\cot\beta} + \frac{1+\cos\beta}{s'}\right) = \frac{\Gamma_0}{2\pi s'}[1+\sec\beta]$$
$$= \frac{2\Gamma_0}{\pi^2 s}(1+\sec\beta)$$

Now by using the Kutta–Zhukovsky theorem, Eqn (4.10) and downwash angle

$$\varepsilon = \frac{w_P}{V}$$

$$\varepsilon = \frac{2C_L VS}{\pi^3 s^2 V}(1+\sec\beta)$$

or

$$\varepsilon = \frac{8C_L}{\pi^3 (AR)}(1+\sec\beta)$$

The derivative

$$\frac{\partial\varepsilon}{\partial\alpha} = \frac{\partial\varepsilon}{\partial C_L}\frac{\partial C_L}{\partial\alpha} = a_1\frac{\partial\varepsilon}{\partial C_L}$$

Thus

$$\frac{\partial\varepsilon}{\partial\alpha} = \frac{8a_1}{\pi^3 (AR)}(1+\sec\beta) \tag{5.19}$$

For cases when the distribution is non-elliptic or the tailplane is above or below the wing centre of pressure, the arithmetic of the problem is altered from that above, which applies only to this restricted problem. Again the mid-span point is taken as representative of the whole tailplane.

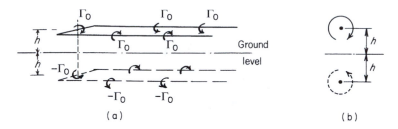

Fig. 5.17

5.3.3 Ground effects

In this section, the influence of solid boundaries on aeroplane (or model) perform-ance is estimated and once again the wing is replaced by the equivalent simplified horseshoe vortex.

Since this is a linear problem, the method of superposition may be used in the following way. If (Fig. 5.17b) a point vortex is placed at height h above a horizontal plane, and an equal but opposite vortex is placed at depth h below the plane, the vertical velocity component induced at any point on the plane by one of the vortices is equal and opposite to that due to the other. Thus the net vertical velocity, induced at any point on the plane, is zero. This shows that the superimposition of the image vortex is equivalent in effect to the presence of a solid boundary. In exactly the same way, the effect of a solid boundary on the horseshoe vortex can be modelled by means of an image horseshoe vortex (Fig. 5.17a). In this case, the boundary is the level ground and its influence on an aircraft h above is the same as that of the 'inverted' aircraft flying 'in formation' h below the ground level (Figs 5.17a and 5.18).

Before working out a particular problem, it is clear from the figure that the image system reduces the downwash on the wing and hence the drag and power required, as well as materially changing the downwash angle at the tail and hence the overall pitching equilibrium of the aeroplane.

Example 5.2 An aeroplane of weight W and span $2s$ is flying horizontally near the ground at altitude h and speed V. Estimate the reduction in drag due to ground effect. If $W = 22 \times 10^4 \, \text{N}$, $h = 15.2 \, \text{m}$, $s = 13.7 \, \text{m}$, $V = 45 \, \text{m s}^{-1}$, calculate the reduction in Newtons.
(U of L)

With the notation of Fig. 5.18 the change in downwash at y along the span is $\Delta w \uparrow$ where

$$\Delta w \uparrow = \frac{\Gamma_0}{4\pi r_1} \cos\theta_1 + \frac{\Gamma_0}{4\pi r_2} \cos\theta_2$$

$$\Delta w = \frac{\Gamma_0}{4\pi} \left(\frac{s' + y}{r_1^2} + \frac{s' - y}{r_2^2} \right)$$

On a strip of span δy at y from the centre-line,

$$\text{lift } l = \rho V \Gamma_0 \, \delta y$$

and change in vortex drag

$$\Delta d_v = \frac{l \Delta w}{V}$$

$$= \frac{\rho V \Gamma_0 \delta y \Delta w}{V} \qquad (5.20)$$

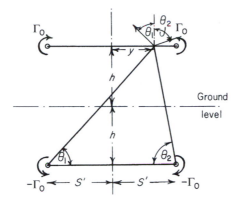

Fig. 5.18

Total change in drag ΔD_v across the span is the integral of Eqn (5.20) from $-s'$ to s' (or twice that from 0 to s'). Therefore

$$-\Delta D_v = 2 \int_0^{s'} \frac{\rho \Gamma_0^2}{4\pi} \left(\frac{s' + y}{r_1^2} + \frac{s' - y}{r_2^2} \right) dy$$

From the geometry, $r_1^2 = 4h^2 + (s' + y)^2$ and $r_2^2 = 4h^2 + (s' - y)^2$. Making these substitutions and evaluating the integral

$$-\Delta D_v = \frac{\rho \Gamma_0^2}{4\pi} \left[\ln \frac{4h^2 + (s' + y)^2}{4h^2 + (s' - y)^2} \right]_0^{s'}$$

$$= \frac{\rho \Gamma_0^2}{4\pi} \ln \left[1 + \left(\frac{s'}{h} \right)^2 \right]$$

With $W = \rho V \Gamma_0 \pi s$ i.e. and $s' = (\pi/4)s$ (assuming elliptic distribution):

$$\Delta D_v = \frac{W^2 \times 2}{\rho V^2 s^2 \pi^3} \ln \left(1 + \frac{\pi^2}{16} \frac{s^2}{h^2} \right)$$

and substituting the values given

$$\Delta D_v = 1390 \, \text{N}$$

A simpler approach is to assume that mid-span conditions are typical of the whole wing. With this the case

$$\theta_1 = \theta_2 = \theta = \arccos \frac{s'}{\sqrt{s'^2 + 4h^2}}$$

and the change in drag is to be 1524 N (a difference of about 10% from the first answer).

5.4 Vortex sheets

To estimate the influence of the near wake on the aerodynamic characteristics of a lifting wing it is useful to investigate the 'hypothetical' bound vortex in greater detail. For this the wing is replaced for the purposes of analysis by a sheet of vortex

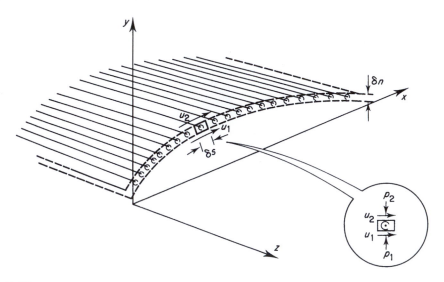

Fig. 5.19

filaments. In order to satisfy Helmholtz's second theorem (Section 5.2.1) each filament must either be part of a closed loop or form a horseshoe vortex with trailing vortex filaments running to infinity. Even with this restriction there are still infinitely many ways of arranging such vortex elements for the purposes of modelling the flow field associated with a lifting wing. For illustrative purposes consider the simple arrangement where there is a sheet of vortex filaments passing in the spanwise direction through a given wing section (Fig. 5.19). It should be noted, however, that at two, here unspecified, spanwise locations each of these filaments must be turned back to form trailing vortex filaments.

Consider the flow in the vicinity of a sheet of fluid moving irrotationally in the xy plane, Fig. 5.19. In this stylized figure the 'sheet' is seen to have a section curved in the xy plane and to be of thickness δn, and the vorticity is represented by a number of vortex filaments normal to the xy plane. The circulation around the element of fluid having sides δs, δn is, by definition, $\Delta\Gamma = \zeta\delta s \cdot \delta n$ where ζ is the vorticity of the fluid within the area $\delta s\,\delta n$.

Now for a sheet $\delta n \to 0$ and if ζ is so large that the product $\zeta\delta n$ remains finite, the sheet is termed a vortex sheet of strength $k = \zeta\,\delta n$. The circulation around the element can now be written

$$\Delta\Gamma = k\,\delta s \qquad (5.21)$$

An alternative way of finding the circulation around the element is to integrate the tangential flow components. Thus

$$\Delta\Gamma = (u_2 - u_1)\delta s \qquad (5.22)$$

Comparison of Eqns (5.21) and (5.22) shows that the local strength k of the vortex sheet is the tangential velocity jump through the sheet.

Alternatively, a flow situation in which the tangential velocity changes discontinuously in the normal direction may be mathematically represented by a vortex sheet of strength proportional to the velocity change.

The vortex sheet concept has important applications in wing theory.

5.4.1   The use of vortex sheets to model the lifting effects of a wing

In Section 4.3, it was shown that the flow around a thin wing could be regarded as a superimposition of a circulatory and a non-circulatory flow. In a similar fashion the same can be established for the flow around a thin wing. For a wing to be classified as thin the following must hold:

- The maximum thickness-to-chord ratio, usually located at mid-span, must be much less than unity.
- The camber lines of all wing sections must only deviate slightly from the corresponding chord-line.
- The wing may be twisted but the angles of incidence of all wing sections must remain small and the rate of change of twist must be gradual.
- The rate of change of wing taper must be gradual.

These conditions would be met for most practical wings. If they are satisfied then the velocities at any point over the wing only differ by a small amount from that of the oncoming flow.

For the thin aerofoil the non-circulatory flow corresponds to that around a symmetrical aerofoil at zero incidence. Similarly for the thin wing it corresponds to that around an untwisted wing, having the same planform shape as the actual wing, but with symmetrical sections at zero angle of incidence. Like its two-dimensional counterpart in aerofoil theory this so-called *displacement* (or *thickness*) *effect* makes no contribution to the lifting characteristics of the wing. The circulatory flow – the so-called *lifting effect* – corresponds to that around an infinitely thin, cambered and possibly twisted, plate at an angle of attack. The plate takes the same planform shape as the mid-plane of the actual wing. This circulatory part of the flow is modelled by a vortex sheet. The lifting characteristics of the wing are determined solely by this component of the flow field. Consequently, the lifting effect is of much greater practical interest than the displacement effect. Accordingly much of this chapter will be devoted to the former. First, however, the displacement effect is briefly considered.

Displacement effect

In Section 4.9, it was shown how the non-circulatory component of the flow around an aerofoil could be modelled by a distribution of sources and sinks along the chord line. Similarly, in the case of the wing, this component of the flow can be modelled by distributing sources and sinks over the entire mid-plane of the wing (Fig. 5.20). In much the same way as Eqn (4.103) was derived (referring to Fig. 5.20 for the geometric notation) it can be shown that the surface pressure coefficient at point (x_1, y_1) due to the thickness effect is given by

$$C_p = -2\frac{u'}{U} = \underbrace{\frac{1}{\pi} \int_{-s}^{s} \int_{x_l(z)}^{x_l(z)+c(z)} \frac{dy_t}{dx}(x,z) \frac{x - x_1}{[(x - x_1)^2 + (z - z_1)^2]^{3/2}} dx\,dz}_{I_1} \qquad (5.23)$$

where $x_l(z)$ denotes the leading edge of the wing.

In general, Eqn (5.23) is fairly cumbersome and nowadays modern computational techniques like the panel method (see Section 5.8) are used. In the special case of

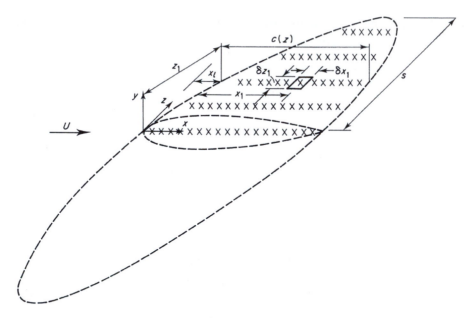

Fig. 5.20 Modelling the displacement effect by a distribution of sources

wings having high aspect ratio, intuition would suggest that the flow over most of the wing behaves as if it were two-dimensional. Plainly this will not be a good approximation near the wing-tips where the formation of the trailing vortices leads to highly three-dimensional flow. However, away from the wing-tip region, Eqn (5.23) reduces approximately to Eqn (4.103) and, to a good approximation, the C_p distributions obtained for symmetrical aerofoils can be used for the wing sections. For completeness this result is demonstrated formally immediately below. However, if this is not of interest go directly to the next section.

Change the variables in Eqn (5.23) to $\bar{x} = (x - x_1)/c$, $\bar{z}_1 = z_1/c$ and $\bar{z} = (z - z_1)/c$. Now provided that the non-dimensional shape of the wing-section does not change along the span, or, at any rate, only changes very slowly $S_t \equiv d(y_t/c)/d\bar{x}$ does not vary with $\bar{z}$ and the integral I_1 in Eqn (5.23) becomes

$$I_1 = \frac{1}{c} \int_0^1 S_t(\bar{x})\bar{x} \underbrace{\int_{-(s+z_1)/c}^{(s-z_1)/c} \frac{d\bar{z}}{(\bar{x}^2 + \bar{z}^2)^{3/2}}}_{I_2} d\bar{x}$$

To evaluate the integral I_2 change variable to $\chi = 1/\bar{z}$ so that

$$I_2 = -\int_{-c/(s+z_1)}^{-\infty} \frac{\chi}{(\bar{z}^2\chi^2 + 1)^{3/2}} d\chi - \int_{\infty}^{c/(s-z_1)} \frac{\chi}{(\bar{z}^2\chi^2 + 1)^{3/2}} d\chi$$

$$= \frac{1}{\bar{z}^2} \left[-\frac{1}{\sqrt{\left(\frac{\bar{z}_1 c}{s+z}\right)^2 + 1}} - \frac{1}{\sqrt{\left(\frac{\bar{z}_1 c}{s-z}\right)^2 + 1}} \right]$$

For large aspect ratios $s \gg c$, so provided z_1 is not close to $\pm s$, i.e. near the wing-tips,

$$\left(\frac{\bar{z}_1 c}{s + z_1}\right)^2 \ll 1 \quad \text{and} \quad \left(\frac{\bar{z}_1 c}{s - z_1}\right)^2 \ll 1$$

giving

$$I_2 \simeq -\frac{2}{z_1^2}$$

Thus Eqn (5.23) reduces to the two-dimensional result, Eqn (4.103), i.e.

$$C_p \simeq -\frac{2}{\pi} \int_{x_l}^{c+x_l} \frac{\mathrm{d}y_t}{\mathrm{d}x} \frac{1}{x - x_1} \mathrm{d}x \qquad (5.24)$$

Lifting effect

To understand the fundamental concepts involved in modelling the lifting effect of a vortex sheet, consider first the simple rectangular wing depicted in Fig. 5.21. Here the vortex sheet is constructed from a collection of horseshoe vortices located in the $y = 0$ plane.

From Helmholtz's second theorem (Section 5.2.1) the strength of the circulation round any section of the vortex sheet (or wing) is the sum of the strengths of the

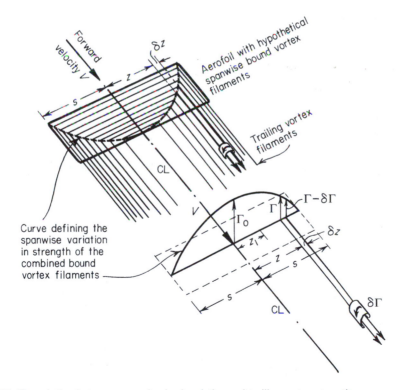

Fig. 5.21 The relation between spanwise load variation and trailing vortex strength

vortex filaments cut by the section plane. As the section plane is progressively moved outwards from the centre section to the tips, fewer and fewer bound vortex filaments are left for successive sections to cut so that the circulation around the sections diminishes. In this way, the spanwise change in circulation round the wing is related to the spanwise lengths of the bound vortices. Now, as the section plane is moved outwards along the bound bundle of filaments, and as the strength of the bundle decreases, the strength of the vortex filaments so far shed must increase, as the overall strength of the system cannot diminish. Thus the change in circulation from section to section is equal to the strength of the vorticity shed between these sections.

Figure 5.21 shows a simple rectangular wing shedding a vortex trail with each pair of trailing vortex filaments completed by a spanwise bound vortex. It will be noticed that a line joining the ends of all the spanwise vortices forms a curve that, assuming each vortex is of equal strength and given a suitable scale, would be a curve of the total strengths of the bound vortices at any section plotted against the span. This curve has been plotted for clarity on a spanwise line through the centre of pressure of the wing and is a plot of (chordwise) circulation (Γ) measured on a vertical ordinate, against spanwise distance from the centre-line (CL) measured on the horizontal ordinate. Thus at a section z from the centre-line sufficient hypothetical bound vortices are cut to produce a chordwise circulation around that section equal to Γ. At a further section $z + \delta z$ from the centre-line the circulation has fallen to $\Gamma - \delta\Gamma$, indicating that between sections z and $z + \delta z$ trailing vorticity to the strength of $\delta\Gamma$ has been shed.

If the circulation curve can be described as some function of z, $f(z)$ say then the strength of circulation shed

$$\delta\Gamma = -\frac{\mathrm{d}f(z)}{\mathrm{d}z}\delta z \tag{5.25}$$

Now at any section the lift per span is given by the Kutta–Zhukovsky theorem Eqn (4.10)

$$l = \rho V \Gamma$$

and for a given flight speed and air density, Γ is thus proportional to l. But l is the local intensity of lift or lift grading, which is either known or is the required quantity in the analysis.

The substitution of the wing by a system of bound vortices has not been rigorously justified at this stage. The idea allows a relation to be built up between the physical load distribution on the wing, which depends, as shall be shown, on the wing geometric and aerodynamic parameters, and the trailing vortex system.

Figure 5.21 illustrates two further points:

(a) It will be noticed from the leading sketch that the trailing filaments are closer together when they are shed from a rapidly diminishing or changing distribution curve. Where the filaments are closer the strength of the vorticity is greater. Near the tips, therefore, the shed vorticity is the most strong, and at the centre where the distribution curve is flattened out the shed vorticity is weak to infinitesimal.

(b) A wing infinitely long in the spanwise direction, or in two-dimensional flow, will have constant spanwise loading. The bundle will have filaments all of equal length and none will be turned back to form trailing vortices. Thus there is no trailing vorticity associated with two-dimensional wings. This is capable of deduction by a more direct process, i.e. as the wing is infinitely long in the spanwise direction the lower surface (high) and upper surface (low) pressures cannot tend to equalize by spanwise components of velocity so that the streams of air meeting at the trailing edge after sweeping under and over the wing have no opposite spanwise motions but join up in symmetrical flow in the direction of motion. Again no trailing vorticity is formed.

A more rigorous treatment of the vortex-sheet modelling is now considered. In Section 4.3 it was shown that, without loss of accuracy, for thin aerofoils the vortices could be considered as being distributed along the chord-line, i.e. the x axis, rather than the camber line. Similarly, in the present case, the vortex sheet can be located on the (x, z) plane, rather than occupying the cambered and possibly twisted mid-surface of the wing. This procedure greatly simplifies the details of the theoretical modelling.

One of the infinitely many ways of constructing a suitable vortex-sheet model is suggested by Fig. 5.21. This method is certainly suitable for wings with a simple planform shape, e.g. a rectangular wing. Some wing shapes for which it is not at all suitable are shown in Fig. 5.22. Thus for the general case an alternative model is required. In general, it is preferable to assign an individual horseshoe vortex of strength $k\,(x, z)$ per unit chord to each element of wing surface (Fig. 5.23). This method of constructing the vortex sheet leads to certain mathematical difficulties

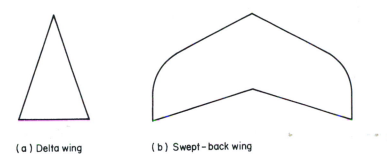

(a) Delta wing   (b) Swept-back wing

Fig. 5.22

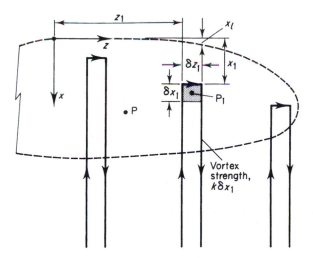

Fig. 5.23 Modelling the lifting effect by a distribution of horseshoe vortex elements

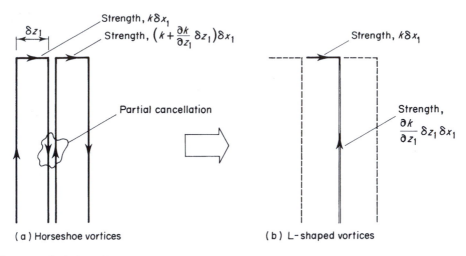

Fig. 5.24 Equivalence between distributions of (a) horseshoe and (b) L-shaped vortices

when calculating the induced velocity. These problems can be overcome by recombining the elements in the way depicted in Fig. 5.24. Here it is recognized that partial cancellation occurs for two elemental horseshoe vortices occupying adjacent spanwise positions, z and $z + \delta z$. Accordingly, the horseshoe-vortex element can be replaced by the L-shaped vortex element shown in Fig. 5.24. Note that although this arrangement appears to violate Helmholtz's second theorem, it is merely a mathematically convenient way of expressing the model depicted in Fig. 5.23 which fully satisfies this theorem.

5.5 Relationship between spanwise loading and trailing vorticity

It is shown below in Section 5.5.1 how to calculate the velocity induced by the elements of the vortex sheet that notionally replace the wing. This is an essential step in the development of a general wing theory. Initially, the general case is considered. Then it is shown how the general case can be very considerably simplified in the special case of wings of high aspect ratio. The general case is then dropped, to be taken up again in Section 5.8, and the assumption of large aspect ratio is made for Section 5.6 and the remainder of the present section. Accordingly, some readers may wish to pass over the material immediately below and go directly to the alternative derivation of Eqn (5.32) given at the end of the present section.

5.5.1 Induced velocity (downwash)

Suppose that it is required to calculate the velocity induced at the point $P_1(x_1, z_1)$ in the $y = 0$ plane by the L-shaped vortex element associated with the element of wing surface located at point $P(x, z)$ now relabelled A (Fig. 5.25).

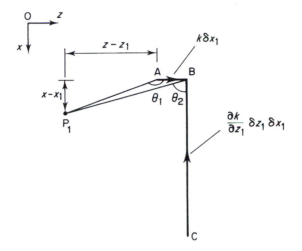

Fig. 5.25 Geometric notation for L-shaped vortex element

Making use of Eqn (5.9) it can be seen that this induced velocity is perpendicular to the $y = 0$ plane and can be written as

$$\delta v_{\mathrm{i}}(x_1, z_1) = (\delta v_{\mathrm{i}})_{\mathrm{AB}} + (\delta v_{\mathrm{i}})_{\mathrm{BC}}$$
$$= -\frac{k\,\delta x}{4\pi(x - x_1)}\left[\cos\theta_1 - \cos\left(\theta_2 + \frac{\pi}{2}\right)\right] + \frac{1}{4\pi}\frac{\partial k}{\partial z}\delta z\,\delta x\frac{(1 + \cos\theta_2)}{(z + \delta z - z_1)} \quad (5.26)$$

From the geometry of Fig. 5.25 the various trigonometric expressions in Eqn (5.26) can be written as

$$\cos\theta_1 = \frac{z - z_1}{\sqrt{(x - x_1)^2 + (z - z_1)^2}}$$

$$\cos\theta_2 = -\frac{x - x_1}{\sqrt{(x - x_1)^2 + (z + \delta z - z_1)^2}}$$

$$\cos\left(\theta_2 + \frac{\pi}{2}\right) = -\sin\theta_2 = \frac{z + \delta z - z_1}{\sqrt{(x - x_1)^2 + (z + \delta z - z_1)^2}}$$

The binomial expansion, i.e.

$$(a + b)^n = a^n + na^{n-1}b + \cdots,$$

can be used to expand some of the terms, for example

$$[(x - x_1)^2 + (z + \delta z - z_1)^2]^{-1/2} = \frac{1}{r} - \frac{(z - z_1)}{r^3}\delta z + \cdots$$

where $r = \sqrt{(x - x_1)^2 + (z - z_1)^2}$. In this way, the trigonometric expressions given above can be rewritten as

$$\cos\theta_1 = \frac{z - z_1}{r} \tag{5.27}$$

$$\cos\theta_2 = -\frac{x - x_1}{r} + \frac{(x - x_1)(z - z_1)}{r^3}\delta z + \cdots \tag{5.28}$$

$$\cos\left(\theta_2 + \frac{\pi}{2}\right) = \frac{z - z_1}{r} + \left[\frac{1}{r} - \frac{(z - z_1)^2}{r^3}\right]\delta z + \cdots \tag{5.29}$$

Equations (5.27 to 5.29) are now substituted into Eqn (5.26), and terms involving $(\delta z)^2$ and higher powers are ignored, to give

$$\delta v_i = \frac{k}{4\pi}\delta x\,\delta z\frac{(x - x_1)}{r^3} + \frac{1}{4\pi}\frac{\partial k}{\partial z}\delta x\,\delta z\left[\frac{1}{z - z_1} - \frac{x - x_1}{r(z - z_1)}\right] \tag{5.30}$$

In order to obtain the velocity induced at P_1 due to all the horseshoe vortex elements, δv_i is integrated over the entire wing surface projected on to the (x, z) plane. Thus using Eqn (5.30) leads to

$$v_i(x_1, z_1) = \frac{1}{4\pi}\int_{-s}^{s}\int_{x_l}^{x_l + c}\left\{\frac{\partial k}{\partial z}\underbrace{\left[\underbrace{\frac{1}{z - z_1}}_{(a)} - \underbrace{\frac{x - x_1}{r(z - z_1)}}_{(b)}\right]} + \underbrace{k\frac{(x - x_1)}{r^3}}_{(c)}\right\}dx\,dz \tag{5.31}$$

The induced velocity at the wing itself and in its wake is usually in a downwards direction and accordingly, is often called the *downwash*, w, so that $w = -v_i$.

It would be a difficult and involved process to develop wing theory based on Eqn (5.31) in its present general form. Nowadays, similar vortex-sheet models are used by the panel methods, described in Section 5.8, to provide computationally based models of the flow around a wing, or an entire aircraft. Accordingly, a discussion of the theoretical difficulties involved in using vortex sheets to model wing flows will be postponed to Section 5.8. The remainder of the present section and Section 5.6 is devoted solely to the special case of unswept wings having high aspect ratio. This is by no means unrealistically restrictive, since aerodynamic considerations tend to dictate the use of wings with moderate to high aspect ratio for low-speed applications such as gliders, light aeroplanes and commuter passenger aircraft. In this special case Eqn (5.31) can be very considerably simplified.

This simplification is achieved as follows. For the purposes of determining the aerodynamic characteristics of the wing it is only necessary to evaluate the induced velocity at the wing itself. Accordingly the ranges for the variables of integration are given by $-s \leq z \leq s$ and $0 \leq x \leq (c)_{max}$. For high aspect ratios $s/c \gg 1$ so that $|x - x_1| \ll r$ over most of the range of integration. Consequently, the contributions of terms (b) and (c) to the integral in Eqn (5.31) are very small compared to that of term (a) and can therefore be neglected. This allows Eqn (5.31) to be simplified to

$$v_i(z_1) = -w(z_1) = \frac{1}{4\pi}\int_{-s}^{s}\frac{d\Gamma}{dz}\frac{1}{z - z_1}dz \tag{5.32}$$

where, as explained in Section 5.4.1, owing to Helmholtz's second theorem

$$\Gamma(z) = \int_{x_l}^{c(z) + x_l} k(x, z)dx \tag{5.33}$$

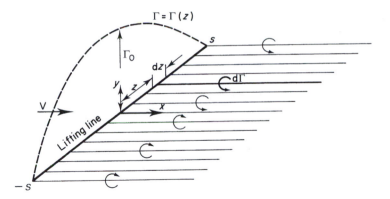

Fig. 5.26 Prandtl's lifting line model

is the total circulation due to all the vortex filaments passing through the wing section at z. Physically the approximate theoretical model implicit in Eqn (5.32) and (5.33) corresponds to replacing the wing by a single bound vortex having variable strength Γ, the so-called *lifting line* (Fig. 5.26). This model, together with Eqns (5.32) and (5.33), is the basis of Prandtl's general wing theory which is described in Section 5.6. The more involved theories based on the full version of Eqn (5.31) are usually referred to as *lifting surface* theories.

Equation (5.32) can also be deduced directly from the simple, less general, theoretical model illustrated in Fig. 5.21. Consider now the influence of the trailing vortex filaments of strength $\delta\Gamma$ shed from the wing section at z in Fig. 5.21. At some other point z_1 along the span, according to Eqn (5.11), an induced velocity equal to

$$\delta v_i(z_1) = \frac{1}{4\pi(z - z_1)} \frac{\mathrm{d}f}{\mathrm{d}z} \delta z$$

will be felt in the downwards direction in the usual case of positive vortex strength. All elements of shed vorticity along the span add their contribution to the induced velocity at z_1 so that the total influence of the trailing system at z_1 is given by Eqn (5.32).

5.5.2 The consequences of downwash – trailing vortex drag

The induced velocity at z_1 is, in general, in a downwards direction and is sometimes called downwash. It has two very important consequences that modify the flow about the wing and alter its aerodynamic characteristics.

Firstly, the downwash that has been obtained for the particular point z_1 is felt to a lesser extent ahead of z_1 and to a greater extent behind (see Fig. 5.27), and has the effect of tilting the resultant oncoming flow at the wing (or anywhere else within its influence) through an angle

$$\varepsilon = \tan^{-1}\frac{w}{V} \simeq \frac{w}{V}$$

where w is the local downwash. This reduces the effective incidence so that for the same lift as the equivalent infinite wing or two-dimensional wing at incidence α_∞ an incidence $\alpha = \alpha_\infty + \varepsilon$ is required at that section on the finite wing. This is illustrated in Fig. 5.28, which in addition shows how the two-dimensional lift L_∞ is normal to

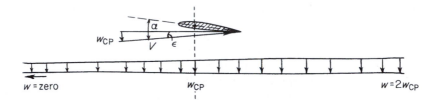

Fig. 5.27 Variation in magnitude of downwash in front of and behind wing

the resultant velocity V_R and is, therefore, tilted back against the actual direction of motion of the wing V. The two-dimensional lift L_∞ is resolved into the aerodynamic forces L and D_v respectively, normal to and against the direction of the forward velocity of the wing. Thus the second important consequence of downwash emerges. This is the generation of a drag force D_v. This is so important that the above sequence will be explained in an alternative way.

A section of a wing generates a circulation of strength Γ. This circulation superimposed on an apparent oncoming flow velocity V produces a lift force $L_\infty = \rho V \Gamma$ according to the Kutta–Zhukovsky theorem (4.10), which is normal to the apparent oncoming flow direction. The apparent oncoming flow felt by the wing section is the resultant of the forward velocity and the downward induced velocity arising from the trailing vortices. Thus the aerodynamic force L_∞ produced by the combination of Γ and V appears as a lift force L normal to the forward motion and a drag force D_v against the normal motion. This drag force is called *trailing vortex drag*, abbreviated to *vortex drag* or more commonly *induced drag* (see Section 1.5.7).

Considering for a moment the wing as a whole moving through air at rest at infinity, two-dimensional wing theory suggests that, taking air as being of small to negligible viscosity, the static pressure of the free stream ahead is recovered behind the wing. This means roughly that the kinetic energy induced in the flow is converted back to pressure energy and zero drag results. The existence of a thin boundary layer and narrow wake is ignored but this does not really modify the argument.

In addition to this motion of the airstream, a finite wing spins the airflow near the tips into what eventually becomes two trailing vortices of considerable core size. The generation of these vortices requires a quantity of kinetic energy that is not recovered

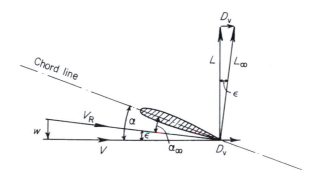

Fig. 5.28 The influence of downwash on wing velocities and forces: w = downwash; V = forward speed of wing; V_R = resultant oncoming flow at wing; α = incidence; ε = downwash angle = w/V; $\alpha_\infty = (\alpha - \varepsilon)$ = equivalent two-dimensional incidence; L_∞ = two-dimensional lift; L = wing lift; D_v = trailing vortex drag

by the wing system and that in fact is lost to the wing by being left behind. This constant expenditure of energy appears to the wing as the induced *drag*. In what follows, a third explanation of this important consequence of downwash will be of use. Figure 5.29 shows the two velocity components of the apparent oncoming flow superimposed on the circulation produced by the wing. The forward flow velocity produces the lift and the downwash produces the vortex drag per unit span.

Thus the lift per unit span of a finite wing (*l*) (or the load grading) is by the Kutta–Zhukovsky theorem:

$$l = \rho V \Gamma$$

the total lift being

$$L = \int_{-s}^{s} \rho V \Gamma \, dz \qquad (5.34)$$

The induced drag per unit span (d_v), or the induced drag grading, again by the Kutta–Zhukovsky theorem is

$$d_v = \rho w \Gamma \qquad (5.35)$$

and by similar integration over the span

$$D_v = \int_{-s}^{s} \rho w \Gamma \, dz \qquad (5.36)$$

This expression for D_v shows conclusively that if w is zero all along the span then D_v is zero also. Clearly, if there is no trailing vorticity then there will be no induced drag. This condition arises when a wing is working under two-dimensional conditions, or if all sections are producing zero lift.

As a consequence of the trailing vortex system, which is produced by the basic lifting action of a (finite span) wing, the wing characteristics are considerably modified, almost always adversely, from those of the equivalent two-dimensional wing of the same section. Equally, a wing with flow systems that more nearly approach the two-dimensional case will have better aerodynamic characteristics than one where

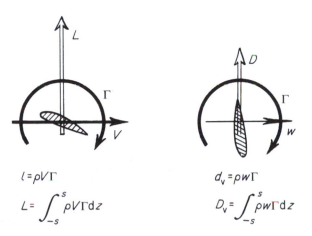

Fig. 5.29 Circulation superimposed on forward wind velocity and downwash to give lift and vortex drag (induced drag) respectively

the end-effects are more dominant. It seems therefore that a wing that is large in the spanwise dimension, i.e. large aspect ratio, is a better wing – nearer the ideal – than a short span wing of the same section. It would thus appear that a wing of large aspect ratio will have better aerodynamic characteristics than one of the same section with a lower aspect ratio. For this reason, aircraft for which aerodynamic efficiency is paramount have wings of high aspect ratio. A good example is the glider. Both the man-made aircraft and those found in nature, such as the albatross, have wings with exceptionally high aspect ratios.

In general, the induced velocity also varies in the chordwise direction, as is evident from Eqn (5.31). In effect, the assumption of high aspect ratio, leading to Eqn (5.32), permits the chordwise variation to be neglected. Accordingly, the lifting character-istics of a section from a wing of high aspect ratio at a local angle of incidence $\alpha(z)$ are identical to those for a two-dimensional wing at an effective angle of incidence $\alpha(z) - \varepsilon$. Thus Prandtl's theory shows how the two-dimensional aerofoil character-istics can be used to determine the lifting characteristics of wings of finite span. The calculation of the *induced angle of incidence* ε now becomes the central problem. This poses certain difficulties because ε depends on the circulation, which in turn is closely related to the lift per unit span. The problem therefore, is to some degree circular in nature which makes a simple direct approach to its solution impossible. The required solution procedure is described in Section 5.6.

Before passing to the general theory in Section 5.6, whereby the spanwise circula-tion distribution must be determined as part of the overall process, the much simpler inverse problem of a specified spanwise circulation distribution is considered in some detail in the next subsection. Although this is a special case it nevertheless leads to many results of practical interest. In particular, a simple quantitative result emerges that reinforces the qualitative arguments given above concerning the greater aero-dynamic efficiency of wings with high aspect ratio.

5.5.3 The characteristics of a simple symmetric loading – elliptic distribution

In order to demonstrate the general method of obtaining the aerodynamic charac-teristics of a wing from its loading distribution the simplest load expression for symmetric flight is taken, that is a semi-ellipse. In addition, it will be found to be a good approximation to many (mathematically) more complicated distributions and is thus suitable for use as first predictions in performance estimates.

The spanwise variation in circulation is taken to be represented by a semi-ellipse having the span ($2s$) as major axis and the circulation at mid-span (Γ_0) as the semi-minor axis (Fig. 5.30). From the general expression for an ellipse

$$\frac{\Gamma^2}{\Gamma_0^2} + \frac{z^2}{s^2} = 1$$

or

$$\Gamma = \Gamma_0 \sqrt{1 - \left(\frac{z}{s}\right)^2} \tag{5.37}$$

This expression can now be substituted in Eqns (5.32), (5.34) and (5.36) to find the lift, downwash and vortex drag on the wing.

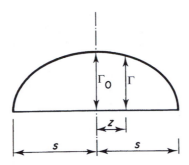

Fig. 5.30 Elliptic loading

Lift for elliptic distribution

From Eqn (5.34)

$$L = \int_{-s}^{s} \rho V \Gamma \, \mathrm{d}z = \int_{-s}^{s} \rho V \Gamma_0 \sqrt{1 - \left(\frac{z}{s}\right)^2} \, \mathrm{d}z$$

i.e.

$$L = \rho V \Gamma_0 \pi \frac{s}{2} \tag{5.38}$$

whence

$$\Gamma_0 = \frac{L}{\left(\frac{1}{2}\rho V \pi s\right)}$$

or introducing

$$L = C_L \frac{1}{2} \rho V^2 S$$

$$\Gamma_0 = \frac{C_L V S}{(\pi s)} \tag{5.39}$$

giving the mid-span circulation in terms of the overall aerofoil lift coefficient and geometry.

Downwash for elliptic distribution

Here

$$\frac{\mathrm{d}\Gamma}{\mathrm{d}z} = -\Gamma_0 \frac{z}{s^2} \left[1 - \left(\frac{z}{s}\right)^2\right]^{-1/2} = -\frac{\Gamma_0}{s} \frac{z}{\sqrt{s^2 - z^2}}$$

Substituting this in Eqn (5.32)

$$w_{z_1} = \frac{\Gamma_0}{4\pi s} \int_{-s}^{s} \frac{z \, \mathrm{d}z}{\sqrt{s^2 - z^2}(z - z_1)}$$

Writing the numerator as $(z - z_1) + z_1$:

$$w_{z_1} = \frac{\Gamma_0}{4\pi s} \int_{-s}^{s} \frac{(z - z_1) + z_1}{\sqrt{s^2 - z^2}(z - z_1)} dz$$

$$= \frac{\Gamma_0}{4\pi s} \left[\int_{-s}^{s} \frac{dz}{\sqrt{s^2 - z^2}} + z_1 \int_{-s}^{s} \frac{dz}{\sqrt{s^2 - z^2}(z - z_1)} \right]$$

Evaluating the first integral which is standard and writing I for the second

$$w_{z_1} = \frac{\Gamma_0}{4\pi s} [\pi + z_1 I] \tag{5.40}$$

Now as this is a symmetric flight case, the shed vorticity is the same from each side of the wing and the value of the downwash at some point z_1 is identical to that at the corresponding point $- z_1$ on the other wing.

So substituting for $\pm z_1$ in Eqn (5.40) and equating:

$$w_{\pm z_1} = \frac{\Gamma_0}{4\pi s} [\pi + z_1 I] = \frac{\Gamma_0}{4\pi s} [\pi - z_1 I]$$

This identity is satisfied only if $I = 0$, so that for any point $z - z_1$ along the span

$$w = \frac{\Gamma_0}{4s} \tag{5.41}$$

This important result shows that the downwash is constant along the span.

Induced drag (vortex drag) for elliptic distribution

From Eqn (5.36)

$$D_V = \int_{-s}^{s} \rho w \Gamma dz = \int_{-s}^{s} \rho \frac{\Gamma_0}{4s} \Gamma_0 \sqrt{1 - \left(\frac{z}{s}\right)^2} dz$$

whence

$$D_V = \frac{\pi}{8} \rho \Gamma_0^2 \tag{5.42}$$

Introducing

$$D_V = C_{D_V} \frac{1}{2} \rho V^2 S$$

and from Eqn (5.39)

$$\Gamma_0 = \frac{C_L V S}{\pi s}$$

Eqn (5.42) gives

$$C_{D_V} \frac{1}{2} \rho V^2 S = \frac{\pi}{8} \rho \left(\frac{C_L V S}{\pi s}\right)^2$$

or

$$C_{D_V} = \frac{C_L^2}{[\pi(AR)]} \tag{5.43}$$

since

$$\frac{4s^2}{S} = \frac{\text{span}^2}{\text{area}} = \text{aspect ratio}(AR)$$

Equation (5.43) establishes quantitatively how C_{D_V} falls with a rise in (AR) and confirms the previous conjecture given above, Eqn (5.36), that at zero lift in symmetric flight C_{D_V} is zero and the other condition that as (AR) increases (to infinity for two-dimensional flow) C_{D_V} decreases (to zero).

5.5.4 The general (series) distribution of lift

In the previous section attention was directed to distributions of circulation (or lift) along the span in which the load is assumed to fall symmetrically about the centre-line according to a particular family of load distributions. For steady symmetric manoeuvres this is quite satisfactory and the previous distribution formula may be arranged to suit certain cases. Its use, however, is strictly limited and it is necessary to seek further for an expression that will satisfy every possible combination of wing design parameter and flight manoeuvre. For example, it has so far been assumed that the wing was an isolated lifting surface that in straight steady flight had a load distribution rising steadily from zero at the tips to a maximum at mid-span (Fig. 5.31a). The general wing, however, will have a fuselage located in the centre sections that will modify the loading in that region (Fig. 5.31b), and engine nacelles or other excrescences may deform the remainder of the curve locally.

The load distributions on both the isolated wing and the general aeroplane wing will be considerably changed in anti-symmetric flight. In rolling, for instance, the upgoing wing suffers a large decrease in lift, which may become negative at some incidences (Fig. 5.31c). With ailerons in operation the curve of spanwise loading for a wing is no longer smooth and symmetrical but can be rugged and distorted in shape (Fig. 5.31d).

It is clearly necessary to find an expression that will accommodate all these various possibilities. From previous work the formula $l = \rho V \Gamma$ for any section of span is familiar. Writing l in the form of the non-dimensional lift coefficient and equating to $\rho V \Gamma$:

$$\Gamma = \frac{C_L}{2} Vc \tag{5.44}$$

is easily obtained. This shows that for a given steady flight state the circulation at any section can be represented by the product of the forward velocity and the local chord.

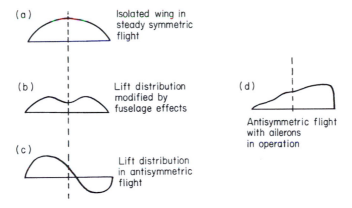

(a) Isolated wing in steady symmetric flight

(b) Lift distribution modified by fuselage effects

(c) Lift distribution in antisymmetric flight

(d) Antisymmetric flight with ailerons in operation

Fig. 5.31 Typical spanwise distributions of lift

Now in addition the local chord can be expressed as a fraction of the semi-span s, and with this fraction absorbed in a new number and the numeral 4 introduced for later convenience, Γ becomes:

$$\Gamma = 4C_\Gamma s$$

where C_Γ is dimensionless circulation which will vary similarly to Γ across the span. In other words, C_Γ is the shape parameter or variation of the Γ curve and being dimensionless it can be expressed as the Fourier sine series $\Sigma_1^\infty A_n \sin n\theta$ in which the coefficients A_n represent the amplitudes, and the sum of the successive harmonics describes the shape. The sine series was chosen to satisfy the end conditions of the curve reducing to zero at the tips where $y = \pm s$. These correspond to the values of $\theta = 0$ and π. It is well understood that such a series is unlimited in angular measure but the portions beyond 0 and π can be disregarded here. Further, the series can fit any shape of curve but, in general, for rapidly changing distributions as shown by a rugged curve, for example, many harmonics are required to produce a sum that is a good representation.

In particular the series is simplified for the symmetrical loading case when the even terms disappear (Fig. 5.32 (II)). For the symmetrical case a maximum or minimum must appear at the mid-section. This is only possible for sines of odd values of $\pi/2$. That is, the symmetrical loading must be the sum of symmetrical harmonics. Odd

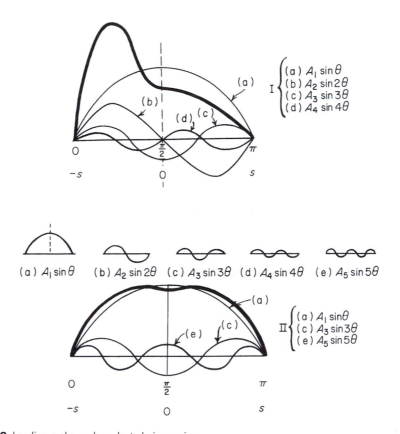

Fig. 5.32 Loading make-up by selected sine series

harmonics are symmetrical. Even harmonics, on the other hand, return to zero again at $\pi/2$ where in addition there is always a change in sign. For any asymmetry in the loading one or more even harmonics are necessary.

With the number and magnitude of harmonics effectively giving all possibilities the general spanwise loading can be expressed as

$$\Gamma = 4sV \sum_1^\infty A_n \sin n\theta \tag{5.45}$$

It should be noted that since $l = \rho V \Gamma$ the spanwise lift distribution can be expressed as

$$l = 4\rho V^2 s \sum_1^\infty A_n \sin n\theta \tag{5.46}$$

The aerodynamic characteristics for symmetrical general loading are derived in the next subsection. The case of asymmetrical loading is not included. However, it may be dealt with in a very similar manner, and in this way expressions derived for such quantities as rolling and yawing moment.

5.5.5 Aerodynamic characteristics for symmetrical general loading

The operations to obtain lift, downwash and drag vary only in detail from the previous cases.

Lift on the wing

$$L = \int_{-s}^{s} \rho V \Gamma \, dz$$

and changing the variable $z = -s \cos \theta$,

$$L = \int_0^\pi \rho V \Gamma s \sin \theta \, d\theta$$

and substituting for the general series expression

$$L = \int_0^\pi \rho V s^2 \sum A_n \sin n\theta \sin \theta \, d\theta$$

$$= 4s^2 \rho V^2 \int_0^\pi \sum A_n [\cos(n-1)\theta - \cos(n+1)\theta] d\theta$$

$$= 4s^2 \rho V^2 \frac{1}{2} \left[\sum A_n \left(\frac{\sin(n-1)\theta}{n-1} - \frac{\sin(n+1)\theta}{n+1} \right) \right]_0^\pi$$

The sum within the squared bracket equals zero for all values of n other than unity when it becomes

$$\left[\lim_{(n-1) \to 0} A_1 \frac{\sin(n-1)\theta}{n-1} \right]_0^\pi = A_1 \pi$$

Thus

$$L = A_1 \pi \frac{1}{2} \rho V^2 4s^2 = C_L \frac{1}{2} \rho V^2 S$$

and writing aspect ratio $(AR) = 4s^2/S$ gives

$$C_L = \pi A_1 (AR) \qquad (5.47)$$

This indicates the rather surprising result that the lift depends on the magnitude of the coefficient of the first term only, no matter how many more may be present in the series describing the distribution. This is because the terms $A_3 \sin 3\theta$, $A_5 \sin 5\theta$, etc., provide positive lift on some sections and negative lift on others so that the overall effect of these is zero. These terms provide the characteristic variations in the spanwise distribution but do not affect the total lift of the whole which is determined solely from the amplitude of the first harmonic. Thus

$$C_L = \pi(AR)A_1 \qquad \text{and} \qquad L = 2\pi\rho V^2 s^2 A_1 \qquad (5.47a)$$

Downwash

Changing the variable and limits of Eqn (5.32), the equation for the downwash is

$$w_{\theta_1} = \frac{1}{4\pi s} \int_0^\pi \frac{\frac{d\Gamma}{d\theta} \, d\theta}{\cos\theta - \cos\theta_1}$$

In this case $\Gamma = 4sV \sum A_n \sin n\theta$ and thus on differentiating

$$\frac{d\Gamma}{d\theta} = 4sV \sum nA_n \cos n\theta$$

Introducing this into the integral expression gives

$$w_{\theta_1} = \frac{4sV}{4\pi s} \int_0^\pi \frac{\sum nA_n \cos n\theta}{\cos\theta - \cos\theta_1} \, d\theta$$

$$= \frac{V}{\pi} \sum nA_n G_n$$

and writing in $G_n = \pi \sin n\theta_1 / \sin\theta_1$ from Appendix 3, and reverting back to the general point θ:

$$w = V \frac{\sum nA_n \sin n\theta}{\sin\theta} \qquad (5.48)$$

This involves all the coefficients of the series, and will be symmetrically distributed about the centre line for odd harmonics.

Induced drag (vortex drag)

The drag grading is given by $d_v = \rho w\Gamma$. Integrating gives the total induced drag

$$D_v = \int_{-s}^{s} \rho w\Gamma \, dz$$

or in the polar variable

$$D_v = \int_0^\pi \rho \underbrace{\frac{V\sum nA_n \sin n\theta}{\sin\theta}}_{w} \underbrace{4sV\sum A_n \sin n\theta}_{\Gamma} \underbrace{s\sin\theta \, d\theta}_{dz}$$

$$= \rho V^2 s^2 \int_0^\pi \sum nA_n \sin\theta \sum A_n \sin n\theta \, d\theta$$

The integral becomes

$$\frac{\pi}{2}\sum nA_n^2$$

This can be demonstrated by multiplying out the first three (say) odd harmonics, thus:

$$I = \int_0^\pi (A_1 \sin\theta + 3A_3 \sin 3\theta + 5A_5 \sin 5\theta)(A_1 \sin\theta + A_3 \sin 3\theta + A_5 \sin\theta)d\theta$$

$$= \int_0^\pi \{A_1^2 \sin^2\theta + 3A_3^2 \sin^2\theta + 5A_5^2 \sin^2\theta + [A_1 A_3 \sin\theta \sin 3\theta \text{ and}$$

other like terms which are products of different multiples of θ]$\}\, d\theta$

On carrying out the integration from 0 to π all terms other than the squared terms vanish leaving

$$I = \int_0^\pi (A_1^2 \sin^2\theta + 3A_3^2 \sin^2 3\theta + 5A_5^2 \sin^2 5\theta + \cdots)d\theta$$

$$= \frac{\pi}{2}[A_1^2 + 3A_3^2 + 5A_5^2 + \cdots] = \frac{\pi}{2}\sum nA_n^2$$

This gives

$$D_V = 4\rho V^2 s^2 \frac{\pi}{2}\sum nA_n^2 = C_{Dv}\frac{1}{2}\rho V^2 S$$

whence

$$C_{Dv} = \pi(AR)\sum nA_n^2 \tag{5.49}$$

From Eqn (5.47)

$$A_1^2 = \frac{C_L^2}{\pi^2(AR)^2}$$

and introducing this into Eqn (5.49)

$$C_{D_v} = \frac{C_L^2}{\pi(AR)}\sum n\left(\frac{A_n}{A_1}\right)^2$$

$$= \frac{C_L^2}{\pi(AR)}\left[1 + \left(\frac{3A_3^2}{A_1^2} + \frac{5A_5^2}{A_1^2} + \frac{7A_7^2}{A_1^2} + \cdots\right)\right]$$

Writing the symbol δ for the term $\left(\frac{3A_3^2}{A_1^2} + \frac{5A_5^2}{A_1^2} + \frac{7A_7^2}{A_1^2} + \cdots\right)$

$$C_{Dv} = \frac{C_L^2}{\pi(AR)}[1 + \delta] \tag{5.50}$$

Plainly δ is always a positive quantity because it consists of squared terms that must always be positive. C_{D_v} can be a minimum only when $\delta = 0$. That is when $A_3 = A_5 = A_7 = \ldots = 0$ and the only term remaining in the series is $A_1 \sin \theta$.

Minimum induced drag condition

Thus comparing Eqn (5.50) with the induced-drag coefficient for the elliptic case (Eqn (5.43)) it can be seen that modifying the spanwise distribution away from the elliptic increases the drag coefficient by the fraction δ that is always positive. It follows that for the induced drag to be a minimum δ must be zero so that the distribution for minimum induced drag is the semi-ellipse. It will also be noted that the minimum drag distribution produces a constant downwash along the span whereas all other distributions produce a spanwise variation in induced velocity. This is no coincidence. It is part of the physical explanation of why the elliptic distribution should have minimum induced drag.

To see this, consider two wings (Fig. 5.33a and b), of equal span with spanwise distributions in downwash velocity $w = w_0 = \text{constant}$ along (a) and $w = f(z)$ along (b). Without altering the latter downwash variation it can be expressed as the sum of two distributions w_0 and $w_1 = f_1(z)$ as shown in Fig. 5.33c.

If the lift due to both wings is the same under given conditions, the rate of change of vertical momentum in the flow is the same for both. Thus for (a)

$$L \propto \int_{-s}^{s} \dot{m}\, w_0 \, dz \tag{5.51}$$

and for (b)

$$L \propto \int_{-s}^{s} \dot{m}(w_0 + f_1(z)) \, dz \tag{5.52}$$

where $\dot{m}$ is a representative mass flow meeting unit span. Since L is the same on each wing

$$\int_{-s}^{s} \dot{m}\, f_1(z) \, dz = 0 \tag{5.53}$$

Now the energy transfer or rate of change of the kinetic energy of the representative mass flows is the induced drag (or vortex drag). For (a):

$$D_{v(a)} \propto \frac{1}{2}\dot{m} \int_{-s}^{s} w_0^2 \, dz \tag{5.54}$$

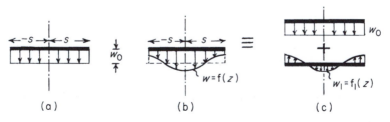

(a) (b) (c)

Fig. 5.33 (a) Elliptic distribution gives constant downwash and minimum drag. (b) Non-elliptic distribution gives varying downwash. (c) Equivalent variation for comparison purposes

For (b):

$$D_{V(b)} \propto \frac{1}{2}\dot{m} \int_{-s}^{s} (w_0 + f_1(z))^2 dz$$

$$\propto \frac{1}{2}\dot{m} \int_{-s}^{s} [w_0^2 + 2w_0 f_1(z) + (f_1(z))^2] dz$$

and since $\int_{-s}^{s} \dot{m} f_1(z) = 0$ in Eqn (5.53)

$$D_{V(b)} \propto \left[\frac{1}{2}\dot{m} \int_{-s}^{s} w_0^2 dz + \frac{1}{2}\dot{m} \int_{-s}^{s} (f_1(z))^2 dz \right] \tag{5.55}$$

Comparing Eqns (5.54) and (5.55)

$$D_{V(b)} = D_{V(a)} + \frac{1}{2}\dot{m} \int_{-s}^{s} f_1(z)^2 dz$$

and since $f_1(z)$ is an explicit function of z,

$$\int_{-s}^{s} (f_1(z))^2 dz > 0$$

since $(f_1(z))^2$ is always positive whatever the sign of $f_1(z)$. Hence $D_{V(b)}$ is always greater than $D_{V(a)}$.

5.6 Determination of the load distribution on a given wing

This is the direct problem broadly facing designers who wish to predict the performance of a projected wing before the long and costly process of model tests begin. This does not imply that such tests need not be carried out. On the contrary, they may be important steps in the design process towards a production aircraft.

The problem can be rephrased to suggest that the designers would wish to have some indication of how the wing characteristics vary as, for example, the geometric parameters of the project wing are changed. In this way, they can balance the aerodynamic effects of their changing ideas against the basic specification – provided there is a fairly simple process relating the changes in design parameters to the aerodynamic characteristics. Of course, this is stating one of the design problems in its baldest and simplest terms, but as in any design work, plausible theoretical processes yielding reliable predictions are very comforting.

The loading on the wing has already been described in the most general terms available and the overall characteristics are immediately to hand in terms of the coefficients of the loading distribution (Section 5.5). It remains to relate the coefficients (or the series as a whole) to the basic aerofoil parameters of planform and aerofoil section characteristics.

5.6.1 The general theory for wings of high aspect ratio

A start is made by considering the influence of the end effect, or downwash, on the lifting properties of an aerofoil section at some distance z from the centre-line of the wing. Figure 5.34 shows the lift-versus-incidence curve for an aerofoil section of

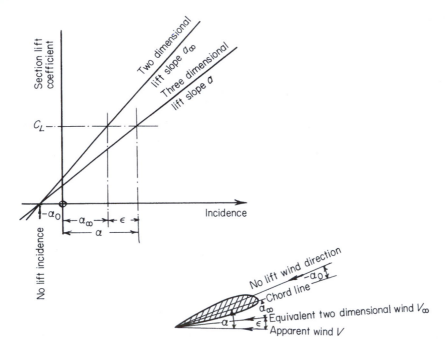

Fig. 5.34 Lift-versus-incidence curve for an aerofoil section of a certain profile, working two-dimensionally and working in a flow regime influenced by end effects, i.e. working at some point along the span of a finite lifting wing

a certain profile working two-dimensionally and working in a flow regime influenced by end effects, i.e. working at some point along the span of a finite lifting wing.

Assuming that both curves are linear over the range considered, i.e. the working range, and that under both flow regimes the zero-lift incidence is the same, then

$$C_L = a_\infty[\alpha_\infty - \alpha_0] = a[\alpha - \alpha_0] \tag{5.56}$$

Taking the first equation with $\alpha_\infty = \alpha - \varepsilon$

$$C_L = a_\infty[(\alpha - \alpha_0) - \varepsilon] \tag{5.57}$$

But equally from Eqn (4.10)

$$C_L = \frac{\text{lift per unit span}}{\frac{1}{2}\rho V^2 c} = \frac{l}{\frac{1}{2}\rho V^2 c}$$

$$= \frac{\rho V \Gamma}{\frac{1}{2}\rho V^2 c}$$

$$C_L = \frac{2\Gamma}{Vc} \tag{5.58}$$

Equating Eqn (5.57) and (5.58) and rearranging:

$$\frac{2\Gamma}{ca_\infty} = V[(\alpha - \alpha_0) - \varepsilon]$$

and since

$$V\varepsilon = w = -\frac{1}{4\pi} \int_{-s}^{s} \frac{(\mathrm{d}\Gamma/\mathrm{d}z)}{z - z_1} \mathrm{d}z \text{ from Eqn (5.32)}$$

$$\frac{2\Gamma}{ca_\infty} = V(\alpha - \alpha_0) + \frac{1}{4\pi} \int_{-s}^{s} \frac{(\mathrm{d}\Gamma/\mathrm{d}z)}{z - z_1} \mathrm{d}z \tag{5.59}$$

This is Prandtl's integral equation for the circulation Γ at any section along the span in terms of all the aerofoil parameters. These will be discussed when Eqn (5.59) is reduced to a form more amenable to numerical solution. To do this the general series expression (5.45) for Γ is taken:

$$\Gamma = 4sV \sum A_n \sin n\theta$$

The previous section gives Eqn (5.48):

$$w = \frac{V \Sigma n A_n \sin n\theta}{\sin \theta}$$

which substituted in Eqn (5.59) gives together

$$2 \frac{4sV \Sigma A_n \sin n\theta}{ca_\infty} = V(\alpha - \alpha_0) - \frac{V \sum n A_n \sin n\theta}{\sin \theta}$$

Cancelling V and collecting $ca_\infty/8s$ into the single parameter μ this equation becomes:

$$\mu(\alpha - \alpha_0) = \sum_{n=1}^{\infty} A_n \sin n\theta \left(1 + \frac{\mu n}{\sin \theta}\right) \tag{5.60}$$

The solution of this equation cannot in general be found analytically for all points along the span, but only numerically at selected spanwise stations and at each end.

5.6.2 General solution of Prandtl's integral equation

This will be best understood if a particular value of θ, or position along the span, be taken in Eqn (5.60). Take for example the position $z = -0.5s$, which is midway between the mid-span sections and the tip. From

$$z = -s \cos \theta, \quad \theta = \cos^{-1} \left(+\frac{1}{2}\right) = 60°$$

Then if the value of the parameter μ is μ_1 and the incidence from no lift is $(\alpha_1 - \alpha_{01})$ Eqn (5.60) becomes

$$\mu_1(\alpha_1 - \alpha_{01}) = A_1 \sin 60° \left[1 + \frac{\mu_1}{\sin 60°}\right] + A_2 \sin 120° \left[1 + \frac{2\mu_1}{\sin 60°}\right]$$

$$+ A_3 \sin 180° \left[1 + \frac{3\mu_1}{\sin 60°}\right] + \text{etc.}$$

This is obviously an equation with A_1, A_2, A_3, A_4, etc. as the only unknowns.

Other equations in which A_1, A_2, A_3, A_4, etc., are the unknowns can be found by considering other points z along the span, bearing in mind that the value of μ and of $(\alpha - \alpha_0)$ may also change from point to point. If it is desired to use, say, four terms in the series, an equation of the above form must be obtained at each of four values of θ, noting that normally the values $\theta = 0$ and π, i.e. the wing-tips, lead to the trivial equation $0 = 0$ and are, therefore, useless for the present purpose. Generally four coefficients are sufficient in the symmetrical case to produce a spanwise distribution that is insignificantly altered by the addition of further terms. In the case of symmetric flight the coefficients would be A_1, A_3, A_5, A_7, since the even harmonics do not appear. Also the arithmetic need only be concerned with values of θ between 0 and $\pi/2$ since the curve is symmetrical about the mid-span section.

If the spanwise distribution is irregular, more harmonics are necessary in the series to describe it adequately, and more coefficients must be found from the integral equation. This becomes quite a tedious and lengthy operation by 'hand', but being a simple mathematical procedure the simultaneous equations can be easily programmed for a computer.

The aerofoil parameters are contained in the expression

$$\mu = \frac{\text{chord} \times \text{two-dimensional lift slope}}{8 \times \text{semi-span}}$$

and the absolute incidence $(\alpha - \alpha_0)$. μ clearly allows for any spanwise variation in the chord, i.e. change in plan shape, or in the two-dimensional slope of the aerofoil profile, i.e. change in aerofoil section. α is the local geometric incidence and will vary if there is any geometric twist present on the wing. α_0, the zero-lift incidence, may vary if there is any aerodynamic twist present, i.e. if the aerofoil section is changing along the span.

Example 5.3 Consider a tapered aerofoil. For completeness in the example every parameter is allowed to vary in a linear fashion from mid-span to the wing-tips.

Mid-span data		Wing-tip data
3.048	Chord m	1.524
5.5	$\left(\dfrac{\partial C_L}{\partial \alpha}\right)_\infty$ per radian	5.8
5.5	absolute incidence $\alpha°$	3.5

Total span of wing is 12.192 m

Obtain the aerofoil characteristics of the wing, the spanwise distribution of circulation, comparing it with the equivalent elliptic distribution for the wing flying straight and level at 89.4 m s^{-1} at low altitude.

From the data:

$$\text{Wing area } S = \frac{3.048 + 1.524}{2} \times 12.192 = 27.85\,\text{m}^2$$

$$\text{Aspect ratio } (AR) = \frac{\text{span}^2}{\text{area}} = \frac{12.192^2}{27.85} = 5.333$$

At any section z from the centre-line [θ from the wing-tip]

$$\text{chord } c = 3.048\left[1 - \frac{3.048 - 1.524}{3.048}\left(\frac{z}{s}\right)\right] = 3.048[1 + 0.5\cos\theta]$$

$$\left(\frac{\partial C_L}{\partial \alpha}\right)_\infty = a = 5.5\left[1 + \frac{5.5 - 5.8}{5.5}\left(\frac{z}{s}\right)\right] = 5.5[1 - 0.054\,55\cos\theta]$$

$$\alpha° = 5.5\left[1 - \frac{5.5 - 3.5}{5.5}\left(\frac{z}{s}\right)\right] = 5.5[1 + 0.363\,64\cos\theta]$$

Table 5.1

θ	$\sin\theta$	$\sin 3\theta$	$\sin 5\theta$	$\sin 7\theta$	$\cos\theta$
$\pi/8$	0.382 68	0.923 88	0.923 88	0.382 68	0.923 88
$\pi/4$	0.707 11	0.707 11	−0.707 11	−0.707 11	0.707 11
$3\pi/8$	0.923 88	−0.382 68	−0.382 68	0.923 88	0.382 68
$\pi/2$	1.000 00	−1.000 00	1.000 00	−1.000 00	0.000 00

This gives at any section:

$$\mu = \frac{ca_\infty}{8s} = 0.34375(1 + 0.5\cos\theta)(1 - 0.054\,55\cos\theta)$$

and

$$\mu\alpha = 0.032\,995(1 + 0.5\cos\theta)(1 - 0.054\,55\cos\theta)(1 + 0.363\,64\cos\theta)$$

where α is now in radians. For convenience Eqn (5.60) is rearranged to:

$$\mu\alpha\,\sin\theta = A_1\sin\theta(\sin\theta + \mu) + A_3\sin 3\theta(\sin\theta + 3\mu) + A_5\sin 5\theta(\sin\theta + 5\mu)$$
$$+ A_7\sin 7\theta(\sin\theta + 7\mu)$$

and since the distribution is symmetrical the odd coefficients only will appear. Four coefficients will be evaluated and because of symmetry it is only necessary to take values of θ between 0 and $\pi/2$, i.e. $\pi/8$, $\pi/4$, $3\pi/8$, $\pi/2$.

Table 5.1 gives values of $\sin\theta$, $\sin n\theta$, and $\cos\theta$ for the above angles and these substituted in the rearranged Eqn (5.60) lead to the following four simultaneous equations in the unknown coefficients.

$$0.0047\,39 = 0.220\,79\ A_1 + 0.892\,02\ A_3 + 1.251\,00\ A_5 + 0.666\,88\ A_7$$
$$0.0116\,37 = 0.663\,19\ A_1 + 0.989\,57\ A_3 - 1.315\,95\ A_5 - 1.642\,34\ A_7$$
$$0.0216\,65 = 1.115\,73\ A_1 - 0.679\,35\ A_3 - 0.896\,54\ A_5 + 2.688\,78\ A_7$$
$$0.0329\,98 = 1.343\,75\ A_1 - 2.031\,25\ A_3 - 2.718\,75\ A_5 - 3.406\,25\ A_7$$

These equations when solved give

$$A_1 := 0.020\,329,\ A_3 = -0.000\,955,\ A_5 = 0.001\,029,\ A_7 = -0.000\,2766$$

Thus

$$\Gamma = 4sV\{0.020\,329\sin\theta - 0.000\,955\sin 3\theta + 0.001\,029\sin 5\theta - 0.000\,2766\sin 7\theta\}$$

and substituting the values of θ taken above, the circulation takes the values of:

θ	0	$\pi/8$	$\pi/4$	$3\pi/8$	$\pi/2$
z/s	1	0.924	0.707	0.383	0
$\Gamma\text{m}^2\text{s}^{-1}$	0	16.85	28.7	40.2	49.2
Γ/Γ_0	0	0.343	0.383	0.82	1.0

As a comparison, the equivalent elliptic distribution with the same coefficient of lift gives a series of values

| $\Gamma m^2 s^{-1}$ | 0 | 14.9 | 27.6 | 36.0 | 38.8 |

The aerodynamic characteristics follow from the equations given in Section 5.5.4. Thus:

$$C_L = \pi(AR)A_1 = 0.3406$$

$$C_D = \frac{C_L^2}{\pi(AR)}[1 + \delta] = 0.007\,068$$

since

$$\delta = 3\left(\frac{A_3}{A_1}\right)^2 + 5\left(\frac{A_5}{A_1}\right)^2 + 7\left(\frac{A_7}{A_1}\right)^2 = 0.020\,73$$

i.e. the induced drag is 2% greater than the minimum.
    For completeness the total lift and drag may be given

$$\text{Lift} = C_L \frac{1}{2}\rho V^2 S = 0.3406 \times 139\,910 = 47.72\,\text{kN}$$

$$\text{Drag (induced)} = C_{D_v} \frac{1}{2}\rho V^2 S = 0.007\,068 \times 139\,910 = 988.82\,\text{N}$$

Example 5.4 A wing is untwisted and of elliptic planform with a symmetrical aerofoil section, and is rigged symmetrically in a wind-tunnel at incidence α_1 to a wind stream having an axial velocity V. In addition, the wind has a small uniform angular velocity ω, about the tunnel axis. Show that the distribution of circulation along the wing is given by

$$\Gamma = 4sV[A_1 \sin\theta + A_2 \sin 2\theta]$$

and determine A_1 and A_2 in terms of the wing parameters. Neglect wind-tunnel constraints.

(CU)

    From Eqn (5.60)

$$\mu(\alpha - \alpha_0) = \Sigma A_n \sin n\theta \left(1 + \frac{\mu n}{\sin\theta}\right)$$

In this case $\alpha_0 = 0$ and the effective incidence at any section z from the centre-line

$$\alpha = \alpha_1 + z\frac{\omega}{V} = \alpha_1 - \frac{\omega}{V}s\cos\theta$$

Also since the planform is elliptic and untwisted $\mu = \mu_0 \sin\theta$ (Section 5.5.3) and the equation becomes for this problem

$$\mu_0 \sin\theta \left[\alpha_1 - \frac{\omega}{V}s\cos\theta\right] = \Sigma A_n \sin n\theta \left(1 + \frac{\mu_0 n \sin\theta}{\sin\theta}\right)$$

Expanding both sides:

$$\mu_0 \alpha_1 \sin\theta - \frac{\mu_0 \omega s}{V}\frac{\sin 2\theta}{2} = A_1 \sin\theta(1 + \mu_0) + A_2 \sin 2\theta(1 + 2\mu_0) + A_3 \sin 3\theta(1 + 3\mu_0) + \text{etc.}$$

Equating like terms:

$$\mu_0 \alpha_1 \sin \theta = A_1(1 + \mu_0) \sin \theta$$

$$\frac{\mu_0 s \omega}{2V} \sin 2\theta = A_2(1 + 2\mu_0) \sin 2\theta$$

$$0 = A_3(1 + 3\mu_0) \sin 3\theta \text{ etc.}$$

Thus the spanwise distribution for this case is

$$\Gamma = 4sV[A_1 \sin \theta + A_2 \sin 2\theta]$$

and the coefficients are

$$A_1 = \left(\frac{\mu_0}{1 + \mu_0}\right)\alpha_1$$

and

$$A_2 = \left(\frac{\mu_0}{2(1 + 2\mu_0)}\right)\frac{\omega s}{V}$$

5.6.3 Load distribution for minimum drag

Minimum induced drag for a given lift will occur if C_D is a minimum and this will be so only if δ is zero, since δ is always a positive quantity. Since δ involves squares of all the coefficients other than the first, it follows that the minimum drag condition coincides with the distribution that provides $A_3 = A_5 = A_7 = A_n = 0$. Such a distribution is $\Gamma = 4sVA_1 \sin \theta$ and substituting $z = -s \cos \theta$

$$\Gamma = 4sVA_1\sqrt{1 - \left(\frac{z}{s}\right)^2}$$

which is an elliptic spanwise distribution. These findings are in accordance with those of Section 5.5.3. This elliptic distribution can be pursued in an analysis involving the general Eqn (5.60) to give a far-reaching expression. Putting $A_n = 0$, $n \neq 1$ in Eqn (5.60) gives

$$\mu(\alpha - \alpha_0) = A_1 \sin \theta \left(1 + \frac{\mu}{\sin \theta}\right)$$

and rearranging

$$A_1 = \frac{\mu}{\sin \theta + \mu}(\alpha - \alpha_0) \tag{5.61}$$

Now consider an untwisted wing producing an elliptic load distribution, and hence minimum induced drag. By Section 5.5.3 the downwash is constant along the span and hence the equivalent incidence $(\alpha - \alpha_0 - w/V)$ anywhere along the span is constant. This means that the lift coefficient is constant. Therefore in the equation

$$\text{lift per unit span } l = \rho V \Gamma = C_L \frac{1}{2}\rho V^2 c \tag{5.62}$$

as l and Γ vary elliptically so must c, since on the right-hand side $C_L \frac{1}{2}\rho V^2$ is a constant along the span. Thus

$$c = c_0\sqrt{1 - \left(\frac{z}{s}\right)^2} = c_0 \sin\theta$$

and the general inference emerges that for a spanwise elliptic distribution an untwisted wing will have an elliptic chord distribution, though the planform may not be a true ellipse, e.g. the one-third chord line may be straight, whereas for a true ellipse, the mid-chord line would be straight (see Fig. 5.35).

It should be noted that an elliptic spanwise variation can be produced by varying the other parameters in Eqn (5.62), e.g. Eqn (5.62) can be rearranged as

$$\Gamma = C_L \frac{V}{2} c$$

and putting

$$C_L = a_\infty[(\alpha - \alpha_0) - \varepsilon] \text{ from Eqn (5.57)}$$
$$\Gamma \propto c a_\infty[(\alpha - \alpha_0) - \varepsilon]$$

Thus to make Γ vary elliptically, geometric twist (varying $(\alpha - \alpha_0)$) or change in aerofoil section (varying a_∞ and/or α_0) may be employed in addition to, or instead of, changing the planform.

Returning to an untwisted elliptic planform, the important expression can be obtained by including $c = c_0 \sin\theta$ in μ to give

$$\mu = \mu_0 \sin\theta \quad \text{where} \quad \mu_0 = \frac{c_0 a_\infty}{8s}$$

Then Eqn (5.61) gives

$$A_1 = \frac{\mu_0}{1 + \mu_0}(\alpha - \alpha_0) \tag{5.63}$$

But

$$A_1 = \frac{C_L}{\pi(AR)} \quad \text{from Eqn (5.47)}$$

Now

$$\frac{C_L}{(\alpha - \alpha_0)} = a = \text{three-dimensional lift slope}$$

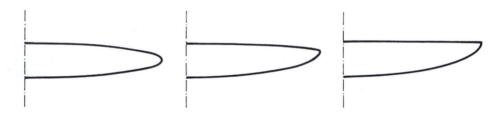

Fig. 5.35 Three different wing planforms with the same elliptic chord distribution

and

$$\mu_0 = \frac{c_0 a_\infty}{8s} = \frac{a_\infty}{\pi(AR)}$$

for an elliptic chord distribution, so that on substituting in Eqn (5.63) and rearranging

$$a = \frac{a_\infty}{1 + [a_\infty/\pi(AR)]} \qquad (5.64)$$

This equation gives the lift–curve slope a for a given aspect ratio (AR) in terms of the two-dimensional slope of the aerofoil section used in the aerofoil. It has been derived with regard to the particular case of an elliptic planform producing minimum drag conditions and is strictly true only for this case. However, most practical aerofoils diverge so little from the elliptic in this respect that Eqn (5.64) and its inverse

$$a_\infty = \frac{a}{1 - [a/\pi(AR)]}$$

can be used with confidence in performance predictions, forecasting of wind-tunnel results and like problems.

Probably the most famous elliptically shaped wing belongs to the Supermarine Spitfire – the British World War II fighter. It would be pleasing to report that the wing shape was chosen with due regard being paid to aerodynamic theory. Unfortunately it is extremely doubtful whether the Spitfire's chief designer, R.D. Mitchell, was even aware of Prandtl's theory. In fact, the elliptic wing was a logical way to meet the structural demands arising from the requirement that four big machine guns be housed in the wings. The elliptic shape allowed the wings to be as thin as possible. Thus the true aerodynamic benefits were rather more indirect than wing theory would suggest. Also the elliptic shape gave rise to considerable manufacturing problems, greatly reducing the rate at which the aircraft could be made. For this reason, the Spitfire's elliptic wing was probably not a good engineering solution when all the relevant factors were taken into account.*

5.7   Swept and delta wings

Owing to the dictates of modern flight many modern aircraft have sweptback or slender delta wings. Such wings are used for the benefits they confer in high-speed flight – see Section 6.8.2. Nevertheless, aircraft have to land and take off. Accordingly, a text on aerodynamics should contain at least a brief discussion of the low-speed aerodynamics of such wings.

5.7.1   Yawed wings of infinite span

For a sweptback wing of fairly high aspect ratio it is reasonable to expect that away from the wing-tips the flow would be similar to that over a yawed (or sheared) wing of infinite span (Fig. 5.36). In order to understand the fundamentals of such flows it is helpful to use the coordinate system (x', y, z'), see Fig. 5.36. In this coordinate system the free stream has two components, namely $U_\infty \cos \Lambda$ and $U_\infty \sin \Lambda$, perpendicular and parallel respectively to the leading edge of the wing. As the flow

*L. Deighton (1977) *Fighter* Jonathan Cape Ltd.

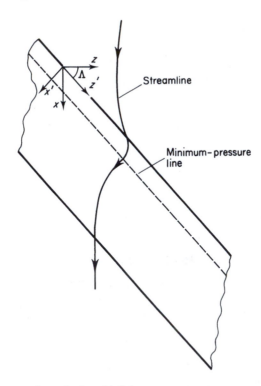

Fig. 5.36 Streamline over a sheared wing of infinite span

approaches the wing it will depart from the freestream conditions. The total velocity field can be thought of as the superposition of the free stream and a perturbation field $(u', v', 0)$ corresponding to the departure from freestream conditions. Note that the velocity perturbation, $w' \equiv 0$ because the shape of the wing remains constant in the z' direction.

An immediate consequence of using the above method to construct the velocity field is that it can be readily shown that, unlike for infinite-span straight wings, the streamlines do not follow the freestream direction in the x–z plane. This is an important characteristic of swept wings. The streamline direction is determined by

$$\left(\frac{\mathrm{d}x'}{\mathrm{d}z'}\right)_{SL} = \frac{U_\infty \cos \Lambda + u'}{U_\infty \sin \Lambda} \tag{5.65}$$

When $u' = 0$, downstream of the trailing edge and far upstream of the leading edge, the streamlines follow the freestream direction. As the flow approaches the leading-edge the streamlines are increasingly deflected in the outboard direction reaching a maximum deflection at the fore stagnation point (strictly a stagnation line) where $u' = U_\infty$. Thereafter the flow accelerates rapidly over the leading edge so that u' quickly becomes positive, and the streamlines are then deflected in the opposite direction – the maximum being reached on the line of minimum pressure.

Another advantage of the (x', y, z') coordinate system is that it allows the theory and data for two-dimensional aerofoils to be applied to the infinite-span yawed wing. So, for example, the lift developed by the yawed wing is given by adapting Eqn (4.43) to read

$$L = \frac{1}{2}\rho(U_\infty \cos \Lambda)^2 S\left(\frac{dC_L}{d\alpha}\right)_{2D}(\alpha_n - \alpha_{0n}) \tag{5.66}$$

where α_n is the angle of incidence defined with respect to the x' direction and α_{0n} is the corresponding angle of incidence for zero lift. Thus

$$\alpha_n = \alpha/\cos \Lambda \tag{5.67}$$

so the lift-curve slope for the infinite yawed wing is given by

$$\frac{dC_L}{d\alpha} = \left(\frac{dC_L}{d\alpha}\right)_{2D}\cos \Lambda \simeq 2\pi \cos \Lambda \tag{5.68}$$

and

$$L \propto \cos \Lambda \tag{5.69}$$

5.7.2 Swept wings of finite span

The yawed wing of infinite span gives an indication of the flow over part of a swept wing, provided it has a reasonably high aspect ratio. But, as with unswept wings, three-dimensional effects dominate near the wing-tips. In addition, unlike straight wings, for swept wings three-dimensional effects predominate in the mid-span region. This has highly significant consequences for the aerodynamic characteristics of swept wings and can be demonstrated in the following way. Suppose that the simple lifting-line model shown in Fig. 5.26, were adapted for a swept wing by merely making a kink in the bound vortex at the mid-span position. This approach is illustrated by the broken lines in Fig. 5.37. There is, however, a crucial difference between straight and kinked bound-vortex lines. For the former there is no self-induced velocity or downwash, whereas for the latter there is, as is readily apparent from Eqn (5.7). Moreover, this self-induced downwash approaches infinity near the kink at mid-span. Large induced velocities imply a significant loss in lift.

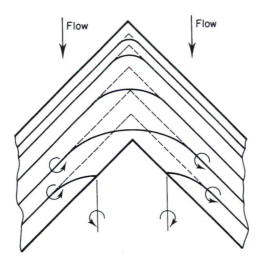

Fig. 5.37 Vortex sheet model for a swept wing

Nature does not tolerate infinite velocities and a more realistic vortex-sheet model is also shown in Fig. 5.37 (full lines). It is evident from this figure that the assumptions leading to Eqn (5.32) cannot be made in the mid-span region even for high aspect ratios. Thus for swept wings simplified vortex-sheet models are inadmissible and the complete expression Eqn (5.31) must be used to evaluate the induced velocity. The bound-vortex lines must change direction and curve round smoothly in the mid-span region. Some may even turn back into trailing vortices before reaching mid-span. All this is likely to occur within about one chord from the mid-span. Further away conditions approximate those for an infinite-span yawed wing. In effect, the flow in the mid-span region is more like that for a wing of low aspect ratio. Accordingly, the generation of lift will be considerably impaired in that region. This effect is evident in the comparison of pressure coefficient distributions over straight and swept wings shown in Fig. 5.38. The reduction in peak pressure over the mid-span region is shown to be very pronounced.

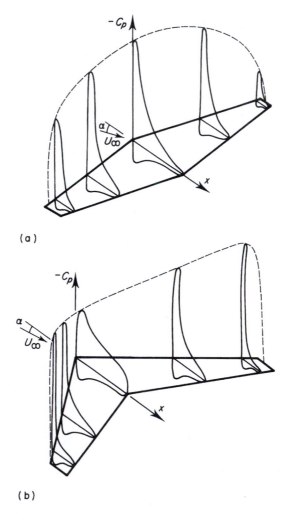

Fig. 5.38 A comparison between the pressure distributions over straight and swept-back wings

The pressure variation depicted in Fig. 5.38b has important consequences. First, if it is borne in mind that suction pressure is plotted in Fig. 5.38, it can be seen that there is a pronounced positive pressure gradient outward along the wing. This tends to promote flow in the direction of the wing-tips which is highly undesirable. Secondly, since the pressure distributions near the wing-tips are much peakier than those further inboard, flow separation leading to wing stall tends to occur first near the wing-tips. For straight wings, on the other hand, the opposite situation prevails and stall usually first occurs near the wing root – a much safer state of affairs. The difficulties briefly described above make the design of swept wings a considerably more challenging affair compared to that of straight wings.

5.7.3 Wings of small aspect ratio

For the wings of large aspect ratio considered in Sections 5.5 and 5.6 above it was assumed that the flow around each wing section is approximately two-dimensional. Much the same assumption is made at the opposite extreme of small aspect ratio. The crucial difference is that now the wing sections are taken as being in the spanwise direction, see Fig. 5.39. Let the velocity components in the (x, y, z) directions be separated into free stream and perturbation components, i.e.

$$(U_\infty \cos \alpha + u', U_\infty \sin \alpha + v', w') \tag{5.70}$$

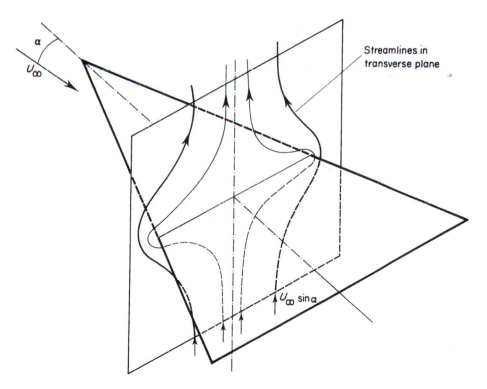

Fig. 5.39 Approximate flow in the transverse plane of a slender delta wing from two-dimensional potential flow theory

Let the velocity potential associated with the perturbation velocities be denoted by φ'. For slender-wing theory φ' corresponds to the two-dimensional potential flow around the spanwise wing-section, so that

$$\frac{\partial^2 \varphi'}{\partial y^2} + \frac{\partial^2 \varphi'}{\partial z^2} = 0 \qquad (5.71)$$

Thus for an infinitely thin uncambered wing this is the flow around a two-dimensional flat plate which is perpendicular to the oncoming flow component $U_\infty \sin \alpha$. The solution to this problem can be readily obtained by means of the potential flow theory described above in Chapter 3. On the surface of the plate the velocity potential is given by

$$\varphi' = \pm U_\infty \sin \alpha \sqrt{(b/2)^2 - z^2} \qquad (5.72)$$

where the plus and minus signs correspond to the upper and lower surfaces respectively.

As previously with thin wing theory, see Eqn (4.103) for example, the coefficient of pressure depends only on $u' = \partial\varphi'/\partial x$. x does not appear in Eqn (5.71), but it does appear in parametric form in Eqn (5.72) through the variation of the wing-section width b.

Example 5.5 Consider the slender delta wing shown in Fig. 5.39. Obtain expressions for the coefficients of lift and drag using slender-wing theory.

From Eqn (5.72) assuming that b varies with x

$$u' = \frac{\partial\varphi'}{\partial x} = \pm \frac{U_\infty \sin \alpha}{2} \frac{b}{\sqrt{b^2 - 4z^2}} \frac{db}{dx} \qquad (5.73)$$

From the Bernoulli equation the surface pressure is given by

$$p = p_0 - \frac{1}{2}\rho(U_\infty + u' + v' + w')^2 \simeq p_\infty - \rho U_\infty u' + O(u'^2)$$

So the pressure difference acting on the wing is given by

$$\Delta p = \rho U_\infty^2 \frac{\sin \alpha}{2} \frac{b}{\sqrt{b^2 - 4z^2}} \frac{db}{dx}$$

The lift is obtained by integrating Δp over the wing surface and resolving perpendicularly to the freestream. Thus, changing variables to $\zeta = 2z/b$, the lift is given by

$$L = \frac{1}{2}\rho U_\infty^2 \sin \alpha \cos \alpha \int_0^c b\frac{db}{dx}\int_{-1}^1 \frac{d\zeta}{\sqrt{1-\zeta^2}}dx \qquad (5.74)$$

Evaluating the inner integral first

$$\int_{-1}^1 \frac{d\zeta}{\sqrt{1-\zeta^2}} = \sin^{-1}(1) - \sin^{-1}(-1) = \pi$$

Therefore Eqn (5.74) becomes

$$L = \frac{\pi}{2}\sin \alpha \cos \alpha \rho U_\infty^2 \int_0^c b\frac{db}{dx}dx \qquad (5.75)$$

For the delta wing $b = 2x \tan \Lambda$ so that

$$\int_0^c b \frac{db}{dx} dx = 4 \tan^2 \Lambda \int_0^c x\, dx = 2c^2 \tan^2 \Lambda$$

Eqn (5.75) then gives

$$C_L = \frac{L}{\frac{1}{2}\rho U_\infty^2 c^2 \tan \Lambda} = 2\pi \tan \Lambda \sin \alpha \cos \alpha \qquad (5.76)$$

The drag is found in a similar fashion except that now the pressure force has to be resolved in the direction of the free stream, so that $C_D \propto \sin \alpha$ whereas $C_L \propto \cos \alpha$ therefore

$$C_D = C_L \tan \alpha \qquad (5.77)$$

For small α, $\sin \alpha \simeq \tan \alpha \simeq \alpha$. Note also that the aspect ratio $(AR) = 4 \tan \Lambda$ and that for small α Eqn (5.76) can be rearranged to give

$$\alpha \simeq \frac{C_L}{2\pi \tan \Lambda}$$

Thus for small α Eqn (5.77) can also be written in the form

$$C_D = \alpha C_L = \frac{2C_L^2}{\pi (AR)^2} \qquad (5.78)$$

Note that this is exactly twice the corresponding drag coefficient given in Eqn (5.43) for an elliptic wing of high aspect ratio.

At first sight the procedure outlined above seems to violate d'Alembert's Law (see Section 4.1) that states that no net force is generated by a purely potential flow around a body. For aerofoils and wings it has been found necessary to introduce circulation in order to generate lift and induced drag. Circulation has not been introduced in the above procedure in any apparent way. However, it should be noted that although the flow around each spanwise wing section is assumed to be non-circulatory potential flow, the integrated effect of summing the contributions of each wing section will not, necessarily, approximate the non-circulatory potential flow around the wing as a whole. In fact, the purely non-circulatory potential flow around a chordwise wing section, at the centre-line for example, will look something like that shown in Fig. 4.1a above. By constructing the flow around the wing in the way described above it has been ensured that there is no flow reversal at the trailing edge and, in fact, a kind of Kutta condition has been implicitly imposed, implying that the flow as a whole does indeed possess circulation.

The so-called *slender wing theory* described above is of limited usefulness because for wings of small aspect ratio the 'wing-tip' vortices tend to roll up and dominate the flow field for all but very small angles of incidence. For example, see the flow field around a slender delta wing as depicted in Fig. 5.40. In this case, the flow separates from the leading edges and rolls up to form a pair of stable vortices over the upper surface. The vortices first appear at the apex of the wing and increase in strength on moving downstream, becoming fully developed by the time the trailing edge is reached. The low pressures generated by these vortices contribute much of the lift.

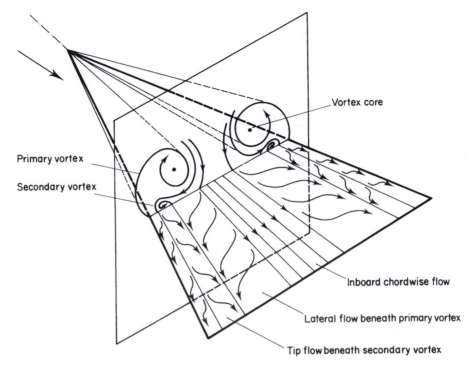

Fig. 5.40 Real flow field around a slender delta wing, showing vortex structure and surface flow pattern

Pohlhamus* offered a simple way to estimate the contribution of the vortices to lift on slender deltas (see Figs 5.41 and 5.42). He suggested that at higher angles of incidence the potential-flow pattern of Fig. 5.39, be replaced by a separated flow pattern similar to that found for real flow around a flat plate oriented perpendicular to the oncoming flow. So, in effect, this transverse flow generates a 'drag force' (per unit chord) of magnitude

$$\frac{1}{2}\rho U_\infty^2 \sin^2\alpha \, bC_{DP}$$

where C_{DP} has the value appropriate to real flow past a flate plate of infinite span placed perpendicular to the free stream (i.e. $C_{DP} \simeq 1.95$). Now this force acts perpendicularly to the wing and the lift is the component perpendicular to the actual free stream, so that

$$L = \frac{1}{2}\rho U_\infty^2 \sin^2\alpha \cos\alpha \, bC_{DP} \int_0^c b\,\mathrm{d}x, \quad \text{or} \quad C_L = C_{DP}\sin^2\alpha \cos\alpha \tag{5.79}$$

This component of the lift is called the *vortex* lift and the component given in Eqn (5.76) is called the *potential flow* lift.

---

* Pohlhamus, E.C. (1966), 'A Concept of the Vortex Lift of Sharp-Edge Delta Wings Based on a Leading-Edge-Suction Analogy', *NASA TN D-3767*; See also 'Applying Slender Wing Benefits to Military Aircraft', *AIAA J. Aircraft*, **21**, 545–559, 1984.

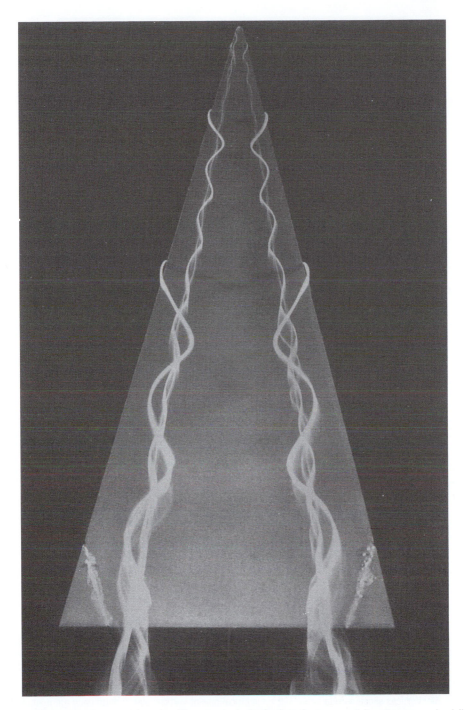

Fig. 5.41 Vortices above a delta wing: The symmetrical pair of vortices over a delta wing are made visible by the use of dye in water flow. The wing is made of thin plate and has a semi-vertex angle of 15°. The angle of attack is 20° and the Reynolds number is 20 000 based on chord. The flow direction is from top to bottom. See also Fig. 5.40 on page 264. (*The photograph was taken by H. Werlé at ONERA, France.*)

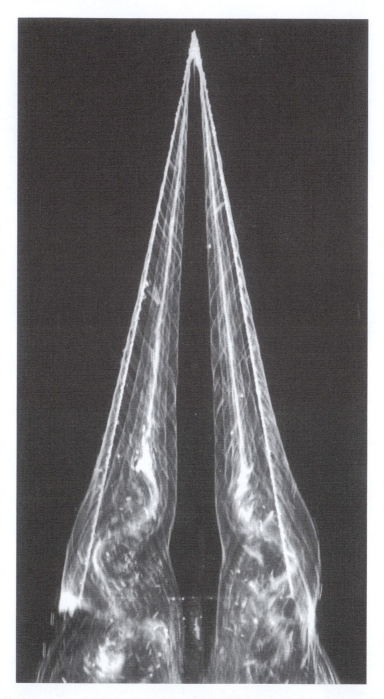

Fig. 5.42 Vortices above a delta wing: The symmetrical pair of vortices over a delta wing are made visible by the use of electrolysis in water flow. The wing is made of thin plate and has a semi-vertex angle of 10°. The angle of attack is 35° and the Reynolds number is 3000 based on chord. The flow direction is from top to bottom. Vortex breakdown occurs at about 0.7 maximum chord. See also Fig. 5.40 on page 264. (*The photograph was taken by J.-L. Solignac at ONERA, France.*)

The total lift acting on a slender delta wing is assumed to be the sum of the vortex and potential flow lifts. Thus

$$C_L = \underbrace{K_P \sin\alpha\cos\alpha}_{\text{Potential flow lift}} + \underbrace{K_V \sin^2\alpha\cos\alpha}_{\text{Vortex lift}} \tag{5.80}$$

where K_P and K_V are coefficients which are given approximately by $2\pi\tan\Lambda$ and 1.95 respectively, or alternatively can be determined from experimental data. The potential-flow term dominates at small angles of incidence and the vortex lift at higher incidence. The mechanism for generating the vortex lift is probably nonlinear to a significant extent, so there is really no theoretical justification for simply summing the two effects. Nevertheless, Eqn (5.80) fits the experimental data reasonably well as shown in Fig. 5.43 where the separate contributions of potential-flow lift and vortex lift are plotted.

It can be seen from Fig. 5.43 that there is not a conventional stalling phenomenon for a slender delta in the form of a sudden catastrophic loss of lift when a certain angle of incidence is reached. Rather there is a gradual loss of lift at around $\alpha = 35°$. This phenomenon is not associated directly with boundary-layer separation, but is caused by the vortices bursting at locations that move progressively further upstream as the angle of incidence is increased. The phenomenon of vortex breakdown is illustrated in Fig. 5.45 (see also Figs 5.42 and 5.44).

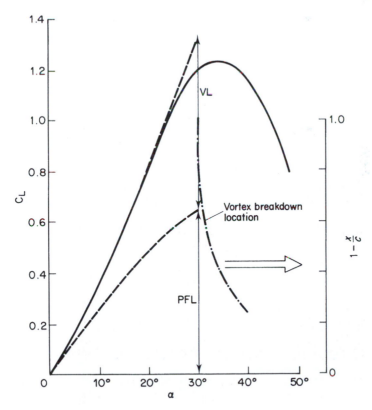

Fig. 5.43 Typical variation of lift coefficient with angle of incidence for a slender delta wing. PFL and VL denote respectively the contributions from the first and second terms on the right-hand side of Eqn (5.80)

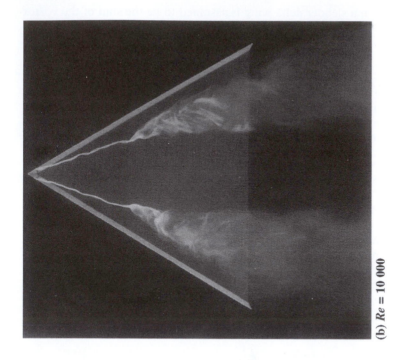

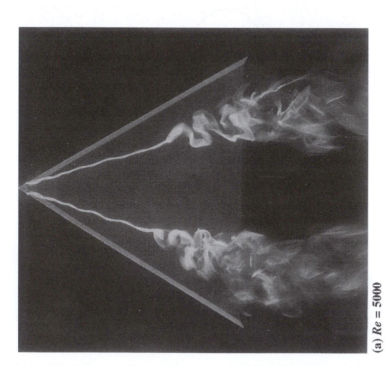

(a) $Re = 5000$

(b) $Re = 10\,000$

Fig. 5.44 Vortex breakdown above a delta wing: The wing is made of thin plate and its planform is an equilateral triangle. The vortices are made visible by the use of dye filaments in water flow. The angle of attack is 20°. In (a) where the Reynolds number based on chord is 5000 the laminar vortices that form after separation from the leading edge abruptly thicken and initially describe a larger-scale spiral motion which is followed by turbulent flow. For (b) the Reynolds number based on chord is 10 000. At this higher Reynolds number the vortex breakdown moves upstream and appears to change form. The flow direction is from top to bottom. See also Fig. 5.42 on page 266. (*The photographs were taken by H. Werlé at ONERA, France.*)

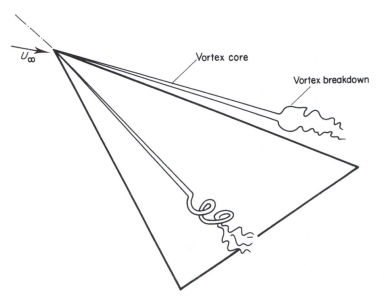

Fig. 5.45 A schematic view of the vortex breakdown over a slender delta wing, showing both the axisymmetric and spiral forms

5.8 Computational (panel) methods for wings

The application of the panel method described in Sections 3.5 and 4.10 above, to whole aircraft leads to additional problems and complexities. For example, it can be difficult to define the trailing edge precisely at the wing-tips and roots. In some more unconventional lifting-body configurations there may well be more widespread difficulties in identifying a trailing edge for the purposes of applying the Kutta condition. In most conventional aircraft configurations, however, it is a relatively straightforward matter to divide the aircraft into lifting and non-lifting portions – see Fig. 5.46. This allows most of the difficulties to be readily overcome and the computation of whole-aircraft aerodynamics is now routine in the aircraft industry.

In Section 4.10, the bound vorticity was modelled by means of either internal or surface vortex panels, see Fig. 4.22. Analogous methods have been used for the three-dimensional wings. There are, however, certain difficulties in using vortex panels. For example, it can often be difficult to avoid violating Helmholtz's theorem (see Section 5.2.1) when constructing vortex panelling. For this and other reasons most modern methods are based on source and doublet distributions. Such methods have a firm theoretical basis since Eqn (3.89b) can be generalized to lifting flows to read

$$\phi = Ux + \iint_{\text{Wing}} \left[\frac{1}{r}\sigma + \frac{\partial}{\partial n}\left(\frac{1}{r}\right)\mu \right] \mathrm{d}S + \iint_{\text{Wake}} \frac{\partial}{\partial n}\left(\frac{1}{r}\right)\mu \mathrm{d}S \qquad (5.81)$$

where n denotes the local normal to the surface and σ and μ are the source and doublet strengths respectively.

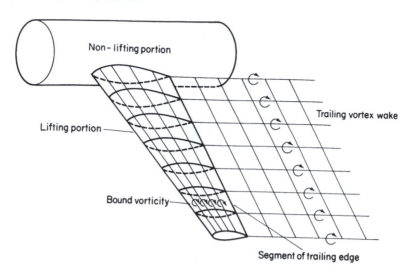

Fig. 5.46 Panel method applied to a wing-body combination

For a given application there is no unique mix of sources and doublets. For many methods* in common use each panel of the lifting surface is assigned a distribution of constant-strength sources. The doublet distribution must now be such that it provides one additional independent parameter for each segment of the trailing edge. Once the doublet strength is known at the trailing edge then the doublet strength on the panels comprising the trailing vorticity is determined. The initially unknown doublet strength at the trailing edge segments represents the spanwise load distribution of the wing. With this arrangement each chordwise segment of wing comprises N (say) panels and 1 trailing-edge segment. There are therefore N unknown source strengths and one unknown doublet parameter. Thus for each chordwise segment the $N + 1$ unknowns are determined by satisfying the N zero-normal-velocity conditions at the collocation points of the panels on the wing, plus the Kutta condition.

As in Section 4.10 the Kutta condition may be implemented either by adding an additional panel at the trailing edge or by requiring that the pressure be the same for the upper and lower panels defining the trailing edge – see Fig. 4.23. The former method is much less accurate since in the three-dimensional case the streamline leaving the trailing edge does not, in general, follow the bisector of the trailing edge. On the other hand, in the three-dimensional case equating the pressures on the two trailing-edge panels leads to a nonlinear system of equations because the pressure is related by Bernoulli equation to the square of the velocity. Nevertheless this method is still to be preferred if computational inaccuracy is to be avoided.

Exercises

1 An aeroplane weighing 73.6 kN has elliptic wings 15.23 m in span. For a speed of 90 m/s in straight and level flight at low altitude find (a) the induced drag; (b) the circulation around sections halfway along the wings.   (*Answer*: 1.37 kN, 44 m²/s)

---

* See B. Hunt (1978) 'The panel method for subsonic aerodynamic flows: A survey of mathematical formulations and numerical models and an outline of the new British Aerospace scheme', in *Computational Fluid Dynamics*, ed. by W. Kollmann, Hemisphere Pub. Corp., 100–165; and a review by J.L. Hess (1990) 'Panel methods in computational fluid dynamics', *Ann. Rev. Fluid Mech*, **22**, 255–274.

2 A glider has wings of elliptical planform of aspect ratio 6. The total drag is given by $C_D = 0.02 + 0.06 C_L^2$. Find the change in minimum angle of glide if the aspect ratio is increased to 10.

3 Discuss the statement that minimum induced drag of a wing is associated with elliptic loading, and plot a curve of induced drag coefficient against lift coefficient for a wing of aspect ratio 7.63.

4 Obtain an expression for the downward induced velocity behind a wing of span $2s$ at a point at distance y from the centre of span, the circulation around the wing at any point y being denoted by Γ. If the circulation is parabolic, i.e.

$$\Gamma = \Gamma_0 \left(1 - \frac{y^2}{s^2} \right)$$

calculate the value of the induced velocity w at mid-span, and compare this value with that obtained when the same lift is distributed elliptically.

5 For a wing with modified elliptic loading such that at distance y from the centre of the span, the circulation is given by

$$\Gamma = \Gamma_0 \left(1 + \frac{1}{6} \frac{y^2}{s^2} \right) \sqrt{1 - \frac{y^2}{s^2}}$$

where s is the semi-span, show that the downward induced velocity at y is

$$\frac{\Gamma_0}{4s} \left(\frac{11}{12} + \frac{y^2}{2s^2} \right)$$

Also prove that for such a wing of aspect ratio (A_R) the induced drag coefficient at lift coefficient C_L is

$$C_{D_0} = \frac{628}{625} \frac{C_L^2}{\pi A_R}$$

6 A rectangular, untwisted, wing of aspect ratio 3 has an aerofoil section for which the lift-curve slope is 6 in two-dimensional flow. Take the distribution of circulation across the span of a wing to be given by

$$\Gamma = 4sU \sum A_n \sin(n\theta)$$

and use the general theory for wings of high aspect ratio to determine the approximate circulation distribution in terms of angle of incidence by retaining only two terms in the above expression for circulation and satisfying the equation at $\theta = \pi/4$ and $\pi/2$.
(*Answer:* $A_1 = 0.372\alpha$, $A_2 = 0.0231\alpha$)

7 A wing of symmetrical cross-section has an elliptical planform and is twisted so that when the incidence at the centre of the span is $2°$ the circulation Γ at a distance y from the wing root is given by

$$\Gamma = \Gamma_0 [1 - (y/s)^2]^{3/2}$$

Find a general expression for the downwash velocity along the span and determine the corresponding incidence at the wing-tips. The aspect ratio is 7 and the lift-curve slope for the aerofoil section in two-dimensional flow is 5.8.
(*Answer:* $\alpha_{tip} = 0.566°$)

8 A straight wing is elliptic and untwisted and is installed symmetrically in a wind-tunnel with its centre-line along the tunnel axis. If the air in the wind-tunnel has an axial velocity V and also has a small uniform angular velocity ω about its axis, show that the distribution of circulation along the wing is given by

$$\Gamma = 4sA_2 \sin(2\theta)$$

and determine A_2 in terms of ω and the wing parameters. (The wind-tunnel wall corrections should be ignored.)

9 The spanwise distribution of circulation along an untwisted rectangular wing of aspect ratio 5 can be written in the form:

$$\Gamma = 4sv\alpha[0.023\,40 \sin(\theta) + 0.002\,68 \sin(3\theta) + 0.000\,72 \sin(5\theta) + 0.000\,10 \sin(7\theta)]$$

Calculate the lift and induced drag coefficients when the incidence α measured to no lift is $10°$. (*Answer*: $C_L = 0.691$, $C_{D_i} = 0.0317$)

10 An aeroplane weighing 250 kN has a span of 34 m and is flying at 40 m/s with its tailplane level with its wings and at height 6.1 m above the ground. Estimate the change due to ground effect in the downwash angle at the tailplane which is 18.3 m behind the centre of pressure of the wing. (*Answer*: $3.83°$)

11 Three aeroplanes of the same type, having elliptical wings of an aspect ratio of 6, fly in vee formation at 67 m/s with $C_L = 1.2$. The followers keep a distance of one span length behind the leader and also the same distance apart from one another. Estimate the percentage saving in induced drag due to flying in this formation. (*Answer*: 22%)

12 An aeroplane weighing 100 kN is 24.4 m in span. Its tailplane, which has a symmetrical section and is located 15.2 m behind the centre of pressure of the wing, is required to exert zero pitching moment at a speed of 67 m/s. Estimate the required tail-setting angle assuming elliptic loading on the wings. (*Answer*: $1.97°$)

13 Show that the downwash angle at the centre span of the tailplane is given to a good approximation by

$$\varepsilon = \text{constant} \times \frac{C_L}{A_R},$$

where A_R is the aspect ratio of the wing. Determine the numerical value of the constant for a tailplane located at $2s/3$ behind the centre of pressure, s being the semi-span. (*Answer*: 0.723 for angle in radians)

14 An aeroplane weighing 100 kN has a span of 19.5 m and a wing-loading of $1.925\,\text{kN/m}^2$. The wings are rather sharply tapered having around the centre of span a circulation 10% greater than that for elliptic wings of the same span and lift. Determine the downwash angle one-quarter of the span behind the centre of pressure, which is located at the quarter-chord point. The air speed is 67 m/s. Assume the trailing vorticity to be completely rolled up just behind the wings. (*Answer*: $4.67°$)

6

Compressible flow

Preamble

Hitherto in this volume the study of aerodynamics has almost exclusively been restricted to incompressible flow. This is really only suitable for the aerodynamics of low-speed flight and similar applications. For incompressible flow the density and temperature of the fluid are assumed invariant throughout the flow field. As the flow speeds rise the changes in pressure become greater, leading to the compression of fluid elements, causing in turn the internal energy, and therefore the temperature, to rise. The generation and transfer of heat due to viscous effects and heat conduction are also significant in the boundary layer. But these and other viscous effects are not considered in this chapter.

The chapter begins with the study of (quasi-) one-dimensional flow. This is an approximate approach that is suitable for flows through ducts and nozzles when the changes in the cross-sectional area are gradual. Under this circumstance the flow variables can be assumed uniform across a cross-section so that they only vary in the streamwise direction. Despite its apparently restrictive nature one-dimensional flow theory is applicable to a wide range of practical problems. It also serves as a good introduction to the concepts and phenomena of compressible flow, such as the development of shock waves when the air is accelerated through and beyond the speed of sound.

The chapter continues with a description of the formation of Mach and shock waves in two-dimensional flow. An important application of this theory is the study of wing aerodynamics. The nature of the flow around wings is greatly affected when the local flow speed exceeds the speed of sound. The flight speed at which this first occurs is called the critical Mach number and methods of estimating this quantity for specified wing sections are demonstrated. The (inviscid) equations of motion governing high-speed flows change their character so that their solutions become wavelike when the local Mach number exceeds unity. The behaviour of the Mach and shock waves in two-dimensional flow is described in some detail. In general, the equations of motion are non-linear in form and not amenable to analytical solution. Special approximate approaches exist for pure subsonic or supersonic flows. For example, the assumption of small perturbations to the freestream flow can be exploited to obtain approximate analytical solutions for both subsonic and supersonic flows around wings. Other approximate methods are also explored. The chapter closes with a short description of compressible flow around wings of finite span.

6.1 Introduction

In previous chapters the study of aerodynamics has been almost exclusively restricted to incompressible flow. This theoretical model is really only suitable for the aerodynamics of low-speed flight and similar applications. For incompressible flow the air density and temperature are assumed to be invariant throughout the flow field. But as flight speeds rise, greater pressure changes are generated, leading to the compression of fluid elements, causing in turn a rise in internal energy and, in consequence, temperature. The resulting variation of these flow variables throughout the flow field makes the results obtained from incompressible flow theory less and less accurate as flow speeds rise. For example, in Section 2.3.4 we showed how use of the incompressibility assumption led to errors in estimating the stagnation-pressure coefficient of 2% at $M = 0.3$, rising to 6% at $M = 0.5$, and 28% at $M = 1$.

But these errors in estimating pressures and other flow variables are not the most important disadvantage of using the incompressible flow model. Far more significant is the marked qualitative changes to the flow field that take place when the local flow speed exceeds the speed of sound. The formation of shock waves is a particularly important phenomenon and is a consequence of the propagation of sound through the air. In incompressible flow the fluid elements are not permitted to change in volume as they pass through the flow field. And, since sound waves propagate by alternately compressing and expanding the medium (see Section 1.2.7), this is tantamount to assuming an infinite speed of sound. This has important consequences when a body like a wing moves through the air otherwise at rest (or, equivalently, a uniform flow of air approaches the body). The presence of the body is signalled by sound waves propagating in all directions. If the speed of sound is infinite the presence of the body is instantly propagated to the farthest extent of the flow field and the flow instantly begins to adjust to the presence of the body.

The consequences of a finite speed of sound for the flow field are illustrated in Fig. 6.11(p.308). Figure 6.11b depicts the wave pattern generated when a source of disturbances (e.g. part of a wing) moves at subsonic speed into still air. It can be seen that the wave fronts are closer together in the direction of flight. But, otherwise, the flow field is qualitatively little different from the one (analogous to incompressible flow) corresponding to the stationary source shown in Fig. 6.11a. In both cases the sound waves eventually reach all parts of the flow field (instantly in the case of incompressible flow). Contrast this with the case, depicted in Fig. 6.11c, of a source moving at supersonic speed. Now the waves propagating in the forward direction line up to make planar wave fronts. The flow field remains undisturbed outside the regions reached by these planar wave fronts, and waves no longer propagate to all parts of the flow field. These planar wave fronts are formed from a superposition of many sound waves and are therefore much stronger than an individual sound wave. In many cases they correspond to shock waves, across which the flow variables change almost discontinuously. At supersonic speeds the flow field is fundamentally wavelike in character, meaning that information is propagated from one part of the flow field to another along wave fronts. Whereas in subsonic flow fields, which are not wavelike in character, information is propagated to all parts of the flow field.

This wavelike character of supersonic flow fields makes them qualitiatively different from the low-speed flow fields studied in previous chapters. Furthermore, the existence of shock waves brings about additional drag and many other undesirable changes from the viewpoint of wing aerodynamics. As a consequence, the effects of flow compressibility has a strong influence on wing design for high-speed flight even at subsonic flight speeds. It might at first be assumed that shock waves only affect wing aerodynamics at supersonic flight speeds. This is not so. It should be recalled that the local flow speeds near the point of minimum pressure over a wing are substantially greater than the free-stream flow speed. The local flow speed first reaches the speed of sound at a free-stream flow speed termed the critical flow speed. So, at flight speeds above critical, regions of supersonic flow appear over the wing, and shock waves are generated. This leads to wave drag and other undesirable effects. It is to postpone the onset of these effects that swept-back wings are used for high-speed subsonic aircraft. It is also worth pointing out that typically for such aircraft, wave drag contributes 20 to 30% of the total.

In recent decades great advances have been made in obtaining computational solutions of the equations of motion for compressible flow. This gives the design engineer much greater freedom to explore a wide range of possible configurations. It might also be thought that the ready availability of such computational solutions makes a knowledge of approximate analytical solutions unnecessary. Up to a point there is some truth in this view. There is certainly no longer any need to learn complex and involved methods of approximation. Nevertheless, approximate analytical methods will continue to be of great value. First and foremost, the study of relatively simple model flows, such as the one-dimensional flows described in Sections 6.2 and 6.3, enables the essential flow physics to be properly understood. In addition, these relatively simple approaches offer approximate methods that can be used to give reasonable estimates within a few minutes. They also offer a valuable way of checking the reliability of computer-generated solutions.

6.2 Isentropic one-dimensional flow

For many applications in aeronautics the viscous effects can be neglected to a good approximation and, moreover, no significant heat transfer occurs. Under these circumstances the thermodynamic processes are termed adiabatic. Provided no other irreversible processes occur we can also assume that the entropy will remain unchanged, such processes are termed *isentropic*. We can, therefore, refer to isentropic flow. At this point it is convenient to recall the special relationships between the main thermodynamic and flow variables that hold when the flow processes are isentropic.

In Section 1.2.8 it was shown that for isentropic processes $p = k\rho^\gamma$ (Eqn (1.24)), where k is a constant. When this relationship is combined with the equation of state for a perfect gas (see Eqn (1.12)), namely $p/(\rho T) = R$, where R is the gas constant, we can write the following relationships linking the variables at two different states (or stations) of an isentropic flow:

$$\frac{p_1}{\rho_1 T_1} = \frac{p_2}{\rho_2 T_2}, \qquad \frac{p_1}{\rho_1^\gamma} = \frac{p_2}{\rho_2^\gamma} \tag{6.1}$$

From these it follows that

$$\frac{T_2}{T_1} = \left(\frac{\rho_2}{\rho_1}\right)^{\gamma-1} = \left(\frac{p_2}{p_1}\right)^{(\gamma-1)/\gamma} \tag{6.2}$$

A useful, special, simplifed model flow is *one-dimensional*, or more precisely *quasi-one-dimensional flow*. This is an internal flow through ducts or passages having slowly varying cross-sections so that to a good approximation the flow is uniform at each cross-section and the flow variables only vary with x in the streamwise direction. Despite the seemingly restrictive nature of these assumptions this is a very useful model flow with several important applications. It also provides a good way to learn about the fundamental features of compressible flow.

The equations of conservation and state for quasi-one-dimensional, adiabatic flow in differential form become

$$\frac{d(\rho u A)}{dx} = 0 \quad \text{(for conservation of mass)} \tag{6.3}$$

where u is the streamwise, and only non-negligible, velocity component.

$$\frac{d(\rho u^2 A)}{dx} + A \frac{dp}{dx} = 0 \quad \text{(for momentum)} \tag{6.4}$$

$$\frac{d(c_p T + u^2/2)}{dx} = 0 \quad \text{(for energy)} \tag{6.5}$$

$$\frac{d(p/\rho T)}{dx} = 0 \quad \text{(for the equation of state)} \tag{6.6}$$

Expanding Eqn (6.3) and rearranging,

$$\frac{d\rho}{\rho} + \frac{du}{u} + \frac{dA}{A} = 0 \tag{6.7}$$

Similarly, for Eqn (6.6)

$$\frac{dp}{p} - \frac{d\rho}{\rho} - \frac{dT}{T} = 0 \tag{6.8}$$

From Eqn (6.4), using eqn (6.3)

$$\rho u A \, du + A \, dp = 0 \tag{6.9}$$

which, on dividing through by $u^2 A$ and using the identity $M^2 = u^2/a^2 = \rho u^2/(\gamma p)$,* using Eqn (1.6d) for the speed of sound in **isentropic** flow becomes

$$\frac{dp}{p} = -\gamma M^2 \frac{du}{u} \tag{6.10}$$

Likewise the energy Eqn (6.5), with $c_p T = a^2/(\gamma - 1)$ found by combining Eqns (1.15) and (1.6d), becomes

$$\frac{dT}{T} = -(\gamma - 1)M^2 \frac{du}{u} \tag{6.11}$$

Then combining Eqns (6.7) and (6.8) to eliminate $d\rho/\rho$ and substituting for dp/p and dT/T gives

$$(M^2 - 1)\frac{du}{u} = \frac{dA}{A} \tag{6.12}$$

---

* M is the symbol for the Mach number, that is defined as the ratio of the flow speed to the speed of sound at a point in a fluid flow and is named after the Austrian physicist Ernst Mach. The Mach number of an aeroplane in flight is the ratio of the flight speed to the speed of sound in the surrounding atmosphere (see also Section 1.4.2).

Equation (6.12) indicates the way in which the cross-sectional area of the stream tube must change to produce a change in velocity for a given mass flow. It will be noted that a change of sign occurs at $M = 1$.

For subsonic flow dA must be negative for an increase, i.e. a positive change, in velocity. At $M = 1$, dA is zero and a throat appears in the tube. For acceleration to supersonic flow a positive change in area is required, that is, the tube diverges from the point of minimum cross-sectional area.

Eqn (6.12) indicates that a stream tube along which gas speeds up from subsonic to supersonic velocity must have a converging–diverging shape. For the reverse process, the one of slowing down, a similar change in tube area is theoretically required but such a deceleration from supersonic flow is not possible in practice.

Other factors also control the flow in the tube and a simple convergence is not the only condition required. To investigate the change of other parameters along the tube it is convenient to consider the model flow shown in Fig. 6.1. In this model the air expands from a high-pressure reservoir (where the conditions may be identified by suffix 0), to a low-pressure reservoir, through a constriction, or throat, in a convergent–divergent tube. Denoting conditions at two separate points along the tube by suffices 1 and 2, respectively, the equations of state, continuity, motion and energy become

$$\frac{p_1}{\rho_1 T_1} = \frac{p_2}{\rho_2 T_2} \tag{6.13}$$

$$\rho_1 u_1 A_1 = \rho_2 u_2 A_2 \tag{6.14}$$

$$\rho_1 u_1^2 A_1 - \rho_2 u_2^2 A_2 + p_1 A_1 - p_2 A_2 + \tfrac{1}{2}(p_1 + p_2)(A_2 - A_1) = 0 \tag{6.15}$$

$$c_p T_1 + \frac{u_1^2}{2} = c_p T_2 + \frac{u_2^2}{2} \tag{6.16}$$

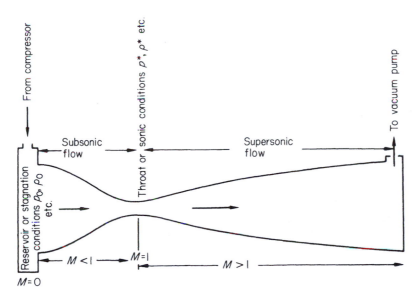

Fig. 6.1 One-dimensional isentropic expansive flow

The last of these equations, on taking account of the various ways in which the acoustic speed can be expressed in isentropic flow (see Eqn (1.6c,d)), i.e.

$$a = \sqrt{\frac{\gamma p}{\rho}} = \sqrt{\gamma RT} = \sqrt{(\gamma - 1)c_p T} = u/M,$$

may be rewritten in several forms for one-dimensional isentropic flow:

or

or

$$\left.\begin{array}{c} \dfrac{u_1^2}{2} + \dfrac{\gamma}{\gamma - 1}\dfrac{p_1}{\rho_1} = \dfrac{u_2^2}{2} + \dfrac{\gamma}{\gamma - 1}\dfrac{p_2}{\rho_2} \\[3ex] \dfrac{u_1^2}{2} + \dfrac{a_1^2}{\gamma - 1} = \dfrac{u_2^2}{2} + \dfrac{a_2^2}{(\gamma - 1)} \\[3ex] M_1^2 + \dfrac{2}{\gamma - 1} = \left(M_2^2 + \dfrac{2}{\gamma - 1}\right)\left(\dfrac{a_2}{a_1}\right)^2 \end{array}\right\} \quad (6.17)$$

6.2.1 Pressure, density and temperature ratios along a streamline in isentropic flow

Occasionally, a further manipulation of Eqn (6.17) is of more use. Rearrangement gives successively

$$\frac{u_2^2 - u_1^2}{2} = \frac{\gamma}{\gamma - 1}\left(\frac{p_1}{\rho_1} - \frac{p_2}{\rho_2}\right) = \frac{\gamma}{\gamma - 1}\frac{p_2}{\rho_2}\left[\left(\frac{p_1}{p_2}\right)^{(\gamma-1)/\gamma} - 1\right]$$

since it follows from the relationship (6.1) for isentropic processes that $p_1/p_2 = (\rho_1/\rho_2)^\gamma$.

Finally, with $a_2^2 = (\gamma p_2/\rho_2)$ this equation can be rearranged to give,

$$\frac{p_1}{p_2} = \left[1 + \frac{\gamma - 1}{2}\frac{u_2^2 - u_1^2}{a_2^2}\right]^{\gamma/(\gamma-1)} \quad (6.18)$$

If conditions 1 refer to stagnation or reservoir conditions, $u_1 = 0$, $p_1 = p_0$, the pressure ratio is

$$\frac{p_0}{p} = \left[1 + \frac{\gamma - 1}{2}M^2\right]^{\gamma/(\gamma-1)} = \left[1 + \frac{M^2}{5}\right]^{7/2} \quad \text{for air} \quad (6.18a)$$

where the quantity without suffix refers to any point in the flow. This ratio is plotted on Fig. 6.2 over the Mach number range 0–4. More particularly, taking the ratio between the pressure in the reservoir and the throat, where $M = M^* = 1$,

$$\frac{p_0}{p^*} = \left[\frac{\gamma + 1}{2}\right]^{\gamma/(\gamma-1)} = 1.89 \quad \text{for air flow} \quad (6.18b)$$

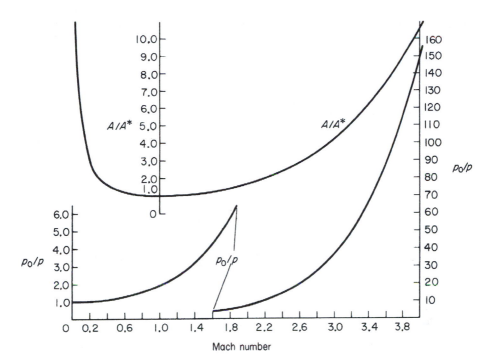

Fig. 6.2

Note that this is the minimum pressure ratio that will permit sonic flow. A greater value is required to produce supersonic flow. The ratios of the other parameters follow from Eqns (6.18) and (6.2):

$$\frac{\rho_1}{\rho_2} = \left(\frac{p_1}{p_2}\right)^{1/\gamma} = \left[1 + \frac{\gamma - 1}{2}\frac{u_2^2 - u_1^2}{a_2^2}\right]^{1/(\gamma-1)} \tag{6.19}$$

$$\frac{\rho_0}{\rho} = \left[1 + \frac{\gamma - 1}{2}M^2\right]^{1/(\gamma-1)} \tag{6.19a}$$

$$\frac{\rho_0}{\rho^*} = \left[\frac{\gamma + 1}{2}\right]^{1/(\gamma-1)} = 1.58 \quad \text{for airflow} \tag{6.19b}$$

and

$$\frac{T_1}{T_2} = \left(\frac{p_1}{p_2}\right)^{(\gamma-1)/\gamma} = 1 + \frac{\gamma - 1}{2}\frac{u_2^2 - u_1^2}{a_2^2} \tag{6.20}$$

$$\frac{T_0}{T} = 1 + \frac{\gamma - 1}{2}M^2 \tag{6.20a}$$

$$\frac{T_0}{T^*} = \frac{\gamma + 1}{2} = 1.2 \quad \text{for airflow} \tag{6.20b}$$

Example 6.1   In streamline airflow near the upper surface of an aeroplane wing the velocity just outside the boundary layer changes from $257\,\mathrm{km\,h^{-1}}$ at a point A near the leading edge to $466\,\mathrm{km\,h^{-1}}$ at a point B to the rear of A. If the temperature at A is $281\,\mathrm{K}$ calculate the temperature at B. Take $\gamma = 1.4$. Find also the value of the local Mach number at the point B.   (LU)

Assume that the flow outside the boundary layer approximates closely to quasi-one-dimensional, isentropic flow.
Then

$$\frac{T_B}{T_A} = 1 + \frac{\gamma - 1}{2}\frac{u_A^2 - u_B^2}{a_A^2}$$

$$= 1 + \frac{1}{5}\frac{u_A^2 - u_B^2}{a_A^2}$$

$$a_A = 72.4\sqrt{T_A} \quad \text{and} \quad T_A = 8 + 273 = 281\,\mathrm{K}$$

$$a_A = 1215\,\mathrm{km\,h^{-1}}$$

$$M_A = \frac{u_A}{a_A} = \frac{257}{1215} = 0.212$$

$$\frac{u_B}{a_A} = \frac{466}{1215} = 0.385$$

$$\frac{T_B}{T_A} = 1 + \frac{1}{5}[0.212^2 - 0.385^2] = 0.979 = 1 - 0.021$$

Therefore

$$T_B = 0.979 \times 281 = 275\,\mathrm{K} = \text{temperature at B}$$

$$a_B = 72.4\sqrt{275} = 1200\,\mathrm{km\,h^{-1}}$$

$$M_B = \frac{466}{1200} = 0.386$$

Example 6.2   An aerofoil is tested in a high-speed wind tunnel at a Mach number of 0.7 and at a point on the upper surface the pressure drop is found to be numerically equal to twice the stagnation pressure of the undisturbed stream. Calculate from first principles the Mach number found just outside the boundary layer at the point concerned. Take $\gamma = 1.4$.   (LU)

Let suffix ∞ refer to the undisturbed stream. Then, from above,

$$\frac{\gamma}{\gamma - 1}\frac{p_\infty}{\rho_\infty} + \frac{u_\infty^2}{2} = \frac{\gamma}{\gamma - 1}\frac{p}{\rho} + \frac{u^2}{2}$$

with $\gamma = 1.4$, this becomes

$$5a_\infty^2 + u_\infty^2 = 5a^2 + u^2$$

Dividing by a^2,

$$M^2 + 5 = \frac{M_\infty^2 + 5}{(a/a_\infty)^2} \tag{i}$$

but

$$(a/a_\infty)^2 = \frac{p}{\rho}\bigg/\frac{p_\infty}{\rho_\infty}, \quad \left(\frac{p}{p_\infty}\right)^{(\gamma - 1)/\gamma} = \left(\frac{p}{p_\infty}\right)^{2/7}$$

with the data given,

$$\frac{p}{p_\infty} = 1 - \gamma M_\infty^2 = 0.314$$

and Eqn (i) gives

$$M^2 + 5 = \frac{M_\infty^2 + 5}{(1 - \gamma M_\infty^2)^{2/7}} = \frac{5.49}{0.314^{2/7}} = 7.635$$

$$M^2 = 2.635 \text{ giving } M = 1.63$$

6.2.2 The ratio of areas at different sections of the stream tube in isentropic flow

It is necessary to introduce the mass flow ($\dot{m}$) and the equation of continuity, Eqn (6.14). Thus $\dot{m} = \rho u A$ for the general section, i.e. without suffix. Introducing again the reservoir or stagnation conditions and using Eqn (6.1):

$$\dot{m} = \frac{\rho}{\rho_0} \rho_0 A u = \rho_0 A u \left(\frac{p}{p_0}\right)^{1/\gamma} \tag{6.21}$$

Now the energy equation (6.17) gives the pressure ratio (6.18) above, which when referred to the appropriate sections of flow is rearranged to

$$u = a_0 \sqrt{\left[1 - \left(\frac{p}{p_0}\right)^{(\gamma-1)/\gamma}\right] \frac{2}{\gamma - 1}}$$

Substituting $\sqrt{\gamma p_0/\rho_0}$ for a_0 and introducing both into Eqn (6.21), the equation of continuity gives

$$\frac{\dot{m}}{A} = \left(\frac{p}{p_0}\right)^{1/\gamma} \sqrt{\frac{2\gamma}{\gamma - 1} p_0 \rho_0 \left[1 - \left(\frac{p}{p_0}\right)^{(\gamma-1)/\gamma}\right]} \tag{6.22}$$

Now, if the general section be taken to be the particular section at the throat, where in general usage conditions are identified by an asterisk (*), the equation of continuity (6.22) becomes

$$\frac{\dot{m}}{A^*} = \left(\frac{p^*}{p_0}\right)^{1/\gamma} \sqrt{\frac{2\gamma}{\gamma - 1} p_0 \rho_0 \left[1 - \left(\frac{p^*}{p_0}\right)^{(\gamma-1)/\gamma}\right]} \tag{6.23}$$

But from Eqn (6.18b) the ratio p^*/p_0 has the explicit value

$$\frac{p^*}{p_0} = \left[\frac{\gamma + 1}{2}\right]^{-\gamma/(\gamma-1)}$$

and hence

$$\frac{\dot{m}}{A^*} = \sqrt{\gamma p_0 \rho_0 \left(\frac{2}{\gamma+1}\right)^{(\gamma+1)/(\gamma-1)}} \tag{6.24}$$

If now the constant quantities $\dot{m}$, p_0, ρ_0 be eliminated from Eqns (6.22) and (6.24), the area ratio becomes

$$\frac{A}{A^*} = \left(\frac{p_0}{p}\right)^{1/\gamma} \sqrt{\frac{\gamma-1}{2}\left(\frac{2}{\gamma+1}\right)^{(\gamma+1)/(\gamma-1)} \left[1 - \left(\frac{p}{p_0}\right)^{(\gamma-1)/\gamma}\right]^{-1}}$$

and, substituting from Eqns (6.20) and (6.20a),

$$\left(\frac{p_0}{p}\right)^{-\gamma/(\gamma-1)} = 1 + \frac{\gamma-1}{2}M^2$$

from Eqn (6.18a), the expression reduces to

$$\frac{A}{A^*} = \frac{1}{M}\left[\frac{1+\frac{\gamma-1}{2}M^2}{\frac{\gamma+1}{2}}\right]^{(\gamma+1)/[2(\gamma-1)]} \tag{6.25}$$

Fig. 6.2 shows this ratio plotted against Mach number over the range $0 < M < 4$.

Example 6.3 Derive from first principles the following expression for the rate of change of stream-tube area with Mach number in the isentropic flow of air, with $\gamma = 1.4$.

$$\frac{dA}{dM} = \frac{5A}{M}\frac{M^2-1}{M^2+5}$$

At the station where $M = 1.4$ the area of the stream tube is increased by 1%. Find the corresponding change in pressure. (LU)

From Eqn (6.25)

$$\left(\frac{A}{A^*}\right)^2 = \frac{1}{M^2}\left[\frac{2}{\gamma+1}\left(1+\frac{\gamma-1}{2}M^2\right)\right]^{(\gamma+1)/(\gamma-1)}$$

For $\gamma = 1.4$:

$$\left(\frac{A}{A^*}\right)^2 = \left(\frac{5+M^2}{6}\right)^6 \frac{1}{M^2}$$

or

$$\frac{A}{A^*} = \frac{\left(\frac{5+M^2}{6}\right)^3}{M} \tag{i}$$

Differentiating this expression with respect to M:

$$\frac{1}{A^*}\frac{dA}{dM} = \left(\frac{5+M^2}{6}\right)^3 \frac{5}{M^2}\frac{M^2-1}{M^2+5} = \frac{A}{A^*}\frac{5}{M}\frac{M^2-1}{M^2+5}$$

which rearranged gives

$$\frac{dA}{dM} = \frac{5A}{M} \frac{M^2 - 1}{M^2 + 5} \tag{ii}$$

Similarly from Eqn (6.18a), with $\gamma = 1.4$:

$$\frac{p}{p_0} = \left[\frac{6}{M^2 + 5}\right]^{7/2}$$

$$\frac{1}{p_0} \frac{dp}{dM} = -7\left(\frac{6}{M^2 + 5}\right)^{7/2} \frac{M}{M^2 + 5} = -\frac{p}{p_0} \frac{7M}{M^2 + 5} \tag{iii}$$

Thus

$$\frac{dp}{p} = \frac{dM}{M} \frac{-7M^2}{M^2 + 5} \tag{iv}$$

From (ii) above:

$$\frac{dM}{M} = \frac{dA}{A} \frac{M^2 + 5}{5(M^2 - 1)} \tag{v}$$

and this substituted in (iv) gives the non-dimensional pressure change in terms of the Mach number and area change, i.e.

$$\frac{dp}{p} = \frac{dA}{A} \frac{-7M^2}{5(M^2 - 1)} \tag{vi}$$

In the question above, $M = 1.4$, $M^2 = 1.96$, $dA/A = 0.01$, so

$$\frac{dp}{p} = -0.0286$$

6.2.3 Velocity along an isentropic stream tube

The velocity at any point may best be expressed as a ratio of either the critical speed of sound a^* or the ultimate velocity c, both of which may be taken as flow parameters.

The critical speed of sound a^* is the local acoustic speed at the throat, i.e. where the local Mach number is unity. Thus the local velocity is equal to the local speed of sound. This can be expressed in terms of the reservoir conditions by applying the energy equation (6.17) between reservoir and throat. Thus:

$$\frac{\gamma}{\gamma - 1} \frac{p_0}{\rho_0} = c_p T_0 = \frac{u^{*2}}{2} + \frac{a^{*2}}{\gamma - 1}$$

which with $u^* = a^*$ yields

$$a^{*2} = \frac{2(\gamma - 1)}{\gamma + 1} c_p T_0 = \frac{2\gamma}{\gamma + 1} \frac{p_0}{\rho_0} \tag{6.26}$$

The ultimate velocity c is the maximum speed to which the flow can accelerate from the given reservoir conditions. It indicates a flow state in which all the energy of the gas is converted to kinetic energy of linear motion. It follows from the definition that this state has zero pressure and zero temperature and thus is not practically attainable.

Again applying the energy Eqn (6.17) between reservoir and ultimate conditions

$$\frac{\gamma}{\gamma - 1}\frac{p_0}{\rho_0} = c_p T_0 = \frac{c^2}{2},$$

so the ultimate, or maximum possible, velocity

$$c^2 = \frac{2\gamma}{\gamma - 1}\frac{p_0}{\rho_0} = 2c_p T_0 \qquad (6.27)$$

Expressing the velocity as a ratio of the ultimate velocity and introducing the Mach number:

$$\frac{u^2}{c^2} = \left(\frac{Ma}{c}\right)^2 = M^2\frac{(\gamma - 1)c_p T}{2c_p T_0}$$

or

$$\frac{u}{c} = M\sqrt{\frac{\gamma - 1}{2}}\sqrt{\frac{T}{T_0}}$$

and substituting Eqn (6.20a) for T/T_0:

$$\frac{u}{c} = M\sqrt{\frac{\gamma - 1}{2 + (\gamma - 1)M^2}} \qquad (6.28)$$

6.2.4   Variation of mass flow with pressure

Consider a converging tube (Fig. 6.3) exhausting a source of air at high stagnation pressure p_0 into a large receiver at some lower pressure. The mass flow induced in the nozzle is given directly by the equation of continuity (Eqn (6.22)) in terms of pressure ratio p/p_0 and the area of exit of the tube A, i.e.

$$\frac{\dot{m}}{A} = \left(\frac{p}{p_0}\right)^{1/\gamma}\sqrt{\frac{2\gamma}{\gamma - 1}p_0\rho_0\left[1 - \left(\frac{p}{p_0}\right)^{(\gamma - 1)/\gamma}\right]} \qquad (Eqn(6.22))$$

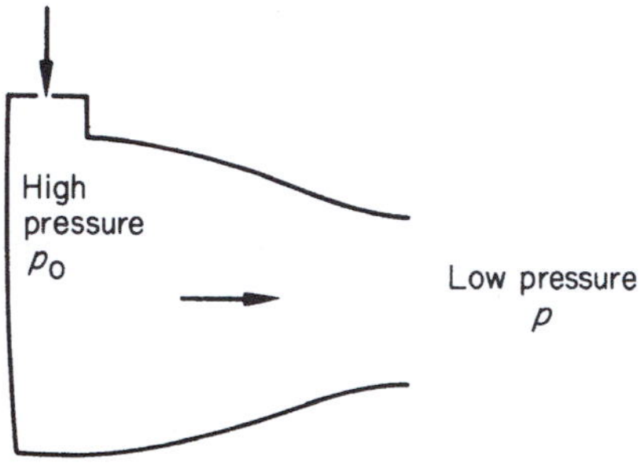

High
pressure
p_0 Low pressure
                                  p

Fig. 6.3

A slight rearrangement allows the mass flow, in a non-dimensional form, to be expressed solely in terms of the pressure ratio, i.e.

$$\left(\frac{\dot{m}}{A\sqrt{p_0\rho_0}}\right)^2 = \frac{2\gamma}{\gamma-1}\left[\left(\frac{p}{p_0}\right)^{2/\gamma} - \left(\frac{p}{p_0}\right)^{(\gamma+1)/\gamma}\right] \qquad (6.29)$$

Inspection of Eqn (6.29), or Eqn (6.22), reveals the obvious fact that $\dot{m} = 0$ when $p/p_0 = 1$, i.e. no flow takes place for zero pressure difference along the duct. Further inspection shows that $\dot{m}$ is also apparently zero when $p/p_0 = 0$, i.e. under maximum pressure drop conditions. This apparent paradox may be resolved by considering the behaviour of the flow as p is gradually decreased from the value p_0. As p is lowered the mass flow increases in magnitude until a condition of maximum mass flow occurs.

The maximum condition may be found by the usual differentiation process, i.e. from Eqn (6.29):

$$\frac{\mathrm{d}}{\mathrm{d}\left(\frac{p}{p_0}\right)}\left[\left(\frac{p}{p_0}\right)^{2/\gamma} - \left(\frac{p}{p_0}\right)^{(\gamma+1)/\gamma}\right] = 0 \quad \text{when} \quad \left(\frac{\dot{m}}{A\sqrt{p_0\rho_0}}\right)^2 \text{ is a maximum,}$$

i.e.

$$\frac{2}{\gamma}\left(\frac{p}{p_0}\right)^{(2/\gamma)-1} - \frac{\gamma+1}{\gamma}\left(\frac{p}{p_0}\right)^{[(\gamma+1)/\gamma]-1} = 0$$

which gives

$$\frac{p}{p_0} = \left(\frac{2}{\gamma+1}\right)^{\gamma/(\gamma-1)} \qquad (6.30)$$

It will be recalled that this is the value of the pressure ratio for the condition $M = 1$ and thus the maximum mass flow occurs when the pressure drop is sufficient to produce sonic flow at the exit.

Decreasing the pressure further will not result in a further increase of mass flow, which retains its maximum value. When these conditions occur the nozzle is said to be choked. The pressure at the exit section remains that given by Eqn (6.30) and as the pressure is further lowered the gas expands from the exit in a supersonic jet.

From previous considerations the condition for sonic flow, which is the condition for maximum mass flow, implies a throat, or section of minimum area, in the stream. Further expansion to a lower pressure and acceleration to supersonic flow will be accompanied by an increase in section area of the jet. It is impossible for the pressure ratio in the exit section to fall below that given by Eqn (6.30), and solutions of Eqn (6.29) have no physical meaning for values of

$$\frac{p}{p_0} < \left(\frac{2}{\gamma+1}\right)^{\gamma/(\gamma-1)}$$

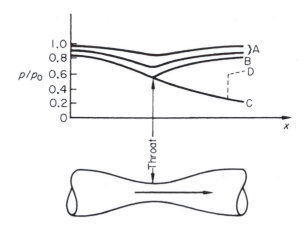

Fig. 6.4

Equally it is necessary for the convergent–divergent tube of Fig. 6.1 to be choked before the divergent portion will maintain supersonic conditions. If this condition is not realized, the flow will accelerate to a maximum value in the throat that is less than the local sonic speed, and then decelerate again in the divergent portion, accompanied by a pressure recovery. This condition can be schematically shown by the curves A in Fig. 6.4 that are plots of p/p_0 against tube length for increasing mass flow magnitudes. Curves B and C result when the tube is carrying its maximum flow. Branch B indicates the pressure recovery resulting from the flow that has just reached sonic conditions in the throat and then has been retarded to subsonic flow again in the divergent portion. Branch B is the limiting curve for subsonic flow in the duct and for mass flows less than the maximum or choked value. The curve C represents the case when the choked flow is accelerated to supersonic velocities downstream of the throat.

Considerations dealt with so far would suggest from the sketch that pressure ratios of a value between those of curves B and C are unattainable at a given station downstream of the throat. This is in fact the case if isentropic flow conditions are to be maintained. To arrive at some intermediate value D between B and C implies that a recompression from some point on the supersonic branch C is required. This is not compatible with isentropic flow and the equations dealt with above no longer apply. The mechanism required is called shock recompression.

Example 6.4 A wind-tunnel has a smallest section measuring 1.25 m × 1 m, and a largest section of 4 m square. The smallest is vented, so that it is at atmospheric pressure. A pressure tapping at the largest section is connected to an inclined tube manometer, sloped at 30° to the horizontal. The manometer reservoir is vented to the atmosphere, and the manometer liquid has a relative density of 0.85. What will be the manometer reading when the speed at the smallest section is (i) 80 m s^{-1} and (ii) 240 m s^{-1}? In the latter case, assume that the static temperature in the smallest section is 0 °C, (273 K).

Denote conditions at the smallest section by suffix 2, and the largest section by suffix 1. Since both the smallest section and the reservoir are vented to the same pressure, the reservoir may be regarded as being connected directly to the smallest section.

Area of smallest section $A_2 = 1.25\,\text{m}^2$
Area of largest section $A_1 = 16\,\text{m}^2$

(i) Since the maximum speed is $80\,\mathrm{m\,s^{-1}}$ the flow may be regarded as incompressible. Then

$$v_1 A_1 = v_2 A_2$$

i.e.

$$v_1 \times 16 = 80 \times 1.25$$

giving

$$v_1 = 6.25\,\mathrm{m\,s^{-1}}$$

By Bernoulli's equation, and assuming standard temperature and pressure:

$$p_1 + \frac{1}{2}\rho v_1^2 = p_2 + \tfrac{1}{2}\rho v_2^2$$

Then

$$p_1 - p_2 = \frac{1}{2}\rho(v_2^2 - v_1^2) = 0.613(80^2 - 6.25^2)$$
$$= 0.613 \times 86.25 \times 73.25$$
$$= 3900\,\mathrm{N\,m^{-2}}$$

This is the pressure across the manometer and therefore

$$\Delta p = \rho_\mathrm{m} g \Delta h$$

where Δh is the head of liquid and ρ_m the manometric fluid density, i.e.

$$3900 = (1000 \times 0.85) \times 9.807 \times \Delta h$$

This gives

$$\Delta h = 0.468\,\mathrm{m}$$

But

$$\Delta h = r \sin \theta$$

where r is the manometer reading and θ is the manometer slope. Then

$$0.468 = r \sin 30° = \frac{1}{2}r$$

and therefore

$$r = 0.936\,\mathrm{m}$$

(ii) In this case the speed is well into the range where compressibility becomes important, and it will be seen how much more complicated the solution becomes. At the smallest section, $T_2 = 0\,°\mathrm{C} = 273\,\mathrm{K}$

$$a_2 = (1.4 \times 287.1 \times 273)^{\frac{1}{2}} = 334\,\mathrm{m\,s^{-1}}$$

From the equation for conservation of mass

$$\rho_1 A_1 v_1 = \rho_2 A_2 v_2$$

i.e.

$$\frac{\rho_1}{\rho_2} = \frac{A_2 v_2}{A_1 v_1}$$

Also, from the isentropic flow relation Eqn (6.19) for compressible flow:

$$\frac{\rho_1}{\rho_2} = \left(1 - \frac{1}{5}\frac{v_1^2 - v_2^2}{a_2^2}\right)^{2.5}$$

Equating these expressions for ρ_1/ρ_2, and putting in the known values for A_1, A_2, v_2 and a_2

$$\frac{1.25 \times 240}{16v_1} = \left[1 - \frac{1}{5}\frac{v_1^2 - (240)^2}{(334)^2}\right]^{2.5}$$

or

$$\frac{18.75}{v_1} = \left[1.1035 - \frac{v_1^2}{557\,780}\right]^{2.5}$$

A first approximation to v_1 can be obtained by assuming incompressible flow, for which

$$v_1 = 240 \times 1.25/16 = 18.75\,\text{m s}^{-1}$$

With this value, $v_1^2/557\,780 \simeq 0.0008$. Therefore the second term within the brackets on the right-hand side can be ignored, and

$$18.75/v_1 = (1.1035)^{2.5} = 1.278$$

Therefore

$$v_1 = 14.7\,\text{m s}^{-1}$$

which value makes the ignored term even smaller.
    Further

$$\rho_1/\rho_2 = 18.75/v_1 = 1.278$$

and therefore

$$\frac{p_1}{p_2} = \left(\frac{\rho_1}{\rho_2}\right)^\gamma = (1.278)^{1.4} = 1.410$$

Therefore

$$p_1 - p_2 = p_2\left(\frac{p_1}{p_2} - 1\right)$$
$$= 101\,325 \times 0.410$$
$$= 41\,500\,\text{N m}^{-2}$$

Then the reading of the manometer is given by

$$r = \frac{\Delta p}{\rho_\text{m}g \sin\theta} = \frac{41\,500 \times 2}{1000 \times 0.85 \times 9.807}$$
$$= 9.95\,\text{m}$$

This result for the manometer reading shows that for speeds of this order a manometer using a low-density liquid is unsuitable. In practice it is probable that mercury would be used, when the reading would be reduced to $9.95 \times 0.85/13.6 = 0.62\,\text{m}$, a far more manageable figure. The use of a suitable transducer that converts the pressure into an electrical signal is even more probable in a modern laboratory.

Example 6.5 The reading of the manometer in Example 6.4 at a certain tunnel speed is 710 mm. Another manometer tube is connected at its free end to a point on an aerofoil model in the smallest section of the tunnel, while a third tube is connected to the total pressure tube of a Pitôt-static tube. If the liquid in the second tube is 76 mm above the zero level, calculate the pressure coefficient and the speed of flow at the point on the model. Calculate also the reading, including sense, of the third tube.

(i) To find speed of flow at smallest section:

$$\text{Manometer reading} = 0.710 \, \text{m}$$

Therefore

$$\text{pressure difference} = 1000 \times 0.85 \times 9.807 \times 0.71 \times \tfrac{1}{2}$$
$$= 2960 \, \text{N} \, \text{m}^{-2}$$

But

$$p_1 - p_2 = \tfrac{1}{2}\rho_0(v_2^2 - v_1^2)$$

and

$$v_1 = 1.25 v_2/16 = 5 v_2/64$$

Therefore

$$2960 = 0.613 v_2^2 \left[1 - \left(\frac{5}{64} \right)^2 \right]$$

$$= 0.613 v_2^2 \left[\frac{4071}{4096} \right]$$

Therefore

$$v_2^2 = \frac{2960 \times 4096}{0.613 \times 4071} = 4860 \, (\text{m} \, \text{s}^{-1})^2$$
$$v_2 = 69.7 \, \text{m} \, \text{s}^{-1}$$

Hence, dynamic pressure at smallest section

$$= \tfrac{1}{2}\rho_0 v_2^2 = 0.613 \, v_2^2$$
$$= 2980 \, \text{N} \, \text{m}^{-2}$$

(ii) Pressure coefficient:
Since static pressure at smallest section = atmospheric pressure, then pressure difference between aerofoil and tunnel stream = pressure difference between aerofoil and atmosphere. This pressure difference is 76 mm on the manometer, or

$$\Delta p = 1000 \times 0.85 \times 9.807 \times 0.076 \times \tfrac{1}{2} = 317.5 \, \text{N} \, \text{m}^{-2}$$

Now the manometer liquid has been drawn upwards from the zero level, showing that the pressure on the aerofoil is less than that of the undisturbed tunnel stream, and therefore the pressure coefficient will be negative, i.e.

$$C_p = \frac{p - p_0}{\tfrac{1}{2}\rho V^2} = \frac{-317.5}{2980} = -0.1068$$

Now

$$C_p = 1 - \left(\frac{q}{v}\right)^2$$

$$-0.1068 = 1 - \left(\frac{q}{v_2}\right)^2$$

Hence

$$q = v_2(1 - C_p)^{1/2} = 69.7(1.1068)^{1/2}$$
$$= 73.2\,\mathrm{m\,s^{-1}}$$

(iii) The total pressure is equal to stream static pressure plus the dynamic pressure and, therefore, pressure difference corresponding to the reading of the third tube is $(p_0 + \frac{1}{2}\rho v_2^2) - p_0$, i.e. is equal to $\frac{1}{2}\rho v_2^2$. Therefore, if the reading is r_3

$$\tfrac{1}{2}\rho v_2^2 = \rho_m g r_3 \sin\theta$$

$$2980 = 1000 \times 0.85 \times 9.807 \times r_3 \times \tfrac{1}{2}$$

whence

$$r_3 = 0.712\,\mathrm{m}$$

Since the total head is greater than the stream static pressure and, therefore, greater than atmospheric pressure, the liquid in the third tube will be depressed below the zero level, i.e. the reading will be $-0.712\,\mathrm{m}$.

Example 6.6 An aircraft is flying at 6100 m, where the pressure, temperature and relative density are $46\,500\,\mathrm{N\,m^{-2}}$, $-24.6\,°C$, and 0.533 respectively. The wing is vented so that its internal pressure is uniform and equal to the ambient pressure. On the upper surface of the wing is an inspection panel 150 mm square. Calculate the load tending to lift the inspection panel and the air speed over the panel under the following conditions:

(i) Mach number = 0.2, mean C_p over panel = -0.8
(ii) Mach number = 0.85, mean C_p over panel = -0.5.

(i) Since the Mach number of 0.2 is small, it is a fair assumption that, although the speed over the panel will be higher than the flight speed, it will still be small enough for compressibility to be ignored. Then, using the definition of coefficient of pressure (see Section 1.5.3)

$$C_{p_1} = \frac{p_1 - p}{0.7p\,M^2}$$

$$p_1 - p = 0.7p\,M^2\,C_{p_1} = 0.7 \times 46\,500 \times (0.2)^2 \times (-0.8)$$
$$= -1041\,\mathrm{N\,m^{-2}}$$

The load on the panel = pressure difference × area
$$= 1041 \times (0.15)^2$$
$$= 23.4\,\mathrm{N}$$

Also

$$C_{p_1} = 1 - \left(\frac{q}{v}\right)^2$$

i.e.

$$-0.8 = 1 - \left(\frac{q}{v}\right)^2$$

whence

$$\left(\frac{q}{v}\right)^2 = 1.8 \text{ giving } \frac{q}{v} = 1.34$$

Now speed of sound $= 20.05 \, (273 - 24.6)^{1/2} = 318 \, \text{m s}^{-1}$
Therefore, true flight speed $= 0.2 \times 318 = 63.6 \, \text{m s}^{-1}$
Therefore, air speed over panel, $q = 63.6 \times 1.34 = 85.4 \, \text{m s}^{-1}$
(ii) Here the flow is definitely compressible. As before,

$$C_{p_1} = \frac{p_1 - p}{0.7p \, M^2}$$

and therefore

$$p_1 - p = 0.7 \times 46 \, 500 \times (0.85)^2 \times (-0.5)$$
$$= -11 \, 740 \, \text{N m}^{-2}$$

Therefore, load on panel $= 11 \, 740 \times (0.15)^2 = 264 \, \text{N}$
There are two ways of calculating the speed of flow over the panel from Eqn (6.18):

(a)

$$\frac{p_1}{p} = \left[1 - \frac{1}{5}\frac{q^2 - v^2}{a^2}\right]^{3.5}$$

where a is the speed of sound in the free stream, i.e.

$$\frac{p_1}{p} = \left[1 - \frac{1}{5}\left\{\left(\frac{q}{a}\right)^2 - M^2\right\}\right]^{3.5}$$

Now

$$p_1 - p = -11 \, 740 \, \text{N m}^{-2}$$

and therefore

$$p_1 = 46 \, 500 - 11 \, 740 = 34 \, 760 \, \text{N m}^{-2}$$

Thus substituting in the above equation the known values $p = 46 \, 500 \, \text{N m}^{-2}$, $p_1 = 34 \, 760 \, \text{N m}^{-2}$ and $M = 0.85$ leads to

$$\left(\frac{q}{a}\right)^2 = 1.124 \text{ giving } \frac{q}{a} = 1.06$$

Therefore

$$q = 1.06a = 1.06 \times 318 = 338 \, \text{m s}^{-1}$$

It is also possible to calculate the Mach number of the flow over the panel, as follows. The local temperature T is found from

$$\frac{T_1}{T} = \left(\frac{p_1}{p}\right)^{1/3.5} = \left(\frac{34 \, 760}{46 \, 500}\right)^{1/3.5} = 0.920$$

giving

$$T_1 = 0.920\,T$$

and

$$a_1 = a(0.920)^{1/2} = 318(0.920)^{1/2} = 306\,\mathrm{m\,s}^{-1}$$

Therefore, Mach number over panel $= 338/306 = 1.103$.

(b) The alternative method of solution is as follows, with the total pressure of the flow denoted by p_0:

$$\frac{p_0}{p} = \left[1 + \frac{1}{5}M^2\right]^{3.5} = \left[1 + \frac{(0.85)^2}{5}\right]^{3.5}$$

$$= (1.1445)^{3.5} = 1.605$$

Therefore

$$p_0 = 46\,500 \times 1.605 = 74\,500\,\mathrm{N\,m}^{-2}$$

As found in method (a)

$$p_1 - p = -11\,740\,\mathrm{N\,m}^{-2}$$

and

$$p_1 = 34\,760\,\mathrm{N\,m}^{-2}$$

Then

$$\frac{p_0}{p_1} = \frac{74\,500}{34\,760} = 2.15 = \left[1 + \frac{1}{5}M_1^2\right]^{3.5}$$

giving

$$M_1^2 = 1.22,\, M_1 = 1.103$$

which agrees with the result found in method (a).

The total temperature T_0 is given by

$$\frac{T_0}{T} = \left(\frac{p_0}{p}\right)^{1/3.5} = 1.1445$$

Therefore

$$T_0 = 1.1445 \times 248.6 = 284\,\mathrm{K}$$

Then

$$\frac{T_0}{T_1} = (2.15)^{1/3.5} = 1.244$$

giving

$$T_1 = \frac{284}{1.244} = 228\,\mathrm{K}$$

and the local speed of sound over the panel, a_1, is

$$a_1 = 20.05(228)^{1/2} = 305 \, \mathrm{m\,s^{-1}}$$

Therefore, flow speed over the panel

$$q = 305 \times 1.103 = 338 \, \mathrm{m\,s^{-1}}$$

which agrees with the answer obtained by method (a).

An interesting feature of this example is that, although the flight speed is subsonic ($M = 0.85$), the flow over the panel is supersonic. This fact was used in the 'wing-flow' method of transonic research. The method dates from about 1940, when transonic wind-tunnels were unsatisfactory. A small model was mounted on the upper surface of the wing of an aeroplane, which then dived at near maximum speed. As a result the model experienced a flow that was supersonic locally. The method, though not very satisfactory, was an improvement on other methods available at that time.

Example 6.7 A high-speed wind-tunnel consists of a reservoir of compressed air that discharges through a convergent-divergent nozzle. The temperature and pressure in the reservoir are $200\,^{\circ}\mathrm{C}$ and $2\,\mathrm{MN\,m^{-2}}$ gauge respectively. In the test section the Mach number is to be 2.5. If the test section is to be 125 mm square, what should be the throat area? Calculate also the mass flow, and the pressure, temperature, speed, dynamic and kinematic viscosity in the test section.

$$\frac{A}{A^*} = \frac{1}{M}\left(\frac{5 + M^2}{6}\right)^3 = \frac{1}{2.5}\left(\frac{5 + 6.25}{6}\right)^3 = 2.64$$

Therefore, throat area $= \dfrac{(125)^2}{2.64} = 5920 \, (\mathrm{mm})^2$

Since the throat is choked, the mass flow may be calculated from Eqn (6.24), is

$$\text{mass flow} = 0.0404\left(\frac{p_0}{\sqrt{T_0}}\right)A^*$$

Now the reservoir pressure is $2\,\mathrm{MN\,m^{-2}}$ gauge, or $2.101\,\mathrm{MN\,m^{-2}}$ absolute, while the reservoir temperature is $200\,^{\circ}\mathrm{C} = 473\mathrm{K}$. Therefore

$$\text{mass flow} = 0.0404 \times 2.101 \times 10^6 \times 5920 \times 10^{-6}/(473)^{1/2}$$
$$= 23.4 \, \mathrm{kg\,s^{-1}}$$

In the test section

$$1 + \frac{1}{5}M^2 = 1 + \frac{6.25}{5} = 2.25$$

Therefore

$$p_0/p_1 = (2.25)^{3.5} = 17.1$$

Therefore

$$\text{pressure in test section} = \frac{2.101 \times 10^6}{17.1} = 123 \, \mathrm{kN\,m^{-2}}$$

Also

$$\frac{T_0}{T_1} = 2.25$$

Therefore:

$$\text{Temperature in test section} = \frac{473}{2.25} = 210 \text{ K} = -63\,^{\circ}\text{C}$$

$$\text{Density in test section} = \frac{123\,000}{287.3 \times 210} = 2.042 \,\text{kg}\,\text{m}^{-3}$$

$$\text{Speed of sound in test section} = (1.4 \times 287.1 \times 210)^{\frac{1}{2}} = 293 \,\text{m}\,\text{s}^{-1}$$

$$\text{Air speed in test section} = 2.5 \times 293 = 732 \,\text{m}\,\text{s}^{-1}$$

Using the approximation given in Section 1.4.2 (Example 1.3) for the variation of viscosity with temperature

$$\mu = 1.71 \times 10^{-5} \left(\frac{210}{273}\right)^{3/4} = 1.50 \times 10^{-5} \,\text{kg}\,\text{m}^{-1}\,\text{s}^{-1}$$

$$v = \frac{\mu}{\rho} = \frac{1.50 \times 10^{-5}}{2.042} = 0.735 \times 10^{-5} \,\text{m}^2\text{s}^{-1}$$

As a check, the mass flow may be calculated from the above results. This gives

$$\text{Mass flow} = \rho v A = 2.042 \times 732 \times 15\,625 \times 10^{-6}$$
$$= 23.4 \,\text{kg}\,\text{s}^{-1}$$

6.3   One-dimensional flow: weak waves

To a certain extent the results of this section have already been assumed in that certain expressions for the speed of sound propagation have been used. Pressure disturbances in gaseous and other media are propagated in longitudinal waves and appeal is made to elementary physics for an understanding of the phenomenon.

Consider the air in a stream tube to be initially at rest and, as a simplification, divided into layers 1, 2, 3, etc., normal to the possible direction of motion. A small pressure impulse felt on the face of the first layer moves the layer towards the right and it acquires a kinetic energy of uniform motion in so doing. At the same time, since layers 1, 2, 3 have inertia, layer 1 converts some kinetic energy of translational motion into molecular kinetic energy associated with heat, i.e. it becomes compressed. Eventually all the relative motion between layers 1 and 2 is absorbed in the pressure inequality between them and, in order to ease the pressure difference, the first layer acquires motion in the reverse direction. At the same time the second layer acquires kinetic energy due to motion from left to right and proceeds to react on layer 3 in a like manner. In the expansive condition, again due to its inertia, it moves beyond the position it previously occupied. The necessary kinetic energy is acquired from internal conditions so that its pressure falls below the original. Reversion to the *status quo* demands that the kinetic energy of motion to the left be transferred back to the conditions of pressure and temperature obtaining before the impulse was felt, with the fluid at rest and not displaced relative to its surroundings.

A first observation of this sequence of events is that the gas has no resultant mean displacement velocity or pressure different from that of the initial conditions, and it serves only to transmit the pressure pulse throughout its length. Secondly, the

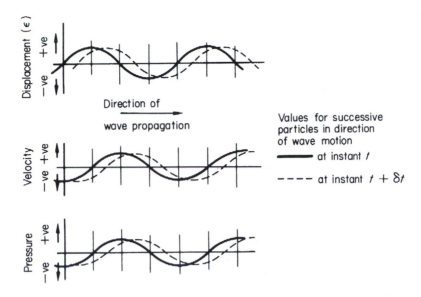

Fig. 6.5

displacement, and hence velocity, pressure, etc., of an individual particle of gas is changing continuously while it is under the influence of the passing impulse.

A more graphic way of expressing the gas conditions in the tube is to plot those of successive particles in the direction of movement of the impulse, at a given instant of time while the impulse is passing. Another curve of the particles' velocities at an instant later shows how individual particles behave.

Fig. 6.5 shows a typical set of curves for the passage of small pressure impulses, and a matter of immediate interest is that an individual particle moves in the direction of the wave propagation when its pressure is above the mean, and in the reverse direction in the expansive phase.

6.3.1   The speed of sound (acoustic speed)

The changing conditions imposed on individual particles of gas as the pressure pulse passes is now considered. As a first simple approach to defining the pulse and its speed of propagation, consider the stream tube to have a velocity such that the pulse is stationary, Fig. 6.6a. The flow upstream of the pulse has velocity u, density ρ and pressure p, while the exit flow has these quantities changed by infinitesimal amounts to $u + \delta u$, $\rho + \delta\rho$, $p + \delta p$.

The flow situation now to be considered is quasi-steady, assumed inviscid and adiabatic (since the very small pressure changes take place too rapidly for heat transfer to be significant), takes place in the absence of external forces, and is one-dimensional, so that the differential equations of continuity and motion are respectively

$$u\frac{\partial\rho}{\partial x} + \rho\frac{\partial u}{\partial x} = 0 \qquad (6.31)$$

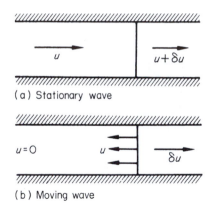

(a) Stationary wave

(b) Moving wave

Fig. 6.6

and

$$u \frac{\partial u}{\partial x} = -\frac{1}{\rho} \frac{\partial p}{\partial x} \tag{6.32}$$

Eliminating $\partial u / \partial x$ from these equations leaves

$$u^2 = \frac{\partial p}{\partial \rho} \tag{6.33}$$

This implies the speed of flow in the stream tube that is required to maintain a stationary pulse of weak strength, is uniquely the speed given by $\sqrt{\partial p / \partial \rho}$ (see Section 1.2.7, Eqn (1.6c)).

The problem is essentially unaltered if the pulse advances at speed $u = \sqrt{\partial p / \partial \rho}$ through stationary gas and, since this is the (ideal) model of the propagation of weak pressure disturbances that are commonly sensed as sounds, the unique speed $\sqrt{\partial p / \partial \rho}$ is referred to as the acoustic speed a. When the pressure density relation is isentropic (as assumed above) this velocity becomes (see Eqn (1.6d))

$$a = \sqrt{\frac{\partial p}{\partial \rho}} = \sqrt{\frac{\gamma p}{\rho}} \tag{6.34}$$

It will be recalled that this is the speed the gas attains in the throat of a choked stream tube and it follows that weak pressure disturbances will not propagate upstream into a flow where the velocity is greater than a, i.e. $u > a$ or $M > 1$.

6.4   One-dimensional flow: plane normal shock waves

In the previous section the behaviour of gas when acting as a transmitter of waves of infinitesimal amplitude was considered and the waves were shown to travel at an (acoustic) speed of $a = \sqrt{\partial p / \partial \rho}$ relative to the gas, while the gas properties of pressure, density etc. varied in a continuous manner.

If a disturbance of large amplitude, e.g. a rapid pressure rise, is set up there are almost immediate physical limitations to its continuous propagation. The accelerations of individual particles required for continuous propagation cannot be sustained and a pressure front or discontinuity is built up. This pressure front is known as a *shock wave* which travels through the gas at a speed, always in excess of the acoustic speed, and together with the pressure jump, the density, temperature and entropy of the gas increases suddenly while the normal velocity drops.

Useful and quite adequate expressions for the change of these flow properties across the shock can be obtained by assuming that the shock front is of zero thickness. In fact the shock wave is of finite thickness being a few molecular mean free path lengths in magnitude, the number depending on the initial gas conditions and the intensity of the shock.

6.4.1 One-dimensional properties of normal shock waves

Consider the flow model shown in Fig. 6.7a in which a plane shock advances from right to left with velocity u_1 into a region of still gas. Behind the shock the velocity is suddenly increased to some value u in the direction of the wave. It is convenient to superimpose on the system a velocity of u_1 from left to right to bring the shock stationary relative to the walls of the tube through which gas is flowing undisturbed at u_1 (Fig. 6.7b). The shock becomes a stationary discontinuity into which gas flows with uniform conditions, p_1, ρ_1, u_1, etc., and from which it flows with uniform conditions, p_2, ρ_2, u_2, etc. It is assumed that the gas is inviscid, and non-heat conducting, so that the flow is adiabatic up to and beyond the discontinuity.

The equations of state and conservation for unit area of shock wave are:

State

$$\frac{p_1}{\rho_1 T_1} = \frac{p_2}{\rho_2 T_2} \tag{6.35}$$

Mass flow

$$\dot{m} = \rho_1 u_1 = \rho_2 u_2 \tag{6.36}$$

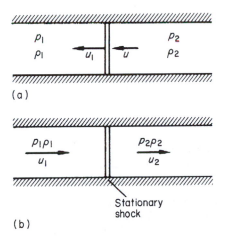

(a)

(b)

Stationary shock

Fig. 6.7

Momentum, in the absence of external and dissipative forces

$$p_1 + \rho_1 u_1^2 = p_2 + \rho_2 u_2^2 \tag{6.37}$$

Energy

$$c_p T_1 + \frac{u_1^2}{2} = c_p T_2 + \frac{u_2^2}{2} = c_p T_0 \tag{6.38}$$

6.4.2 Pressure–density relations across the shock

Eqn (6.38) may be rewritten (from e.g. Eqn (6.27)) as

$$\frac{u_1^2}{2} + \frac{\gamma}{\gamma - 1}\frac{p_1}{\rho_1} = c_p T_0 = \frac{u_2^2}{2} + \frac{\gamma}{\gamma - 1}\frac{p_2}{\rho_2}$$

which on rearrangement gives

$$\frac{\gamma}{\gamma - 1}\left(\frac{p_1}{\rho_1} - \frac{p_2}{\rho_2}\right) = \tfrac{1}{2}(u_2 - u_1)(u_2 + u_1) \tag{6.39}$$

From the continuity equation (6.36):

$$u_2 + u_1 = \dot{m}\left(\frac{1}{\rho_2} + \frac{1}{\rho_1}\right)$$

and from the momentum equation (6.37):

$$u_2 - u_1 = \frac{1}{\dot{m}}(p_1 - p_2)$$

Substituting for both of these in the rearranged energy equation (6.39)

$$\frac{\gamma}{\gamma - 1}\left(\frac{p_1}{\rho_1} - \frac{p_2}{\rho_2}\right) = \tfrac{1}{2}(p_1 - p_2)\left(\frac{1}{\rho_1} + \frac{1}{\rho_2}\right) \tag{6.40}$$

and this, rearranged by isolating the pressure and density ratios respectively, gives the *Rankine–Hugoniot relations*:

$$\frac{p_2}{p_1} = \frac{\dfrac{\gamma + 1}{\gamma - 1}\dfrac{\rho_2}{\rho_1} - 1}{\dfrac{\gamma + 1}{\gamma - 1} - \dfrac{\rho_2}{\rho_1}} \tag{6.41}$$

$$\frac{\rho_2}{\rho_1} = \frac{\dfrac{\gamma + 1}{\gamma - 1}\dfrac{p_2}{p_1} + 1}{\dfrac{\gamma + 1}{\gamma - 1} + \dfrac{p_2}{p_1}} \tag{6.42}$$

Taking $\gamma = 1.4$ for air, these equations become:

$$\frac{p_2}{p_1} = \frac{6\frac{\rho_2}{\rho_1} - 1}{6 - \frac{\rho_2}{\rho_1}} \qquad (6.41a)$$

and

$$\frac{\rho_2}{\rho_1} = \frac{6\frac{p_2}{p_1} + 1}{6 + \frac{p_2}{p_1}} \qquad (6.42a)$$

Eqns (6.42) and (6.42a) show that, as the value of ρ_2/ρ_1 tends to $(\gamma + 1)/(\gamma - 1)$ (or 6 for air), p_2/p_1 tends to infinity, which indicates that the maximum possible density increase through a shock wave is about six times the undisturbed density.

6.4.3 Static pressure jump across a normal shock

From the equation of motion (6.37) using Eqn (6.36):

$$\frac{p_2 - p_1}{p_1} = \frac{\rho_1 u_1^2 - \rho_2 u_2^2}{p_1} = \frac{\rho_1 u_1^2}{p_1}\left[1 - \frac{u_2}{u_1}\right]$$

or

$$\frac{p_2}{p_1} - 1 = \gamma M_1^2\left[1 - \frac{u_2}{u_1}\right]$$

but from continuity $u_2/u_1 = \rho_1/\rho_2$, and from the Rankine–Hugoniot relations ρ_2/ρ_1 is a function of (p_2/p_1). Thus, by substitution:

$$\frac{p_2}{p_1} - 1 = \gamma M_1^2\left[1 - \left\{\left(\frac{\gamma+1}{\gamma-1}\frac{p_1}{p_2} + 1\right)\bigg/\left(\frac{\gamma+1}{\gamma-1} + \frac{p_1}{p_2}\right)\right\}\right]$$

Isolating the ratio p_2/p_1 and rearranging gives

$$\frac{p_2}{p_1} = \frac{2\gamma M_1^2}{\gamma + 1} - \frac{\gamma - 1}{\gamma + 1} \qquad (6.43)$$

Note that for air

$$\frac{p_2}{p_1} = \frac{7M_1^2 - 1}{6} \qquad (6.43a)$$

Expressed in terms of the downstream or exit Mach number M_2, the pressure ratio can be derived in a similar manner (by the inversion of suffices):

$$\frac{p_1}{p_2} = \frac{2\gamma M_2^2 - (\gamma - 1)}{\gamma + 1} \qquad (6.44)$$

or

$$\frac{p_1}{p_2} = \frac{7M_2^2 - 1}{6} \quad \text{for air} \tag{6.44a}$$

6.4.4 Density jump across the normal shock

Using the previous results, substituting for p_2/p_1 from Eqn (6.43) in the Rankine–Hugoniot relations Eqn (6.42):

$$\frac{p_2}{\rho_1} = \frac{\left(\frac{\gamma+1}{\gamma-1}\left[\frac{2\gamma M_1^2 - (\gamma-1)}{\gamma+1}\right] + 1\right)}{\left(\frac{\gamma+1}{\gamma-1} + \left[\frac{2\gamma M_1^2 - (\gamma-1)}{\gamma+1}\right]\right)}$$

or rearranged

$$\frac{\rho_2}{\rho_1} = \frac{(\gamma+1)M_1^2}{2 + (\gamma-1)M_1^2} \tag{6.45}$$

For air $\gamma = 1.4$ and

$$\frac{\rho_2}{\rho_1} = \frac{6M_1^2}{5 + M_1^2} \tag{6.45a}$$

Reversed to give the ratio in terms of the exit Mach number

$$\frac{\rho_1}{\rho_2} = \frac{(\gamma+1)M_2^2}{2 + (\gamma-1)M_2^2} \tag{6.46}$$

For air

$$\frac{\rho_1}{\rho_2} = \frac{6M_2^2}{5 + M_2^2} \tag{6.46a}$$

6.4.5 Temperature rise across the normal shock

Directly from the equation of state and Eqns (6.43) and (6.45):

$$\frac{T_2}{T_1} = \frac{p_2}{p_1} \bigg/ \frac{\rho_2}{\rho_1}$$

$$\frac{T_2}{T_1} = \left(\frac{2\gamma M_1^2 - (\gamma-1)}{\gamma+1}\right)\left(\frac{2 + (\gamma-1)M_1^2}{(\gamma+1)M_1^2}\right) \tag{6.47}$$

For air

$$\frac{T_2}{T_1} = \frac{7M_1^2 - 5/M_1^2 + 34}{36} \tag{6.47a}$$

Since the flow is non-heat conducting the total (or stagnation) temperature remains constant.

6.4.6 Entropy change across the normal shock

Recalling the basic equation (1.32)

$$e^{\Delta S/c_V} = \left(\frac{T_2}{T_1}\right)^{\gamma}\left(\frac{p_1}{p_2}\right)^{\gamma-1} = \left(\frac{p_2}{p_1}\right)\left(\frac{\rho_1}{\rho_2}\right)^{\gamma} \quad \text{from the equation of state}$$

which on substituting for the ratios from the sections above may be written as a sum of the natural logarithms:

$$\frac{\Delta S}{c_V} = \ln\frac{2\gamma M_1^2 - (\gamma-1)}{\gamma+1} + \gamma\ln\left(2 + (\gamma-1)M_1^2\right) - \gamma\ln M_1^2$$

These are rearranged in terms of the new variable $(M_1^2 - 1)$

$$\frac{\Delta S}{c_V} = \ln\left[1 + \frac{2\gamma}{\gamma+1}(M_1^2 - 1)\right] + \gamma\ln\left[1 + \frac{\gamma-1}{\gamma+1}(M_1^2 - 1)\right]$$
$$- \gamma\ln\left[1 + (M_1^2 - 1)\right]$$

On expanding these logarithms and collecting like terms, the first and second powers of $(M_1^2 - 1)$ vanish, leaving a converging series commencing with the term

$$\frac{\Delta S}{c_V} = \frac{2\gamma(\gamma-1)}{(\gamma+1)^2}\frac{(M_1^2 - 1)^3}{3} \quad (6.48)$$

Inspection of this equation shows that: (a) for the second law of thermodynamics to apply, i.e. ΔS to be positive, M_1 must be greater than unity and an expansion shock is not possible; (b) for values of M_1 close to (but greater than) unity the values of the change in entropy are small and rise only slowly for increasing M_1. Reference to the appropriate curve in Fig. 6.9 below shows that for quite moderate supersonic Mach numbers, i.e. up to about $M_1 = 2$, a reasonable approximation to the flow conditions may be made by assuming an isentropic state.

6.4.7 Mach number change across the normal shock

Multiplying the above pressure (or density) ratio equations together gives the Mach number relationship directly:

$$\frac{p_2}{p_1} \times \frac{p_1}{p_2} = \frac{2\gamma M_1^2 - (\gamma-1)}{\gamma+1} \times \frac{2\gamma M_2^2 - (\gamma-1)}{\gamma+1} = 1$$

Rearrangement gives for the exit Mach number:

$$M_2^2 = \frac{(\gamma-1)M_1^2 + 2}{2\gamma M_1^2 - (\gamma-1)} \quad (6.49)$$

For air

$$M_2^2 = \frac{M_1^2 + 5}{7M_1^2 - 1} \tag{6.49a}$$

Inspection of these last equations shows that M_2 has upper and lower limiting values:

For $M_1 \to \infty$ $M_2 \to \sqrt{\frac{\gamma - 1}{2\gamma}} = (1/\sqrt{7} = 0.378$ for air)

For $M_1 \to 1$ $M_2 \to 1$

Thus the exit Mach number from a normal shock wave is always subsonic and for air has values between 1 and 0.378.

6.4.8 Velocity change across the normal shock

The velocity ratio is the inverse of the density ratio, since by continuity $u_2/u_1 = \rho_1/\rho_2$. Therefore, directly from Eqns (6.45) and (6.45a):

$$\frac{u_2}{u_1} = \frac{2 + (\gamma - 1)M_1^2}{(\gamma + 1)M_1^2} \tag{6.50}$$

or for air

$$\frac{u_2}{u_1} = \frac{5 + M_1^2}{6M_1^2} \tag{6.50a}$$

Of added interest is the following development. From the energy equations, with $c_p T$ replaced by $[\gamma/(\gamma - 1)]p/\rho$, p_1/ρ_1 and p_2/ρ_2 are isolated:

$$\frac{p_1}{\rho_1} = \frac{\gamma - 1}{\gamma}\left(c_p T_0 - \frac{u_1^2}{2}\right) \quad \text{ahead of the shock}$$

and

$$\frac{p_2}{\rho_2} = \frac{\gamma - 1}{\gamma}\left(c_p T_0 - \frac{u_2^2}{2}\right) \quad \text{downstream of the shock}$$

The momentum equation (6.37) is rearranged with $\rho_1 u_1 = \rho_2 u_2$ from the equation of continuity (6.36) to

$$u_1 - u_2 = \frac{p_2}{\rho_2 u_2} - \frac{p_1}{\rho_1 u_1}$$

and substituting from the preceding line

$$u_1 - u_2 = \frac{\gamma - 1}{\gamma}\left[(u_1 - u_2)\left(\frac{1}{2} + \frac{c_p T_0}{u_1 u_2}\right)\right]$$

Disregarding the uniform flow solution of $u_1 = u_2$ the conservation of mass, motion and energy apply for this flow when

$$u_1 u_2 = \frac{2(\gamma - 1)}{\gamma + 1} c_p T_0 \qquad (6.51)$$

i.e. the product of normal velocities through a shock wave is a constant that depends on the stagnation conditions of the flow and is independent of the strength of the shock. Further it will be recalled from Eqn (6.26) that

$$\frac{2(\gamma - 1)}{\gamma + 1} c_p T_0 = a^{*2}$$

where a^* is the critical speed of sound and an alternative parameter for expressing the gas conditions. Thus, in general across the shock wave:

$$u_1 u_2 = a^{*2} \qquad (6.52)$$

This equation indicates that $u_1 > a^* > u_2$ or *vice versa* and appeal has to be made to the second law of thermodynamics to see that the second alternative is inadmissible.

6.4.9   Total pressure change across the normal shock

From the above sections it can be seen that a finite entropy increase occurs in the flow across a shock wave, implying that a degradation of energy takes place. Since, in the flow as a whole, no heat is acquired or lost the total temperature (total enthalpy) is constant and the dissipation manifests itself as a loss in total pressure. Total pressure is defined as the pressure obtained by bringing gas to rest isentropically.

Now the model flow of a uniform stream of gas of unit area flowing through a shock is extended upstream, by assuming the gas to have acquired the conditions of suffix 1 by expansion from a reservoir of pressure p_{01} and temperature T_0, and downstream, by bringing the gas to rest isentropically to a total pressure p_{02} (Fig. 6.8)

Isentropic flow from the upstream reservoir to just ahead of the shock gives, from Eqn (6.18a):

$$\frac{p_{01}}{p_1} = \left[1 + \frac{\gamma - 1}{2} M_1^2 \right]^{\gamma/(\gamma - 1)} \qquad (6.53)$$

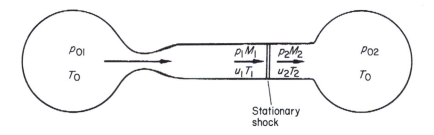

Fig. 6.8

and from just behind the shock to the downstream reservoir:

$$\frac{p_{02}}{p_2} = \left[1 + \frac{\gamma - 1}{2} M_2^2\right]^{\gamma/(\gamma-1)} \tag{6.54}$$

Eqn (6.43) recalled is

$$\frac{p_2}{p_1} = \frac{2\gamma M_1^2 - (\gamma - 1)}{\gamma + 1}$$

and Eqn (6.49) is

$$M_2^2 = \frac{(\gamma - 1)M_1^2 + 2}{2\gamma M_1^2 - (\gamma - 1)}$$

These four expressions, by division and substitution, give successively

$$\frac{p_{01}}{p_{02}} = \frac{p_1}{p_2} \frac{\left[1 + \dfrac{\gamma - 1}{2} M_1^2\right]^{\gamma/(\gamma-1)}}{\left[1 + \dfrac{\gamma - 1}{2} M_2^2\right]}$$

$$= \frac{\gamma + 1}{2\gamma M_1^2 - (\gamma - 1)} \left[\frac{2 + (\gamma - 1)M_1^2}{2 + (\gamma - 1)\left[\dfrac{(\gamma - 1)M_1^2 + 2}{2\gamma M_1^2 - (\gamma - 1)}\right]}\right]^{\gamma/(\gamma-1)}$$

$$= \left[\frac{2\gamma M_1^2 - (\gamma - 1)}{\gamma + 1}\right]^{1/(\gamma-1)} \left[\frac{2 + (\gamma - 1)M_1^2}{(\gamma + 1)M_1^2}\right]^{\gamma/(\gamma-1)}$$

Rewriting in terms of $(M_1^2 - 1)$:

$$\frac{p_{01}}{p_{02}} = \left[1 + \frac{2\gamma}{\gamma + 1}(M_1^2 - 1)\right]^{1/(\gamma-1)} \left[1 + \frac{\gamma - 1}{\gamma + 1}(M_1^2 - 1)\right]^{\gamma/(\gamma-1)}$$

$$\times [1 + (M_1^2 - 1)]^{-\gamma/(\gamma-1)}$$

Expanding each bracket and multiplying through gives the series

$$\frac{p_{01}}{p_{02}} = 1 + \frac{2\gamma}{(\gamma + 1)^2} \frac{(M_1^2 - 1)^3}{3} + \cdots$$

For values of Mach number close to unity (but greater than unity) the sum of the terms involving M_1^2 is small and very close to the value of the first term shown, so that the proportional change in total pressure through the shock wave is

$$\frac{\Delta p_0}{p_{01}} = \frac{p_{01} - p_{02}}{p_{01}} \simeq \frac{2\gamma}{(\gamma + 1)^2} \frac{(M_1^2 - 1)^3}{3} \tag{6.55}$$

It can be deduced from the curve (Fig. 6.9) that this quantity increases only slowly from zero near $M_1 = 1$, so that the same argument for ignoring the entropy increase (Section 6.4.6) applies here. Since from entropy considerations $M_1 > 1$, Eqn (6.55) shows that the total pressure always drops through a shock wave. The two phenomena,

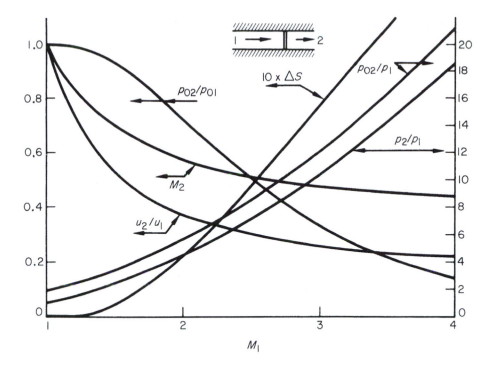

Fig. 6.9

i.e. total pressure drop and entropy increase, are in fact related, as may be seen in the following.

Recalling Eqn (1.32) for entropy:

$$e^{\Delta S/c_v} = \frac{p_2}{p_1}\left(\frac{\rho_1}{\rho_2}\right)^\gamma = \frac{p_{02}}{p_{01}}\left(\frac{\rho_{01}}{\rho_{02}}\right)^\gamma$$

since

$$\frac{p_1}{\rho_1^\gamma} = \frac{p_{01}}{\rho_{01}^\gamma}, \quad \text{etc.}$$

But across the shock T_0 is constant and, therefore, from the equation of state $p_{01}/\rho_{01} = p_{02}/\rho_{02}$ and entropy becomes

$$e^{\Delta S/c_v} = \left(\frac{p_{01}}{p_{02}}\right)^{\gamma-1}$$

$$\ln\frac{p_{01}}{p_{02}} = \frac{\Delta S}{c_V(\gamma-1)} \tag{6.56}$$

and substituting for ΔS from Eqn (6.48):

$$\ln\frac{p_{01}}{p_{02}} = \frac{2\gamma}{(\gamma+1)^2}\frac{(M_1^2-1)^3}{3} + \cdots = \beta \text{ (say)}$$

Now for values of M_1 near unity $\beta \ll 1$ and

$$\frac{\Delta p_0}{p_{01}} = \frac{p_{01} - p_{02}}{p_{01}} = 1 - e^{-\beta}$$

$$\frac{\Delta p_0}{p_{01}} \simeq \frac{2\gamma}{(\gamma + 1)^2} \frac{(M_1^2 - 1)^3}{3} \quad \text{(as before, Eqn (6.55))}$$

6.4.10 Pitôt tube equation

The pressure registered by a small open-ended tube facing a supersonic stream is effectively the 'exit' (from the shock) total pressure p_{02}, since the bow shock wave may be considered normal to the axial streamline, terminating in the stagnation region of the tube. That is, the axial flow into the tube is assumed to be brought to rest at pressure p_{02} from the subsonic flow p_2 behind the wave, after it has been compressed from the supersonic region p_1 ahead of the wave, Fig. 6.10. In some applications this pressure is referred to as the *static pressure* of the free or undisturbed supersonic stream p_1 and evaluated in terms of the free stream Mach number, hence providing a method of determining the undisturbed Mach number, as follows.

From the normal shock static-pressure ratio equation (6.43)

$$\frac{p_2}{p_1} = \frac{2\gamma M_1^2 - (\gamma - 1)}{\gamma + 1}$$

From isentropic flow relations,

$$\frac{p_{02}}{p_2} = \left[1 + \frac{\gamma - 1}{2} M_2^2\right]^{\gamma/(\gamma-1)}$$

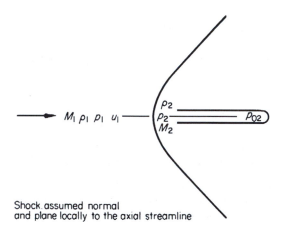

Shock assumed normal
and plane locally to the axial streamline

Fig. 6.10

Dividing these expressions and recalling Eqn (6.49), as follows:

$$M_2^2 = \frac{(\gamma - 1)M_1^2 + 2}{2\gamma M_1^2 - (\gamma - 1)}$$

the required pressure ratio becomes

$$\frac{p_{02}}{p_1} = \frac{\left[\dfrac{\gamma + 1}{2} M_1^2\right]^{\gamma/(\gamma-1)}}{\left[\dfrac{2\gamma M_1^2 - (\gamma - 1)}{\gamma + 1}\right]^{1/(\gamma-1)}} \qquad (6.57)$$

This equation is sometimes called *Rayleigh's supersonic Pitôt tube equation.*

The observed curvature of the detached shock wave on supersonic Pitôt tubes was once thought to be sufficient to bring the assumption of plane-wave theory into question, but the agreement with theory reached in the experimental work was well within the accuracy expected of that type of test and was held to support the assumption of a normal shock ahead of the wave.*

6.5 Mach waves and shock waves in two-dimensional flow

A small deflection in supersonic flow always takes place in such a fashion that the flow properties are uniform along a front inclined to the flow direction, and their only change is in the direction normal to the front. This front is known as a wave and for small flow changes it sets itself up at the Mach angle (μ) appropriate to the upstream flow conditions.

For finite positive or compressive flow deflections, that is when the downstream pressure is much greater than that upstream, the (shock) wave angle is greater than the Mach angle and characteristic changes in the flow occur (see Section 6.4). For finite negative or expansive flow deflections where the downstream pressure is less, the turning power of a single wave is insufficient and a fan of waves is set up, each inclined to the flow direction by the local Mach angle and terminating in the wave whose Mach angle is that appropriate to the downstream condition.

For small changes in supersonic flow deflection both the compression shock and expansion fan systems approach the character and geometrical properties of a Mach wave and retain only the algebraic sign of the change in pressure.

6.6 Mach waves

Figure 6.11 shows the wave pattern associated with a point source P of weak pressure disturbances: (a) when stationary; and (b) and (c) when moving in a straight line.

(a) In the stationary case (with the surrounding fluid at rest) the concentric circles mark the position of successive wave fronts, at a particular instant of time. In three-dimensional flow they will be concentric spheres, but a close analogy to the

*D.N. Holder *et al.*, *ARCR and M*, 2782, 1953.

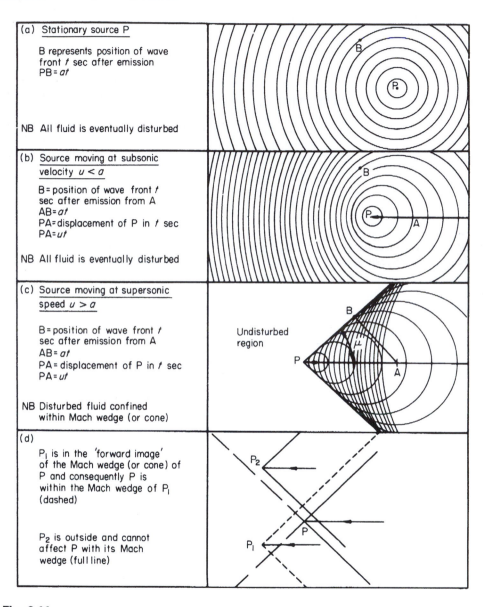

Fig. 6.11

two-dimensional case is the appearance of the ripples on the still surface of a pond from a small disturbance. The wave fronts emanating from P advance at the acoustic speed a and consequently the radius of a wave t seconds after its emission is at. If t is large enough the wave can traverse the whole of the fluid, which is thus made aware of the disturbance.

(b) When the intermittent source moves at a speed u less than a in a straight line, the wave fronts adopt the different pattern shown in Fig. 6.11b. The individual waves remain circular with their centres on the line of motion of the source and are eccentric but non-intersecting. The point source moves through a distance ut in the time the wave moves through the greater distance at. Once again the waves signalling the pressure disturbance will move through the whole region of fluid, ahead of and behind the moving source.

(c) If the steady speed of the source is increased beyond that of the acoustic speed the individual sound waves (at any one instant) are seen in Fig. 6.11c to be eccentric intersecting circles with their centres on the line of motion. Further the circles are tangential to two symmetrically inclined lines (a cone in three dimensions) with their apex at the point source P.

While a wave has moved a distance at, the point P has moved ut and thus the semi-vertex angle

$$\mu = \text{arc sin} \frac{at}{ut} = \text{arc sin} \frac{1}{M} \tag{6.58}$$

M, the Mach number of the speed of the point P relative to the undisturbed stream, is the ratio u/a, and the angle μ is known as the *Mach angle*. Were the disturbance continuous, the inclined lines (or cone) would be the envelope of all the waves produced and are then known as *Mach waves* (or cones).

It is evident that the effect of the disturbance does not proceed beyond the Mach lines (or cone) into the surrounding fluid, which is thus unaware of the disturbance. The region of fluid outside the Mach lines (or cone) has been referred to as the zone of silence or more dramatically as the zone of forbidden signals.

It is possible to project an image wedge (or cone) forward from the apex P, Fig. 6.11d, and this contains the region of the flow where any disturbance P_1, say, ahead would have an effect on P, since a disturbance P_2 outside it would exclude P from its Mach wedge (or cone); providing always that P_1 and P_2 are moving at the same Mach number.

If a uniform supersonic stream M is superimposed from left to right on the flow in Fig. 6.11c the system becomes that of a uniform stream of Mach number $M > 1$ flowing past a weak disturbance. Since the flow is symmetrical, the axis of symmetry may represent the surface of a flat plate along which an inviscid supersonic stream flows. Any small disturbance caused by a slight irregularity, say, will be communicated to the flow at large along a Mach wave. Figure 6.12 shows the Mach wave emanating from a disturbance which has a net effect on the flow similar to a pressure pulse that leaves the downstream flow unaltered. If the pressure change across the Mach wave is to be permanent, the downstream flow direction must change. The converse is also true.

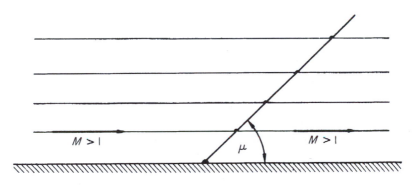

Fig. 6.12

It is shown above that a slight pressure change in supersonic flow is propagated along an oblique wave inclined at μ to the flow direction. The pressure difference is across, or normal to, the wave and the gas velocity will alter, as a consequence, in its component perpendicular to the wave front. If the downstream pressure is less, the flow velocity component normal to the wave increases across the wave so that the resultant downstream flow is inclined at a greater angle to the wave front, Fig. 6.13a. Thus the flow has been expanded, accelerated and deflected away from the wave front. On the other hand, if the downstream pressure is greater, Fig. 6.13b, the flow component across the wave is reduced, as is the net outflow velocity, which is now inclined at an angle less than μ to the wave front. The flow has been compressed, retarded and deflected towards the wave.

Quantitatively the turning power of a wave may be obtained as follows: Figure 6.14 shows the slight expansion round a small deflection δv_p from flow conditions p, ρ, M, q, etc., across a Mach wave set at μ to the initial flow direction. Referring to the velocity components normal and parallel to the wave, it may be recalled that the final velocity $q + \delta q$ changes only by virtue of a change in the normal velocity component u to $u + \delta u$ as it crosses the wave, since the tangential velocity remains uniform throughout the field. Then, from the velocity diagram after the wave:

$$(q + \delta q)^2 = (u + \delta u)^2 + v^2$$

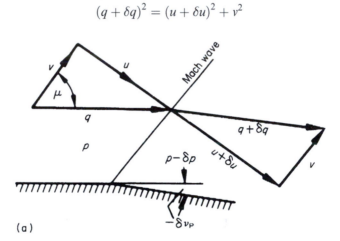

(a)

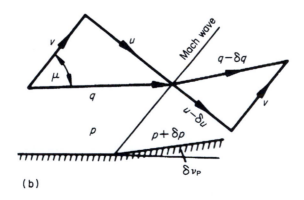

(b)

Fig. 6.13

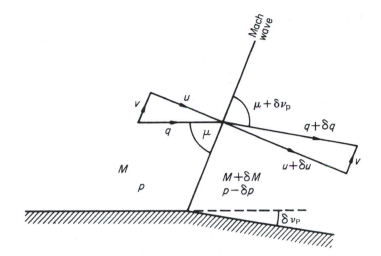

Fig. 6.14 Expansion round an infinitesimal deflection through a Mach wave

on expanding

$$q + 2q\delta q + (\delta q)^2 = u^2 + 2u\delta u + (\delta u)^2 + v^2$$

and in the limit, ignoring terms of the second order, and putting $u^2 + v^2 = q^2$:

$$q\mathrm{d}q = u\mathrm{d}u \tag{6.59}$$

Equally, from the definition of the velocity components:

$$\mu = \arctan\frac{u}{v} \quad \text{and} \quad \mathrm{d}\mu = \frac{1}{1 + (u/v)^2}\frac{\mathrm{d}u}{v} = \frac{v}{q^2}\mathrm{d}u$$

but the change in deflection angle is the incremental change in Mach angle. Thus

$$\mathrm{d}v_\mathrm{P} = \mathrm{d}\mu = \frac{v}{q^2}\mathrm{d}u \tag{6.60}$$

Combining Eqns (6.59) and (6.60) yields

$$\frac{\mathrm{d}q}{\mathrm{d}v_\mathrm{P}} = q\frac{u}{v} \quad \text{since} \quad \frac{u}{v} = \arctan\mu = \frac{1}{\sqrt{M^2 - 1}}$$

$$\frac{\mathrm{d}q}{q} = \pm\frac{\mathrm{d}v_\mathrm{P}}{\sqrt{M^2 - 1}} = \pm\mathrm{d}v_\mathrm{P}\tan\mu \tag{6.61}$$

where q is the flow velocity inclined at v_P to some datum direction. It follows from Eqn (6.10), with q substituted for μ, that

$$\frac{\mathrm{d}p}{p} = \pm\mathrm{d}v_\mathrm{P}\frac{\gamma M^2}{\sqrt{M^2 - 1}} \tag{6.62}$$

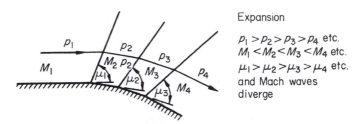

Expansion

$p_1 > p_2 > p_3 > p_4$ etc.
$M_1 < M_2 < M_3 < M_4$ etc.
$\mu_1 > \mu_2 > \mu_3 > \mu_4$ etc.
and Mach waves
diverge

Fig. 6.15

or in pressure-coefficient form

$$C_p = \pm \frac{2\,\mathrm{d}v_P}{\sqrt{M^2 - 1}} \tag{6.63}$$

The behaviour of the flow in the vicinity of a single weak wave due to a small pressure change can be used to study the effect of a larger pressure change that may be treated as the sum of a number of small pressure changes. Consider the expansive case first. Figure 6.15 shows the expansion due to a pressure decrease equivalent to three incremental pressure reductions to a supersonic flow initially having a pressure p_1 and Mach number M_1. On expansion through the wavelets the Mach number of the flow successively increases due to the acceleration induced by the successive pressure reductions and the Mach angle ($\mu = \arcsin 1/M$) successively decreases. Consequently, in such an expansive regime the Mach waves spread out or diverge, and the flow accelerates smoothly to the downstream conditions. It is evident that the number of steps shown in the figure may be increased or the generating wall may be continuous without the flow mechanism being altered except by the increased number of wavelets. In fact the finite pressure drop can take place abruptly, for example, at a sharp corner and the flow will continue to expand smoothly through a fan of expansion wavelets emanating from the corner. This case of two-dimensional expansive supersonic flow, i.e. round a corner, is known as the Prandtl–Meyer expansion and has the same physical mechanism as the one-dimensional isentropic supersonic accelerating flow of Section 6.2. In the Prandtl–Meyer expansion the streamlines are turned through the wavelets as the pressure falls and the flow accelerates. The flow velocity, angular deflection (from some upstream datum), pressure etc. at any point in the expansion may be obtained, with reference to Fig. 6.16.

Algebraic expressions for the wavelets in terms of the flow velocity be obtained by further manipulation of Eqn (6.61) which, for convenience, is recalled in the form:

$$\frac{1}{q}\frac{\mathrm{d}q}{\mathrm{d}v_P} = +\tan\mu$$

Introduce the velocity component $v = q\cos\mu$ along or tangential to the wave front (Fig. 6.13). Then

$$\mathrm{d}v = \mathrm{d}q\cos\mu - q\sin\mu\,\mathrm{d}\mu = q\sin\mu\left(\frac{1}{q}\frac{\mathrm{d}q}{\tan\mu} - \mathrm{d}\mu\right) \tag{6.64}$$

It is necessary to define the lower limiting or datum condition. This is most conveniently the sonic state where the Mach number is unity, $a = a^*$, $v_p = 0$, and the

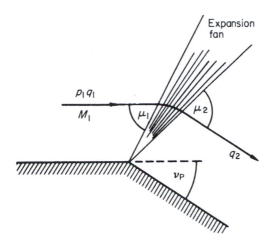

Fig. 6.16 Prandtl–Meyer expansion with finite deflection angle

wave angle $\mu = \pi/2$. In the general case, the datum (sonic) flow may be inclined by some angle α to the coordinate in use. Substitute dv_P for $(1/q)dq/\tan\mu$ from Eqn (6.61) and, since $q\sin\mu = a$, Eqn (6.64) becomes $dv_P - d\mu = dv/a$. But from the energy equation, with $c = $ ultimate velocity, $a^2/(\gamma - 1) + (q^2/2) = (c^2/2)$ and with $q^2 = (v^2 + a^2)$ (Eqn (6.17)):

$$a^2 = \frac{\gamma - 1}{\gamma + 1}(c^2 - v^2) \tag{6.65}$$

which gives the differential equation

$$dv_P - d\mu = \frac{dv}{\left(\dfrac{\gamma - 1}{\gamma + 1}(c^2 - v^2)\right)^{1/2}} \tag{6.66}$$

Equation (6.66) may now be integrated. Thus

$$\int_{v_P=\alpha}^{v_P} dv_P - \int_{\mu=\pi/2}^{\mu} d\mu = \left[\sqrt{\frac{\gamma + 1}{\gamma - 1}}\sin^{-1}\frac{v}{c}\right]_0^v$$

or

$$(v_P - \alpha) - \left(\mu - \frac{\pi}{2}\right) = \sqrt{\frac{\gamma + 1}{\gamma - 1}}\sin^{-1}\frac{v}{c} \tag{6.67}$$

From Eqn (6.65)

$$\sin^{-1}\frac{v}{c} = \tan^{-1}\sqrt{\frac{\gamma - 1}{\gamma + 1}}\frac{v}{a} = \tan^{-1}\sqrt{\frac{\gamma - 1}{\gamma + 1}}\cot\mu$$

which allows the flow deflection in Eqn (6.67) to be expressed as a function of Mach angle, i.e.

$$v_P - \alpha = \mu + \sqrt{\frac{\gamma - 1}{\gamma + 1}} \tan^{-1} \sqrt{\frac{\gamma - 1}{\gamma + 1}} \cot \mu - \frac{\pi}{2} \tag{6.68}$$

or

$$v_P - \alpha = f(\mu) \tag{6.68a}$$

In his original paper Meyer* used the complementary angle to the Mach wave $(\psi) = [(\pi/2) - \mu]$ and expressed the function

$$\sqrt{\frac{\gamma + 1}{\gamma - 1}} \tan^{-1} \sqrt{\frac{\gamma - 1}{\gamma + 1}} \cot \mu$$

as the angle ϕ to give Eqn (6.68a) in the form

$$v_P - \alpha = \phi - \psi \tag{6.68b}$$

The local velocity may also be expressed in terms of the Mach angle μ by rearranging the energy equation as follows:

$$\frac{q^2}{2} + \frac{a^2}{\gamma - 1} = \frac{c^2}{2}$$

but $a^2 = q^2 \sin^2 \mu$. Therefore

$$q^2 \left(\frac{1}{2} + \frac{\sin^2 \mu}{\gamma - 1} \right) = \frac{c^2}{2}$$

or

$$q^2 = \frac{c^2}{\left(1 + \frac{2}{\gamma - 1} \sin^2 \mu \right)} \tag{6.69}$$

Equations (6.68) and (6.69) give expressions for the flow velocity and direction at any point in a turning supersonic flow in terms of the local Mach angle μ and hence the local Mach number M.

Values of the deflection angle from sonic conditions $(v_P - \alpha)$, the deflection of the Mach angle from its position under sonic conditions ϕ, and velocity ratio q/c for a given Mach number may be computed once and for all and used in tabular form thereafter. Numerous tables of these values exist but most of them have the Mach number as dependent variable. It will be recalled that the turning power of a wave is a significant property and a more convenient tabulation has the angular deflection $(v_P - \alpha)$ as the dependent variable, but it is usual of course to give α the value of zero for tabular purposes.[†]

---

* Th. Meyer, *Über zweidimensionale Bewegungsvorgänge in einem Gas das mit Überschallgeschwindigkeit strömt*, 1908.

[†] See, for example, E.L. Houghton and A.E. Brock, *Tables for the Compressible Flow of Dry Air*, 3rd Edn, Edward Arnold, 1975.

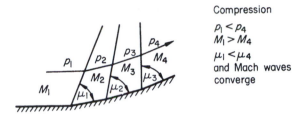

Compression

$p_1 < p_4$
$M_1 > M_4$
$\mu_1 < \mu_4$
and Mach waves
converge

Fig. 6.17

Compression flow through three wavelets springing from the points of flow deflection are shown in Fig. 6.17. In this case the flow velocity is reducing, M is reducing, the Mach angle increases, and the compression wavelets converge towards a region away from the wall. If the curvature is continuous the large number of wavelets reinforce each other in the region of the convergence, to become a finite disturbance to form the foot of a shock wave which is propagated outwards and through which the flow properties change abruptly. If the finite compressive deflection takes place abruptly at a point, the foot of the shock wave springs from the point and the initiating system of wavelets does not exist. In both cases the presence of boundary layers adjacent to real walls modifies the flow locally, having a greater effect in the compressive case.

6.6.1 Mach wave reflection

In certain situations a Mach wave, generated somewhere upstream, may impinge on a solid surface. In such a case, unless the surface is bent at the point of contact, the wave is reflected as a wave of the same sign but at some other angle that depends on the geometry of the system. Figure 6.18 shows two wavelets, one expansive and the other compressive, each of which, being generated somewhere upstream, strikes a plane wall at P along which the supersonic stream flows, at the Mach angle

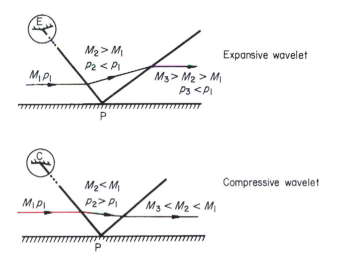

Fig. 6.18 Impingement and reflection of plane wavelets on a plane surface

appropriate to the upstream flow. Behind the wave the flow is deflected away from the wave (and wall) in the expansive case and towards the wave (and wall) in the compressive case, with appropriate increase and decrease respectively in the Mach number of the flow.

The physical requirement of the reflected wave is contributed by the wall downstream of the point P that demands the flow leaving the reflected wave parallel to the wall. For this to be so, the reflected wave must turn the flow away from itself in the former case, expanding it further to $M_3 > M_1$, and towards itself in the compressive case, thus additionally compressing and retarding its downstream flow.

If the wall is bent in the appropriate sense at the point of impingement at an angle of sufficient magnitude for the exit flow from the impinging wave to be parallel to the wall, then the wave is absorbed and no reflection takes place, Fig. 6.19. Should the wall be bent beyond this requirement a wavelet of the opposite sign is generated.

A particular case arises in the impingement of a compressive wave on a wall if the upstream Mach number is not high enough to support a supersonic flow after the two compressions through the impinging wave and its reflection. In this case the impinging wave bends to meet the surface normally and the reflected wave forks from the incident wave above the normal part away from the wall, Fig. 6.20. The resulting wave system is Y-shaped.

On reflection from an *open boundary* the impinging wavelets change their sign as a consequence of the physical requirement of pressure equality with the free atmosphere through which the supersonic jet is flowing. A sequence of wave reflections is shown in Fig. 6.21 in which an adjacent solid wall serves to reflect the wavelets onto the jet boundary. As in a previous case, an expansive wavelet arrives from upstream and is reflected from the point of impingement P_1 while the flow behind it is expanded to the ambient pressure p and deflected away from the wall. Behind the reflected wave from P_1 the flow is further expanded to p_3 in the fashion discussed above, to bring the streamlines back parallel to the wall.

On the reflection from the free boundary in Q_1 the expansive wavelet P_1Q_1 is required to compress the flow from p_3 back to p again along Q_1P_2. This compression

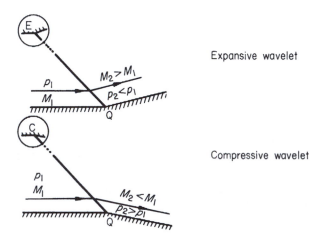

Fig. 6.19 Impingement and absorption of plane wavelets at bent surfaces

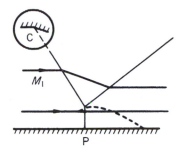

Fig. 6.20

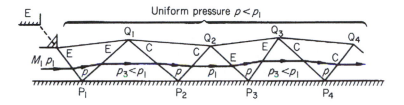

Fig. 6.21 Wave reflection from an open boundary

deflects the flow towards the wall where the compressive reflected wave from the wall (P_2Q_2) is required to bring the flow back parallel to the wall and in so doing increases its pressure to p_1 (greater than p). The requirement of the reflection of P_2Q_2 in the open boundary is thus expansive wavelet Q_2P_3 which brings the pressure back to the ambient value p again. And so the cycle repeats itself.

The solid wall may be replaced by the axial streamline of a (two-dimensional) supersonic jet issuing into gas at a uniformly (slightly) lower pressure. If the ambient pressure were (slightly) greater than that in the jet, the system would commence with a compressive wave and continue as above (Q_1P_2) onwards.

In the complete jet the diamonds are seen to be regions where the pressure is alternately higher or lower than the ambient pressure but the streamlines are axial, whereas when they are outside the diamonds, in the region of pressure equality with the boundary, the streamlines are alternately divergent or convergent.

The simple model discussed here is considerably different from that of the flow in a real jet, mainly on account of jet entrainment of the ambient fluid which affects the reflections from the open boundary, and for a finite pressure difference between the jet and ambient conditions the expansive waves are systems of fans and the compressive waves are shock waves.

6.6.2 Mach wave interference

Waves of the same character and strength intersect one another with the same configuration as those of reflections from the plane surface discussed above, since the surface may be replaced by the axial streamline, Fig. 6.22a and b. When the intersecting wavelets are of opposite sign the axial streamline is bent at the point of intersection in a direction away from the expansive wavelet. This is shown in Fig. 6.22c. The streamlines are also changed in direction at the intersection of waves of the same sign but of differing turning power.

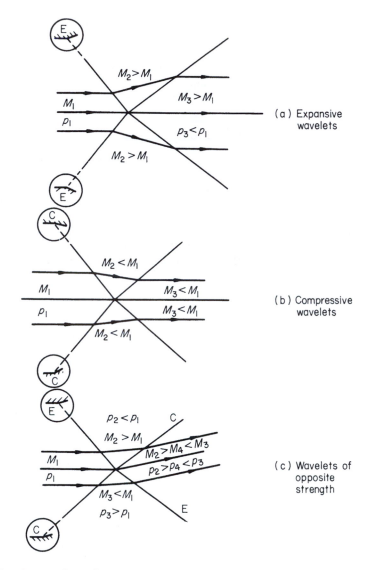

Fig. 6.22 Interference of wavelets

6.7 Shock waves

The generation of the flow discontinuity called a shock wave has been discussed in Section 6.4 in the case of one-dimensional flow. Here the treatment is extended to plane oblique and curved shocks in two-dimensional flows. Once again, the thickness of the shock wave is ignored, the fluid is assumed to be inviscid and non-heat-conducting. In practice the (thickness) distance in which the gas stabilizes its properties of state from the initial to the final conditions is small but finite. Treating a curved shock as consisting of small elements of plane oblique shock wave is reasonable only as long as its radius of curvature is large compared to the thickness.

With these provisos, the following exact, but relatively simple, extension to the one-dimensional shock theory will provide a deeper insight into those problems of shock waves associated with aerodynamics.

6.7.1 Plane oblique shock relations

Let a datum be fixed relative to the shock wave and angular displacements measured from the free-stream direction. Then the model for general oblique flow through a plane shock wave may be taken, with the notation shown in Fig. 6.23, where V_1 is the incident flow and V_2 the exit flow from the shock wave. The shock is inclined at an angle β to the direction of V_1 having components normal and tangential to the wave front of u_1 and v_1 respectively. The exit velocity V_2 (normal u_2, tangential v_2 components) will also be inclined to the wave but at some angle other than β. Relative to the incident flow direction the exit flow is deflected through δ. The equation of continuity for flow normal to the shock gives

$$\rho_1 u_1 = \rho_2 u_2 \qquad (6.70)$$

Conservation of linear momentum parallel to the wave front yields

$$\rho_1 u_1 v_1 = \rho_2 u_2 v_2 \qquad (6.71)$$

i.e. since no tangential force is experienced along the wave front, the product of the mass entering the wave per unit second and its tangential velocity at entry must equal

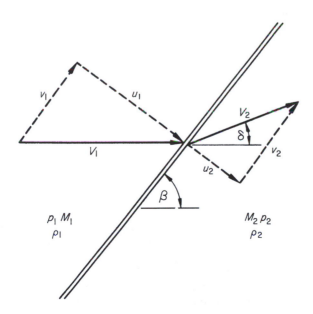

Fig. 6.23

the product of the mass per second leaving the wave and the exit tangential velocity. From continuity, Eqn (6.71) yields

$$v_1 = v_2 \tag{6.72}$$

Thus the velocity component along the wave front is unaltered by the wave and the model reduces to that of the one-dimensional flow problem (cf. Section 6.4.1) on which is superimposed a uniform velocity parallel to the wave front.

Now the normal component of velocity decreases abruptly in magnitude through the shock, and a consequence of the constant tangential component is that the exit flow direction, as well as magnitude, changes from that of the incident flow, and the change in the direction is towards the shock front. From this it emerges that the oblique shock is a mechanism for turning the flow inwards as well as compressing it. In the expansive mechanism for turning a supersonic flow (Section 6.6) the angle of inclination to the wave increases.

Since the tangential flow component is unaffected by the wave, the wave properties may be obtained from the one-dimensional flow case but need to be referred to datum conditions and direction are different from the normal velocities and directions. In the present case:

$$M_1 = \frac{V_1}{a_1} = \frac{u_1}{a_1} \frac{V_1}{u_1}$$

or

$$M_1 = \frac{\dfrac{u_1}{a_1}}{\sin \beta} \tag{6.73}$$

Similarly

$$M_2 = \frac{\dfrac{u_2}{a_2}}{\sin(\beta - \delta)} \tag{6.74}$$

The results of Section 6.4.2 may now be used directly, but with M_1 replaced by $M_1 \sin \beta$, and M_2 by $M_2 \sin (\beta - \delta)$. The following ratios pertain:

Static pressure jump from Eqn (6.43):

$$\frac{p_2}{p_1} = \frac{2\gamma M_1^2 \sin^2 \beta - (\gamma - 1)}{\gamma + 1} \tag{6.75}$$

or as inverted from Eqn (6.44):

$$\frac{p_1}{p_2} = \frac{2\gamma M_2^2 \sin^2(\beta - \delta) - (\gamma - 1)}{\gamma + 1} \tag{6.76}$$

Density jump from Eqn (6.45):

$$\frac{\rho_2}{\rho_1} = \frac{(\gamma + 1)M_1^2 \sin^2 \beta}{2 + (\gamma - 1)M_1^2 \sin^2 \beta} \tag{6.77}$$

or from Eqn (6.46)

$$\frac{\rho_1}{\rho_2} = \frac{(\gamma + 1)M_2^2 \sin^2(\beta - \delta)}{2 + (\gamma - 1)M_2^2 \sin^2(\beta - \delta)} \tag{6.78}$$

Static temperature change from Eqn (6.47):

$$\frac{T_2}{T_1} = \frac{2\gamma M_1^2 \sin^2 \beta - (\gamma - 1)}{\gamma + 1} \frac{2 + (\gamma - 1)M_1^2 \sin^2 \beta}{(\gamma + 1)M_1^2 \sin^2 \beta} \tag{6.79}$$

Mach number change from Eqn (6.49):

$$M_2^2 \sin^2(\beta - \delta) = \frac{(\gamma - 1)M_1^2 \sin^2 \beta + 2}{2\gamma M_1^2 \sin^2 \beta - (\gamma - 1)} \tag{6.80}$$

The equations above contain one or both of the additional parameters β and δ that must be known for the appropriate ratios to be evaluated.

An expression relating the incident Mach number M_1, the wave angle β and flow deflection δ may be obtained by introducing the geometrical configuration of the flow components, i.e.

$$\frac{u_1}{v_1} = \tan \beta, \quad \frac{u_2}{v_2} = \tan(\beta - \delta)$$

but

$$v_1 = v_2 \quad \text{and} \quad \frac{u_1}{u_2} = \frac{\rho_2}{\rho_1}$$

by continuity. Thus

$$\frac{\rho_2}{\rho_1} = \frac{\tan \beta}{\tan(\beta - \delta)} \tag{6.81}$$

Equations (6.77) and (6.81) give the different expressions for ρ_2/ρ_1, therefore the right-hand sides may be set equal, to give:

$$\frac{\tan \beta}{\tan(\beta - \delta)} = \frac{(\gamma + 1)M_1^2 \sin^2 \beta}{2 + (\gamma - 1)M_1^2 \sin^2 \beta} \tag{6.82}$$

Algebraic rearrangement gives

$$\tan \delta = \cot \beta \frac{M_1^2 \sin^2 \beta - 1}{\frac{\gamma + 1}{2} M_1^2 - (M_1^2 \sin^2 \beta - 1)} \tag{6.83}$$

Plotting values of β against δ for various Mach numbers gives the carpet of graphs shown in Fig. 6.24.

It can be seen that all the curves are confined within the $M_1 = \infty$ curve, and that for a given Mach number a certain value of deflection angle δ up to a maximum value δ_M may result in a smaller (weak) or larger (strong) wave angle β. To solve Eqn (6.83) algebraically, i.e. to find β for a given M_1 and δ, is very difficult. However, Collar* has shown that the equation may be expressed as the cubic

$$x^3 - Cx^2 - Ax + (B - AC) = 0 \tag{6.84}$$

* A.R. Collar, *J R Ae S*, Nov 1959.

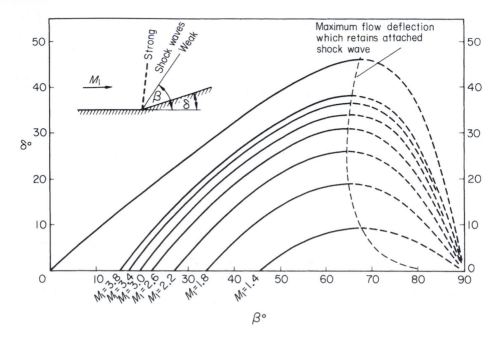

Fig. 6.24

where

$$x = \cot \beta, \qquad A = M^2 - 1, \qquad B = \frac{\gamma+1}{2} M_1^4 \tan \delta$$

and

$$C = \left(\frac{\gamma+1}{2} M_1^2 + 1 \right) \tan \delta$$

He further showed that the first root may be obtained from the iterative process

$$x_{n+1} = \sqrt{A - \frac{B}{x_n - C}} \tag{6.85}$$

and a suitable first approximation is $x_1 = \sqrt{M_1^2 - 1}$.

The iteration completed yields the root $x_0 = \cot \beta_w$ where β_w is the wave angle corresponding to the weak wave, i.e. β_w is the smaller value of wave angle shown graphically above (Fig. 6.24). Extracting this root (x_0) as a factor from the cubic equation (6.84) gives the quadratic equation

$$x^2 + (C + x_0)x + [x_0(C + x_0) - A] = 0 \tag{6.86}$$

having the formal solution

$$x = \tfrac{1}{2}[-(C + x_0) \pm \sqrt{(C + x_0)(C - 3x_0) + 4A}] \tag{6.87}$$

Now $x_0 = \cot \beta_w$ is one of the positive roots of the cubic equations and one of the physically possible solutions. The other physical solution, corresponding to the strong shock wave, is given by the positive root of the quadratic equation (6.87).

It is thus possible to obtain both physically possible values of the wave angle providing the deflection angle $\delta < \delta_{max}$. δ_{max} may be found in the normal way by differentiating Eqn (6.83) with reference to β, with M_1 constant and equating to zero. This gives, for the maximum value of $\tan \delta$:

$$\sin^2 \beta_{max} = \frac{1}{\gamma M_1^2} \left[\frac{\gamma+1}{4} M_1^2 - 1 + \sqrt{(\gamma+1)\left(1 + \frac{\gamma-1}{2} M_1^2 + \frac{\gamma+1}{16} M_1^4\right)} \right] \quad (6.88)$$

Substituting back in Eqn (6.82) gives a value for $\tan \delta_{max}$.

6.7.2 The shock polar

Although in practice plane-shock-wave data are used in the form of tables and curves based upon the shock relationships of the previous section, the study of shock waves is considerably helped by the use of a hodograph or velocity polar diagram set up for a given free-stream Mach number. This curve is the exit velocity vector displacement curve for all possible exit flows downstream of an attached plane shock in a given undisturbed supersonic stream, and to plot it out requires rearrangement of the equations of motion in terms of the exit velocity components and the inlet flow conditions.

Reference to Fig. 6.25 shows the exit component velocities to be used. These are q_t and q_n, the radial and tangential polar components with respect to the free stream V_1 direction taken as a datum. It is immediately apparent that the exit flow direction is given by $\arctan(q_t/q_n)$. For the wave angle β (recall the additional notation of

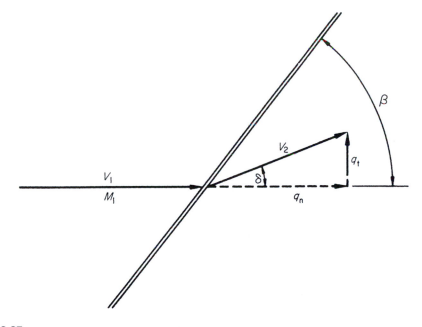

Fig. 6.25

Fig. 6.23), linear conservation of momentum along the wave front, Eqn (6.72), gives $v_1 = v_2$, or, in terms of geometry:

$$V_1 \cos \beta = V_2 \cos(\beta - \delta) \qquad (6.89)$$

Expanding the right-hand side and dividing through:

$$V_1 = V_2[\cos \delta + \tan \beta \sin \delta] \qquad (6.90)$$

or, in terms of the polar components:

$$V_1 = V_2 \left(\frac{q_n}{V_2} + \tan \beta \frac{q_t}{V_2} \right)$$

which rearranged gives the wave angle

$$\beta = \arctan \frac{V_1 - q_n}{q_t} \qquad (6.91)$$

To express the conservation of momentum normal to the wave in terms of the polar velocity components, consider first the flow of unit area normal to the wave, i.e.

$$p_1 + \rho_1 u_1^2 = p_2 + \rho_2 u_2^2 \qquad (6.92)$$

Then successively, using continuity and the geometric relations:

$$p_2 = p_1 + \rho_1 V_1 \sin \beta [V_1 \sin \beta - q_n \sin \beta + q_n \cos \beta \tan \delta]$$
$$p_2 = p_1 + \rho_1 V_1 \sin \beta [(V_1 - q_n) \sin \beta + q_t \cos \beta]$$

and, using Eqn (6.89):

$$p_2 = p_1 + \rho_1 V_1 (V_1 - q_n) \qquad (6.93)$$

Again from continuity (expressed in polar components):

$$\rho_1 V_1 \sin \beta = \rho_2 V_2 \sin(\beta - \delta) = \rho_2 q_n (\sin \beta - \cos \beta \tan \delta)$$

or

$$\rho_1 V_1 = \rho_2 q_n \left(1 - \frac{q_t}{q_n \tan \beta} \right) \qquad (6.94)$$

Divide Eqn (6.93) by Eqn (6.94) to isolate pressure and density:

$$\frac{p_2}{\rho_2 q_n} \frac{1}{\left(1 - \dfrac{q_t}{q_n \tan \beta} \right)} = \frac{p_1}{\rho_1 V_1} + (V_1 - q_n)$$

Again recalling Eqn (6.91) to eliminate the wave angle and rearranging:

$$\frac{p_2}{\rho_2} \frac{V_1 - q_n}{V_1 q_n - q_n^2 - q_t^2} = \frac{p_1}{\rho_1 V_1} + (V_1 - q_n) \qquad (6.95)$$

Finally from the energy equation expressed in polar velocity components: up to the wave

$$\frac{p_1}{\rho_1} = \frac{\gamma - 1}{2}\left(\frac{\gamma + 1}{\gamma - 1}a^{*^2} - V_1^2\right) \qquad (6.96)$$

and downstream from the wave

$$\frac{p_2}{\rho_2} = \frac{\gamma - 1}{2}\left(\frac{\gamma + 1}{\gamma - 1}a^{*^2} - q_n^2 - q_t^2\right) \qquad (6.97)$$

Substituting for these ratios in Eqn (6.93) and isolating the exit tangential velocity component gives the following equation:

$$q_t^2 = \frac{(V_1 - q_n)^2(V_1 q_n - a^{*^2})}{\left(\dfrac{2}{\gamma + 1}\right)V_1^2 - V_1 q_n + a^{*^2}} \qquad (6.98)$$

that is a basic form of the shock-wave-polar equation.

To make Eqn (6.98) more amenable to graphical analysis it may be made non-dimensional. Any initial flow parameters, such as the critical speed of sound a^*, the ultimate velocity c, etc., may be used but here we follow the originator A. Busemann* and divide through by the undisturbed acoustic speed a_1:

$$\hat{q}_t^2 = \frac{(M_1 - \hat{q}_n)^2\left(M_1\hat{q}_n - \dfrac{a^{*^2}}{a_1^2}\right)}{\dfrac{2}{\gamma + 1}M_1^2 - \left(M_1\hat{q}_n - \dfrac{a^{*^2}}{a_1^2}\right)} \qquad (6.99)$$

where $\hat{q}_t^2 = (q_t/a_1)^2$, etc. This may be further reduced to

$$\hat{q}_t^2 = (M_1 - \hat{q}_n)^2\frac{\hat{q}_n - \overline{M}_1}{\dfrac{2}{\gamma + 1}M_1 - (\hat{q}_n - \overline{M}_1)} \qquad (6.100)$$

where

$$\overline{M}_1 = \frac{a^{*^2}}{a_1^2 M_1} = \frac{\gamma - 1}{\gamma + 1}M_1 + \frac{2}{\gamma + 1}\frac{1}{M_1}$$

$$= M_1 - \frac{2}{\gamma + 2}\left(M_1 - \frac{1}{M_1}\right) = \overline{M}(M_1) \qquad (6.101)$$

Inspection of Eqn (6.99) shows that the curve of the relationship between $\hat{q}_t$ and $\hat{q}_n$ is uniquely determined by the free-stream conditions (M_1) and conversely one shock-polar curve is obtained for each free stream Mach number. Further, since

*A. Busemann, *Stodola Festschrift*, Zürich, 1929.

the non-dimensional tangential component $\hat{q}_t$ appears in the expression as a squared term, the curve is symmetrical about the q_n axis.

Singular points will be given by setting $\hat{q}_t = 0$ and ∞. For $\hat{q}_t = 0$,

$$(M_1 - \hat{q}_n)^2(\hat{q}_n^2 - \overline{M}_1) = 0$$

giving intercepts of the $\hat{q}_n$ axis at A:

$$\hat{q}_n = M_1 \quad \text{(twice)} \tag{6.102}$$

at B

$$\hat{q}_n = \overline{M}_1 = M_1 - \frac{2}{\gamma + 1}\left(M_1 - \frac{1}{M_1}\right) \tag{6.103}$$

For $\hat{q}_t = \infty$, at C,

$$\hat{q}_n = \frac{2}{\gamma + 1}M_1 + \overline{M}_1 = M_1 + \frac{2}{(\gamma + 1)M_1} \tag{6.104}$$

For a shock wave to exist $M_1 > 1$. Therefore the three points B, A and C of the q_n axis referred to above indicate values of $q_n < M_1$, $= M_1$, and $> M_1$ respectively. Further, as the exit flow velocity cannot be greater than the inlet flow velocity for a shock wave the region of the curve between A and C has no physical significance and attention need be confined only to the curve between A and B.

Plotting Eqn (6.98) point by point confirms the values A, B and C above. Fig. 6.26 shows the shock polar for the undisturbed flow condition of $M_1 = 3$. The upper

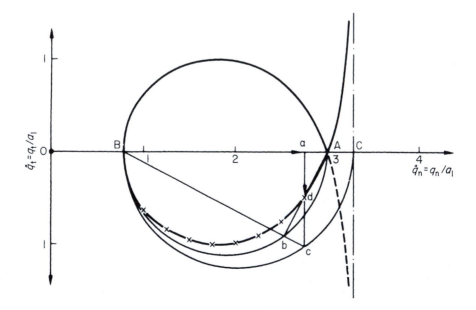

Fig. 6.26 Construction of shock polar March 3

branch of the curve in Fig. 6.26 is plotted point by point for the case of flow at a free stream of $M_1 = 3$. The lower half, which is the image of the upper reflected in the $\hat{q}_n$ axis shows the physically significant portion, i.e. the closed loop, obtained by a simple geometrical construction. This is as follows:

(i) Find and plot points A, B and C from the equations above. They are all explicitly functions of M_1.
(ii) Draw semi-circles (for a half diagram) with AB and CB as diameters.
(iii) At a given value $\hat{q}_n$ (Oa) erect ordinates to meet the larger semi-circle in c.
(iv) Join c to B intersecting the smaller semi-circle at b.
(v) The required point d is the intercept of bA and ac.

Geometrical proof   Triangles Aad and acB are similar. Therefore

$$\frac{ad}{aB} = \frac{aA}{ac}$$

i.e.

$$(ad)^2 = (aA)^2 \left(\frac{aB}{ac}\right)^2 \tag{6.105}$$

Again, from geometrical properties of circles,

$$(ac)^2 = aB \times aC$$

which substituted in Eqn (6.105) gives

$$(ad)^2 = (aA)^2 \frac{aB}{aC} \tag{6.106}$$

Introduction of the scaled values, ad = $\hat{q}_t$,

$$aB = Oa - OB = \hat{q}_n - \overline{M}_1, \qquad aA = OA - Oa = M_1 - \hat{q}_n,$$
$$aC = OC - Oa = [2/(\gamma+1)]M_1 + \overline{M}_1 - \hat{q}_n$$

reveals Eqn (6.100), i.e.

$$q_t^2 = \frac{(M_1 - \hat{q}_n)^2(\hat{q}_n - \overline{M}_1)}{\left(\dfrac{2}{\gamma+1}\right)M_1 - (\hat{q}_n - \overline{M}_1)} \quad \text{(Eqn (6.100))}$$

Consider the physically possible flows represented by various points on the closed portion of the shock polar diagram shown in Fig. 6.27. Point A is the upper limiting value for the exit flow velocity and is the case where the free stream is subjected only to an infinitesimal disturbance that produces a Mach wave inclined at μ to the free stream but no deflection of the stream and no change in exit velocity. The Mach wave angle is given by the inclination of the tangent of the curve at A to the vertical and this is the limiting case of the construction required to find the wave angle in general.

Point D is the second point at which a general vector emanating from the origin cuts the curve (the first being point E). The representation means that in going through a certain oblique shock the inlet stream of direction and magnitude given by OA is deflected through an angle δ and has magnitude and direction given by vector OD (or Od in the lower half diagram). The ordinates of OD give the normal and tangential exit velocity components.

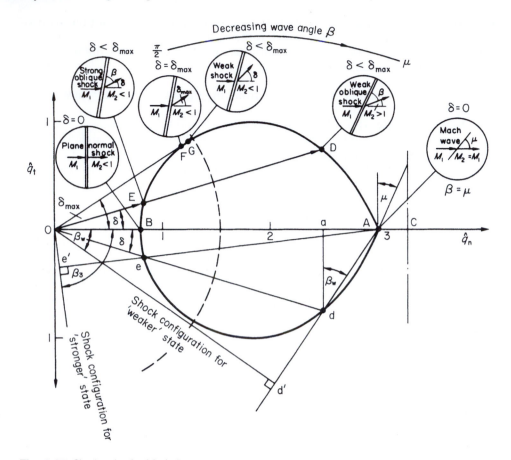

Fig. 6.27 Shock polar for Mach 3

The appropriate wave angle β_w is determined by the geometrical construction shown in the lower half of the curve, i.e. by the angle Ada. To establish this recall Eqn (6.99):

$$\tan \beta_W = \frac{V_1 - q_n}{q_t} = \frac{M_1 - \hat{q}_n}{\hat{q}_t} = \frac{(OA) - (Oa)}{(ad)} = \frac{(aA)}{(ad)}$$

i.e. $\beta_W = \widehat{adA}$.

The wave angle may be seen in better juxtaposition to the deflection δ, by a small extension to the geometrical construction. Produce Ad to meet the perpendicular from O in d'. Since $\Delta aAd|||\Delta Ad'O$,

$$\widehat{Aod'} = \widehat{Ada} = \beta_W$$

Of the two intercepts of the curve the point D yields the weaker shock wave, i.e. the shock wave whose inclination, characteristics, etc., are closer to the Mach wave at A. The other physically possible shock to produce the deflection δ is represented by the point E.

Point E; by a similar construction the wave angle appropriate to this shock condition is found (see Fig. 6.27), i.e. by producing Ae to meet the perpendicular from O in e'. Inspection shows that the wave is nearly normal to the flow, the velocity drop to OE from OA is much greater than the previous velocity drop OD for the same flow deflection and the shock is said to be the stronger shock.

As drawn, OE is within the sonic line, which is an arc of centre O and radius $[\hat{q}_n]_{M_2} = 1$, i.e. of radius

$$\frac{a^*}{q_1} = \sqrt{\frac{(\gamma-1)M_1^2 + 2}{\gamma+1}}$$

Point F is where the tangent to the curve through the origin meets the curve, and the angle so found by the tangent line and the $\hat{q}_n$ axis is the maximum flow deflection possible in the given supersonic stream that will still retain an attached shock wave. For deflections less than this maximum the curve is intersected in two physically real points as shown above in D and E. Of these two the exit flows OE corresponding to the strong shock wave case are always subsonic. The exit flows OD due to the weaker shocks are generally *supersonic* but a few deflection angles close to δ_{max} allow of weak shocks with subsonic exit flows. These are represented by point G. In practice weaker waves are experienced in uniform flows with plane shocks. When curved detached shocks exist, their properties may be evaluated locally by reference to plane-shock theory, and for the near-normal elements the strong shock representation OE may be used. Point B is the lower (velocity) limit to the polar curve and represents the normal (strongest) shock configuration in which the incident flow of velocity OA is compressed to the exit flow of velocity OB.

There is no flow deflection through a normal shock wave, which has the maximum reduction to subsonic velocity obtainable for the given undisturbed conditions.

6.7.3 Two-dimensional supersonic flow past a wedge

This can be described bearing the shock polar in mind. For an attached plane wave the wedge semi-vertex angle Δ, say, Fig. 6.28, must be less than or equal to the maximum deflection angle δ_{max} given by point F of the polar. The shock wave then sets itself up at the angle given by the weaker-shock case. The exit flow is uniform and parallel at a lower, in general supersonic, Mach number but with increased entropy compared with the undisturbed flow.

If the wedge angle Δ is increased, or the free-stream conditions altered to allow $\Delta > \beta_{max}$, the shock wave stands detached from the tip of the wedge and is curved

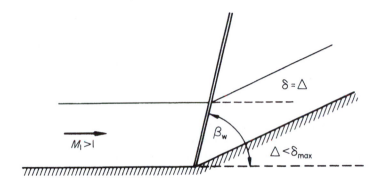

Fig. 6.28

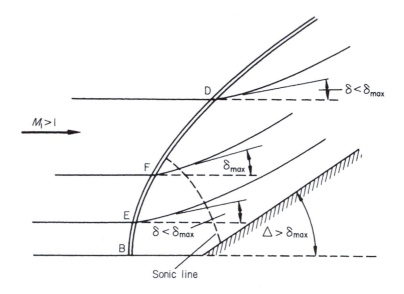

Fig. 6.29

from normal to the flow at the dividing streamline, to an angle approaching the Mach angle a long way from the axial streamline (Fig. 6.29).

All the conditions indicated by the closed loop of the shock polar can be identified:

B – on the axis the flow is undeflected but compressed through an element of normal shock to a subsonic state.

E – a little way away from the axis the stream deflection through the shock is less than the maximum possible but the exit flow is still subsonic given by the stronger shock.

F – further out the flow deflection through the shock wave reaches the maximum possible for the free-stream conditions and the exit flow is still subsonic. Beyond this point elements of shock wave behave in the weaker fashion giving a supersonic exit for streamlines meeting the shock wave beyond the intersection with the broken sonic line.

D – this point corresponds to the weaker shock wave. Further away from the axis the inclination of the wave approaches the Mach angle (the case given by point A in Fig. 6.27).

It is evident that a significant variable along the curved wave front is the product $M_1 \sin \beta$, where β is the inclination of the wave locally to the incoming streamline. Uniform undisturbed flow is assumed for simplicity, but is not a necessary restriction. Now $\mu < \beta < \pi/2$ and $M_1 \sin \beta$, the Mach number of the normal to the wave inlet component velocity, is thus the effective variable, that is a maximum on the axis, reducing to a minimum at the extremes of the wave (Fig. 6.30). Likewise all the other properties of the flow across the curved wave that are functions of $M_1 \sin \beta$, will vary along the shock front. In particular, the entropy jump across the shock, which from Eqn (6.48) is

$$\Delta S = c_V \frac{2\gamma(\gamma - 1)}{(\gamma + 1)^2} \frac{(M_1^2 \sin^2 \beta - 1)^3}{3} \tag{6.107}$$

will vary from streamline to streamline behind the shock wave.

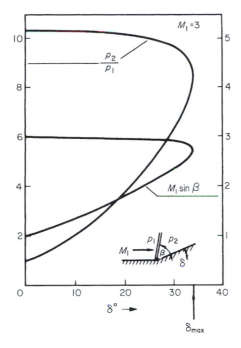

Fig. 6.30

An entropy gradient in the flow is associated with rotational flow and thus a curved shock wave produces a flow in which vorticity exists away from the surface of the associated solid body. At low initial Mach numbers or with waves of small curvature the same approximations as those that are the consequences of assuming $\Delta S \simeq 0$ may be made. For highly curved strong shock waves, such as may occur at hypersonic speed, the downstream flow contains shock-induced vorticity, or the entropy wake as it is sometimes called, which forms a large and significant part of the flow in the immediate vicinity of the body associated with the wave.

6.8 Wings in compressible flow

In this section the compressible-flow equations in their various forms are considered in order to predict the behaviour of aerofoil sections in high sub- and supersonic flows. Except in the descriptive portions the effects of viscosity are largely neglected.

6.8.1 Transonic flow, the critical Mach number

When the air flows past a body, or vice versa, e.g. a symmetrical aerofoil section at low incidence, the local airspeed adjacent to the surface just outside the boundary layer is higher or lower than the free-stream speed depending on whether the local static pressure is less or greater than the ambient. In such a situation the value of the velocity somewhere on the aerofoil exceeds that of the free stream. Thus as the free-stream flow speed rises the Mach number at a point somewhere adjacent to the surface reaches sonic conditions before the free stream. This point is usually the minimum-pressure point which in this case is on the upper surface. The value of the free-stream Mach number (M_∞) at which the flow somewhere on the surface first reaches $M = 1$ is called the *critical Mach number*, M_c. Typically for a slender wing section at low incidence M_c may be about 0.75. Below that critical Mach number the flow is subsonic throughout.

Above the critical Mach number the flow is mixed, part supersonic part subsonic. As M_∞ is increased progressively from low numbers to M_c the aerodynamic characteristics of the aerofoil section undergo progressive and generally smooth changes, and for thin aerofoil shapes at low incidences these changes may be predicted by the small-perturbation or linearized theory outlined below due to Prandtl and Glauert.

As M_∞ is increased progressively beyond M_c a limited region in which the flow is supersonic develops from the point where the flow first became sonic and grows outwards and downstream, terminating in a shock wave that is at first approximately normal to the surface. As M_∞ increases the shock wave becomes stronger, longer and moves rearwards. At some stage, at a value of $M_\infty > M_c$, the velocity somewhere on the lower surface approaches and passes the sonic value, a supersonic region terminating in a shock wave appears on the lower surface, and that too grows stronger and moves back as the lower supersonic region increases.

Eventually at a value of M_∞ close to unity the upper and lower shock waves reach the trailing edge. In their rearward movement the shock waves approach the trailing edge in general at different rates, the lower typically starting later and ahead of the upper, but moving more rapidly and overtaking the upper before they reach the trailing edge. When the free-stream Mach number has reached unity a bow shock wave appears at a small stand-off distance from the rounded leading edge. For higher Mach numbers the extremes of the bow and trailing waves incline rearwards to approach the Mach angle. For round-nosed aerofoils or bodies the bow wave is a 'strong' wave and always stands off, and a small subsonic region exists around the front stagnation point. The sequence is shown in Fig. 6.31. For sharp leading edges the bow shock waves are plane, and usually 'weak', with the downstream flow still, at a lower Mach number, supersonic. This case is dealt with separately below.

The effect on the aerofoil characteristics of the flow sequence described above is dramatic. The sudden loss of lift, increase in drag and rapid movement in centre of pressure are similar in flight to those experienced at the stall and this flight regime became known as the shock stall. Many of the effects can be minimized or delayed by design methods that are beyond the scope of the present volume.

To appreciate why the aerofoil characteristics begin to change so dramatically we must recall the properties of shock waves the first appearance of which signals the start of the change. Across the shock wave, which is the only mechanism for a finite pressure increase in supersonic flow, the pressure rise is large and sudden. Moreover the shock wave is a process accompanied by an entropy change which manifests itself as an immediate rise in drag, i.e. an irreversible conversion of mechanical energy to heat (which is dissipated) takes place and sustaining this loss results in the drag. The drag increase is directly related to the strengths of the shock waves which in turn depend on the magnitude of the supersonic regions ahead. Another contribution to the drag will occur if the boundary layer at the foot of the shock separates as a consequence of accommodating the sudden pressure rise.

The lift on the other hand continues to rise smoothly with increase in $M_\infty > M_c$ as a consequence of the increased low-pressure area on the upper surface. The sequence is seen in Fig. 6.32. The lift does not begin to decrease significantly until the low-pressure area on the lower surface becomes appreciably extensive owing to the

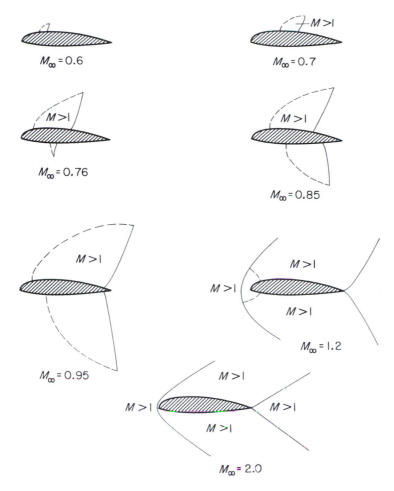

Fig. 6.31 Flow development on two-dimensional aerofoil as M_∞ increases beyond M_{crit}; $M_{crit} = 0.58$

growth of the supersonic region there; Fig. 6.32c. The presence of the shock wave can be seen by the sharp vertical pressure recovery terminating the supersonic regions (shaded areas in Fig. 6.32). It is apparent that the marked effect on the lift is associated more with the growth of the shock wave on the lower surface. Movement of the centre of pressure also follows as a consequence of the varying pressure distributions and is particularly marked as the lower shock wave moves behind the upper at the higher Mach numbers approaching unity.

It may also be noted from the pressure distributions (e) and (f) that the pressure recovery at the trailing edge is incomplete. This is due to the flow separation at the feet of the shock waves. This will lead to buffeting of any control surface near the trailing edge. It is also worth noting that even if the flow remained attached the pressure information that needs to be propagated to the pressure distribution by a control movement (say) cannot be propagated upstream through the supersonic region so that the effectiveness of a trailing-edge control surface is much reduced. As the free-stream Mach number M_∞ becomes supersonic the flow over the aerofoil,

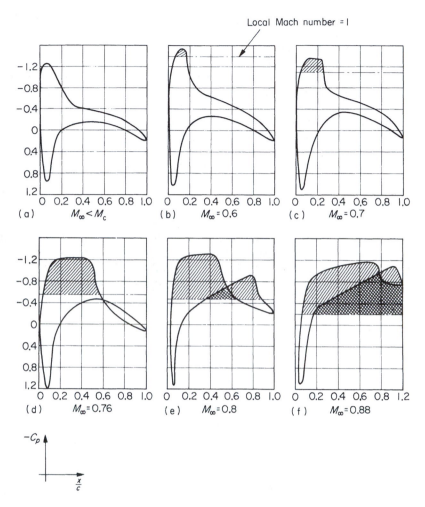

Fig. 6.32 Pressure distribution on two-dimensional aerofoil ($M_{crit} = 0.57$) as M_∞ increases through M_{crit}

except for the small region near the stagnation point, is supersonic, and the shock system stabilizes to a form similar to the supersonic case shown last in Fig. 6.31.

6.8.2 Subcritical flow, small perturbation theory (Prandtl–Glauert rule)

In certain cases of compressible flow, notably in supersonic flow, exact solutions to the equations of motion may be found (always assuming the fluid to be inviscid) and when these are applied to the flow in the vicinity of aerofoils they have acquired the soubriquet of *exact* theories. As described, aerofoils in motion near the speed of sound, in the transonic region, have a mixed-flow regime, where regions of subsonic and supersonic flow exist side by side around the aerofoil. Mathematically the analysis of this regime involves the solution of a set of nonlinear differential equations, a task that demands either advanced computational techniques or some degree of approximation.

The most sweeping approximations, producing the simplest solutions, are made in the present case and result in the transformation of the equations into soluble linear differential equations. This leads to the expression *linearized theory* associated with aerofoils in, for example, high subsonic or low supersonic flows. The approximations come about mainly from assuming that all disturbances are *small disturbances* or *small perturbations* to the free-stream flow conditions and, as a consequence, these two terms are associated with the development of the theory.

Historically, H. Glauert was associated with the early theoretical treatment of the compressibility effects on aerofoils approaching the speed of sound, and developed what are, in essence, the linearized equations for subsonic compressible flow, in *R and M*, 305 (1927), a note previously published by the Royal Society. In this, mention is made of the same results being quoted by Prandtl in 1922. The significant compressibility effect in subsonic flow has subsequently been given the name of the *Prandtl–Glauert rule* (or law).

Furthermore, although the theory takes no account of viscous drag or the onset of shock waves in localized regions of supersonic flow, the relatively crude experimental results of the time (obtained from the analysis of tests on an airscrew) did indicate the now well-investigated critical region of flight where the theory breaks down. Glauert suggested that the critical speed at which the lift falls off depends on the shape and incidence of the aerofoil, and this has since been well-substantiated.

In what follows, attention is paid to the approximate methods of satisfying the equations of motion for an inviscid compressible fluid. These depend on the simultaneous solution of the fundamental laws of conservation and of state. Initially a single equation is desired that will combine all the physical requirements. The complexity of this equation and whether it is amenable to solution will depend on the nature of the particular problem and those quantities that may be conveniently minimized.

The equations of motion of a compressible fluid

The equation of continuity may be recalled in Cartesian coordinates for two-dimensional flow in the form

$$\frac{\partial \rho}{\partial t} + \frac{\partial(\rho u)}{\partial x} + \frac{\partial(\rho v)}{\partial y} = 0 \tag{6.108}$$

since, in what follows, analysis of two-dimensional conditions is sufficient to demonstrate the method and derive valuable equations. The equations of motion may also be recalled in similar notation as

$$\left.\begin{array}{l} \dfrac{\partial u}{\partial t} + u\dfrac{\partial u}{\partial x} + v\dfrac{\partial u}{\partial y} = -\dfrac{1}{\rho}\dfrac{\partial p}{\partial x} \\[2mm] \dfrac{\partial v}{\partial t} + u\dfrac{\partial v}{\partial x} + v\dfrac{\partial v}{\partial y} = -\dfrac{1}{\rho}\dfrac{\partial p}{\partial y} \end{array}\right\} \tag{6.109}$$

and for steady flow

$$\left.\begin{array}{l} -\dfrac{1}{\rho}\dfrac{\partial p}{\partial x} = u\dfrac{\partial u}{\partial x} + v\dfrac{\partial u}{\partial y} \\[2mm] -\dfrac{1}{\rho}\dfrac{\partial p}{\partial y} = u\dfrac{\partial v}{\partial x} + v\dfrac{\partial v}{\partial y} \end{array}\right\} \tag{6.110}$$

For adiabatic flow (since the assumption of negligible viscosity has already been made, the further stipulations of adiabatic compression and expansion imply isentropic flow):

$$p = k\rho^\gamma, \qquad \frac{\partial p}{\partial \rho} = a^2 = \frac{\gamma p}{\rho} \tag{6.111}$$

For steady flow, Eqn (6.108) may be expanded to

$$u \frac{\partial \rho}{\partial x} + v \frac{\partial \rho}{\partial y} + \rho \frac{\partial u}{\partial x} + \rho \frac{\partial v}{\partial y} = 0 \tag{6.112}$$

But

$$\frac{\partial \rho}{\partial x} = \frac{\partial p}{\partial x} \frac{\partial \rho}{\partial p} = \frac{1}{a^2} \frac{\partial p}{\partial x}, \quad \text{etc.}$$

so that Eqn (6.112) becomes

$$\frac{u}{a^2} \frac{\partial p}{\partial x} + \frac{v}{a^2} \frac{\partial p}{\partial y} + \rho \frac{\partial u}{\partial x} + \rho \frac{\partial v}{\partial y} = 0 \tag{6.113}$$

Substitution in Eqn (6.113) for $\partial p/\partial x$, $\partial p/\partial y$ from Eqn (6.110) and cancelling ρ gives

$$-\frac{u^2}{a^2} \frac{\partial u}{\partial x} - \frac{uv}{a^2} \frac{\partial u}{\partial y} - \frac{vu}{a^2} \frac{\partial v}{\partial x} - \frac{v^2}{a^2} \frac{\partial v}{\partial y} + \frac{\partial u}{\partial x} + \frac{\partial v}{\partial y} = 0$$

or, collecting like terms:

$$\left(1 - \frac{u^2}{a^2}\right) \frac{\partial u}{\partial x} - \frac{uv}{a^2} \left(\frac{\partial v}{\partial x} + \frac{\partial u}{\partial y}\right) + \left(1 - \frac{v^2}{a^2}\right) \frac{\partial v}{\partial y} = 0 \tag{6.114}$$

For irrotational flow $\partial v/\partial x = \partial u/\partial y$, and a velocity potential ϕ_1 (say) exists, so that

$$\left(1 - \frac{u^2}{a^2}\right) \frac{\partial u}{\partial x} - \frac{2uv}{a^2} \frac{\partial u}{\partial y} + \left(1 - \frac{v^2}{a^2}\right) \frac{\partial v}{\partial y} = 0 \tag{6.114a}$$

and since $u = \partial\phi_1/\partial x$, $v = \partial\phi_1/\partial y$, Eqn (6.114a) can be written

$$\left(1 - \frac{u^2}{a^2}\right) \frac{\partial^2 \phi_1}{\partial x^2} - \frac{2uv}{a^2} \frac{\partial^2 \phi_1}{\partial x \partial y} + \left(1 - \frac{v^2}{a^2}\right) \frac{\partial^2 \phi_1}{\partial y^2} = 0 \tag{6.114b}$$

Finally the energy equation provides the relation between a, u, v, and the acoustic speed. Thus:

$$\frac{u^2 + v^2}{2} + \frac{a^2}{\gamma - 1} = \text{constant} \tag{6.115}$$

or

$$\left(\frac{\partial\phi_1}{\partial x}\right)^2 + \left(\frac{\partial\phi_1}{\partial y}\right)^2 + \frac{2}{\gamma - 1} a^2 = \text{constant} \tag{6.115a}$$

Combining Eqns (6.114b) and (6.115a) gives an expression in terms of the local velocity potential.

Even without continuing the algebra beyond this point it may be noted that the resulting nonlinear differential equation in ϕ_1 is not amenable to a simple closed solution and that further restrictions on the variables are required. Since all possible restrictions on the generality of the flow properties have already been made, it is necessary to consider the nature of the component velocities themselves.

Small disturbances

So far it has been tacitly assumed that the flow is steady at infinity, and the local flow velocity has components u and v parallel to coordinate axes x and y respectively, the origin of coordinate axes furnishing the necessary datum. Let the equations now refer to a class of flows in which the velocity changes only slightly from its steady value at infinity and the velocity gradients themselves are small (thin wings at low incidence, etc). Further, identify the x axis with the undisturbed flow direction (see Fig. 6.33). The local velocity components u and v can now be written:

$$u = U_\infty + u', \qquad v = v'$$

where u' and v' are small compared to the undisturbed stream velocity, and are termed the perturbation or disturbance velocities. These may be expressed non-dimensionally in the form

$$\frac{u'}{U_\infty} \ll 1, \frac{v'}{U_\infty} \ll 1$$

Similarly, $\partial u'/\partial x$, $\partial v'/\partial y$ are small.

Making this substitution, Eqn (6.115) becomes

$$\frac{(U_\infty + u')^2 + v'^2}{2} + \frac{a^2}{\gamma - 1} = \frac{U_\infty^2}{2} + \frac{a_\infty^2}{\gamma - 1}$$

When the squares of small quantities are neglected this equation simplifies to

$$U_\infty u' = \frac{a_\infty^2 - a^2}{\gamma - 1}$$

similarly

$$a^2 = a_\infty^2 - (\gamma - 1)U_\infty u'$$

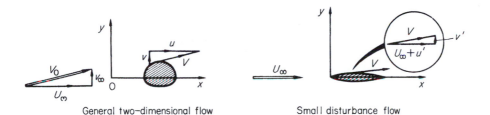

General two-dimensional flow   Small disturbance flow

Fig. 6.33

Thus the coefficient terms of Eqn (6.114a) become:

$$1 - \frac{u^2}{a^2} = 1 - \left(\frac{U_\infty + u'}{a}\right)^2 = 1 - \frac{U_\infty^2 + 2U_\infty u'}{a_\infty^2 - (\gamma - 1)U_\infty u'}$$

Putting $U_\infty/a_\infty = M_\infty$ the free stream Mach number:

$$1 - \left(\frac{u}{a}\right)^2 = 1 - M_\infty^2 \left(\frac{1 + \dfrac{2u'}{U_\infty}}{1 - (\gamma - 1)M_\infty^2 (u'/U_\infty)}\right)$$

$$= 1 - M_\infty^2 \left(1 - \frac{u'[2 + (\gamma - 1)M_\infty^2]}{U_\infty[1 - (\gamma - 1)M_\infty^2(u'/U_\infty)]}\right)$$

and

$$\frac{2uv}{a^2} = \frac{2(U_\infty + u')}{a^2} v' = \frac{2U_\infty v'}{a_\infty^2 - (\gamma - 1)U_\infty u'}$$

$$= M_\infty^2 \frac{\dfrac{2v'}{U_\infty}}{1 - [(\gamma - 1)M_\infty^2(u'/U_\infty)]}$$

Also, $1 - (v'/U_\infty)^2 \simeq 1$ from the small disturbance assumption.

Now, if the velocity potential ϕ_1 is expressed as the sum of a velocity potential due to the flow at infinity plus a velocity potential due to the disturbance, i.e. $\phi_1 = \phi_\infty + \phi$, Eqn (6.114b) becomes, with slight re-arrangement:

$$(1 - M_\infty^2)\frac{\partial^2 \phi}{\partial x^2} + \frac{\partial^2 \phi}{\partial y^2} = \frac{M_\infty^2}{1 - (\gamma - 1)M_\infty^2 u'/U_\infty}$$

$$\times \left[[2 + (\gamma - 1)M_\infty^2]\frac{u'}{U_\infty}\frac{\partial^2 \phi}{\partial x^2} + \frac{2v'}{U_\infty}\frac{\partial^2 \phi}{\partial x \partial y}\right] \qquad (6.116)$$

where ϕ is the disturbance potential and $u' = \partial\phi/\partial x$, $v' = \partial\phi/\partial y$, etc.

The right-hand side of Eqn (6.116) vanishes when $M_\infty = 0$ and the coefficient of the first term becomes unity, so that the equation reduces to the Laplace equation, i.e. when $M_\infty = 0$, Eqn (6.116) becomes

$$\frac{\partial^2 \phi}{\partial x^2} + \frac{\partial^2 \phi}{\partial y^2} = 0 \qquad (6.117)$$

Since velocity components and their gradients are of the same small order their products can be neglected and the bracketed terms on the right-hand side of Eqn (6.116) will be negligibly small. This will control the magnitude of the right-hand side, which can therefore be assumed essentially zero unless the remaining quantity outside the bracket becomes large or indeterminate. This will occur when

$$\frac{M_\infty^2}{\left[1 - (\gamma - 1)M_\infty^2 \dfrac{u'}{U_\infty}\right]} \to \infty$$

i.e. when $M_\infty^2(\gamma - 1)u'/U_\infty \to 1$. It can be seen that by assigning reasonable values to u'/U_∞ and γ the equality will be made when $M_\infty \simeq 5$, i.e. put $u'/U_\infty = 0.1$, $\gamma = 1.4$, then $M_\infty^2 \simeq 25$.

Within the limitations above the equation of motion reduces to the linear equation

$$(1 - M_\infty^2)\frac{\partial^2 \phi}{\partial x^2} + \frac{\partial^2 \phi}{\partial y^2} = 0 \tag{6.118}$$

A further limitation in application of Eqn (6.116) occurs when M_∞ has a value in the vicinity of unity, i.e. where the flow regime may be described as transonic. Inspection of Eqn (6.116) will also show a fundamental change in form as M_∞ approaches and passes unity, i.e. the quantity $(1 - M_\infty^2)$ changes sign, the equation changing from an elliptic to a hyperbolic type.

As a consequence of these restrictions the further application of the equations finds its most use in the high subsonic region where $0.4 < M_\infty < 0.8$, and in the supersonic region where $1.2 < M_\infty < 5$.

To extend theoretical investigation to transonic or hypersonic Mach numbers requires further development of the equations that is not considered here.

Prandtl–Glauert rule – the application of linearized theories of subsonic flow

Consider the equation (6.118) in the subsonic two-dimensional form:

$$(1 - M_\infty^2)\frac{\partial^2 \phi}{\partial x^2} + \frac{\partial^2 \phi}{\partial y^2} = 0 \quad \text{(Eqn(6.118))}$$

For a given Mach number M_∞ this equation can be written

$$B^2\frac{\partial^2 \phi}{\partial x^2} + \frac{\partial^2 \phi}{\partial y^2} = 0 \tag{6.119}$$

where B is a constant. This bears a superficial resemblance to the Laplace equation:

$$\frac{\partial^2 \Phi}{\partial \xi^2} + \frac{\partial^2 \Phi}{\partial \eta^2} = 0 \tag{6.120}$$

and if the problem expressed by Eqn (6.118), that of finding ϕ for the subsonic compressible flow round a thin aerofoil, say, could be transformed into an equation such as (6.120), its solution would be possible by standard methods.

Figure 6.34 shows the thin aerofoil occupying, because it is thin, in the definitive sense, the Ox axis in the *real* or *compressible* plane, where the velocity potential ϕ exists in the region defined by the xy ordinates. The corresponding aerofoil in the *Laplace* or *incompressible* $\xi\eta$ plane has a velocity potential Φ. If the simple relations:

$$\Phi = A\phi, \qquad \xi = Cx \qquad \text{and} \qquad \eta = Dy \tag{6.121}$$

are assumed, where A, C and D are constants, the transformation can proceed.

The boundary conditions on the aerofoil surface demand that the flow be locally tangential to the surface so that in each plane respectively

$$v'_c = U_\infty\frac{\mathrm{d}y}{\mathrm{d}x} = \left(\frac{\partial \phi}{\partial y}\right)_{y=0} \tag{6.122}$$

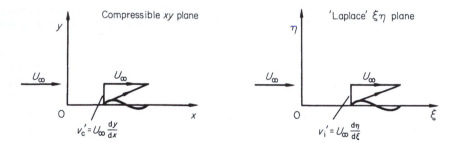

Fig. 6.34

and

$$v'_i = U_\infty \frac{\mathrm{d}\eta}{\mathrm{d}\xi} = \left(\frac{\partial\Phi}{\partial\eta}\right)_{\eta=0} \tag{6.123}$$

where the suffices c and i denote the compressible and incompressible planes respectively.

Using the simple relationships of Eqn (6.121) gives

$$\frac{\partial\Phi}{\partial\xi} = \frac{\partial(A\phi)}{\partial(Cx)} = \frac{A}{C}\frac{\partial\phi}{\partial x}, \qquad \frac{\partial^2\Phi}{\partial\xi^2} = \frac{A}{C^2}\frac{\partial^2\phi}{\partial x^2}$$

and

$$\frac{\partial\Phi}{\partial\eta} = \frac{A}{D}\frac{\partial\phi}{\partial y}, \qquad \frac{\partial^2\Phi}{\partial\eta^2} = \frac{A}{D^2}\frac{\partial^2\phi}{\partial y^2}$$

Thus Eqn (6.120) by substitution and rearrangement of constants becomes

$$\frac{\partial^2\Phi}{\partial\xi^2} + \frac{\partial^2\Phi}{\partial\eta^2} = \frac{A}{D^2}\left[\left(\frac{D^2}{C^2}\right)\frac{\partial^2\phi}{\partial x^2} + \frac{\partial^2\phi}{\partial y^2}\right] = 0 \tag{6.124}$$

Comparison of Eqns (6.124) and (6.119) shows that a solution to Eqn (6.120) (the left-hand part of Eqn (6.124)) provides a solution to Eqn (6.119) (the right-hand part of Eqn (6.119)) if the bracketed constant can be identified as the B^2, i.e. when

$$\frac{D}{C} = B = \sqrt{1 - M_\infty^2} \tag{6.125}$$

Without generalizing further, two simple procedures emerge from Eqn (6.125). These are followed by making C or D unity when $D = B$ or $1/C = B$ respectively. Since C and D control the spatial distortion in the Laplace plane the two procedures reduce to the distortion of one or the other of the two ordinates.

Constant chordwise ordinates

If the aerofoil is thin, and by definition this must be so for the small-disturbance conditions of the theory from which Eqn (6.118) is derived, the implication of this restriction is that the aerofoils are of similar shape in both planes. Take first the case of $C = 1$, i.e. $\xi = x$. This gives $D = B = \sqrt{1 - M_\infty^2}$ from Eqn (6.125). A solution of

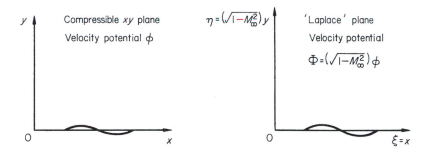

Fig. 6.35

Eqns (6.118) or (6.119) is found by applying the transformation $\eta = \sqrt{1 - M^2}\, y$ (see Fig. 6.35). Then Eqns (6.122) and (6.123) give

$$v'_c = U_\infty \frac{dy}{dx} = \left[\frac{\partial\phi}{\partial y}\right]_{y=0} = \frac{B}{A}\left[\frac{\partial\Phi}{\partial\eta}\right]_{\eta=0} = \frac{B}{A} U_\infty \frac{d\eta}{d\xi}$$

but $D = B$, since $D/C = B$ and $C = 1$.

For similar aerofoils, it is required that $dy/dx = d\eta/d\xi$ at corresponding points, and, for this to be so:

$$A = B = \sqrt{1 - M_\infty^2} \tag{6.126}$$

The transformed potential is thus

$$\Phi = \sqrt{1 - M_\infty^2}\,\phi \tag{6.127}$$

The horizontal flow perturbations are now easily found:

$$u'_c = \frac{\partial\phi}{\partial x} = \frac{1}{\sqrt{1 - M_\infty^2}} \frac{\partial\Phi}{\partial x} = \frac{u'_i}{\sqrt{1 - M_\infty^2}} \tag{6.128}$$

$$C_{p_c} = \frac{-2u'_c}{U_\infty} = \frac{-1}{\sqrt{1 - M_\infty^2}} \frac{2u'_i}{U_\infty} = \frac{C_{pi}}{\sqrt{1 - M_\infty^2}} \tag{6.129}$$

Since

$$C_L = \frac{1}{c}\oint C_p\, dx$$

the relationship between lift coefficients in compressible and equivalent incompressible flows follows that of the pressure coefficients, i.e.

$$C_L = \frac{C_{L_i}}{\sqrt{1 - M_\infty^2}} \tag{6.130}$$

This simple use of the factor $\sqrt{1 - M_\infty^2}$ is known as the Prandtl–Glauert rule or law and $\sqrt{1 - M_\infty^2}$ is known as Glauert's factor.

Constant normal ordinates (Fig. 6.36)

Glauert,* however, developed the affine transformation implicit in taking a transformed plane distorted in the x-direction. The consequence of this is that, for a thin aerofoil, the transformed section about which the potential Φ exists has its chordwise lengths altered by the factor $1/C$.

With $D = 1$, i.e. $\eta = y$, Eqn (6.131) gives

$$C = \frac{1}{B} = \frac{1}{\sqrt{1 - M_\infty^2}} \tag{6.131}$$

Thus the solution to Eqn (6.118) or (6.119) is found by applying the transformation

$$\xi = \frac{x}{\sqrt{1 - M_\infty^2}}$$

and, for this and the geometrical condition of Eqns (6.122) and (6.123) to apply, A can be found. Eqn (6.122) gives

$$v'_c = U_\infty \frac{\mathrm{d}y}{\mathrm{d}x} = \left[\frac{\partial \phi}{\partial y}\right]_{y=0} = \frac{1}{A}\left[\frac{\partial \Phi}{\partial \eta}\right]_{\eta=0}$$

By substituting $\Phi = A\phi$, Eqn (6.121), $y = \eta$, Eqn (6.131), but from Eqn (6.122)

$$v'_c = \frac{U_\infty}{A}\frac{\mathrm{d}\eta}{\mathrm{d}\zeta} = \frac{U_\infty}{A}\frac{\mathrm{d}y}{\mathrm{d}x}\sqrt{1 - M_\infty^2}$$

To preserve the identity, $A = \sqrt{1 - M_\infty^2}$ and the transformed potential $\Phi = \sqrt{1 - M_\infty^2}\,\phi$, as previously shown in Eqn (6.127). The horizontal flow perturbations, pressure coefficients and lift coefficients follow as before.

Glauert explained the latter transformation in physical terms by appealing to the fact that the flow at infinity in both the original compressible plane and the transformed, ideal or Laplace plane is the same, and hence the overall lifts to the systems are the same. But the chord of the ideal aerofoil is greater (due to the $\xi = x/\sqrt{1 - M_\infty^2}$ distortion) than the equivalent compressible aerofoil and thus for an identical aerofoil in the compressible plane the lift is greater than that in the ideal (or incompressible) case. The ratio L_c/L_i is as before, i.e. $(1 - M_\infty^2)^{-1/2}$.

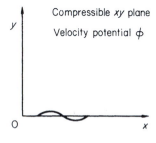
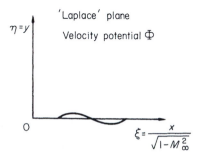

Fig. 6.36

* H. Glauert, The effect of compressibility on aerofoil lift, *ARCR and M*, 1135, 1927.

Critical pressure coefficient

The pressure coefficient of the point of minimum pressure on an aerofoil section, using the notation of Fig. 6.37b, is

$$C_{p_{min}} = \frac{p_{min} - p_\infty}{\frac{1}{2}\rho_\infty V_\infty^2} \qquad (6.132)$$

but since $\frac{1}{2}\rho_\infty V_\infty^2 = \frac{1}{2}\gamma p_\infty M_\infty^2$, Eqn (6.132) may be written

$$C_{p_{min}} = \left[\frac{p_{min}}{p_\infty} - 1\right]\frac{2}{\gamma M_\infty^2} \qquad (6.133)$$

The critical condition is when p_{min} first reaches the sonic pressure p^* and M_∞ becomes M_{crit} (see Fig. 6.37c). $C_{p_{min}}$ is then the critical pressure coefficient of the aerofoil section. Thus

$$C_{p_{crit}} = \left[\frac{p^*}{p_\infty} - 1\right]\frac{2}{\gamma M_{crit}^2} \qquad (6.134)$$

An expression for p^*/p_∞ in terms of M_{crit} may be readily found by recalling the energy equation applied to isentropic flow along a streamline (see Section 6.2) which in the present notation gives

$$\frac{V_\infty^2}{2} + \frac{a_\infty^2}{\gamma - 1} = \frac{V^2}{2} + \frac{a^2}{\gamma - 1}$$

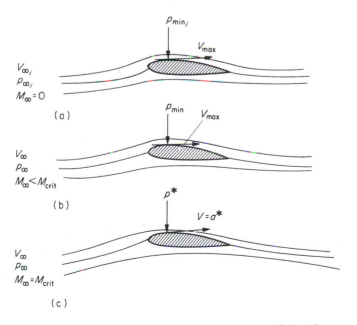

Fig. 6.37 (a) 'Incompressible' flow. (b) Compressible subcritical flow. (c) Critical flow

Divide through by a_∞^2 for the condition when $V = a = a^*$, $M_\infty = M_{crit}$, etc.

$$\frac{M_{crit}^2}{2} + \frac{1}{\gamma - 1} = \left(\frac{a^*}{a_\infty}\right)\left(\frac{1}{2} + \frac{1}{\gamma - 1}\right)$$

On rearranging

$$\left(\frac{a^*}{a_\infty}\right)^2 = M_{crit}^2 \frac{\gamma - 1}{\gamma + 1} + \frac{2}{\gamma + 1}$$

But

$$\frac{p^*}{p_\infty} = \left(\frac{a^*}{a_\infty}\right)^{2\gamma/(\gamma - 1)}$$

and on substituting for p^*/p_∞ in Eqn (6.134)

$$C_{p_{crit}} = \left[\left(\frac{\gamma - 1}{\gamma + 1}M_{crit}^2 + \frac{2}{\gamma + 1}\right)^{\gamma/(\gamma - 1)} - 1\right]\frac{2}{\gamma M_{crit}^2} \qquad (6.135)$$

In this expression M_{crit} is the critical Mach number of the wing, and is the parameter that is often required to be found. $C_{p_{crit}}$ is the pressure coefficient at the point of maximum velocity on the wing when locally sonic conditions are just attained, and is usually also unknown in practice; it has to be predicted from the corresponding minimum pressure coefficient (C_{p_i}) in incompressible flow. C_{p_i} may be obtained from pressure-distribution data from low-speed models or, as previously, from the solution of the Laplace equation of a potential flow.

The approximate relationship between $C_{p_{crit}}$ and C_{p_i} was discussed above for two-dimensional wings. The Prandtl–Glauert rule gives:

$$C_{p_{crit}} = \frac{C_{p_i}}{\sqrt{1 - M_{crit}^2}} \qquad (6.136)$$

A simultaneous solution of Eqns (6.135) and (6.136) with a given C_{p_i} yields values of critical Mach number M_{crit}.

Application to swept wings

In the same way as for the incompressible case (see Section 5.7), the compressible flow over an infinite-span swept (or sheared) wing can be considered to be the superposition of two flows. One component is the flow perpendicular to the swept leading edge. The other is the flow parallel to the leading edge. The free-stream velocity now consists of two components, see Fig. 6.38. For the component perpendicular to the leading edge Eqn (6.118) becomes

$$(1 - M_\infty^2 \cos^2 \Lambda)\frac{\partial^2 \phi}{\partial x^2} + \frac{\partial^2 \phi}{\partial y^2} \qquad (6.137)$$

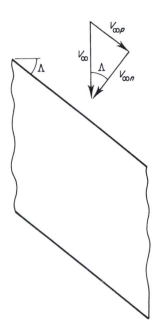

Fig. 6.38

Only the perpendicular component affects the pressure so that Eqns (6.128) and (6.129) become

$$(u'_c)_n = \frac{\partial \phi}{\partial x} = \frac{1}{\sqrt{1 - 1.1^2_\infty \cos^2 \Lambda}} \frac{\partial \Phi}{\partial x} = \frac{(u'_i)_n}{\sqrt{1 - M^2_\infty \cos^2 \Lambda}} \qquad (6.138)$$

$$C_{p_c} = -2 \frac{\cos \Lambda (u'_c)_n}{U_\infty} = -2 \frac{\cos \Lambda (u'_i)_n}{U_\infty \sqrt{1 - M^2_\infty \cos^2 \Lambda}} \qquad (6.139)$$

It follows directly that

$$C_L = \frac{C_{Li}}{\sqrt{1 - M^2_\infty \cos^2 \Lambda}} \qquad (6.140)$$

Example 6.8   For the NACA 4 digit series of symmetrical aerofoil sections in incompressible flow the maximum disturbance velocity $(u'/V_\infty)_{max}$ (corresponding to $(C_p)_{min}$) varies in the following way with thickness-to-chord ratio, t/c:

NACA AEROFOIL DESIGNATION	t/c	$(u'/V_\infty)_{max}$
NACA0006	0.06	0.107
NACA0008	0.08	0.133
NACA0010	0.10	0.158
NACA0012	0.12	0.188
NACA0015	0.15	0.233
NACA0018	0.18	0.278
NACA0021	0.21	0.325
NACA0024	0.24	0.374

Use this data to determine the critical Mach number for

(i) A straight wing of infinite span with a NACA 0010 wing section; and
(ii) An infinite-span wing with a 45° sweepback with the same wing section perpendicular to the leading edge.

All the 4-digit NACA wing sections are essentially the same shape, but with different thickness-to-chord ratios, as denoted by the last two digits. Thus a NACA 0010 wing section at a given freestream Mach number M_∞ is equivalent to a 4-digit NACA series in incompressible flow having a thickness of

$$(t/c)_i = 0.10\sqrt{1 - M_\infty^2}$$

The maximum disturbance velocity, $[(u'/V_\infty)_{max}]_i$ for such a wing section is obtained by using linear interpolation on the data in the table given above. The maximum perturbation velocity in the actual compressible flow at M_∞ is given by

$$(u'/V_\infty)_{max} = \frac{1}{1 - M_\infty^2}[(u'/V_\infty)_{max}]_i \tag{a}$$

The maximum local Mach number is given approximately by

$$M_{max} \simeq \frac{V_\infty + (u_c)_{max}}{a_\infty} = \frac{V_\infty}{a_\infty}\left(1 + \frac{(u_c)_{max}}{V_\infty}\right) = M_\infty\left(1 + \frac{(u_c)_{max}}{V_\infty}\right) \tag{b}$$

Equations (a) and (b) and linear interpolation of the table of data can be used to determine M_{max} for a specified M_∞. The results are set out in the table below

M_∞	$\sqrt{1 - M_\infty^2}$	$\left[\left(\frac{u'}{V_\infty}\right)_{max}\right]_i$	$\left[\left(\frac{u'}{V_\infty}\right)_{max}\right]_c$	M_{max}
0.5	0.866	0.141	0.188	0.594
0.6	0.08	0.133	0.2078	0.725
0.7	0.0714	0.120	0.2353	0.865
0.75	0.066	0.114	0.2606	0.945
0.8	0.06	0.107	0.2972	1.038

Linear interpolation between $M_\infty = 0.75$ and 0.8 gives the critical value of $M_\infty \simeq 0.78$ (i.e. corresponding to $M_{max} = 1.0$).
  For the 45° swept-back wing

$$(t/c)_i = 0.10\sqrt{1 - M_\infty^2\cos^2\Lambda} = 0.10\sqrt{1.0 - 0.5\,M_\infty^2}$$

V_∞ must be replaced by $(V_\infty)_n$, i.e. $V_\infty\cos\Lambda$, so the maximum disturbance velocity is given by

$$\left[\left(\frac{u'}{V_{\infty n}}\right)_{max}\right]_c = \frac{1}{1 + 0.5M_\infty^2}\left[\left(\frac{u'}{V_{\infty n}}\right)_{max}\right]_i$$

The maximum local Mach number is then obtained from

$$M_{max} \simeq M_\infty\left\{1 + \left[\left(\frac{u'}{V_{\infty n}}\right)_{max}\right]_c\right\}$$

Thus in a similar way as for the straight wing the following table is obtained.

M_∞	$\sqrt{1 - M_\infty^2}$	$\left[\left(\dfrac{u'}{V_{\infty n}}\right)_{max}\right]_i$	$\left[\left(\dfrac{u'}{V_{\infty n}}\right)_{max}\right]_c$	M_{max}
0.5	0.0935	0.144	0.116	0.558
0.6	0.0906	0.143	0.123	0.674
0.7	0.0869	0.141	0.132	0.792
0.8	0.0825	0.136	0.141	0.913
0.85	0.0799	0.133	0.147	0.975
0.9	0.0771	0.128	0.152	1.037

Linear interpolation gives a critical Mach number of about 0.87.

It will be noted that although the wing section is the same in both cases the critical Mach number is much higher for the sweptback wing. This is principally because $V_{\infty n}$ is considerably less than V_∞. It is for this reason that aircraft, such as airliners, that are designed to cruise at high subsonic Mach numbers invariably have swept-back wings, in order to keep wave drag to a minimum.

6.8.3   Supersonic linearized theory (Ackeret's rule)

Before proceeding to considerations of solution to the supersonic form of the simplified (small perturbation) equation of motion, Eqn (6.118), i.e. where the Mach number is everywhere greater than unity, it is pertinent to review the early work of Ackeret[*] in this field. Notwithstanding the intrinsic historical value of the work a fresh reading many decades later still has interest in the general development of first-order theory.

Making obvious simplifications, such as assuming thin sharp-edged wings of small camber at low incidence in two-dimensional frictionless shock-free supersonic flow, briefly Ackeret argued that the flow in the vicinity of the aerofoil may be likened directly to that of the Prandtl–Meyer expansion round a corner. With the restrictions imposed above any leading-edge effect will produce two Mach waves issuing from the sharp leading edge (Fig. 6.39) ahead of which the flow is undisturbed. Over the upper surface of the aerofoil the flow may expand according to the two-dimensional solution of the flow equations originated by Prandtl and Meyer (see p. 314). If the same restrictions apply to the leading edge and lower surface, then providing the inclinations are gentle and no shock waves exist the Prandtl–Meyer solution may still be used by employing the following device. Since the undisturbed flow is supersonic it may be assumed to have reached that condition by expanding through the appropriate angle v_p from sonic conditions, then any isentropic compressive deflection δ will lead to flow conditions equivalent to an expansion of $(v_p - \delta)$ from sonic flow conditions.

Thus, providing that nowhere on the surface will any compressive deflections be large, the Prandtl–Meyer values of pressure may be found by reading off the values[†] appropriate to the flow deflection caused by the aerofoil surface, and the aerodynamic forces etc. obtained from pressure integration.

Referring back to Eqn (6.118) with $M_\infty > 1$:

$$(M_\infty^2 - 1)\frac{\partial^2 \phi}{\partial x^2} - \frac{\partial^2 \phi}{\partial y^2} = 0 \tag{6.141}$$

---

[*] J. Ackeret, Z. Flugtech Motorluftschiff, pp. 72–4, 1925. Trans. D.M. Milner in *NAC & TM*, 317.

[†] From suitable tables, e.g. E.L. Houghton and A.E. Brock, *Tables for the Compressible Flow of Dry Air*, 3rd Edn, Edward Arnold, 1975.

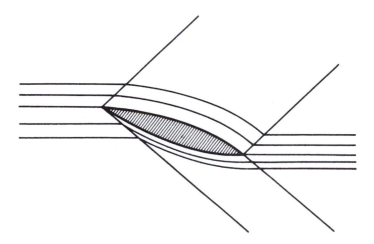

Fig. 6.39

Fig. 6.40 Supersonic flow over as wedge: The schlieren method was used to obtain this flow visualization. A parallel light beam is refracted by the density differences in the flow field. It is then focused on to a knife edge and gives a flow visualization in the image on the photographic film, which takes the form of bright or dark patterns, depending on the direction the beam is bent by refraction. The main features of the flow field are the oblique bow shock wave which is slightly rounded at the nose (see Fig. 7.54, page 479), the expansion fans at the trailing edge, followed by recompression shock waves which form downstream in the wake. These last are slightly curved owing to the interaction with the expansion waves from trailing edge. (*The photograph was taken by D.J. Buckingham at the School of Engineering, University of Exeter, UK.*)

This wave equation has a general solution

$$\phi = F_1(x - \sqrt{M_\infty^2 - 1}\ y) + F_2(x + \sqrt{M_\infty^2 - 1}\ y)$$

where F_1 and F_2 are two independent functions with forms that depend on the boundary conditions of the flow. In the present case physical considerations show that each function exists separately in well-defined regions of the flow (Figs 6.40, 6.41 and 6.42).

By inspection, the solution $\phi = F_1(x - \sqrt{M_\infty^2 - 1}\ y)$ allows constant values of ϕ along the lines $x - \sqrt{M_\infty^2 - 1}\ y = C$, i.e. along the straight lines with an inclination of arc tan $1/\sqrt{M_\infty^2 - 1}$ to the x axis (Fig. 6.42). This means that the disturbance originating on the aerofoil shape (as shown) is propagated into the flow at large, along the straight lines $x = \sqrt{M_\infty^2 - 1}\ y + C$. Similarly, the solution $\phi = F_2(x + \sqrt{M_\infty^2 - 1}\ y)$ allows constant values of ϕ along the straight lines $x = -\sqrt{M_\infty^2 - 1}\ y + C$ with inclinations of

$$\text{arc tan} \left(-\frac{1}{\sqrt{M_\infty^2 - 1}} \right)$$

to the axis.

It will be remembered that Mach lines are inclined at an angle

$$\mu = \text{arc tan} \left(\pm \frac{1}{\sqrt{M_\infty^2 - 1}} \right)$$

to the free-stream direction and thus the lines along which the disturbances are propagated coincide with Mach lines.

Since disturbances cannot be propagated forwards into supersonic flow, the appropriate solution is such that the lines incline downstream. In addition, the effect of the disturbance is felt only in the region between the first and last Mach lines and any flow conditions away from the disturbance are a replica of those adjacent to the body. Therefore within the region in which the disturbance potential exists, taking the positive solution, for example:

$$u' = \frac{\partial \phi}{\partial x} = \frac{\partial F_1}{\partial (x - \sqrt{M_\infty^2 - 1}\ y)} \frac{\partial (x - \sqrt{M_\infty^2 - 1}\ y)}{\partial x}$$

$$u' = F_1' \tag{6.142}$$

and

$$v' = \frac{\partial \phi}{\partial y} = \frac{\partial F_1}{\partial (x - \sqrt{M_\infty^2 - 1}\ y)} \frac{\partial (x - \sqrt{M_\infty^2 - 1}\ y)}{\partial y}$$

$$v' = -\sqrt{M_\infty^2 - 1} F_1' \tag{6.143}$$

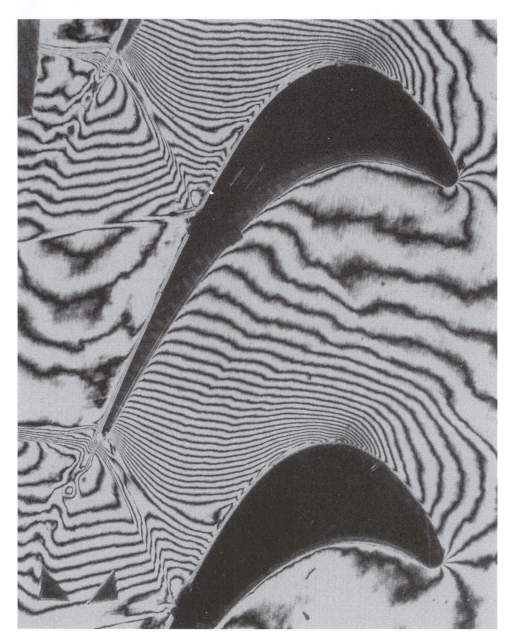

Fig. 6.41 Transonic flow through a turbine cascade: The holographic interferogram shows fringes corresponding to lines of constant density. The flow enters from the right and exits at a Mach number of about 1.3 from the left. The convex and concave surfaces of the turbine blades act as suction and pressure surfaces respectively. Various features of the flow field may be discerned from the interferogram: e.g. the gradual drop in density from inlet to outlet until the formation of a sharp density gradient marking a shock wave where the constant-density lines fold together. The shock formation at the trailing edge may be compared with Fig. 7.51 on page 476. (*The phototgraph was taken by P.J. Bryanston-Cross in the Engineering Department, University of Warwick, UK.*)

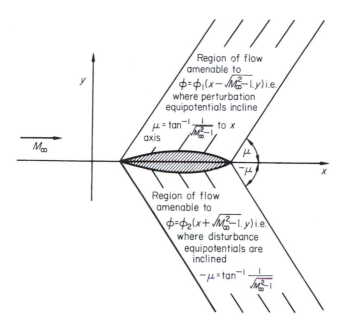

Fig. 6.42

Now the physical boundary conditions to the problem are such that the velocity on the surface of the body is tangential to the surface. This gives an alternative value for v', i.e.

$$v' = U_\infty \frac{\mathrm{d}f(x)}{\mathrm{d}x} \tag{6.144}$$

where $\mathrm{d}f(x)/\mathrm{d}x$ is the local surface slope, $f(x)$ the shape of the disturbing surface and U_∞ the undisturbed velocity. Equating Eqns (6.143) and (6.144) on the surface where $y = 0$:

$$[F_1']_{y=0} = -\frac{U_\infty}{\sqrt{M_\infty^2 - 1}} f'(x)$$

or

$$\left[\frac{\partial F_1}{\partial (x - \sqrt{M_\infty^2 - 1}\, y)} \right]_{y=0} = -\frac{U_\infty}{\sqrt{M_\infty^2 - 1}} \left[\frac{\mathrm{d}f(x - \sqrt{M_\infty^2 - 1}\, y)}{\mathrm{d}(x - \sqrt{M^2 - 1}\, y)} \right]_{y=0}$$

On integrating:

$$\phi = (F_1 =) \frac{-U_\infty}{\sqrt{M_\infty^2 - 1}} f(x - \sqrt{M_\infty^2 - 1}\, y) \tag{6.145}$$

With the value of ϕ (the disturbance potential) found, the horizontal perturbation velocity becomes on the surface, from Eqn (6.142):

$$[u']_{y=0} = \frac{-U_\infty}{\sqrt{M_\infty^2 - 1}} [f'(x - \sqrt{M_\infty^2 - 1}\, y)]_{y=0} = \frac{-U_\infty}{\sqrt{M_\infty^2 - 1}} \frac{\mathrm{d}f\, x}{\mathrm{d}x}$$

The local pressure coefficient which in the linearized form is $C_p = -(2u')/U_\infty$ gives, for this flow:

$$C_p = \frac{2}{\sqrt{M_\infty^2 - 1}} \frac{\mathrm{d}fx}{\mathrm{d}x}$$

But $\mathrm{d}fx/\mathrm{d}x$ is the local inclination of the surface to the direction of motion, i.e. $\mathrm{d}fx/\mathrm{d}x = \varepsilon$ (say). Thus

$$C_p = \frac{2\varepsilon}{\sqrt{M_\infty^2 - 1}} \qquad (6.146)$$

Example 6.9 A shallow irregularity of length l, in a plane wall along which a two-dimensional supersonic stream $M_0 = u_0/a_0$ is flowing, is given approximately by the expression $y = kx[1 - (x/l)]$, where $0 < x < l$ and $k \ll 1$ (see Fig. 6.43). Using Ackeret's theory, prove that the velocity potential due to the disturbance in the flow is

$$\phi = \frac{-u_0}{\sqrt{M_0^2 - 1}} k(x - \sqrt{M_0^2 - 1}\, y) \left[1 - \frac{x - \sqrt{M_0^2 - 1}\, y}{l} \right]$$

and obtain a corresponding expression for the local pressure coefficient anywhere on the irregularity. (LU)

$$(M_0^2 - 1)\frac{\partial^2 \phi}{\partial x^2} - \frac{\partial^2 \phi}{\partial y^2} = 0$$

has the solution applicable here of $\phi = \phi(x - \sqrt{M_0^2 - 1}\, y)$ where ϕ is the disturbance potential function. Local perturbation velocity components on the wall are

$$u = \frac{\partial \phi}{\partial x} = \left[\frac{\partial \phi}{\partial (x - \sqrt{M_0^2 - 1}\, y)} \right]_{y=0}$$

$$v = \frac{\partial \phi}{\partial y} = -\sqrt{M_0^2 - 1} \left[\frac{\partial \phi}{\partial (x - \sqrt{M_0^2 - 1}\, y)} \right]_{y=0}$$

At x from the leading edge the boundary conditions require the flow velocity to be tangential to the surface

$$v = (u_0 + u)\frac{\mathrm{d}y}{\mathrm{d}x} \simeq u_0 \frac{\mathrm{d}y}{\mathrm{d}x} = u_0 k \left[1 - \frac{2x}{l} \right]$$

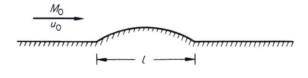

Fig. 6.43

Equating Eqns (6.143) and (6.144) on the surface where $y = 0$, Eqn (6.145) gives:

$$[\partial\phi/\partial y]_{y=0} = \left[\frac{-u_0}{\sqrt{M_0^2 - 1}} k\left(1 - \frac{2x}{l}\right) \mathrm{d}(x - \sqrt{M_0^2 - 1}\,y) \right]_{y=0}$$

$$\phi = \frac{-u_0}{\sqrt{M_0^2 - 1}} k(x - \sqrt{M_0^2 - 1}\,y)\left[1 - \frac{x - \sqrt{M_0^2 - 1}\,y}{l}\right]$$

Also on the surface, with the same assumptions:

$$C_p = -\frac{2u}{U_0} = \left[-\frac{2\frac{\partial\phi}{\partial x}}{u_0} \right]_{y=0} = \left[-\frac{2\frac{\partial\phi}{\partial(x - \sqrt{M_0^2 - 1}\,y)}}{u_0} \right]_{y=0}$$

$$C_p = \frac{2}{\sqrt{M_0^2 - 1}} k\left(1 - \frac{2x}{l}\right)$$

It is now much more convenient to use the pressure coefficient (dropping the suffix ∞):

$$C_p = \frac{2\varepsilon}{\sqrt{M^2 - 1}} \tag{6.147}$$

where ε may be taken as +ve or −ve according to whether the flow is compressed or expanded respectively. Some care is necessary in designating the sign in a particular case, and in the use of this result the angle ε is always measured from the undisturbed stream direction where the Mach number is M, and not from the previous flow direction if different from this.

Symmetrical double wedge aerofoil in supersonic flow

Example 6.10   Plot the pressure distribution over the symmetrical double-wedge, 10 per cent thick supersonic aerofoil shown in Fig. 6.44 when the Mach 2.2 flow meets the upper surface (a) tangentially; and (b) and (c) at incidence 2° above and below this. Estimate also the lift, drag, and pitching moment coefficients for these incidences.

The semi-wedge angle $\varepsilon_0 = \arctan 0.1 = 5.72° = 0.1$ radians

$$M = 2.2; \qquad M^2 = 4.84; \qquad \sqrt{M^2 - 1} = 1.96$$

and for the incidence $\alpha = \varepsilon_0 = 0.1°$. Using Eqn (6.146), the distribution is completed in tabular and graphical forms in Fig. 6.45.

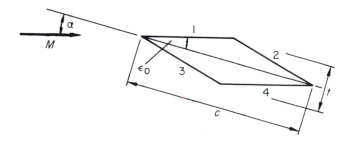

Fig. 6.44

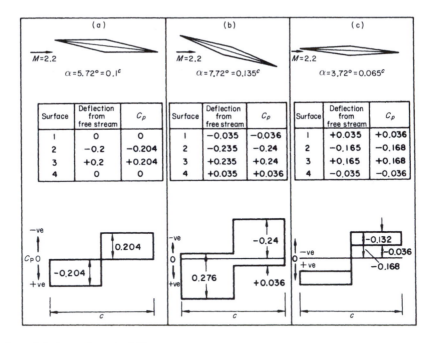

Fig. 6.45 Solution to Example 6.10

For the lift drag and moment a more general approach can be adopted. If a chordwise element δx, x from the leading edge be taken, the net force normal to the chord is

$$(p_L - p_U)\delta x = (C_{pL} - C_{pU})\frac{1}{2}\rho V^2 \delta x$$

Total normal force = lift (since α is small)

$$L = C_L \frac{1}{2}\rho V^2 c = \int_0^c (C_{pL} - C_{pU})\frac{1}{2}\rho V^2 \mathrm{d}x \tag{6.148}$$

In this case Eqn (6.148) integrates to give

$$C_L c = \frac{c}{2}(C_{p_3} - C_{p_1}) + \frac{c}{2}(C_{p_4} - C_{p_2})$$

and on substituting $C_{p_2} = 2\varepsilon_1 \sqrt{M^2 - 1}$, etc.,

$$C_L = \frac{1}{\sqrt{M^2 - 1}}[\varepsilon_3 - \varepsilon_1 + \varepsilon_4 - \varepsilon_2] \tag{6.149}$$

But for the present configuration

$$\varepsilon_1 = \varepsilon_0 - \alpha, \qquad \varepsilon_2 = -\varepsilon_0 - \alpha, \qquad \varepsilon_3 = \varepsilon_0 + \alpha, \qquad \varepsilon_4 = -\varepsilon_0 + \alpha$$

when Eqn (6.149) becomes

$$C_L = \frac{1}{\sqrt{M^2 - 1}}[(\varepsilon_0 + \alpha) - (\varepsilon_0 - \alpha) + (-\varepsilon_0 + \alpha) + (\varepsilon_0 - \alpha)]$$

$$C_L = \frac{4\alpha}{\sqrt{M^2 - 1}} \tag{6.150}$$

In the present example

$$C_{L3.42} = 0.132, \qquad C_{L5.72} = 0.204, \qquad C_{L7.72} = 0.275$$

The contribution to drag due to a chordwise element of lower surface, say, is

$$p_L \varepsilon_L \delta x = C_{pL} \frac{1}{2} \rho V^2 \varepsilon_L \delta x + p_0 \varepsilon_L \delta x$$

where p_0 is the free-stream static pressure which integrates to zero and may be neglected throughout. Again using $C_{pL} = 2\varepsilon_L^2 \sqrt{M^2 - 1}$, the elemental contribution to drag becomes

$$\frac{2\varepsilon_L^2}{\sqrt{M^2 - 1}} \frac{1}{2} \rho V^2 \delta x$$

The corresponding contribution from the upper surface is

$$\frac{2\varepsilon_U^2}{\sqrt{M^2 - 1}} \frac{1}{2} \rho V^2 \delta x$$

The total wave drag becomes

$$D = \frac{2}{\sqrt{M^2 - 1}} \frac{1}{2} \rho V^2 \int_0^c (\varepsilon_L^2 + \varepsilon_U^2) dx = C_D \frac{1}{2} \rho V^2 c$$

$$C_D = \frac{2}{\sqrt{M^2 - 1}} \int_0^1 (\varepsilon_L^2 - \varepsilon_U^2) d\left(\frac{x}{c}\right)$$

In the present case, with $\varepsilon_1 = \varepsilon_0 - \alpha$, $\varepsilon_2 = -(\alpha + \varepsilon_0)$, $\varepsilon_3 = \varepsilon_0 + \alpha$, $\varepsilon_4 = \alpha - \varepsilon_0$:

$$C_D = \frac{2}{\sqrt{M^2 - 1}} \{[(\alpha - \varepsilon_0)^2 + (\varepsilon_0 - \alpha)^2] \frac{1}{2} + [(\alpha + \varepsilon_0)^2 + (\varepsilon_0 + \alpha)^2] \frac{1}{2}\}$$

$$= \frac{4}{\sqrt{M^2 - 1}} [\alpha^2 + \varepsilon_0^2]$$

But $\varepsilon_0^2 = (t/c)^2$, therefore:

$$C_D = \frac{4}{\sqrt{M^2 - 1}} \left[\alpha^2 + \left(\frac{t}{c}\right)^2\right] \tag{6.151}$$

It is now seen that aerofoil thickness contributes to the wave drag, which is a minimum for a wing of zero thickness, i.e. a flat plate. Alternatively, for other than the flat plate, minimum wave drag occurs at zero incidence. This is generally true for symmetrical sections although the magnitude of the minimum wave drag varies. In the present example the required values are

$$C_{D3.72} = 0.029, \qquad C_{D5.72} = 0.0408, \qquad C_{D7.72} = 0.0573$$

Lift to wave drag ratio. Directly from Eqns (6.150), and (6.151):

$$\frac{L}{D} = \frac{4\alpha}{\sqrt{M^2 - 1}} \left\{ 4\left[\alpha^2 + \left(\frac{t}{c}\right)^2\right] \frac{1}{\sqrt{M^2 - 1}} \right\}^{-1} = \frac{\alpha}{\alpha^2 + (t/c)^2} \tag{6.152}$$

Now L/D is a maximum when $D/L = \alpha + (t/c)^2/\alpha$ is a minimum and this occurs when the two terms involved are numerically equal, i.e. in this case, when $\alpha = t/c$. Substituting back gives the maximum L/D ratio as

$$\left[\frac{L}{D}\right]_{\text{max}} = \frac{t/c}{2\left(\frac{t}{c}\right)^2} = \frac{1}{2\left(\frac{t}{c}\right)}$$

For the present example, with $t/c = 0.1$, $[L/D]_{\text{max}} = 5$ occurring at $5.72°$ of incidence, and

$$\left[\frac{L}{D}\right]_{3.72} = 4.55, \qquad \left[\frac{L}{D}\right]_{5.72} = 5.0, \qquad \left[\frac{L}{D}\right]_{7.72} = 4.8$$

Moment about the leading edge Directly from the lift case above, the force normal to the chord from an element δx of chord x from the leading edge is $(C_{p_L} - C_{p_U})\frac{1}{2}\rho V^2 \delta x$ and this produces the negative increment of pitching moment

$$\Delta M = -(C_{p_L} - C_{p_U})x\frac{1}{2}\rho V^2 \delta x$$

Integrating gives the total moment

$$-M = \int_0^c (C_{p_L} - C_{p_U})\frac{1}{2}\rho V^2 x\,dx = -C_M \frac{1}{2}\rho V^2 c^2$$

Making the appropriate substitution for C_p:

$$-C_M = \frac{2}{c^2\sqrt{M^2 - 1}}\int_0^c M(\varepsilon_L - \varepsilon_U)x\,dx$$

which for the profile of the present example gives

$$-C_M = \left(\frac{2}{\sqrt{M^2 - 1}}\right)\frac{1}{c^2}\left\{[\alpha - \varepsilon_0 - (\varepsilon_0 - x)]\frac{c^2}{4} + [(\varepsilon_0 + \alpha) + (\alpha + \varepsilon_0)]\frac{c^2}{4}\right\}$$

$$= \frac{1}{\sqrt{M^2 - 1}}[2x]$$

i.e.

$$-C_M = \frac{2\alpha}{\sqrt{M^2 - 1}} = \frac{C_L}{2} \tag{6.153}$$

Hence

$$C_{M_{3.72}} = -0.066, \qquad C_{M_{6.72}} = -0.102, \qquad C_{M_{7.72}} = -0.138$$

Centre of pressure coefficient k_{CP}:

$$k_{CP} = \frac{-C_M}{C_L} = 0.5 \tag{6.154}$$

and this is independent of Mach number and (for symmetrical sections) of incidence.

Symmetrical biconvex circular arc aerofoil in supersonic flow

While still dealing with symmetrical sections it is of use to consider another class of profile, i.e. one made up of biconvex circular arcs. Much early experimental work

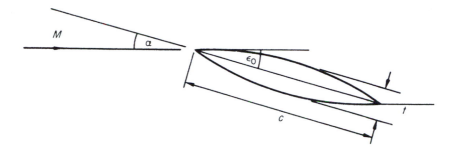

Fig. 6.46

was done on these sections both with symmetrical and cambered profiles and this is readily available to compare with the theory.

Consider the thin symmetrical aerofoil section shown in Fig. 6.46. On the upper surface x from the leading edge and the deflection of the flow from the free-stream direction is ε_U, and $\varepsilon_U = -\alpha + \varepsilon_0[1 - (2x/c)]$,* so that the local pressure coefficient is

$$C_{p_U} = \frac{2}{\sqrt{M^2 - 1}}\left[-\alpha + \varepsilon_0\left(1 - \frac{2x}{c}\right)\right] \tag{6.155}$$

For the lower surface

$$C_{p_L} = \frac{2}{\sqrt{M^2 - 1}}\left[\alpha + \varepsilon_0\left(1 - \frac{2x}{c}\right)\right]$$

Then the contribution to lift of the upper and lower surfaces x from the leading edge is

$$\delta L = \frac{2}{\sqrt{M^2 - 1}}2\alpha\frac{1}{2}\rho V^2\delta x = \delta C_L\frac{1}{2}\rho V^2 c$$

and integrating over the chord gives

$$C_L = \frac{4\alpha}{\sqrt{M^2 - 1}} \tag{6.156}$$

as before. The contributions to wave drag of each of the surfaces x from the leading edge are together

---

* This approximate form of equation for a circular arc is justified for shallow concavities, i.e. large radii of curvature, and follows from simple geometry, i.e. from Fig. 6.47:

$$\varepsilon_1 = \varepsilon_0 - \theta = \varepsilon_0\left(1 - \frac{\theta}{\varepsilon_0}\right), \qquad \theta = \frac{s}{R} = \frac{x}{R}, \qquad \varepsilon_0 = \sin^{-1}\frac{c}{2R} \approxeq \frac{c}{2R}$$

Therefore

$$\frac{\theta}{\varepsilon_0} = \frac{2x}{c} \qquad \text{and} \qquad \varepsilon = \varepsilon_0\left(1 - \frac{2x}{c}\right)$$

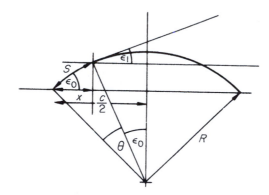

Fig. 6.47

$$\delta D = \frac{2}{\sqrt{M^2 - 1}} \frac{1}{2} \rho V^2 (\varepsilon_U^2 + \varepsilon_L^2) \delta x = \delta C_D \frac{1}{2} \rho V^2 c$$

and integrating gives after substituting for ε_U^2 and ε_L^2

$$C_D = \frac{4}{\sqrt{M^2 - 1}} \left[\alpha^2 + \frac{\varepsilon_0^2}{3} \right]$$

Now, by geometry, and since ε_0 is small, $\varepsilon_0 = 2(t/c)$, giving

$$C_D = \frac{4}{\sqrt{M^2 - 1}} \left[\alpha^2 + \frac{4}{3} \left(\frac{t}{c} \right)^2 \right]$$

The lift/drag ratio is a maximum when, by division, $D/L = \alpha + \left[\frac{4}{3}(t/c)^2 1/\alpha \right]$ is a minimum, and this occurs when

$$\alpha = \frac{4}{3} \left(\frac{t}{c} \right)^2 \frac{1}{\alpha}$$

Then

$$\left[\frac{L}{D} \right]_{max} = \frac{\alpha}{\alpha^2 + \alpha^2} = \frac{1}{2\alpha} = \frac{\sqrt{3}}{4} \frac{c}{t} = \frac{0.433}{t/c}$$

For a 10% thick section $(L/D)_{max} = 4\frac{1}{3}$ at $\alpha = 6.5°$

Moment coefficient and k_{CP}
Directly from previous work, i.e. taking the moment of δL about the leading edge:

$$M = C_M \frac{1}{2} \rho V^2 c^2 = - \int_0^c \frac{4\alpha}{\sqrt{M^2 - 1}} x \frac{1}{2} \rho V^2 \, dx$$

$$C_{M_0} = \frac{-2\alpha}{\sqrt{M^2 - 1}} \qquad (6.157)$$

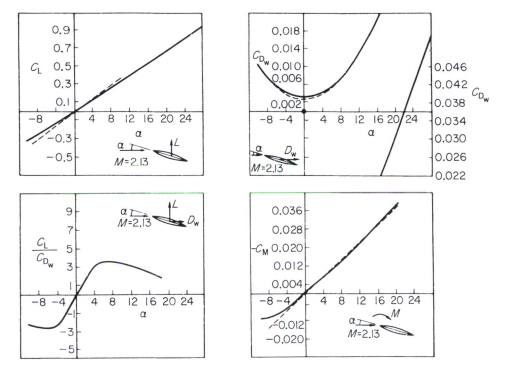

Fig. 6.48

and the centre of pressure coefficient $= -(C_M/C_L) = 0.5$ as before. A series of results of tests on supersonic aerofoil sections published by A. Ferri* serve to compare with the theory. The set chosen here is for a symmetrical bi-convex aerofoil section of $t/c = 0.1$ set in an air flow of Mach number 2.13. The incidence was varied from $-10°$ to 28° and also plotted on the graphs of Fig. 6.48 are the theoretical values of Eqns (6.156) and (6.157).

Examination of Fig. 6.48 shows the close approximation of the theoretical values to the experimental results. The lift coefficient varies linearly with incidence but at some slightly smaller value than that predicted. No significant reduction in C_L, as is common at high incidences in low-speed tests, was found even with incidence $>20°$.

The measured drag values are all slightly higher than predicted which is understandable since the theory accounts for wave drag only. The difference between the two may be attributed to skin-friction drag or, more generally, to the presence of viscosity and the behaviour of the boundary layer. It is unwise, however, to expect the excellent agreement of these particular results to extend to more general aerofoil sections – or indeed to other Mach numbers for the same section, as severe limitations on the use of the theory appear at extreme Mach numbers. Nevertheless, these and other published data amply justify the continued use of the theory.

* A. Ferri, Experimental results with aerofoils tested in the high-speed tunnel at Guidornia, *Atti Guidornia*, No. 17, September 1939.

General aerofoil section

Retaining the major assumptions of the theory that aerofoil sections must be slender and sharp-edged permits the overall aerodynamic properties to be assessed as the sum of contributions due to thickness, camber and incidence. From previous sections it is known that the local pressure at any point on the surface is due to the magnitude and sense of the angular deflection of the flow from the free-stream direction. This deflection in turn can be resolved into components arising from the separate geometric quantities of the section, i.e. from the thickness, camber and chord incidence.

The principle is shown figuratively in the sketch, Fig. 6.49, where the pressure p acting on the aerofoil at a point where the flow deflection from the free stream is ε may be considered as the sum of $p_t + p_c + p_\alpha$. If, as is more convenient, the pressure coefficient is considered, care must be taken to evaluate the sum algebraically. With the notation shown in Fig. 6.49;

$$C_p = C_{p_t} + C_{p_c} + C_{p_\alpha} \qquad (6.158)$$

or

$$\frac{2}{\sqrt{M^2 - 1}}\varepsilon = \frac{2}{\sqrt{M^2 - 1}}(\varepsilon_t + \varepsilon_c + \varepsilon_\alpha) \qquad (6.159)$$

Lift The lift coefficient due to the element of surface is

$$\delta C_L = \frac{-2}{\sqrt{M^2 - 1}}(\varepsilon_t + \varepsilon_c + \varepsilon_\alpha)\frac{\delta x}{c}$$

which is made up of terms due to thickness, camber and incidence. On integrating round the surface of the aerofoil the contributions due to thickness and camber vanish leaving only that due to incidence. This can be easily shown by isolating the contribution due to camber, say, for the upper surface. From Eqn (6.148)

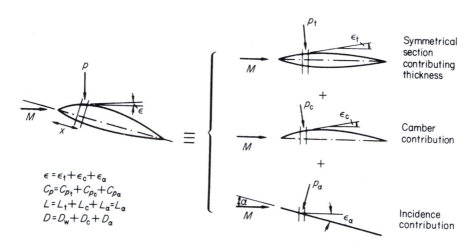

$\varepsilon = \varepsilon_t + \varepsilon_c + \varepsilon_\alpha$
$C_p = C_{p_t} + C_{p_c} + C_{p_\alpha}$
$L = L_t + L_c + L_\alpha = L_\alpha$
$D = D_w + D_c + D_\alpha$

Symmetrical section contributing thickness

Camber contribution

Incidence contribution

Fig. 6.49

$$C_{L_{\text{camber}}} = \frac{-2}{\sqrt{M^2 - 1}} \int_0^c \varepsilon_c \frac{dx}{c}$$

but

$$\int_0^c \varepsilon_c \, dx = \int_0^c \left(\frac{dy}{dx}\right)_c dx = \int_0^c dy_c = [y]_0^c = 0$$

Therefore

$$C_{L_{\text{camber}}} = 0$$

Similar treatment of the lower surface gives the same result, as does consideration of the contribution to the lift due to the thickness.

This result is also borne out by the values of C_L found in the previous examples, i.e.

$$C_L = \frac{-2}{\sqrt{M^2 - 1}} \left(\underset{\substack{\text{upper}\\\text{surface}}}{\int_0^c \varepsilon_\alpha \frac{dx}{c}} - \underset{\substack{\text{lower}\\\text{surface}}}{\int_0^c \varepsilon_\alpha \frac{dx}{c}} \right)$$

Now ε_α (upper surface) $= -\alpha$ and ε_α (lower surface) $= +\alpha$

$$C_L = \frac{-2}{\sqrt{M^2 - 1}} \frac{1}{c} ([-\alpha]_0^c + [\alpha]_c^0)$$

$$C_L = \frac{4\alpha}{\sqrt{M^2 - 1}} \tag{6.160}$$

Drag (wave) The drag coefficient due to the element of surface shown in Fig. 6.49 is

$$\delta C_D = C_p \varepsilon^2 \frac{\delta x}{c}$$

which, on putting $\varepsilon = \varepsilon_t + \varepsilon_c + \varepsilon_\alpha$ etc., becomes

$$\delta C_D = \frac{2}{\sqrt{M^2 - 1}} (\varepsilon_t + \varepsilon_c + \varepsilon_\alpha)^2 \frac{\delta x}{c}$$

On integrating this expression round the contour to find the overall drag, only the integration of the squared terms contributes, since integration of other products vanishes for the same reason as given above for the development leading to Eqn (6.160). Thus

$$C_D = \frac{2}{\sqrt{M^2 - 1}} \oint (\varepsilon_t^2 + \varepsilon_c^2 + \varepsilon_\alpha^2) \frac{dx}{c} \tag{6.161}$$

Now

$$2 \oint \varepsilon_\alpha^2 \, dx = 4\alpha^2 c$$

and for a particular section

$$2 \oint \varepsilon_t^2 \, dx = k_t \left(\frac{t}{c}\right)^2 c$$

and

$$2 \oint \varepsilon_c^2 \, dx = k_c \beta^2 c$$

so that for a given aerofoil profile the drag coefficient becomes in general

$$C_D = \frac{1}{\sqrt{M^2 - 1}} \left(4\alpha^2 + k_t \left(\frac{t}{c}\right)^2 + k_c \beta^2\right) \tag{6.161a}$$

where t/c and β are the thickness chord ratio and camber, respectively, and k_t, k_c are geometric constants.

Lift/wave drag ratio   It follows from Eqns (6.160) and (6.161) that

$$\frac{D}{L} = \alpha + \frac{k_t(t/c)^2 + k_t\beta^2}{4\alpha}$$

which is a minimum when

$$\alpha^2 = \frac{k_t(t/c)^2 + k_c\beta^2}{4}$$

Moment coefficient and centre of pressure coefficient   Once again the moment about the leading edge is generated from the normal contribution and for the general element of surface x from the leading edge

$$\delta C_M = -\left(\frac{2}{\sqrt{M^2 - 1}}\right) \frac{x}{c} \varepsilon \frac{dx}{c}$$

$$C_M = \frac{-2}{\sqrt{M^2 - 1}} \oint (\varepsilon_\alpha + \varepsilon_t + \varepsilon_c) \frac{x}{c} \frac{dx}{c}$$

Now

$$\oint \varepsilon_t \frac{x}{c} \frac{dx}{c}$$

is zero for the general symmetrical thickness, since the pressure distribution due to the section (which, by definition, is symmetrical about the chord) provides neither lift nor moment, i.e. the net lift at any chordwise station is zero. However, the effect of camber is not zero in general, although the overall lift is zero (since the integral of the slope is zero) and the influence of camber is to exert a pitching moment that is negative (nose down for positive camber), i.e. concave downwards. Thus

$$C_M = \frac{-2}{\sqrt{M^2 - 1}} \left(2\frac{\alpha}{2} + k_c\beta\right)$$

The centre of pressure coefficient follows from

$$k_{\text{CP}} = \frac{-C_M}{C_L} = \frac{\dfrac{2}{\sqrt{M^2 - 1}}(\alpha + k_c\beta)}{\dfrac{4\alpha}{\sqrt{M^2 - 1}}}$$

$$k_{\text{CP}} = 0.5\left(1 + \frac{k_c\beta}{\alpha}\right)$$

and this is no longer independent of incidence, although it is still independent of Mach number.

Aerofoil section made up of unequal circular arcs

A convenient aerofoil section to consider as a first example is the biconvex aerofoil used by Stanton* in some early work on aerofoils at speeds near the speed of sound. In his experimental work he used a conventional, i.e. round-nosed, aerofoil (RAF 31a) in addition to the biconvex sharp-edged section at subsonic as well as supersonic speeds, but the only results used for comparison here will be those for the biconvex section at the supersonic speed $M = 1.72$.

Example 6.11 Made up of two unequal circular arcs a profile has the dimensions shown in Fig. 6.50. The exercise here is to compare the values of lift, drag, moment and centre of pressure coefficients found by Stanton* with those predicted by Ackeret's theory. From the geometric data given, the tangent angles at leading and trailing edges are $16° = 0.28$ radians and $7° = 0.12$ radians for upper and lower surfaces respectively. Then, measuring x from the leading edge, the local deflections from the free-stream direction are

$$\varepsilon_U = 0.28\left(1 - 2\frac{x}{c}\right) - \alpha$$

and

$$\varepsilon_L = 0.12\left(1 - 2\frac{x}{c}\right) + \alpha$$

for upper and lower surfaces respectively.

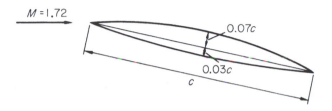

Fig. 6.50 Stanton's biconvex aerofoil section $t/c = 0.1$

---

*T.E. Stanton, A high-speed wind channel for tests on aerofoils, *ARCR and M*, 1130, January 1928.

Lift coefficient

$$C_L = \frac{-2}{c\sqrt{M^2-1}} \left[\int_0^c \varepsilon_U \, dx - \int_0^c \varepsilon_L \, dx \right]$$

$$= \frac{-2}{c\sqrt{M^2-1}} \left[\int_0^c 2\alpha + 0.16\left(1 - 2\frac{x}{c}\right) dx \right]$$

$$C_L = \frac{4\alpha}{\sqrt{M^2-1}}$$

For $M = 1.72$

$$C_L = 2.86\alpha$$

Drag (wave) coefficient

$$C_D = \frac{2}{c\sqrt{M^2-1}} \int_0^c \left[\left(0.28\left(1 - \frac{2x}{c}\right) - \alpha \right)^2 + \left(0.12\left(1 - 2\frac{x}{c}\right) + \alpha \right)^2 \right] dx$$

$$C_D = \frac{(4\alpha^2 + 0.0619)}{\sqrt{M^2-1}}$$

For $M = 1.72$ (as in Stanton's case),

$$C_D = 2.86\alpha^2 + 0.044$$

Moment coefficient (about leading edge)

$$M_{\text{LE}} = C_{M_{\text{LE}}} \frac{1}{2} \rho_0 V^2 c^2 = -\int_0^c \frac{\rho_0 V^2}{\sqrt{M^2-1}} \left[2\alpha - 0.16\left(1 - 2\frac{x}{c}\right) \right] x \, dx$$

or

$$C_{M_{\text{LE}}} = \frac{2}{\sqrt{M^2-1}} [\alpha + 0.0271]$$

For $M = 1.72$:

$$-C_{M_{\text{LE}}} = 1.43\alpha + 0.039$$

Centre-of-pressure coefficient

$$k_{\text{CP}} = \frac{-C_{M_{\text{LE}}}}{C_L} = \frac{2\alpha + 0.054}{4\alpha}$$

$$k_{\text{CP}} = \frac{0.5 + 0.0135}{\alpha}$$

Lift/drag ratio

$$\frac{L}{D} = \frac{\dfrac{4\alpha}{\sqrt{M^2-1}}}{\dfrac{4\alpha^2 + 0.0619}{\sqrt{M^2-1}}} = \frac{\alpha}{\alpha^2 - 0.0155}$$

This is a maximum when $\alpha = \sqrt{0.0155} = 0.125$ rads. $= 8.4°$ giving $(L/D)_{\text{max}} = 4.05$.

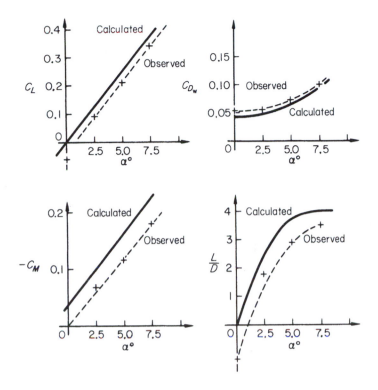

Incidence degrees		0	2·5	5·0	7·5
C_L	calculated	0	0·125	0·25	0·375
	observed	−0·064	0·096	0·203	0·342
C_D	calculated	0·044	0·0495	0·066	0·093
	observed	0·052	0·054	0·070	0·093
$-C_M$	calculated	0·039	0·101	0·164	0·226
	observed	−0·002	0·068	0·114	0·178
k_{CP}	calculated	∞	0·81	0·65	0·60
	observed	0·03	0·69	0·54	0·49
L/D	calculated	0	2·5	3·8	4·0
	observed	−1·2	1·8	2·9	3·5

Fig. 6.51

It will be noted again that the calculated and observed values are close in shape but the latter are lower in value, Fig. 6.51. The differences between theory and experiment are probably explained by the fact that viscous drag is neglected in the theory.

Double wedge aerofoil section

Example 6.12   Using Ackeret's theory obtain expressions for the lift and drag coefficients of the cambered double-wedge aerofoil shown in Fig. 6.52. Hence show that the minimum lift-drag ratio for the uncambered double-wedge aerofoil is $\sqrt{2}$ times that for a cambered one with $h = t/2$. Sketch the flow patterns and pressure distributions around both aerofoils at the incidence for $(L/D)_{max}$.
(U of L)

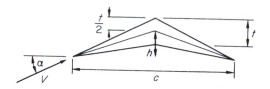

Fig. 6.52

Lift Previous work, Eqn (6.160) has shown that

$$C_L = \frac{4\alpha}{\sqrt{M^2 - 1}}$$

Drag (wave) From Eqn (6.161) on the general aerofoil

$$C_D = \frac{2}{\sqrt{M^2 - 1}} \oint (\varepsilon_t^2 + \varepsilon_c^2 + \varepsilon_\alpha^2) \frac{\mathrm{d}x}{c}$$

Here, as before:

$$2 \oint \varepsilon_\alpha^2 \frac{\mathrm{d}x}{c} = 4\alpha^2 c$$

For the given geometry

$$\oint \varepsilon_t^2 \frac{\mathrm{d}x}{c} = 4 \left[\frac{1}{2} \left(\frac{t}{c} \right)^2 \right]$$

i.e. one equal contribution from each of four flat surfaces, and

$$\oint \varepsilon_c^2 \frac{\mathrm{d}x}{c} = 4 \left[\frac{1}{2} \left(\frac{h}{c/2} \right)^2 \right]$$

i.e. one equal contribution from each of four flat surfaces. Therefore

$$C_D = \frac{4}{\sqrt{M^2 - 1}} \left[\alpha^2 + \left(\frac{t}{c} \right)^2 + 4 \left(\frac{h}{c} \right)^2 \right]$$

Lift–drag ratio

$$\frac{L}{D} = \frac{C_L}{C_D} = \frac{\alpha}{\left[\alpha^2 + \left(\frac{t}{c} \right)^2 + 4 \left(\frac{h}{c} \right)^2 \right]}$$

For the uncambered aerofoil $h = 0$:

$$\left[\frac{L}{D} \right]_{\mathrm{max}} = \left[\frac{\alpha}{\alpha^2 + (t/c)^2} \right]_{\mathrm{max}} \quad \overset{\alpha = t/c}{\Rightarrow} \quad \left[\frac{L}{D} \right]_{\mathrm{max}} = \frac{1}{2(t/c)}$$

For the cambered section, given $h = t/c$:

$$\left[\frac{L}{D} \right]_{\mathrm{max}} = \left[\frac{\alpha}{\alpha^2 + 2(t/c)^2} \right]_{\mathrm{max}} \quad \overset{\alpha + \sqrt{2}t/c}{\Rightarrow} \quad \left[\frac{L}{D} \right]_{\mathrm{max}} = \frac{1}{2\sqrt{2}(t/c)}$$

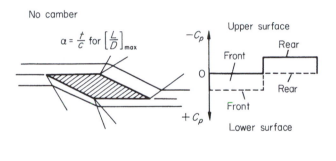

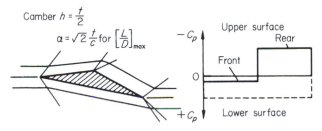

Fig. 6.53 Flow patterns and pressure distributions around both aerofoils at incidence of $[L/D]_{max}$

6.8.4 Other aspects of supersonic wings

The shock-expansion approximation

The supersonic linearized theory has the advantage of giving relatively simple formulae for the aerodynamic characteristics of aerofoils. However, as shown below in Example 6.13 the exact pressure distribution can be readily found for a double-wedge aerofoil. Hence the coefficients of lift and drag can be obtained.

Example 6.13 Consider a symmetrical double-wedge aerofoil at zero incidence, similar in shape to that in Fig. 6.44 above, except that the semi-wedge angle $\varepsilon_0 = 10°$. Sketch the wave pattern for $M_\infty = 2.0$, calculate the Mach number and pressure on each face of the aerofoil, and hence determine C_D. Compare the results with those obtained using the linear theory. Assume the free-stream stagnation pressure, $p_{0\infty} = 1$ bar.

The wave pattern is sketched in Fig. 6.54a. The flow properties in the various regions can be obtained using isentropic flow and oblique shock tables.* In region 1 $M = M_\infty = 2.0$ and $p_{0\infty} = 1$ bar. From the isentropic flow tables $p_{01}/p_1 = 7.83$ leading to $p_1 = 0.1277$ bar. In region 2 the oblique shock-wave tables give $p_2/p_1 = 1.7084$ (leading to $p_2 = 0.2182$ bar), $M_2 = 1.6395$ and shock angle $= 39.33°$. Therefore

$$C_{p2} = \frac{p_2 - p_\infty}{\frac{1}{2}\rho_\infty V_\infty^2} = \frac{(p_2/p_\infty) - 1}{\frac{1}{2}\gamma(p_\infty/\gamma p_\infty)V_\infty^2} = \frac{(p_2/p_\infty) - 1}{\frac{1}{2}\gamma M_\infty^2}$$

$$= \frac{(0.2182/0.1277) - 1}{0.5 \times 1.4 \times 2^2} = 0.253$$

---

* e.g. E.L. Houghton and A.E. Brock, *Tables for the Compressible Flow of Dry Air*, 3rd Edn., Edward Arnold, 1975.

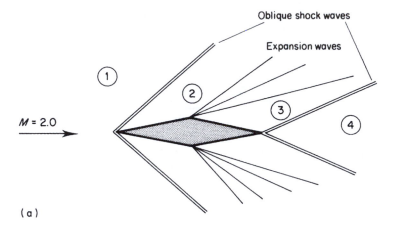

(a)

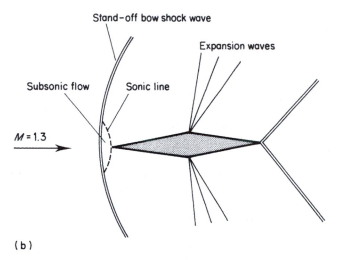

(b)

Fig. 6.54

(Using the linear theory, Eqn (6.145) gives

$$C_{p2} = \frac{2\varepsilon}{\sqrt{M_\infty^2 - 1}} = \frac{2 \times (10\pi/180)}{\sqrt{2^2 - 1}} = 0.202)$$

In order to continue the calculation into region 3 it is first necessary to determine the Prandtl–Meyer angle and stagnation pressure in region 2. These can be obtained as follows using the isentropic flow tables: $p_{02}/p_2 = 4.516$ giving $p_{02} = 4.516 \times 0.2182 = 0.9853$ bar; and Mach angle, $\mu_2 = 37.57°$ and Prandtl–Meyer angle, $v_2 = 16.01°$.

Between regions 2 and 3 the flow expands isentropically through 20° so $v_3 = v_2 + 20° = 36.01°$. From the isentropic flow tables this value of v_3 corresponds to $M_3 = 2.374$, $\mu_3 = 24.9°$ and

$p_{03}/p_3 = 14.03$. Since the expansion is isentropic $p_{03} = p_{02} = 0.9853$ bar so that $p_3 = 0.9853/14.03 = 0.0702$ bar. Thus

$$C_{p3} = \frac{(0.0702/0.1277) - 1}{0.7 \times 2^2} = -0.161$$

(Using the linear theory, Eqn (6.145) gives

$$C_{p3} = \frac{2\varepsilon}{\sqrt{M_\infty^2 - 1}} = \frac{-2 \times (10\pi/180)}{\sqrt{2^2 - 1}} = -0.202)$$

There is an oblique shock wave between regions 3 and 4. The oblique shock tables give $p_4/p_3 = 1.823$ and $M_4 = 1.976$ giving $p_4 = 1.823 \times 0.0702 = 0.128$ bar and a shock angle of $33.5°$.

The drag per unit span acting on the aerofoil is given by resolving the pressure forces, so that

$$D = 2(p_2 - p_3) \times \frac{(c/2)}{\cos(10°)} \times \sin(10°)$$

so

$$C_D = (C_{p2} - C_{p3})\tan(10°) = 0.0703$$

(Using the linear theory, Eqn (6.151) with $\alpha = 0$ gives

$$C_D = \frac{4(t/c)^2}{\sqrt{M_\infty^2 - 1}} = \frac{4\tan^2(10°)}{\sqrt{2^2 - 1}} = 0.072)$$

It can be seen from the calculations above that, although the linear theory does not approximate the value of C_p very accurately, it does yield an accurate estimate of C_D.

When $M_\infty = 1.3$ it can be seen from the oblique shock tables that the maximum compression angle is less than $10°$. This implies that in this case the flow can only negotiate the leading edge by being compressed through a shock wave that stands off from the leading edge and is normal to the flow where it intersects the extension of the chord line. This leads to a region of subsonic flow being formed between the stand-off shock wave and the leading edge. The corresponding flow pattern is sketched in Fig. 6.54b.

A similar procedure to that in Example 6.13 can be followed for aerofoils with curved profiles. In this case, though, the procedure becomes approximate because it ignores the effect of the Mach waves reflected from the bow shock wave – see Fig. 6.55. The so-called shock-expansion approximation is made clearer by the example given below.

Example 6.14 Consider a biconvex aerofoil at zero incidence in supersonic flow at $M_\infty = 2$, similar in shape to that shown in Fig. 6.46 above so that, as before, the shape of the upper surface is given by

$$y = x\varepsilon_0\left(1 - \frac{x}{c}\right) \text{ giving local flow angle } \theta(= \varepsilon) = \arctan\left[\varepsilon_0\left(1 - \frac{2x}{c}\right)\right]$$

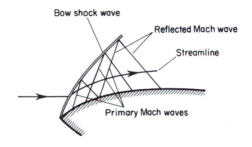

Fig. 6.55

Calculate the pressure and Mach number along the surface as functions of x/c for the case of $\varepsilon_0 = 0.2$. Compare with the results obtained with linear theory. Take the freestream stagnation pressure to be 1 bar.

Region 1 as in Example 6.13, i.e. $M_1 = 2.0$, $p_{01} = 1$ bar and $p_1 = 0.1277$ bar

At $x = 0$ $\theta = \arctan{(0.2)} = 11.31°$. Hence initially the flow is turned by the bow shock through an angle of $11.31°$, so using the oblique shock tables gives $p_2/p_1 = 1.827$ and $M_2 = 1.59$. Thus $p_2 = 1.827 \times 0.1277 = 0.233$ bar. From the isentropic flow tables it is found that $M_2 = 1.59$ corresponds to $p_{02}/p_2 = 4.193$ giving $p_{02} = 0.977$ bar.

Thereafter the pressures and Mach numbers around the surface can be obtained using the isentropic flow tables as shown in the table below.

$\frac{x}{c}$	$\tan\theta$	θ	$\Delta\theta$	ν	M	$\frac{p_0}{p}$	p (bar)	C_p	$(C_p)_{lin}$
0.0	0.2	11.31°	0°	14.54°	1.59	4.193	0.233	0.294	0.228
0.1	0.16	9.09°	2.22°	16.76°	1.666	4.695	0.208	0.225	0.183
0.2	0.12	6.84°	4.47°	19.01°	1.742	5.265	0.186	0.163	0.138
0.3	0.08	4.57°	6.74°	21.28°	1.820	5.930	0.165	0.104	0.092
0.5	0.0	0.0	11.31°	25.85°	1.983	7.626	0.128	0.0008	0
0.7	−0.08	−4.57°	15.88°	30.42°	2.153	9.938	0.098	−0.0831	−0.098
0.8	−0.12	−6.84°	18.15°	32.69°	2.240	11.385	0.086	−0.1166	−0.138
0.9	−0.16	−9.09°	20.40°	34.94°	2.330	13.104	0.075	−0.1474	−0.183
1.0	−0.20	−11.31°	22.62°	37.16°	2.421	15.102	0.065	−0.1754	−0.228

Wings of finite span

When the component of the free-stream velocity perpendicular to the leading edge is greater than the local speed of sound the wing is said to have a *supersonic leading edge*. In this case, as illustrated in Fig. 6.56, there is two-dimensional supersonic flow over much of the wing. This flow can be calculated using supersonic aerofoil theory. For the rectangular wing shown in Fig. 6.56 the presence of a wing-tip can only be communicated within the Mach cone apex which is located at the wing-tip. The same consideration would apply to any inboard three-dimensional effects, such as the 'kink' at the centre-line of a swept-back wing.

The opposite case is when the component of free-stream velocity perpendicular to the leading edge is less than the local speed of sound and the term *subsonic leading edge* is used. A typical example is the swept-back wing shown in Fig. 6.57. In this case the Mach cone generated by the leading edge of the centre section subtends the whole wing. This implies that the leading edge of the outboard portions of the wing influences the oncoming flow just as for subsonic flow. Wings having finite thickness and incidence actually generate a shock cone, rather than a Mach cone, as shown in

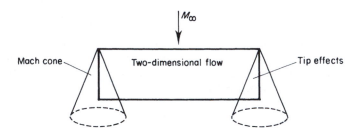

Fig. 6.56 A typical wing with a supersonic leading edge

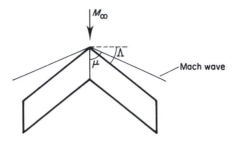

Fig. 6.57 A wing with a subsonic leading edge

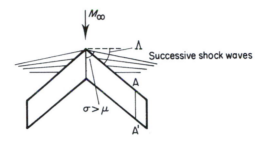

Fig. 6.58

Fig. 6.58. Additional shocks are generated by other points on the leading edge and the associated shock angles will tend to increase because each successive shock wave leads to a reduction in the Mach number. These shock waves progressively decelerate the flow, so that at some section, such as AA', the flow approaching the leading edge will be subsonic. Thus subsonic wing sections would be used over most of the wing.

Wings with subsonic leading edges have lower wave drag than those with supersonic ones. Consequently highly swept wings, e.g. slender deltas, are the preferred configuration at supersonic speeds. On the other hand swept wings with supersonic leading edges tend to have a greater wave drag than straight wings.

Computational methods

Computational methods for compressible flows, particularly transonic flow over wings, have been the subject of a very considerable research effort over the past three decades. Substantial progress has been made, although much still remains to be done. A discussion of these methods is beyond the scope of the present book, save to note that for the linearized compressible potential flow Eqn (6.118) panel methods (see Sections 3.5, 4.10 and 5.8) have been developed for both subsonic and supersonic flow. These can be used to obtain approximate numerical solutions in cases with exceedingly complex geometries. A review of the computational methods developed for the full inviscid and viscous equations of motion is given by Jameson.*

* A. Jameson, 'Full-Potential, Euler and Navier-Stokes Schemes', in *Applied Computational Aerodynamics, Vol. 125 of Prog. in Astronautics and Aeronautics* (ed. By P.A. Henne), 39–88 (1990), AIAA: New York.

Exercises

1 A convergent–divergent duct has a maximum diameter of 150 mm and a pitot-static tube is placed in the throat of the duct. Neglecting the effect of the pitot-static tube on the flow, estimate the throat diameter under the following conditions:

(i) air at the maximum section is of standard pressure and density, pressure difference across the pitot-static tube $\equiv$ 127 mm water;
(ii) pressure and temperature in the maximum section are 101 300 N m^{-2} and 100 °C respectively, pressure difference across pitot-static tube $\equiv$ 127 mm mercury.

(*Answer*: (i) 123 mm; (ii) 66.5 mm)

2 In the wing-flow method of transonic research an aeroplane dives at a Mach number of 0.87 at a height where the pressure and temperature are 46 500 N m^{-2} and −24.6 °C respectively. At the position of the model the pressure coefficient is −0.5. Calculate the speed, Mach number, $0.7p\,M^2$, and the kinematic viscosity of the flow past the model.

(*Answer*: 344 m s^{-1}; $M = 1.133$; $0.7\,pM^2 = 30\,800$ N m^{-2}; $v = 2.64 \times 10^{-3}$ m^{2}s^{-1})

3 What would be the indicated air speed and the true air speed of the aeroplane in Exercise 2, assuming the air-speed indicator to be calibrated on the assumption of incompressible flow in standard conditions, and to have no instrument errors?

(*Answer*: TAS = 274 m s^{-1}; IAS = 219 m s^{-1})

4 On the basis of Bernoulli's equation, discuss the assumption that the compressibility of air may be neglected for low subsonic speeds.

A symmetric aerofoil at zero lift has a maximum velocity which is 10% greater than the free-stream velocity. This maximum increases at the rate of 7% of the free-stream velocity for each degree of incidence. What is the free-stream velocity at which compressibility effects begins to become important (i.e. the error in pressure coefficient exceeds 2%) on the aerofoil surface when the incidence is 5°?

(*Answer*: Approximately 70 m s^{-1}) (U of L)

5 A closed-return type of wind-tunnel of large contraction ratio has air at standard conditions of temperature and pressure in the settling chamber upstream of the contraction to the working section. Assuming isentropic compressible flow in the tunnel estimate the speed in the working section where the Mach number is 0.75. Take the ratio of specific heats for air as $\gamma = 1.4$.   (*Answer*: 242 m s^{-1}) (U of L)

7

Viscous flow and boundary layers*

Preamble

This chapter introduces the concept of the boundary layer, describes the flow phenomena involved, and explains how the Navier–Stokes equations can be simplified for the analysis of boundary-layer flows. Certain useful solutions to the boundary-layer equations are described. There are two phenomena, governed by viscous effects and the behaviour of the boundary layer, that are vitally important for engineering applications of aerodynamics. These are flow separation and transition from laminar to turbulent flow. A section is devoted to each in turn. The momentum-integral form of the boundary-layer equations is derived. Its use for obtaining approximate solutions for laminar, turbulent, and mixed laminar-turbulent boundary layers is explored in detail. Its application for estimating profile drag is also described. These approximate techniques are illustrated with examples chosen to show how to estimate the aerodynamic characteristics, such as drag, that depend on the behaviour of the boundary layer. Computational methods for obtaining numerical solutions to the boundary-layer equations are presented and reviewed. Some of the computational methods are explained in detail. A substantial section is devoted to the flow physics of turbulent boundary layers with illustrations from aeronautical applications. Computational methods for turbulent boundary layers and other flows are also reviewed. As viscosity is the key physical property governing the behaviour of boundary layers and related phenomena, a quantitative treatment of compressible effects is omitted. But the chapter closes with a detailed qualitative description of the influence of compressible effects on boundary layers, particularly their interaction with shock waves.

7.1 Introduction

In the other chapters of this book, the effects of viscosity, which is an inherent property of any real fluid, have, in the main, been ignored. At first sight, it would seem to be a waste of time to study inviscid fluid flow when all practical fluid

---

* This chapter is concerned mainly with incompressible flows. However, the general arguments developed are also applicable to compressible flows.

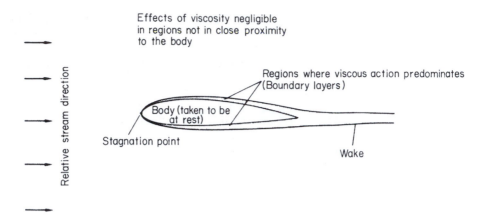

Effects of viscosity negligible
in regions not in close proximity
to the body

Regions where viscous action predominates
(Boundary layers)

Body (taken to be
at rest)

Relative stream direction

Stagnation point

Wake

Fig. 7.1

problems involve viscous action. The purpose behind this study by engineers dates back to the beginning of the previous century (1904) when Prandtl conceived the idea of the *boundary layer*.

In order to appreciate this concept, consider the flow of a fluid past a body of reasonably slender form (Fig. 7.1). In aerodynamics, almost invariably, the fluid viscosity is relatively small (i.e. the Reynolds number is high); so that, unless the transverse velocity gradients are appreciable, the shearing stresses developed [given by Newton's equation $\tau = \mu(\partial u/\partial y)$ (see, for example, Section 1.2.6 and Eqn (2.86))] will be very small. Studies of flows, such as that indicated in Fig. 7.1, show that the transverse velocity gradients are usually negligibly small throughout the flow field except for thin layers of fluid immediately adjacent to the solid boundaries. Within these boundary layers, however, large shearing velocities are produced with consequent shearing stresses of appreciable magnitude.

Consideration of the intermolecular forces between solids and fluids leads to the assumption that at the boundary between a solid and a fluid (other than a rarefied gas) there is a condition of no slip. In other words, the relative velocity of the fluid tangential to the surface is everywhere zero. Since the mainstream velocity at a small distance from the surface may be considerable, it is evident that appreciable shearing velocity gradients may exist within this boundary region.

Prandtl pointed out that these boundary layers were usually very thin, provided that the body was of streamline form, at a moderate angle of incidence to the flow and that the flow Reynolds number was sufficiently large; so that, as a first approximation, their presence might be ignored in order to estimate the pressure field produced about the body. For aerofoil shapes, this pressure field is, in fact, only slightly modified by the boundary-layer flow, since almost the entire lifting force is produced by normal pressures at the aerofoil surface, it is possible to develop theories for the evaluation of the lift force by consideration of the flow field outside the boundary layers, where the flow is essentially inviscid in behaviour. Herein lies the importance of the inviscid flow methods considered previously. As has been noted in Section 4.1, however, no drag force, other than induced drag, ever results from these theories. The drag force is mainly due to shearing stresses at the body surface (see Section 1.5.5) and it is in the estimation of these that the study of boundary-layer behaviour is essential.

The enormous simplification in the study of the whole problem, which follows from Prandtl's boundary-layer concept, is that the equations of viscous motion need be considered only in the limited regions of the boundary layers, where appreciable simplifying assumptions can reasonably be made. This was the major single impetus to the rapid advance in aerodynamic theory that took place in the first half of the twentieth century. However, in spite of this simplification, the prediction of boundary-layer behaviour is by no means simple. Modern methods of computational fluid dynamics provide powerful tools for predicting boundary-layer behaviour. However, these methods are only accessible to specialists; it still remains essential to study boundary layers in a more fundamental way to gain insight into their behaviour and influence on the flow field as a whole. To begin with, we will consider the general physical behaviour of boundary layers.

7.2 The development of the boundary layer

For the flow around a body with a sharp leading edge, the boundary layer on any surface will grow from zero thickness at the leading edge of the body. For a typical aerofoil shape, with a bluff nose, boundary layers will develop on top and bottom surfaces from the front stagnation point, but will not have zero thickness there (see Section 2.10.3).

On proceeding downstream along a surface, large shearing gradients and stresses will develop adjacent to the surface because of the relatively large velocities in the mainstream and the condition of no slip at the surface. This shearing action is greatest at the body surface and retards the layers of fluid immediately adjacent to the surface. These layers, since they are now moving more slowly than those above them, will then influence the latter and so retard them. In this way, as the fluid near the surface passes downstream, the retarding action penetrates farther and farther away from the surface and the boundary layer of retarded or 'tired' fluid grows in thickness.

7.2.1 Velocity profile

Further thought about the thickening process will make it evident that the increase in velocity that takes place along a normal to the surface must be continuous. Let y be the perpendicular distance from the surface at any point and let u be the corresponding velocity parallel to the surface. If u were to increase discontinuously with y at any point, then at that point $\partial u/\partial y$ would be infinite. This would imply an infinite shearing stress [since the shear stress $\tau = \mu(\partial u/\partial y)$] which is obviously untenable.

Consider again a small element of fluid (Fig. 7.2) of unit depth normal to the flow plane, having a unit length in the direction of motion and a thickness δy normal to the flow direction. The shearing stress on the lower face AB will be $\tau = \mu(\partial u/\partial y)$ while that on the upper face CD will be $\tau + (\partial \tau/\partial y)\delta y$, in the directions shown, assuming u to increase with y. Thus the resultant shearing force in the x-direction will be $[\tau + (\partial \tau/\partial y)\delta y] - \tau = (\partial \tau/\partial y)\delta y$ (since the area parallel to the x-direction is unity) but $\tau = \mu(\partial u/\partial y)$ so that the net shear force on the element $= \mu(\partial^2 u/\partial y^2)\delta y$. Unless μ be zero, it follows that $\partial^2 u/\partial y^2$ cannot be infinite and therefore the rate of change of the velocity gradient in the boundary layer must also be continuous.

Also shown in Fig. 7.2 are the streamwise pressure forces acting on the fluid element. It can be seen that the net pressure force is $-(\partial p/\partial x)\delta x$. Actually, owing to the very small total thickness of the boundary layer, the pressure hardly varies at all normal to the surface. Consequently, the net transverse pressure force is zero to a very good approximation and Fig. 7.2 contains all the significant fluid forces. The

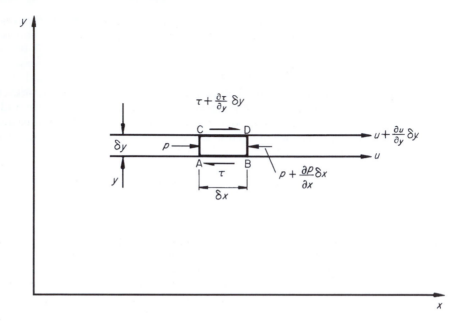

Fig. 7.2

effects of streamwise pressure change are discussed in Section 7.2.6 below. At this stage it is assumed that $\partial p/\partial x = 0$.

If the velocity u is plotted against the distance y it is now clear that a smooth curve of the general form shown in Fig. 7.3a must develop (see also Fig. 7.11). Note that at the surface the curve is not tangential to the u axis as this would imply an infinite gradient $\partial u/\partial y$, and therefore an infinite shearing stress, at the surface. It is also evident that as the shearing gradient decreases, the retarding action decreases, so that

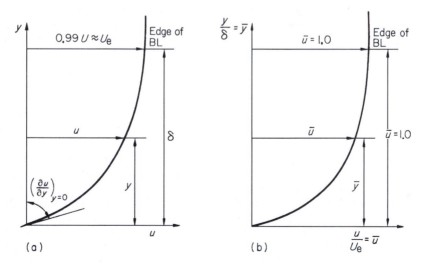

Fig. 7.3

at some distance from the surface, when $\partial u/\partial y$ becomes very small, the shear stress becomes negligible, although theoretically a small gradient must exist out to $y = \infty$.

7.2.2 Boundary-layer thickness

In order to make the idea of a boundary layer realistic, an arbitrary decision must be made as to its extent and the usual convention is that the boundary layer extends to a distance δ from the surface such that the velocity u at that distance is 99% of the local mainstream velocity U_e that would exist at the surface in the absence of the boundary layer. Thus δ is the physical thickness of the boundary layer so far as it needs to be considered and when defined specifically as above it is usually designated the 99%, or general, thickness. Further thickness definitions are given in Section 7.3.2.

7.2.3 Non-dimensional profile

In order to compare boundary-layer profiles of different thickness, it is convenient to express the profile shape non-dimensionally. This may be done by writing $\bar{u} = u/U_e$ and $\bar{y} = y/\delta$ so that the profile shape is given by $\bar{u} = f(\bar{y})$. Over the range $y = 0$ to $y = \delta$, the velocity parameter $\bar{u}$ varies from 0 to 0.99. For convenience when using $\bar{u}$ values as integration limits, negligible error is introduced by using $\bar{u} = 1.0$ at the outer boundary, and considerable arithmetical simplification is achieved. The velocity profile is then plotted as in Fig. 7.3b.

7.2.4 Laminar and turbulent flows

Closer experimental study of boundary-layer flows discloses that, like flows in pipes, there are two different regimes which can exist: laminar flow and turbulent flow. In *laminar flow*, the layers of fluid slide smoothly over one another and there is little interchange of fluid mass between adjacent layers. The shearing tractions that develop due to the velocity gradients are thus due entirely to the viscosity of the fluid, i.e. the momentum exchanges between adjacent layers are on a molecular scale only.

In *turbulent flow* considerable seemingly random motion exists, in the form of velocity fluctuations both along the mean direction of flow and perpendicular to it. As a result of the latter there are appreciable transports of mass between adjacent layers. Owing to these fluctuations the velocity profile varies with time. However, a time-averaged, or mean, velocity profile can be defined. As there is a mean velocity gradient in the flow, there will be corresponding interchanges of streamwise momentum between the adjacent layers that will result in shearing stresses between them. These shearing stresses may well be of much greater magnitude than those that develop as the result of purely viscous action, and the velocity profile shape in a turbulent boundary layer is very largely controlled by these *Reynolds stresses* (see Section 7.9), as they are termed.

As a consequence of the essential differences between laminar and turbulent flow shearing stresses, the velocity profiles that exist in the two types of layer are also different. Figure 7.4 shows a typical laminar-layer profile and a typical turbulent-layer profile plotted to the same non-dimensional scale. These profiles are typical of those on a flat plate where there is no streamwise pressure gradient.

In the laminar boundary layer, energy from the mainstream is transmitted towards the slower-moving fluid near the surface through the medium of viscosity alone and only a relatively small penetration results. Consequently, an appreciable proportion of the boundary-layer flow has a considerably reduced velocity. Throughout the

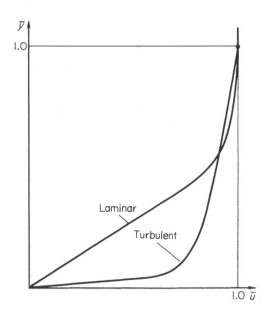

Fig. 7.4

boundary layer, the shearing stress τ is given by $\tau = \mu(\partial u/\partial y)$ and the wall shearing stress is thus $\tau_w = \mu(\partial u/\partial y)_{y=0} = \mu(\partial u/\partial y)_w$ (say).

In the turbulent boundary layer, as has already been noted, large Reynolds stresses are set up owing to mass interchanges in a direction perpendicular to the surface, so that energy from the mainstream may easily penetrate to fluid layers quite close to the surface. This results in the turbulent boundary away from the immediate influence of the wall having a velocity that is not much less than that of the mainstream. However, in layers that are very close to the surface (at this stage of the discussion considered smooth) the velocity fluctuations perpendicular to the wall are evidently damped out, so that in a very limited region immediately adjacent to the surface, the flow approximates to purely viscous flow.

In this *viscous sublayer* the shearing action becomes, once again, purely viscous and the velocity falls very sharply, and almost linearly, within it, to zero at the surface. Since, at the surface, the wall shearing stress now depends on viscosity only, i.e. $\tau_w = \mu(\partial u/\partial y)_w$, it will be clear that the surface friction stress under a turbulent layer will be far greater than that under a laminar layer of the same thickness, since $(\partial u/\partial y)_w$ is much greater. It should be noted, however, that the viscous shear-stress relation is only employed in the viscous sublayer very close to the surface and not throughout the turbulent boundary layer.

It is clear, from the preceding discussion, that the viscous shearing stress at the surface, and thus the surface friction stress, depends only on the slope of the velocity profile at the surface, whatever the boundary-layer type, so that accurate estimation of the profile, in either case, will enable correct predictions of skin-friction drag to be made.

7.2.5   Growth along a flat surface

If the boundary layer that develops on the surface of a flat plate held edgeways on to the free stream is studied, it is found that, in general, a laminar boundary layer starts to

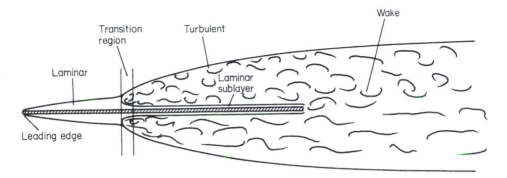

Fig. 7.5 Note: Scale normal to surface of plate is greatly exaggerated

develop from the leading edge. This laminar boundary layer grows in thickness, in accordance with the argument of Section 7.2, from zero at the leading edge to some point on the surface where a rapid transition to turbulence occurs (see Fig. 7.29). This transition is accompanied by a corresponding rapid thickening of the layer. Beyond this transition region, the turbulent boundary layer then continues to thicken steadily as it proceeds towards the trailing edge. Because of the greater shear stresses within the turbulent boundary layer its thickness is greater than for a laminar one. But, away from the immediate vicinity of the transition region, the actual rate of growth along the plate is lower for turbulent boundary layers than for laminar ones. At the trailing edge the boundary layer joins with the one from the other surface to form a wake of retarded velocity which also tends to thicken slowly as it flows away downstream (see Fig. 7.5).

On a flat plate, the laminar profile has a constant shape at each point along the surface, although of course the thickness changes, so that one non-dimensional relationship for $\bar{u} = f(\bar{y})$ is sufficient (see Section 7.3.4). A similar argument applies to a reasonable approximation to the turbulent layer. This constancy of profile shape means that flat-plate boundary-layer studies enjoy a major simplification and much work has been undertaken to study them both theoretically and experimentally.

However, in most aerodynamic problems, the surface is usually that of a stream-line form such as a wing or fuselage. The major difference, affecting the boundary-layer flow in these cases, is that the mainstream velocity and hence the pressure in a streamwise direction is no longer constant. The effect of a pressure gradient along the flow can be discussed purely qualitatively initially in order to ascertain how the boundary layer is likely to react.

7.2.6 Effects of an external pressure gradient

In the previous section, it was noted that in most practical aerodynamic applications the mainstream velocity and pressure change in the streamwise direction. This has a profound effect on the development of the boundary layer. It can be seen from Fig. 7.2 that the net streamwise force acting on a small fluid element within the boundary layer is

$$\frac{\partial \tau}{\partial y} \delta y - \frac{\partial p}{\partial x} \delta x$$

When the pressure decreases (and, correspondingly, the velocity along the edge of the boundary layer increases) with passage along the surface the *external* pressure

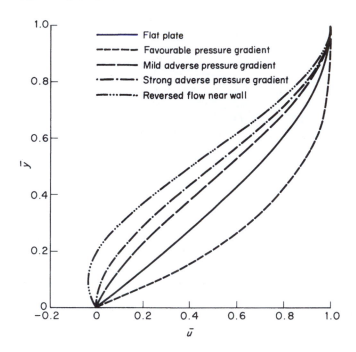

Fig. 7.6 Effect of external pressure gradient on the velocity profile in the boundary layer

gradient is said to be *favourable*. This is because $\partial p/\partial x < 0$ so, noting that $\partial \tau/\partial y < 0$, it can be seen that the streamwise pressure forces help to counter the effects, discussed earlier, of the shearing action and shear stress at the wall. Consequently, the flow is not decelerated so markedly at the wall, leading to a fuller velocity profile (see Fig. 7.6), and the boundary layer grows more slowly along the surface than for a flat plate.

The converse case is when the pressure increases and mainstream velocity decreases along the surface. The external pressure gradient is now said to be *unfavourable* or *adverse*. This is because the pressure forces now reinforce the effects of the shearing action and shear stress at the wall. Consequently, the flow decelerates more markedly near the wall and the boundary layer grows more rapidly than in the case of the flat plate. Under these circumstances the velocity profile is much less full than for a flat plate and develops a point of inflexion (see Fig. 7.6). In fact, as indicated in Fig. 7.6, if the adverse pressure gradient is sufficiently strong or prolonged, the flow near the wall is so greatly decelerated that it begins to reverse direction. Flow reversal indicates that the boundary layer has *separated* from the surface. Boundary-layer separation can have profound consequences for the whole flow field and is discussed in more detail in Section 7.4.

7.3 The boundary-layer equations

To fix ideas it is helpful to think about the flow over a flat plate. This is a particularly simple flow, although like much else in aerodynamics the more one studies the details the less simple it becomes. If we consider the case of infinite Reynolds number, i.e. ignore viscous effects completely, the flow becomes exceedingly simple. The stream-lines are everywhere parallel to the flat plate and the velocity uniform and equal to U_∞, the value in the free stream infinitely far from the plate. There would, of course, be no drag, since the shear stress at the wall would be equivalently zero. (This is a special case of d'Alembert's paradox that states that no force is generated by irrota-tional flow around any body irrespective of its shape.) Experiments on flat plates would confirm that the potential (i.e. inviscid) flow solution is indeed a good approximation at high Reynolds number. It would be found that the higher the Reynolds number, the closer the streamlines become to being everywhere parallel with the plate. Furthermore, the non-dimensional drag, or drag coefficient (see Section 1.4.5), becomes smaller and smaller the higher the Reynolds number becomes, indicating that the drag tends to zero as the Reynolds number tends to infinity.

But, even though the drag is very small at high Reynolds number, it is evidently important in applications of aerodynamics to estimate its value. So, how may we use this excellent infinite-Reynolds-number approximation, i.e. potential flow, to do this? Prandtl's boundary-layer concept and theory shows us how this may be achieved. In essence, he assumed that the potential flow is a good approximation everywhere except in a thin boundary layer adjacent to the surface. Because the boundary layer is very thin it hardly affects the flow outside it. Accordingly, it may be assumed that the flow velocity at the edge of the boundary layer is given to a good approximation by the potential-flow solution for the flow velocity along the surface itself. For the flat plate, then, the velocity at the edge of the boundary layer is U_∞. In the more general case of the flow over a streamlined body like the one depicted in Fig. 7.1, the velocity at the edge of boundary layer varies and is denoted by U_e. Prandtl went on to show, as explained below, how the Navier–Stokes equations may be simplified for application in this thin boundary layer.

7.3.1 Derivation of the laminar boundary-layer equations

At high Reynolds numbers the boundary-layer thickness, δ, can be expected to be very small compared with the length, L, of the plate or streamlined body. (In aeronautical examples, such as the wing of a large aircraft δ/L is typically around 0.01 and would be even smaller if the boundary layer were laminar rather than turbulent.) We will assume that in the hypothetical case of $Re_L \rightarrow \infty$ (where $Re_L = \rho U_\infty L/\mu$), $\delta \rightarrow 0$. Thus if we introduce the small parameter

$$\varepsilon = \frac{1}{Re_L} \tag{7.1}$$

we would expect that $\delta \rightarrow 0$ as $\varepsilon \rightarrow 0$, so that

$$\frac{\delta}{L} \propto \varepsilon^n \tag{7.2}$$

where n is a positive exponent that is to be determined.

Suppose that we wished to estimate the magnitude of velocity gradient within the laminar boundary layer. By considering the changes across the boundary layer along line AB in Fig. 7.7, it is evident that a rough approximation can be obtained by writing

$$\frac{\partial u}{\partial y} \simeq \frac{U_\infty}{\delta} = \frac{U_\infty}{L}\frac{1}{\varepsilon^n}$$

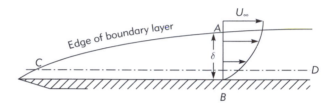

Fig. 7.7

Although this is plainly very rough, it does have the merit of remaining valid as the Reynolds number becomes very high. This is recognized by using a special symbol for the rough approximation and writing

$$\frac{\partial u}{\partial y} = \mathcal{O}\left(\frac{U_\infty}{L}\frac{1}{\varepsilon^n}\right)$$

For the more general case of a streamlined body (e.g. Fig. 7.1), we use x to denote the distance along the surface from the leading edge (strictly from the fore stagnation point) and y to be the distance along the local normal to the surface. Since the boundary layer is very thin and its thickness much smaller than the local radius of curvature of the surface, we can use the cartesian form, Eqns (2.92a,b) and (2.93), of the Navier–Stokes equations. In this more general case, the velocity varies along the edge of the boundary layer and we denote it by U_e, so that

$$\frac{\partial u}{\partial y} = \mathcal{O}\left(\frac{U_e}{\delta}\right) = \mathcal{O}\left(\frac{U_e}{L}\frac{1}{\varepsilon^n}\right) \tag{7.3}$$

where U_e replaces U_∞, so that Eqn (7.3) applies to the more general case of a boundary layer around a streamlined body. Engineers think of $\mathcal{O}(U_e/\delta)$ as meaning order of magnitude of U_e/δ or very roughly a similar magnitude to U_e/δ. To mathematicans $F = \mathcal{O}(1/\varepsilon^n)$ means that $F \propto 1/\varepsilon^n$ as $\varepsilon \to 0$. It should be noted that the order-of-magnitude estimate is the same irrespective of whether the term is negative or positive.

Estimating $\partial u/\partial y$ is fairly straightforward, but what about $\partial u/\partial x$? To estimate this quantity consider the changes along the line CD in Fig. 7.7. Evidently, $u = U_\infty$ at C and $u \to 0$ as D becomes further from the leading edge of the plate. So the total change in u is approximately $U_\infty - 0$ and takes place over a distance $\Delta x \simeq L$. Thus for the general case where the flow velocity varies along the edge of the boundary layer, we deduce that

$$\frac{\partial u}{\partial x} = \mathcal{O}\left(\frac{U_e}{L}\right) \tag{7.4}$$

Finally, in order to estimate second derivatives like $\partial^2 u/\partial y^2$, we again consider the changes along the vertical line AB in Fig. 7.7. At B the estimate (7.3) holds for $\partial u/\partial y$ whereas at A, $\partial u/\partial y \simeq 0$. Therefore, the total change in $\partial u/\partial y$ across the boundary layer is approximately $(U_\infty/\delta) - 0$ and occurs over a distance δ. So, making use of Eqn (7.1), in the general case we obtain

$$\frac{\partial^2 u}{\partial y^2} = \mathcal{O}\left(\frac{U_e}{\delta^2}\right) = \mathcal{O}\left(\frac{U_e}{L^2}\frac{1}{\varepsilon^{2n}}\right) \tag{7.5}$$

In a similar way we deduce that

$$\frac{\partial^2 u}{\partial x^2} = \mathcal{O}\left(\frac{U_e}{L^2}\right) \tag{7.6}$$

We can now use the order-of-magnitude estimates (7.3)–(7.6) to estimate the order of magnitude of each of the terms in the Navier–Stokes equations. We begin with the continuity Eqn (2.93)

$$\underbrace{\frac{\partial u}{\partial x}}_{\mathcal{O}(U_e/L)} + \underbrace{\frac{\partial v}{\partial y}}_{\mathcal{O}\left(v/(\varepsilon^n L)\right)} = 0 \tag{7.7}$$

If both terms are the same order of magnitude we can deduce from Eqn (7.7) that

$$v = \mathcal{O}\left(U_e \frac{\delta}{L}\right) = \mathcal{O}\left(U_e \varepsilon^n\right) \tag{7.8}$$

One might question the assumption that two terms are the same order of magnitude. But, the slope of the streamlines in the boundary layer is equal to v/u by definition and will also be given approximately by δ/L, so Eqn (7.8) is evidently correct.

We will now use Eqns (7.3)–(7.6) and (7.8) to estimate the orders of magnitude of the terms in the Navier–Stokes equations (2.92a,b). We will assume steady flow, ignore the body-force terms, and divide throughout by ρ (noting that the kinematic viscosity $\nu = \mu/\rho$), thus

$$\underbrace{u\frac{\partial u}{\partial x}}_{\mathcal{O}(U_e^2/L)} + \underbrace{v\frac{\partial u}{\partial y}}_{\mathcal{O}(U_e^2/L)} = \underbrace{-\frac{1}{\rho}\frac{\partial p}{\partial x}}_{\mathcal{O}(\text{unknown})} + \underbrace{\nu\left(\frac{\partial^2 u}{\partial x^2}\right)}_{\mathcal{O}(\nu U_e/L^2)=\mathcal{O}(\varepsilon U_e^2/L)} + \underbrace{\nu\left(\frac{\partial^2 u}{\partial y^2}\right)}_{\mathcal{O}(\nu U_e/(L^2\varepsilon^{2n}))=\mathcal{O}(\varepsilon^{1-2n}U_e^2/L)} \tag{7.9}$$

$$\underbrace{u\frac{\partial v}{\partial x}}_{\mathcal{O}(\varepsilon^n U_e^2/L)} + \underbrace{v\frac{\partial v}{\partial y}}_{\mathcal{O}(\varepsilon^n U_e^2/L)} = \underbrace{-\frac{1}{\rho}\frac{\partial p}{\partial y}}_{\mathcal{O}(\text{unknown})} + \underbrace{\nu\left(\frac{\partial^2 v}{\partial x^2}\right)}_{\mathcal{O}(\nu U_e\varepsilon^n/L^2)=\mathcal{O}(\varepsilon^{1+n}U_e^2/L)} + \underbrace{\nu\left(\frac{\partial^2 v}{\partial y^2}\right)}_{\mathcal{O}(\nu U_e/(L^2\varepsilon^n))=\mathcal{O}(\varepsilon^{1-n}U_e^2/L)} \tag{7.10}$$

Now $\varepsilon = 1/Re_L$ is a very small quantity so that a quantity of $\mathcal{O}(\varepsilon U_e^2/L)$ is negligible compared with one of $\mathcal{O}(U_e^2/L)$. It therefore follows that the second term on the right-hand side of Eqn (7.9) can be dropped in comparison with the terms on the left-hand side. What about the third term on the right-hand side of Eqn (7.9)? If $2n = 1$ it will be the same order of magnitude as those on the left-hand side. If $2n < 1$ then this remaining viscous term will be negligible compared with the left-hand side. This cannot be so, because we know that the viscous effects are not negligible within the boundary layer. On the other hand, if $2n > 1$ the terms on the left-hand side of Eqn (7.9) will be negligible in comparison with the remaining viscous term. So, for the flat plate for which $\partial p/\partial x \equiv 0$, Eqn (7.9) reduces to

$$\frac{\partial^2 u}{\partial y^2} = 0$$

This can be readily integrated to give

$$u = f(x)y + g(x)$$

Note that, as it is a partial derivative, arbitrary functions of x, $f(x)$ and $g(x)$, take the place of constants of integration. In order to satisfy the no-slip condition ($u = 0$) at the surface, $y = 0$, $g(x) = 0$, so that $u \propto y$. Evidently this does not conform to the required smooth velocity profile depicted in Figs 7.3 and 7.7. We therefore conclude that the only possiblity that fits the physical requirements is

$$2n = 1 \quad \text{implying} \quad \delta \propto \varepsilon^{1/2} \left(= \frac{1}{\sqrt{Re_L}} \right) \tag{7.11}$$

and Eqn (7.9) simplifies to

$$u \frac{\partial u}{\partial x} + v \frac{\partial u}{\partial y} = -\frac{1}{\rho} \frac{\partial p}{\partial x} + \nu \left(\frac{\partial^2 u}{\partial y^2} \right) \tag{7.12}$$

It is now plain that all the terms in Eqn (7.10) must be $\mathcal{O}(\varepsilon^{1/2} U_e^2 / L)$ or even smaller and are therefore negligibly small compared to the terms retained in Eqn (7.12). We therefore conclude that Eqn (7.10) simplifies drastically to

$$\frac{\partial p}{\partial y} \simeq 0 \tag{7.13}$$

In other words, the pressure does not change across the boundary layer. (In fact, this could be deduced from the fact that the boundary layer is very thin, so that the streamlines are almost parallel with the surface.) This implies that p depends only on x and can be determined in advance from the potential-flow solution. Thus Eqn (7.12) simplifies further to

$$u \frac{\partial u}{\partial x} + v \frac{\partial u}{\partial y} = \underbrace{-\frac{1}{\rho} \frac{dp}{dx}}_{\text{a known function of } x} + \nu \left(\frac{\partial^2 u}{\partial y^2} \right) \tag{7.14}$$

Equation (7.14) plus (7.7) are usually known as the (Prandtl) *boundary-layer equations*.

To sum up, then, the velocity profiles within the boundary layer can be obtained as follows:

(i) Determine the potential flow around the body using the methods described in Chapter 3;
(ii) From this potential-flow solution determine the pressure and the velocity along the surface;
(iii) Solve equations (7.7) and (7.14) subject to the boundary conditions

$$u = v = 0 \quad \text{at} \quad y = 0; \qquad u = U_e \quad \text{at} \quad y = \delta \text{ (or } \infty) \tag{7.15}$$

The boundary condition, $u = 0$, is usually referred to as the *no-slip condition* because it implies that the fluid adjacent to the surface must stick to it. Explanations can be offered for why this should be so, but fundamentally it is an empirical observation. The second boundary condition, $v = 0$, is referred to as the *no-penetration condition* because it states that fluid cannot pass into the wall. Plainly, it will not hold when the surface is porous, as with boundary-layer suction (see Section 8.4.1). The third boundary condition (7.15) is applied at the boundary-layer edge where it requires the flow velocity to be equal to the potential-flow solution. For the approximate methods described in Section 7.7, one usually applies it at $y = \delta$. For accurate solutions of the boundary-layer equations, however, no clear edge can be defined;

the velocity profile is such that u approaches ever closer to U_e the larger y becomes. Thus for accurate solutions one usually chooses to apply the boundary condition at $y = \infty$, although it is commonly necessary to choose a large finite value of y for seeking computational solutions.

7.3.2 Various definitions of boundary-layer thickness

In the course of deriving the boundary-layer equations we have shown in Eqn (7.11) how the boundary-layer thickness varies with Reynolds number. This is another example of obtaining useful practical information from an equation without needing to solve it. Its practical use will be illustrated later in Example 7.1. Notwithstanding such practical applications, however, we have already seen that the boundary-layer thickness is rather an imprecise concept. It is difficult to give it a precise numerical value. In order to do so in Section 7.2.2 it was necessary, rather arbitrarily, to identify the edge of the boundary layer as corresponding to the point where $u = 0.99 U_e$. Partly owing to this rather unsatisfactory vagueness, several more precise definitions of boundary-layer thickness are given below. As will become plain, each definition also has a useful and significant physical interpretation relating to boundary-layer characteristics.

Displacement thickness (δ^*)

Consider the flow past a flat plate (Fig. 7.8a). Owing to the build-up of the boundary layer on the plate surface a stream tube that, at the leading edge, is close to the surface will become entrained into the boundary layer. As a result the mass flow in the streamtube will decrease from ρU_e, in the main stream, to some value ρu, and – to satisfy continuity – the tube cross-section will increase. In the two-dimensional flows considered here, this means that the widths, normal to the plate surface, of the boundary-layer stream tubes will increase, and stream tubes that are in the main-stream will be displaced slightly away from the surface. The effect on the mainstream flow will then be as if, with no boundary layer present, the solid surface had been displaced a small distance into the stream. The amount by which the surface would be displaced under such conditions is termed the boundary-layer displacement thickness (δ^*) and may be calculated as follows, provided the velocity profile $\bar{u} = f(\bar{y})$ (see Fig. 7.3) is known.

At station x (Fig. 7.8c), owing to the presence of the boundary layer, the mass flow rate is reduced by an amount equal to

$$\int_0^\infty (\rho U_e - \rho u)\,\mathrm{d}y$$

corresponding to area OABR. This must equate to the mass flow rate deficiency that would occur at uniform density ρ and velocity U_e through the thickness δ^*, corresponding to area OPQR. Equating these mass flow rate deficiencies gives

$$\int_0^\infty (\rho U_e - \rho u)\,\mathrm{d}y = \rho U_e \delta^*$$

i.e.

$$\delta^* = \int_0^\infty \left(1 - \frac{u}{U_e}\right)\mathrm{d}y \tag{7.16}$$

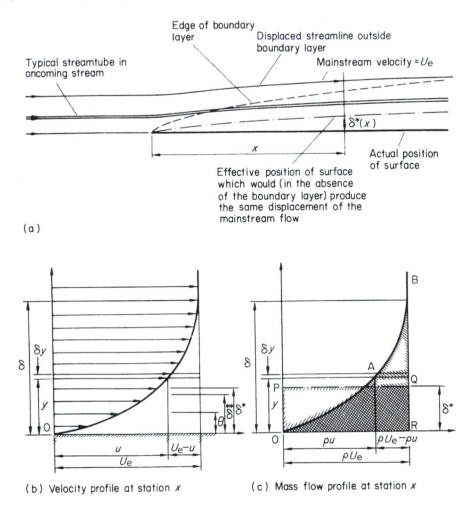

(a)

(b) Velocity profile at station x

(c) Mass flow profile at station x

Fig. 7.8

The idea of displacement thickness has been put forward here on the basis of two-dimensional flow past a flat plate purely so that the concept may be considered in its simplest form. The above definition may be used for any incompressible two-dimensional boundary layer without restriction and will also be largely true for boundary layers over three-dimensional bodies provided the curvature, in planes normal to the free-stream direction, is not large, i.e. the local radius of curvature should be much greater than the boundary-layer thickness. If the curvature is large a displacement thickness may still be defined but the form of Eqn (7.16) will be slightly modified. An example of the use of displacement thickness will be found later in this chapter (Examples 7.2 and 7.3).

Similar arguments to those given above will be used below to define other boundary-layer thicknesses, using either momentum flow rates or energy flow rates.

Momentum thickness (θ)

This is defined in relation to the momentum flow rate within the boundary layer. This rate is less than that which would occur if no boundary layer existed, when the velocity in the vicinity of the surface, at the station considered, would be equal to the mainstream velocity U_e.

For the typical streamtube within the boundary layer (Fig. 7.8b) the rate of momentum defect (relative to mainstream) is $\rho u(U_e - u)\delta y$. Note that the mass flow rate ρu actually within the stream tube must be used here, the momentum defect of this mass being the difference between its momentum based on mainstream velocity and its actual momentum at position x within the boundary layer.

The rate of momentum defect for the thickness θ (the distance through which the surface would have to be displaced in order that, with no boundary layer, the total flow momentum at the station considered would be the same as that actually occurring) is given by $\rho U_e^2 \theta$. Thus:

$$\int_0^\infty \rho u(U_e - u)\mathrm{d}y = \rho U_e^2 \theta$$

i.e.

$$\theta = \int_0^\infty \frac{u}{U_e}\left(1 - \frac{u}{U_e}\right)\mathrm{d}y \qquad (7.17)$$

The momentum thickness concept is used in the calculation of skin friction losses.

Kinetic energy thickness (δ**)

This quantity is defined with reference to kinetic energies of the fluid in a manner comparable with the momentum thickness. The rate of kinetic-energy defect within the boundary layer at any station x is given by the difference between the energy that the element would have at main-stream velocity U_e and that it actually has at velocity u, being equal to

$$\int_0^\infty \frac{1}{2}\rho u(U_e^2 - u^2)\mathrm{d}y$$

while the rate of kinetic-energy defect in the thickness δ^{**} is $\frac{1}{2}\rho U_e^3 \delta^{**}$. Thus

$$\int_0^\infty \rho u(U_e^2 - u^2)\mathrm{d}y = \rho U_e^2 \delta^{**}$$

i.e.

$$\delta^{**} = \int_0^\infty \frac{u}{U_e}\left[1 - \left(\frac{u}{U_e}\right)^2\right]\mathrm{d}y \qquad (7.18)$$

7.3.3 Skin friction drag

The shear stress between adjacent layers of fluid in a laminar flow is given by $\tau = \mu(\partial u/\partial y)$ where $\partial u/\partial y$ is the transverse velocity gradient. Adjacent to the solid surface at the base of the boundary layer, the shear stress in the fluid is due entirely to viscosity and is given by $\mu(\partial u/\partial y)_w$. This statement is true for both laminar and turbulent boundary layers because, as discussed in Section 7.2.4, a viscous sublayer exists at the surface even if the main boundary-layer flow is turbulent. The shear stress in the fluid layer in contact with the surface is essentially the same as the shear stress between that layer and the surface so that for all boundary layers the shear stress at the wall, due to the presence of the boundary layer, is given by

$$\tau_w = \mu \left(\frac{\partial u}{\partial y} \right)_w \tag{7.19}$$

where τ_w is the wall shear stress or surface friction stress, usually known as the skin friction.

Once the velocity profile (laminar or turbulent) of the boundary layer is known, then the surface (or skin) friction can be calculated. The skin-friction stress can be defined in terms of a non-dimensional local skin-friction coefficient, C_f, as follows.

$$\tau_w = C_f \frac{1}{2} \rho U_e^2 \tag{7.20}$$

Of particular interest is the total skin-friction force F on the surface under consideration. This force is obtained by integrating the skin-friction stress over the surface. For a two-dimensional flow, the force F per unit width of surface may be evaluated, with reference to Fig. 7.9, as follows. The skin-friction force per unit width on an elemental length (δx) of surface is

$$\delta F = \tau_w \delta x$$

Therefore the total skin-friction force per unit width on length L is

$$F = \int_0^L \tau_w \mathrm{d}x \tag{7.21}$$

The skin-friction force F may be expressed in terms of a non-dimensional coefficient C_F, defined by

$$F = C_F \frac{1}{2} \rho U_\infty^2 S_w \tag{7.22}$$

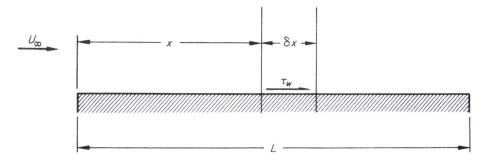

Fig. 7.9

where S_w is the wetted area of the surface under consideration. Similarly for a flat plate or aerofoil section, the total skin-friction drag coefficient C_{D_F} is defined by

$$D_F = C_{D_F} \frac{1}{2} \rho U_\infty^2 S \tag{7.23}$$

where D_F = total skin-friction force on both surfaces resolved in the direction of the free stream, and S = plan area of plate or aerofoil. For a flat plate or symmetrical aerofoil section, at zero incidence, when the top and bottom surfaces behave identically, $D_F = 2F$ and $S = S_w$ (the wetted area for each surface). Thus

$$C_{D_F} = \frac{2F}{\frac{1}{2}\rho U_\infty^2 S} = 2C_F \tag{7.24}$$

When flat-plate flows (at constant pressure) are considered, $U_e = U_\infty$. Except where a general definition is involved, U_e will be used throughout.

Subject to the above condition, use of Eqns (7.20) and (7.22) in Eqn (7.21) leads to

$$C_F = \int_0^1 C_f \, d\left(\frac{x}{L}\right) \tag{7.25}$$

Equation (7.25) is strictly applicable to a flat plate only, but on a slim aerofoil, for which U_e does not vary greatly from U_∞ over most of the surface, the expression will give a good approximation to C_F.

We have seen in Eqn (7.11) how the boundary-layer thickness varies with Reynolds number. This result can also be used to show how skin friction and skin-friction drag varies with Re_L. Using the order-of-magnitude estimate (7.3) it can be seen that

$$\tau_w = \mu \frac{\partial u}{\partial y} \propto \frac{\mu U_e}{\delta} \propto \frac{\mu U_\infty}{L} \sqrt{Re_L}$$

But, by definition, $Re_L = \rho U_\infty L / \mu$, so the above becomes

$$\tau_w \propto \rho U_\infty^2 \frac{\mu}{\rho U_\infty L} \sqrt{Re_L} = \rho U_\infty^2 \frac{1}{\sqrt{Re_L}} \tag{7.26}$$

It therefore follows from Eqns (7.20) and (7.25) that the relationships between the coefficients of skin-friction and skin-friction drag and Reynolds number are identical and given by

$$C_f \propto \frac{1}{\sqrt{Re_L}} \qquad \text{and} \qquad C_{D_f} \propto \frac{1}{\sqrt{Re_L}} \tag{7.27}$$

Example 7.1   Some engineers wish to obtain a good estimate of the drag and boundary-layer thickness at the trailing edge of a miniature wing. The chord and span of the wing are 6 mm and 30 mm respectively. A typical flight speed is 5 m/s in air (kinematic viscosity $= 15 \times 10^{-6}$ m^2/s, density $= 1.2$ kg/m^3). They decide to make a superscale model with chord and span of 150 mm and 750 mm respectively. Measurements on the model in a water channel flowing at 0.5 m/s (kinematic viscosity $= 1 \times 10^{-6}$ m^2/s, density $= 1000$ kg/m^3) gave a drag of 0.19 N and a boundary-layer thickness of 3 mm. Estimate the corresponding values for the prototype.

The Reynolds numbers of model and prototype are given by

$$(Re_L)_m = \frac{0.15 \times 0.5}{1 \times 10^{-6}} = 75\,000 \qquad \text{and} \qquad (Re_L)_p = \frac{0.006 \times 5}{15 \times 10^{-6}} = 2000$$

Evidently, the Reynolds numbers are not the same for model and prototype, so the flows are not dynamically similar. But, as a streamlined body is involved, we can use Eqns (7.11) and (7.27). From Eqn (7.11)

$$\delta_p = \delta_m \times \frac{L_p}{L_m} \times \left(\frac{(Re_L)_m}{(Re_L)_p}\right)^{1/2}$$

$$= 3 \times \frac{6}{150} \times \left(\frac{75\,000}{2000}\right)^{1/2} = 0.735\,\text{mm} = 735\,\mu\text{m}$$

and from Eqn (7.27)

$$(C_{Df})_p = (C_{Df})_m \times \left(\frac{(Re_L)_m}{(Re_L)_p}\right)^{1/2}$$

But $D_f = \frac{1}{2}\rho U_\infty^2 S C_{Df}$. So, if we assume that skin-friction drag is the dominant type of drag and that it scales in the same way as the total drag, the prototype drag is given by

$$D_p = D_m \frac{(\rho U_\infty^2 S)_p}{(\rho U_\infty^2 S)_m} \left(\frac{(Re_L)_m}{(Re_L)_p}\right)^{1/2} = 0.19 \times \frac{1.2 \times 5^2 \times 6 \times 30}{1000 \times 0.5^2 \times 150 \times 750} \times \left(\frac{75\,000}{2000}\right)^{1/2}$$

$$= 0.00022\,\text{N} = 220\,\mu\text{N}$$

7.3.4 Solution of the boundary-layer equations for a flat plate

There are a few special exact solutions of the boundary-layer equations (7.7) and (7.14). The one for the boundary layer in the vicinity of a stagnation point is an exact solution of the Navier–Stokes equations and was described in Section 2.10.3. In this case, we saw that an exact solution was interpreted as meaning that the governing equations are reduced to one or more ordinary differential equations. This same interpretation carries over to the boundary-layer equations. The most famous, and probably most useful, example is the solution for the boundary layer over a flat plate (see Fig. 7.7). This was first derived by one of Prandtl's PhD students, Blasius, in 1908.

A useful starting point is to introduce the stream function, ψ, (see Section 2.5 and Eqns 2.56a,b) such that

$$u = \frac{\partial \psi}{\partial y}, \qquad v = -\frac{\partial \psi}{\partial x} \tag{7.28a,b}$$

This automatically satisfies Eqn (7.7), reducing the boundary-layer equations to a single equation (7.14) that takes the form:

$$\frac{\partial \psi}{\partial y}\frac{\partial^2 \psi}{\partial x \partial y} - \frac{\partial \psi}{\partial x}\frac{\partial^2 \psi}{\partial y^2} = -\frac{1}{\rho}\frac{dp}{dx} + \nu\frac{\partial^3 \psi}{\partial y^3} \tag{7.29}$$

For the flat plate $dp/dx = 0$.

Consider the hypothetical case of an infinitely long flat plate. For practical application we can always assume that the boundary layer at a point $x = L$, say, on an infinitely long plate is identical to that at the trailing edge of a flat plate of length L. But, if the hypothetical plate is infinitely long, we cannot use its length as a reference dimension. In fact, the only length dimension available is ν/U_∞. This strongly suggests that the boundary layer at a point x_1, say, will be identical to that at another point x_2, except that the boundary-layer thicknesses will differ. Accordingly we propose that, as suggested in Fig. 7.7, the velocity profile does not change shape as the boundary layer develops along the plate. That is, we can write

$$\frac{u}{U_\infty} = fn\left(\frac{y}{\delta}\right) \quad \text{only} \tag{7.30}$$

If we replace L by x in Eqn (7.11) we deduce that

$$\frac{\delta}{x} \propto \frac{1}{\sqrt{Re_x}} \quad \text{implying} \quad \delta \propto x^{1/2}$$

So that Eqn (7.30) can be written

$$\frac{u}{U_\infty} = fn(\eta); \qquad \eta \propto \frac{y}{\delta} = \frac{ay}{x^{1/2}}, \quad \text{say} \tag{7.31}$$

Let

$$\psi = bx^m f(\eta) \tag{7.32}$$

where b and m will be determined below.

We now tranform the independent variables

$$(x, y) \quad \longrightarrow \quad \left(\xi = x, \eta = \frac{ay}{x^{1/2}}\right)$$

so that

$$\left(\frac{\partial \psi}{\partial x}\right)_y = \underbrace{\left(\frac{\partial \xi}{\partial x}\right)_y}_{=1} \left(\frac{\partial \psi}{\partial \xi}\right)_\eta + \underbrace{\left(\frac{\partial \eta}{\partial x}\right)_y}_{=-\frac{1}{2}ayx^{-3/2}=-\frac{1}{2}\eta/\xi} \times \left(\frac{\partial \psi}{\partial \eta}\right)_\xi = \left(\frac{\partial \psi}{\partial \xi}\right)_\eta - \frac{\eta}{2\xi}\left(\frac{\partial \psi}{\partial \eta}\right)_\xi \tag{7.33}$$

$$\left(\frac{\partial \psi}{\partial y}\right)_x = \underbrace{\left(\frac{\partial \xi}{\partial y}\right)_x}_{=0} \left(\frac{\partial \psi}{\partial \xi}\right)_\eta + \underbrace{\left(\frac{\partial \eta}{\partial y}\right)_x}_{=a/x^{1/2}=a/\xi^{1/2}} \times \left(\frac{\partial \psi}{\partial \eta}\right)_\xi = a\xi^{-1/2}\left(\frac{\partial \psi}{\partial \eta}\right)_\xi \tag{7.34}$$

From Eqns (7.28a), (7.32) and (7.34) we obtain

$$u = a\xi^{-\frac{1}{2}}\frac{\partial}{\partial \eta}\left(bx^m f(\eta)\right) = abx^{m-\frac{1}{2}}\frac{df}{d\eta}$$

Thus if the form Eqn (7.31) is to hold we must require $ab = U_\infty$ and $m = 1/2$, therefore

$$\psi = \frac{U_\infty}{a}\xi^{\frac{1}{2}}f(\eta) \tag{7.35}$$

Thus

$$u = \frac{\partial \psi}{\partial y} = U_\infty \frac{df}{d\eta} \tag{7.36}$$

Substituting Eqn (7.35) into (7.33) gives

$$\left(\frac{\partial \psi}{\partial x}\right)_y = \frac{U_\infty}{2a}\xi^{-1/2}f(\eta) - \frac{U_\infty \eta}{2a}\xi^{-1/2}\frac{df}{d\eta} \tag{7.37}$$

Likewise, if we replace ψ by $\partial\psi/\partial y$ in Eqn (7.33) and make use of Eqn (7.36), we obtain

$$\frac{\partial^2\psi}{\partial x\partial y} = \frac{\partial}{\partial x}\left(\frac{\partial\psi}{\partial y}\right) = \frac{\partial}{\partial\xi}\left(\frac{\partial\psi}{\partial y}\right) - \frac{\eta}{2\xi}\frac{\partial}{\partial\eta}\left(\frac{\partial\psi}{\partial y}\right) = 0 - \frac{\eta U_\infty}{2\xi}\frac{\mathrm{d}^2 f}{\mathrm{d}\eta^2} \tag{7.38}$$

Similarly using Eqn (7.34) and (7.36)

$$\frac{\partial^2\psi}{\partial y^2} = \frac{\partial}{\partial y}\left(\frac{\partial\psi}{\partial y}\right) = \frac{\partial}{\partial y}\left(U_\infty\frac{\mathrm{d}f}{\mathrm{d}\eta}\right) = a\xi^{-1/2}U_\infty\frac{\mathrm{d}^2 f}{\mathrm{d}\eta^2} \tag{7.39}$$

$$\frac{\partial^3\psi}{\partial y^3} = a^2\xi^{-1}U_\infty\frac{\mathrm{d}^3 f}{\mathrm{d}\eta^3} \tag{7.40}$$

We now substitute Eqns (7.36)–(7.40) into Eqn (7.29) to obtain

$$U_\infty\frac{\mathrm{d}f}{\mathrm{d}\eta}\left(-\frac{\eta}{2\xi}\frac{\mathrm{d}^2 f}{\mathrm{d}\eta^2}\right) - \frac{U_\infty\xi^{-1/2}}{2a}\left(f - \eta\frac{\mathrm{d}f}{\mathrm{d}\eta}\right)a\xi^{-1/2}U_\infty\frac{\mathrm{d}^2 f}{\mathrm{d}\eta^2} = \nu a^2\xi^{-1}U_\infty\frac{\mathrm{d}^3 f}{\mathrm{d}\eta^3}$$

After cancelling like terms, this simplifies to

$$-\frac{U_\infty^2}{2}\xi^{-1}f\frac{\mathrm{d}^2 f}{\mathrm{d}\eta^2} = \nu a^2\xi^{-1}U_\infty\frac{\mathrm{d}^3 f}{\mathrm{d}\eta^3}$$

Then cancelling common factors and rearranging leads to

$$\frac{\mathrm{d}^3 f}{\mathrm{d}\eta^3} + \underbrace{\frac{U_\infty}{2\nu a^2}}_{=1,\text{ say}}f\frac{\mathrm{d}^2 f}{\mathrm{d}\eta^2} = 0 \tag{7.41}$$

As suggested, if we wish to obtain the simplest universal (i.e. independent of the values of U_∞ and ν) form of Eqn (7.41), we should set

$$\frac{U_\infty}{2\nu a^2} = 1 \quad\text{implying}\quad a = \sqrt{\frac{U_\infty}{2\nu}} \tag{7.42}$$

So that Eqn (7.41) reduces to

$$\frac{\mathrm{d}^3 f}{\mathrm{d}\eta^3} + f\frac{\mathrm{d}^2 f}{\mathrm{d}\eta^2} = 0; \qquad \eta = y\sqrt{\frac{U_\infty}{2\nu x}} \tag{7.43}$$

The boundary conditions (7.15) become

$$f = \frac{\mathrm{d}f}{\mathrm{d}\eta} = 0 \quad\text{at } \eta = 0; \quad f \to 1 \quad\text{as}\quad \eta \to \infty \tag{7.44}$$

The ordinary differential equation can be solved numerically for f. The velocity profile $\mathrm{d}f/\mathrm{d}\eta$ thus obtained is plotted in Fig. 7.10 (see also Fig. 7.11).

From this solution the various boundary-layer thicknesses given in Section 7.3.2 can be obtained by evaluating the integrals numerically in the forms:

$$\delta_{0.99} \simeq 5.0\sqrt{\frac{\nu x}{U_\infty}} \tag{7.45}$$

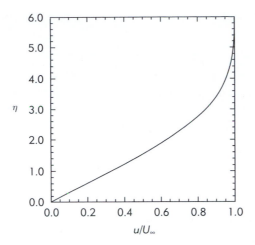

Fig. 7.10

Displacement thickness

$$\delta^* = \int_0^\infty \left(1 - \frac{u}{U_\infty}\right) dy = \int_0^\infty \left(1 - \frac{u}{U_\infty}\right) \underbrace{\frac{dy}{d\eta}}_{\sqrt{\frac{2\nu x}{U_\infty}}} d\eta$$

$$= \sqrt{\frac{2\nu x}{U_\infty}} \underbrace{\int_0^\infty \left(1 - \frac{df}{d\eta}\right) d\eta}_{1.7208} = 1.7208 \sqrt{\frac{\nu x}{U_\infty}} \qquad (7.46)$$

Momentum thickness

$$\theta = \int_0^\infty \frac{u}{U_\infty}\left(1 - \frac{u}{U_\infty}\right) dy = \sqrt{\frac{2\nu x}{U_\infty}} \int_0^\infty \frac{df}{d\eta}\left(1 - \frac{df}{d\eta}\right) d\eta = 0.664 \sqrt{\frac{\nu x}{U_\infty}} \qquad (7.47)$$

Energy thickness

$$\theta = \int_0^\infty \frac{u}{U_\infty}\left(1 - \frac{u^2}{U_\infty^2}\right) dy = \sqrt{\frac{2\nu x}{U_\infty}} \int_0^\infty \frac{df}{d\eta}\left\{1 - \left(\frac{df}{d\eta}\right)^2\right\} d\eta = 1.0444 \sqrt{\frac{\nu x}{U_\infty}} \qquad (7.48)$$

The local wall shear stress and hence the skin-friction drag can also be calculated readily from function $f(\eta)$:

$$\tau_w(x) = \mu \left(\frac{\partial u}{\partial y}\right)_{y=0} = \mu \underbrace{\frac{\partial \eta}{\partial y}}_{\sqrt{\frac{U_\infty}{2\nu x}}} \times \underbrace{\left(\frac{\partial u}{\partial \eta}\right)_{\eta=0}}_{U_\infty \left(\frac{d^2 f}{d\eta^2}\right)_{\eta=0}}$$

$$= \mu U_\infty \sqrt{\frac{U_\infty}{2\nu x}} \underbrace{\left(\frac{d^2 f}{d\eta^2}\right)_{\eta=0}}_{0.332\sqrt{2}} = 0.332\, \mu U_\infty \sqrt{\frac{U_\infty}{\nu x}} \qquad (7.49)$$

Fig. 7.11 Velocity profile in a boundary layer: The velocity profile that forms in a boundary layer along a flat wall is made visible by lines of aluminium powder dropped from a trough on to the flowing fluid surface. The fluid is a dilute solution of wallpaper paste in water. The Reynolds number based on distance along the wall is about 50 000. See also Fig. 7.3 on page 376. (*The photograph was taken by D.J. Buckingham at the School of Engineering, University of Exeter, UK.*)

Thus the skin-friction coefficient is given by

$$C_f(x) = \frac{\tau_w(x)}{\frac{1}{2}\rho U_\infty^2} = \frac{0.664}{\sqrt{Re_x}} \quad \text{where} \quad Re_x = \frac{\rho U_\infty x}{\mu} \tag{7.50}$$

The drag of one side of the plate (spanwise breadth B and length L) is given by

$$D_F = B \int_0^L \tau_w(x)\,\mathrm{d}x \tag{7.51}$$

Thus combining Eqns (7.49) and (7.51) we find that the drag of one side of the plate is given by

$$D_F = 0.332\,\mu B U_\infty \sqrt{\frac{U_\infty}{\nu}} \int_0^L \frac{\mathrm{d}x}{\sqrt{x}} = 0.664\,\mu B U_\infty \sqrt{\frac{L U_\infty}{\nu}} = 0.664\,\mu B U_\infty \sqrt{Re_L} \tag{7.52}$$

Hence the coefficient of skin-friction drag is given by

$$C_{DF} = \frac{D_F}{\frac{1}{2}\rho U_\infty^2 BL} = \frac{1.328}{\sqrt{Re_L}} \tag{7.53}$$

Example 7.2 The Blasius solution for the laminar boundary layer over a flat plate will be used to estimate the boundary-layer thickness and skin-friction drag for the miniature wing of Example 7.1.

The Reynolds number based on length $Re_L = 2000$, so according to Eqns (7.45) and (7.46) the boundary-layer thicknesses at the trailing edge are given by

$$\delta_{0.99} \simeq \frac{5.0L}{Re_L^{1/2}} = \frac{5 \times 6}{\sqrt{2000}} = 0.67\,\text{mm} \qquad \delta^* = \frac{1.7208}{5} \times 0.67 = 0.23\,\text{mm}$$

Remembering that the wing has two sides, an estimate for its skin-friction drag is given by

$$D_F = 2 \times C_{DF} \times \frac{1}{2} U_\infty^2 BL = \frac{1.328}{\sqrt{2000}} \times 1.2 \times 5^2 \times 30 \times 6 \times 10^{-6} = 160\,\mu\text{N}$$

7.3.5 Solution for the general case

The solution of the boundary-layer equations for the flat plate described in Section 7.3.4 is a very special case. Although other *similarity* solutions exist (i.e. cases where the boundary-layer equations reduce to an ordinary differential equation), they are of limited practical value. In general, it is necessary to solve Eqns (7.7) and (7.14) or, equivalently Eqn (7.29), as partial differential equations.

To fix ideas, consider the flow over an aerofoil, as shown in Fig. 7.12. Note that the boundary-layer thickness is greatly exaggerated. The first step is to determine the potential flow around the aerofoil. This would be done computationally by using the panel method described in Section 3.6 for non-lifting aerofoils or Section 4.10 in the case where lift is generated. From this solution for the potential flow the velocity U_e along the surface of the aerofoil can be determined. This will be assumed to be the velocity at the edge of the boundary layer. The location of the fore stagnation point F can also be determined from the solution for U_e. Plainly it corresponds to $U_e = 0$. (For the non-lifting case of a symmetric aerofoil at zero angle of attack the location of the fore stagnation point will be known in advance from symmetry). This point corresponds to $x = 0$. And the development of the boundary layers over the top and bottom of the aerofoil have to be calculated separately, unless they are identical, as in symmetric aerofoils at zero incidence.

Mathematically, the boundary-layer equations are *parabolic*. This means that their solution (i.e. the boundary-layer velocity profile) at an arbitrary point P_1, say, (where $x = x_1$) on the aerofoil depends only on the solutions upstream, i.e. at $x < x_1$. This property allows special efficient numerical methods to be used whereby one begins with the solution at the fore stagnation point and marches step by step around the aerofoil, solving the boundary-layer equations at each value of x in turn. This is very much easier than solving the Navier–Stokes equations that in subsonic steady flow are *elliptic* equations like the Laplace equation. The term elliptic implies that the solution (i.e. the velocity field) at a particular point depends on the solutions at all other points. For elliptic equations the flow field upstream does depend on conditions downstream. How else would the flow approaching the aerofoil sense its presence and begin gradually to deflect from uniform flow in order to flow smoothly

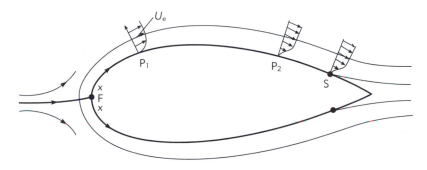

Fig. 7.12 The boundary layer developing around an aerofoil

around the aerofoil? Nevertheless, numerical solution of the boundary-layer equations is not particularly simple. In order to avoid numerical instability, so-called implicit methods are usually required. These are largely beyond the scope of the present book, but are described in a simple treatment given below in Section 7.11.3.

For aerofoils and other bodies with rounded leading edges, the stagnation flow field determined in Section 2.10.3 gives the initial boundary-layer velocity profile in the vicinity of $x = 0$. The velocity U_e along the edge of the boundary layer increases rapidly away from the fore stagnation point, F. The evolving velocity profile in the boundary layer is found by solving the boundary-layer equations step by step by 'marching' around the surface of the aerofoil. At some point U_e will reach a maximum at the point of minimum pressure. From this point onwards the pressure gradient along the surface will change sign to become adverse and begin to slow down the boundary-layer flow (as explained above in Section 7.2.6 and below in Section 7.4). A point of inflexion develops in the velocity profile (e.g. at point P_2 in Fig. 7.12, see also Fig. 7.6) that moves towards the wall as x increases. Eventually, the inflexion point reaches the wall itself, the shear stress at the wall falls to zero, reverse flow occurs (see Fig. 7.6), and the boundary layer separates from the surface of the aerofoil at point S. The boundary-layer equations cease to be valid just before separation (where $\tau_w = \mu(\partial u/\partial y)_w = 0$) and the calculation is terminated.

Overall the same procedures are involved when using the approximate methods described in Section 7.11 below. There a more detailed account of the computation of the boundary layer around an aerofoil will be presented.

7.4   Boundary-layer separation

The behaviour of a boundary layer in a positive pressure gradient, i.e. pressure increasing with distance downstream, may be considered with reference to Fig. 7.13. This shows a length of surface that has a gradual but steady convex curvature, such as the surface of an aerofoil beyond the point of maximum thickness. In such a flow region, because of the retardation of the mainstream flow, the pressure in the mainstream will rise (Bernoulli's equation). The variation in pressure along a normal to the surface through the boundary-layer thickness is essentially zero, so that the pressure at any point in the mainstream, adjacent to the edge of the boundary layer, is transmitted unaltered through the layer to the surface. In the light of this, consider the small element of fluid (Fig. 7.13) marked ABCD. On face AC, the pressure is p, while on face BD the pressure has increased to $p + (\partial p/\partial x)\delta x$. Thus the net pressure force on the element is tending to retard the flow. This retarding force is in addition to the viscous shears that act along AB and CD and it will continuously slow the element down as it progresses downstream.

This slowing-down effect will be more pronounced near the surface where the elements are more remote from the accelerating effect, via shearing actions, of the mainstream, so that successive profile shapes in the streamwise direction will change in the manner shown.

Ultimately, at a point S on the surface, the velocity gradient $(\partial u/\partial y)_w$ becomes zero. Apart from the change in shape of the profile it is evident that the boundary layer must thicken rapidly under these conditions, in order to satisfy continuity within the boundary layer. Downstream of point S, the flow adjacent to the surface will be in an upstream direction, so that a circulatory movement, in a plane normal to the surface, takes place near the surface. A line (shown dotted in Fig. 7.13) may be drawn from the point S such that the mass flow above this

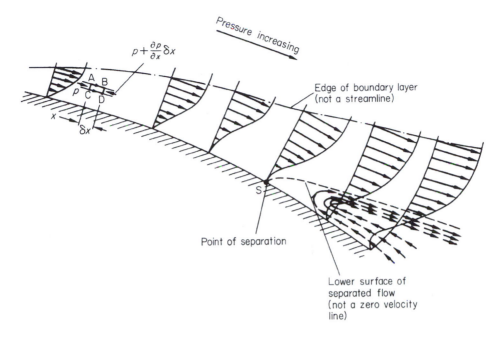

Fig. 7.13

line corresponds to the mass flow ahead of point S. The line represents the continuation of the lower surface of the upstream boundary layer, so that, in effect, the original boundary layer separates from the surface at point S. This is termed the separation point.

Reference to the velocity profiles for laminar and turbulent layers (Fig. 7.4) will make it clear that, owing to the greater extent of lower-energy fluid near the surface in the laminar boundary layer, the effect of a positive pressure gradient will cause separation of the flow much more rapidly than if the flow were turbulent. A turbulent boundary layer is said to stick to the surface better than a laminar one.

The result of separation on the rear half of an aerofoil is to increase the thickness of the wake flow, with a consequent reduction in the pressure rise that should occur near the trailing edge. This pressure rise means that the forward-acting pressure force components on the rear part of the aerofoil do not develop to offset the rearward-acting pressures near the front stagnation point, in consequence the pressure drag of the aerofoil increases. In fact, if there were no boundary layers, there would be a stagnation point at the trailing edge and the boundary-layer pressure drag, as well as the skin-friction drag, would be zero. If the aerofoil incidence is sufficiently large, the separation may take place not far downstream of the maximum suction point, and a very large wake will develop. This will cause such a marked redistribution of the flow over the aerofoil that the large area of low pressure near the upper-surface leading edge is seriously reduced, with the result that the lift force is also greatly reduced. This condition is referred to as the *stall*. A negative pressure gradient will obviously have the reverse effect, since the streamwise pressure forces will cause energy to be added to the slower-moving air near the surface, decreasing any tendency for the layer adjacent to the surface to come to rest.

7.4.1 Separation bubbles

On many aerofoils with relatively large upper-surface curvatures, high local curvature over the forward part of the chord may initiate a laminar separation when the aerofoil is at quite a moderate angle of incidence (Fig. 7.14).

Small disturbances grow much more readily and at low Reynolds numbers in separated, as compared to attached, boundary layers. Consequently, the separated laminar boundary layer may well undergo transition to turbulence with characteristic rapid thickening. This rapid thickening may be sufficient for the lower edge of the, now-turbulent, shear layer to come back into contact with the surface and re-attach as a turbulent boundary layer on the surface. In this way, a bubble of fluid is trapped under the separated shear layer between the separation and re-attachment points. Within the bubble, the boundary of which is usually taken to be the streamline that leaves the surface at the separation point, two regimes exist. In the upstream region a pocket of stagnant fluid at constant pressure extends back some way and behind this a circulatory motion develops as shown in Fig. 7.14, the pressure in this latter region increasing rapidly towards the re-attachment point.

Two distinct types of bubble are observed to occur:

(i) a short bubble of the order of 1 per cent of the chord in length (or 100 separation-point displacement thicknesses*) that exerts negligible effect on the peak suction value just ahead of the bubble.

(ii) a long bubble that may be of almost any length from a few per cent of the chord (10 000 separation displacement thicknesses) up to almost the entire chord, which exerts a large effect on the value of the peak suction near the aerofoil leading edge.

It has been found that a useful criterion, as to whether a short or long bubble is formed, is the value at the separation point of the displacement-thickness Reynolds number $Re_{\delta^*} = U_e \delta^* / \nu$. If $Re_{\delta^*} < 400$ then a long bubble will almost certainly form,

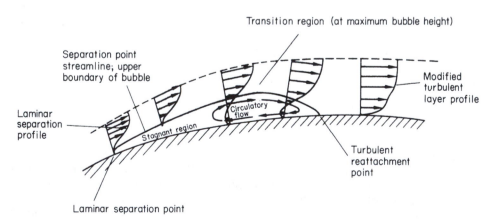

Fig. 7.14

---

* Displacement thickness δ^* is defined in Section 7.3.2.

while for values >550 a short bubble is almost certain. In between these values either type may occur. This is the Owen–Klanfer* criterion.

Short bubbles exert very little influence on the pressure distribution over the aerofoil surface and remain small, with increasing incidence, right up to the stall. They will, in general, move slowly forward along the upper surface as incidence is increased. The final stall may be caused by forward movement of the rear turbulent separation point (trailing-edge stall) or by breakdown of the small bubble at the leading edge owing to failure, at high incidence, of the separated shear flow to re-attach (leading-edge stall).

If a long bubble forms at moderate incidence, its length will rapidly increase with increasing incidence, causing a continuous reduction of the leading-edge suction peak. The bubble may ultimately extend right to the trailing edge or even into the wake downstream, and this condition results in a low lift coefficient and effective stalling of the aerofoil. This type of progressive stall usually occurs with thin aerofoils and is often referred to as thin-aerofoil stall. There are thus three alternative mechanisms that may produce subsonic stalling of aerofoil sections.

7.5   Flow past cylinders and spheres

Some of the properties of boundary layers discussed above help in the explanation of the behaviour, under certain conditions, of a cylinder or sphere immersed in a uniform free stream. So far discussion has been restricted to the flow over bodies of reasonably streamline form, behind which a relatively thin wake is formed. In such cases, the drag forces are largely due to surface friction, i.e. to shear stresses at the base of the boundary layer. When dealing with non-streamlined or bluff bodies, it is found that, because of the adverse effect of a positive pressure gradient on the boundary layer, the flow usually separates somewhere near points at the maximum cross-section, with the formation of a broad wake. As a result, the skin-friction drag is only small, and the major part of the total drag now consists of form drag due to the large area at the rear of the body acted upon by a reduced pressure in the wake region. Experimental observation of the flow past a sphere or cylinder indicates that the drag of the body is markedly influenced by the cross-sectional area of the wake, a broad wake being accompanied by a relatively high drag and vice versa.

The way in which the flow pattern around a bluff body can change dramatically as the Reynolds number is varied may be considered with reference to the flow past a circular cylinder. For the most part the flow past a sphere also behaves in a similar way. At very low Reynolds number,[†] i.e. less than unity, the flow behaves as if it were purely viscous with negligible inertia. Such flow is known as *creeping* or *Stokes* flow. For such flows there are no boundary layers and the effects of viscosity extend an infinite distance from the body. The streamlines are completely symmetrical fore and aft, as depicted in Fig. 7.15a. In appearance the streamline pattern is superficially similar to that for potential flow. For creeping flow, however, the influence of the cylinder on the streamlines extend to much greater distances than for potential flow. Skin-friction drag is the only force generated by the fluid flow on the cylinder. Consequently, the body with the lowest drag for a fixed volume is the sphere.

---

*P.R. Owen and L. Klanfer, *RAE Reports Aero.*, 2508, 1953.

[†] Reynolds number here is defined as $U_\infty D/\nu$, where U_∞ is the free stream velocity, ν is the kinematic viscosity in the free stream and D is the cylinder diameter.

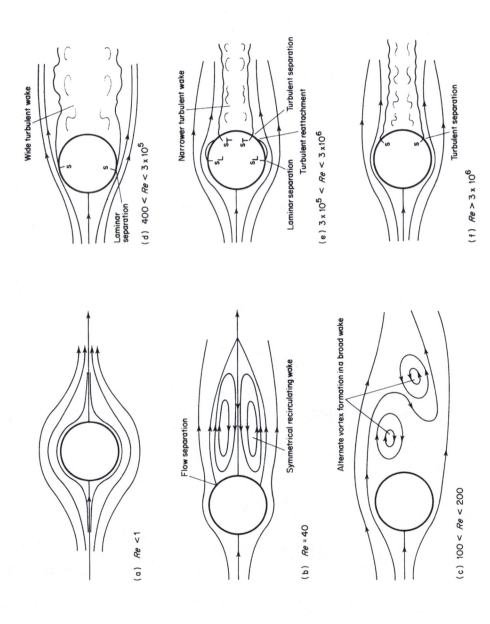

Fig. 7.15 (Note that the Reynolds number limits quoted are only approximate, as they depend appreciably on the free-stream turbulence level)

Perhaps, it is for this reason that microscopic swimmers such as protozoa, bacteria and spermatazoa tend to be near-spherical. In the range $1 < Re < 5$, the streamline pattern remains fairly similar to that of Fig. 7.15a, except that as Re is increased within this range a more and more pronounced asymmetry develops between the fore and aft directions. Nevertheless, the flow remains attached.

When Re exceeds a value of about 5, a much more profound change in the flow pattern occurs. The flow separates from the cylinder surface to form a closed wake of recirculating flow – see Fig. 7.15b. The wake grows progressively in length as Re is increased from 5 up to about 41. The flow pattern is symmetrical about the horizontal axis and is *steady*, i.e. it does not change with time. At these comparatively low Reynolds numbers the effects of viscosity still extend a considerable distance from the surface, so it is not valid to use the concept of the boundary layer, nevertheless the explanation for flow separation occurring is substantially the same as that given in Section 7.4.

When Re exceeds a value of about 41 another profound change occurs; steady flow becomes impossible. In some respects what happens is similar to the early stages of laminar-turbulent transition (see Section 7.9), in that the steady recirculating wake flow, seen in Fig. 7.15b, becomes unstable to small disturbances. In this case, though, the small disturbances develop as vortices rather than waves. Also in this case, the small disturbances do not develop into turbulent flow, but rather a steady laminar wake develops into an unsteady, but stable, laminar wake. The vortices are generated

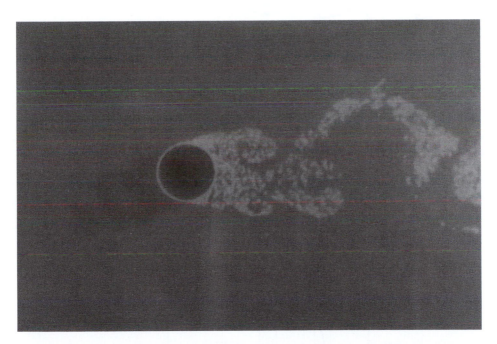

Fig. 7.16 The wake of a circular cylinder at $Re_D \simeq 5000$: Vortices are formed when flow passes over circular cylinders for a wide range of Reynolds numbers – see Fig. 7.15 on page 400. The flow is from left to right. The Reynolds number is sufficiently large for a thin laminar boundary layer to form over the upstream surface of the cylinder. It separates at a point just ahead of maximum thickness and breaks up into a turbulent wake which is dominated by large-scale vortices. Flow visualization is obtained by using aluminium particle tracers on water flow. (*The photograph was taken by D.J. Buckingham at the School of Engineering, University of Exeter, UK.*)

periodically on alternate sides of the horizontal axis through the wake and the centre of the cylinder. In this way, a row of vortices are formed, similar to that shown in Fig. 7.17c. The vortex row persists for a very considerable distance downstream. This phenomenon was first explained theoretically by von Kármán in the first decade of the twentieth century.

For Reynolds numbers between just above 40 and about 100 the vortex street develops from amplified disturbances in the wake. However, as the Reynolds number

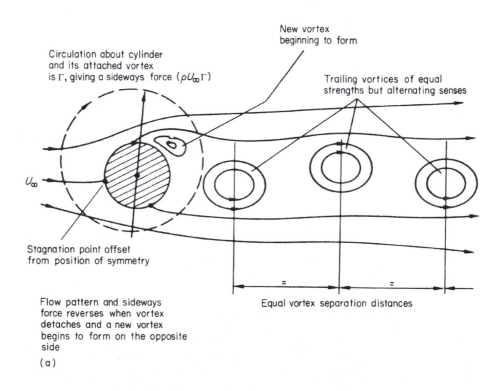

Circulation about cylinder and its attached vortex is Γ, giving a sideways force $(\rho U_\infty \Gamma)$

New vortex beginning to form

Trailing vortices of equal strengths but alternating senses

U_∞

Stagnation point offset from position of symmetry

Flow pattern and sideways force reverses when vortex detaches and a new vortex begins to form on the opposite side

Equal vortex separation distances

(a)

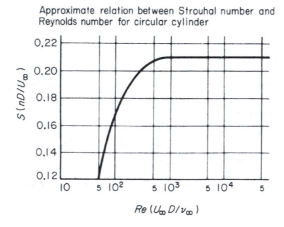

Approximate relation between Strouhal number and Reynolds number for circular cylinder

$S\,(nD/U_\infty)$

$Re\,(U_\infty D/\nu_\infty)$

(b)

Fig. 7.17

rises an identifiable thin boundary layer begins to form on the cylinder surface and the disturbance develops increasingly closer to the cylinder. Finally, above about $Re = 100$ eddies are shed alternately from the laminar separation points on either side of the cylinder (see Fig. 7.16). Thus, a vortex will be generated in the region behind the separation point on one side, while a corresponding vortex on the other side will break away from the cylinder and move downstream in the wake. When the attached vortex reaches a particular strength, it will in turn break away and a new vortex will begin to develop again on the second side and so on.

The wake thus consists of a procession of equal-strength vortices, equally spaced but alternating in sign. This type of wake, which can occur behind all long cylinders of bluff cross-section, including flat plates normal to the flow direction, is termed a von Kármán *vortex street* or *trail* (see Fig. 7.17a). In a uniform stream flowing past a cylinder the vortices move downstream at a speed somewhat less than the free-stream velocity, the reduction in speed being inversely proportional to the streamwise distance separating alternate vortices.

It will be appreciated that, during the formation of any single vortex while it is bound to the cylinder, an increasing circulation will exist about the cylinder, with the consequent generation of a transverse (lift) force. With the development of each successive vortex this force will change sign, giving rise to an alternating transverse force on the cylinder at the same frequency as the vortex shedding. If the frequency happens to coincide with the natural frequency of oscillation of the cylinder, however it may be supported, then appreciable vibration may be caused. This phenomenon is responsible, for example, for the singing of telegraph wires in the wind (Aeolian tones).

A unique relationship is found to exist between the Reynolds number and a dimensionless parameter involving the shedding frequency. This parameter, known as the Strouhal number, is defined by the expression $S = nD/U_\infty$, where n is the frequency of vortex shedding. Figure 7.17b shows the typical variation of S with Re in the vortex street range.

Despite the many other changes, described below, which occur in the flow pattern as Re increases still further, markedly periodic vortex shedding remains a characteristic flow around the circular cylinder and other bluff bodies up to the highest Reynolds numbers. This phenomenon can have important consequences in engineering applications. An example was the Tacoma Narrows Bridge (in Washington State, U.S.A.) A natural frequency of the bridge deck was close to its shedding frequency causing resonant behaviour in moderate winds, although its collapse in 1940 was due to torsional aeroelastic instability excited by stronger winds.

For two ranges of Reynolds number, namely $200 < Re < 400$ and $3 \times 10^5 < Re < 3 \times 10^6$, the regularity of vortex shedding is greatly diminished. In the former range very considerable scatter occurs in values of Strouhal number, while for the latter range all periodicity disappears except very close to the cylinder. The values of Reynolds number marking the limits of these two ranges are associated with pronounced changes in the flow pattern. In the case of $Re \simeq 400$ and 3×10^6 the transitions in flow pattern are such as to restore periodicity.

Below $Re \simeq 200$ the vortex street persists to great distances downstream. Above this Reynolds number, transition to turbulent flow occurs in the wake thereby destroying the periodic vortex wake far downstream. At this Reynolds number the vortex street also becomes unstable to three-dimensional disturbances leading to greater irregularity.

At $Re \simeq 400$ a further change occurs. Transition to turbulence now occurs close to the separation points on the cylinder. Rather curiously, perhaps, this has a stabilizing

(a) $Re_D = 15\,000$

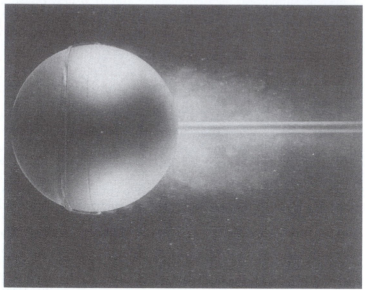

(b) $Re_D = 30\,000$

Fig. 7.18 Flow past a sphere: In both cases the flow is from left to right. $Re_D = 15\,000$ for (a) which uses dye in water to show a laminar boundary layer separating ahead of the equator and remaining laminar for almost one radius before becoming turbulent. This case corresponds to Fig. 7.15d on page 400. Air bubbles in water provide the flow visualization in (b). For this case $Re_D = 30\,000$ and a wire hoop on the downstream surface trips the boundary layer ensuring that transition occurs in the separation bubble leading to reattachment and a final turbulent separation much further rearward. This case corresponds to Fig. 7.15e on page 400. The much reduced wake in (b) as compared with (a) leads to a dramatically reduced drag. The use of a wire hoop to promote transition artificially produces the drag reduction at a much lower Reynolds number than for the smooth sphere. (*The photographs were taken by H. Werlé at ONERA, France.*)

effect on the shedding frequency even though the vortices themselves develop considerable irregular fluctuations. This pattern with laminar boundary-layer separation and a turbulent vortex wake persists until $Re \simeq 3 \times 10^5$, and is illustrated in Figs 7.15d and 7.16. Note that with laminar separation the flow separates at points on the front half of the cylinder, thereby forming a large wake and producing a high-level of form drag. In this case, the contribution of skin-friction drag is all but negligible.

When the Reynolds number reaches a value in the vicinity of 3×10^5 the laminar boundary layer undergoes transition to turbulence almost immediately after separation. The increased mixing re-energizes the separated flow causing it to reattach as a turbulent boundary layer, thereby forming a separation bubble (as described in Section 7.4.1) – see Figs 7.15e and 7.18.

At this critical stage the second and final point of separation, which now takes place in a turbulent layer, moves suddenly downstream, because of the better sticking property of the turbulent layer, and the wake width is very appreciably decreased. This stage is therefore accompanied by a sudden decrease in the total drag of the cylinder. For this reason the value of Re at which this transition in flow pattern occurs is often called the *Critical Reynolds* number. The wake vorticity remains random with no clearly discernible frequency. With further increase in Reynolds number the wake width will gradually increase to begin with, as the turbulent separation points slowly move upstream round the rear surface. The total drag continues to increase steadily in this stage, due to increases in both pressure and skin-friction drag, although the drag coefficient, defined by

$$C_D = \frac{\text{drag per unit span}}{\frac{1}{2}\rho_\infty U_\infty^2 D}$$

tends to become constant, at about 0.6, for values of $Re > 1.3 \times 10^6$. The final change in the flow pattern occurs at $Re \simeq 3 \times 10^6$ when the separation bubble disappears, see Fig. 7.15f. This transition has a stabilizing effect on the shedding frequency which becomes discernible again. C_D rises slowly as the Reynolds number increases beyond 3×10^6.

The actual value of the Reynolds number at the critical stage when the dramatic drag decrease occurs depends, for a smooth cylinder, on the small-scale turbulence level existing in the oncoming free stream (see Fig. 7.18). Increased turbulence, or, alternatively, increased surface roughness, will provoke turbulent reattachment, with its accompanying drag decrease, at a lower Reynolds number. The behaviour of a smooth sphere under similarly varying conditions exhibits the same characteristics as the cylinder, although the Reynolds numbers corresponding to the changes of flow regime are somewhat different. One marked difference in behaviour is that the eddying vortex street, typical of bluff cylinders, does not develop in so regular a fashion behind a sphere. Graphs showing the variations of drag coefficient with Reynolds number for circular cylinders and spheres are given in Fig. 7.19.

7.5.1 Turbulence spheres

The effect of free-stream turbulence on the Reynolds number at which the critical drag decrease occurs was widely used many years ago to ascertain the turbulence level in the airstream of a wind-tunnel working section. In this application, a smooth sphere is mounted in the working section and its drag, for a range of tunnel speeds, is read off on the drag balance. The speed, and hence the Reynolds number, at which the drag suddenly decreases is recorded. Experiments in air of virtually zero small-scale

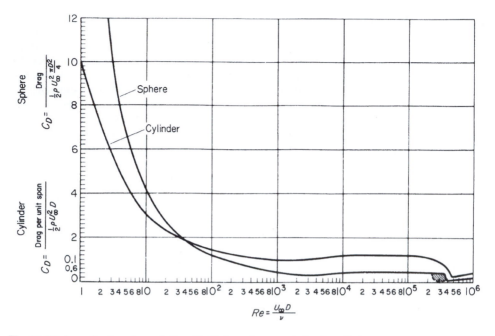

Fig. 7.19 Approximate values of C_D with Re for spheres and circular cylinders

turbulence have indicated that the highest critical sphere Reynolds number attainable is 385 000. A turbulence factor, for the tunnel under test, is then defined as the ratio of 385 000 to the critical Reynolds number of the test tunnel.

A major difficulty in this application is the necessity for extreme accuracy in the manufacture of the sphere, as small variations from the true spherical shape can cause appreciable differences in the behaviour at the critical stage. As a result, this technique for turbulence measurement is not now in favour, and more recent methods, such as hot-wire anemometry, took its place some time ago.

7.5.2   Golf balls

In the early days of the sport, golf balls were made with a smooth surface. It was soon realized, however, that when the surface became worn the ball travelled farther when driven, and subsequently golf balls were manufactured with a dimpled surface to simulate the worn surface. The reason for the increase in driven distance with the rough surface is as follows.

The diameter of a golf ball is about 42 mm, which gives a critical velocity in air, for a smooth ball, of just over $135\,\mathrm{m\,s^{-1}}$ (corresponding to $Re = 3.85 \times 10^5$). This is much higher than the average flight speed of a driven ball. In practice, the critical speed would be somewhat lower than this owing to imperfections in manufacture, but it would still be higher than the usual flight speed. With a rough surface, promoting early transition, the critical Reynolds number may be as low as 10^5, giving a critical speed for a golf ball of about $35\,\mathrm{m\,s^{-1}}$, which is well below the flight speed. Thus, with the roughened surface, the ball travels at above the critical drag speed during its flight and so experiences a smaller decelerating force throughout, with consequent increase in range.

7.5.3 Cricket balls

The art of the seam bowler in cricket is also explainable with reference to boundary-layer transition and separation. The bowling technique is to align the seam at a small angle to the flight path (see Fig. 7.20). This is done by spinning the ball about an axis perpendicular to the plane of the seam, and using the gyroscopic inertia to stabilize this seam position during the trajectory. On the side of the front stagnation point where the boundary layer passes over the seam, it is induced to become turbulent before reaching the point of laminar separation. On this side, the boundary layer remains attached to a greater angle from the fore stagnation point than it does on the other side where no seam is present to trip the boundary layer. The flow past the ball thus becomes asymmetric with a larger area of low pressure on the turbulent side, producing a lateral force tending to move the ball in a direction normal to its flight path. The range of flight speeds over which this phenomenon can be used corresponds to those of the medium to medium-fast pace bowler. The diameter of a cricket ball is between 71 and 72.5 mm. In air, the critical speed for a smooth ball would be about 75 m s^{-1}. However, in practice it is found that transition to turbulence for the seam-free side occurs at speeds in the region of 30 to 35 m s^{-1}, because of inaccuracies in the spherical shape and minor surface irregularities. The critical speed for a rough ball with early transition ($Re \approx 10^5$) is about 20 m s^{-1} and below this speed the flow asymmetry tends to disappear because laminar separation occurs before the transition, even on the seam side.

Thus within the speed range 20 to about 30 m s^{-1}, very approximately, the ball may be made to swing by the skilful bowler. The very fast bowler will produce a flight speed in excess of the upper critical and no swing will be possible. A bowler may make the ball swing late by bowling at a speed just too high for the asymmetric condition to exist, so that as the ball loses speed in flight the asymmetry will develop later in the trajectory. It is obvious that considerable skill and experience is required to know at just what speed the delivery must be made to do this.

It will also be realized that the surface condition, apart from the seam, will affect the possibility of swinging the ball, e.g. a new, smooth-surfaced ball will tend to

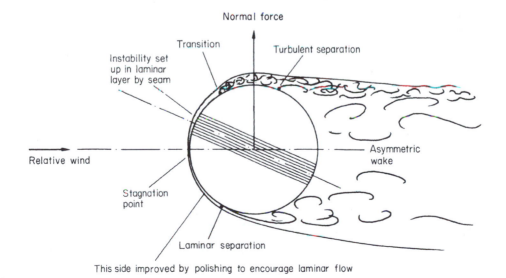

Fig. 7.20

maintain laminar layers up to separation even on the seam side, while a badly worn ball will tend to induce turbulence on the side remote from the seam. The slightly worn ball is best, especially if one side can be kept reasonably polished to help maintain flow on that side only.

7.6   The momentum integral equation

The accurate evaluation of most of the quantities defined above in Sections 7.3.2 and 7.3.3 requires the numerical solution of the differential equations of motion. This will be discussed in Section 7.11. Here an integral form of the equations of motion is derived that allows practical solutions to be found fairly readily for certain engineering problems.

The required momentum integral equation is derived by considering mass and momentum balances on a thin slice of boundary layer of length δx. This slice is illustrated in Fig. 7.21. Remember that in general, quantities vary with x, i.e. along the surface; so it follows from elementary differential calculus that the value of a quantity f, say, on CD (where the distance from the origin is $x + \delta x$) is related to its value on AB (where the distance from the origin is x) in the following way:

$$f(x + \delta x) \simeq f(x) + \frac{df}{dx}\delta x \tag{7.54}$$

First the conservation of mass for an elemental slice of boundary layer will be considered, see Fig. 7.21b. Since the density is assumed to be constant the mass flow balance for slice ABCD states, in words, that

Volumetric flow rate into the slice across AB = Volumetric flow rate out across CD

+ Volumetric flow rate out across AD

+ Volumetric flow rate out across BC

The last item in the volumetric flow balance allows for the possibility of flow due to suction passing through a porous wall. In the usual case of an impermeable wall $V_s = 0$. Expressed mathematically this equation becomes

$$\underbrace{Q_i}_{\text{across AB}} = \underbrace{Q_i + \frac{dQ_i}{dx}\delta x}_{\text{across CD}} + \underbrace{V_e\delta x - U_e\frac{d\delta}{dx}\delta x}_{\text{across AD}} + \underbrace{V_s\delta x}_{\text{across BC}} \tag{7.55}$$

Note that Eqn (7.54) has been used, Q_i replacing f where

$$Q_i \equiv \int_0^\delta u\,dy$$

Cancelling common factors, rearranging Eqn (7.55) and taking the limit $\delta x \to dx$ leads to an expression for the perpendicular velocity component at the edge of the boundary layer, i.e.

$$V_e = -\frac{d}{dx}\int_0^\delta u\,dy + U_e\frac{d\delta}{dx} - V_s$$

The definition of displacement thickness, Eqn (7.16), is now introduced to give

$$V_e = -\frac{d}{dx}(\delta^* U_e) - V_s \tag{7.56}$$

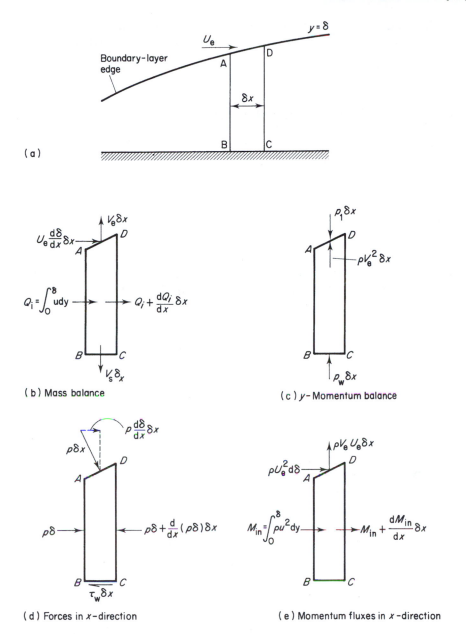

(a)

(b) Mass balance

(c) y-Momentum balance

(d) Forces in x-direction

(e) Momentum fluxes in x-direction

Fig. 7.21 Mass and momentum balances on a thin slice of boundary layer

The y-momentum balance for the slice ABCD of boundary layer is now considered. This is illustrated in Fig. 7.21c. In this case, noting that the y-component of momentum can be carried by the flow across side AD only and that the only force in the y direction is pressure,* the momentum theorem states that the

---

* The force of gravity is usually ignored in aerodynamics.

> Rate at which y-component   The net pressure force in the y direction
> of momentum crosses AD $=$ acting on the slice ABCD

Or in mathematical terms

$$\rho V_e^2 \delta x = (p_w - p_1)\delta x$$

Thus cancelling the common factor δx leads to

$$p_w - p_1 = \rho V_e^2$$

It can readily be shown from this result that the net pressure difference across the boundary layer is negligible, i.e. $p_w \simeq p_1$, as it should be according to boundary-layer theory. For simplicity the case of the boundary layer along an impermeable flat plate when $U_e = U_\infty (=\text{const.})$ and $V_s = 0$ is considered, so that from Eqn (7.56)

$$V_e = U_\infty \frac{\mathrm{d}\delta^*}{\mathrm{d}x} \Rightarrow p_w - p_1 = \rho U_\infty^2 \left(\frac{\mathrm{d}\delta^*}{\mathrm{d}x}\right)^2$$

It must be remembered, however, that the boundary layer is very thin compared with the length of the plate thus $\mathrm{d}\delta^*/\mathrm{d}x \ll 1$, so that its square is negligibly small. This argument can be readily extended to the more general case where U_e varies along the edge of the boundary layer. Thus it can be demonstrated that the assumption of a thin boundary layer implies that the pressure does not vary appreciably across the boundary layer. This is one of the major features of boundary-layer theory (see Section 7.3.1). It also implies that within the boundary layer the pressure p is a function of x only.

Finally, the x-momentum balance for the slice ABCD is considered. This case is more complex since there are both pressure and surface friction forces to be considered, and furthermore the x component of momentum may be carried across AB, CD and AD. The forces involved are depicted in Fig. 7.21d while the momentum fluxes are shown in Fig. 7.21e. In this case, the momentum theorem states that

$$\begin{pmatrix} \text{Rate at which} \\ \text{momentum leaves} \\ \text{across CD and AD} \end{pmatrix} - \begin{pmatrix} \text{Rate at which} \\ \text{momentum enters} \\ \text{across AB} \end{pmatrix} = \begin{pmatrix} \text{Net pressure force} \\ \text{in } x \text{ direction acting} \\ \text{on ABCD} \end{pmatrix} - \begin{pmatrix} \text{Surface friction} \\ \text{force acting} \\ \text{on BC} \end{pmatrix}$$

Using Eqn (7.54) this can be expressed mathematically as:

$$\underbrace{M_{\mathrm{in}} + \frac{\mathrm{d}M_{\mathrm{in}}}{\mathrm{d}x}}_{\text{out across CD}} + \underbrace{\rho V_e U_e \delta x - \rho U_e^2 \frac{\mathrm{d}\delta}{\mathrm{d}x}\delta x}_{\text{out across AD}} \quad \underbrace{-M_{\mathrm{in}}}_{\substack{\text{in across AB}}}$$

$$= \underbrace{p\delta}_{\text{on AB}} - \underbrace{\left(p\delta + \frac{\mathrm{d}}{\mathrm{d}x}(p\delta)\delta x\right)}_{\text{on CD}} + \underbrace{p\frac{\mathrm{d}\delta}{\mathrm{d}x}\delta x}_{\text{on AD}} - \underbrace{\tau_w \delta x}_{\text{on BC}} \tag{7.57}$$

where

$$M_{\mathrm{in}} = \int_0^\delta \rho u^2 \mathrm{d}y.$$

After cancelling common factors, taking the limit $\delta x \to \mathrm{d}x$ and simplification Eqn (7.57) becomes

$$\frac{\mathrm{d}}{\mathrm{d}x}\int_0^\delta \rho u^2 \mathrm{d}y - \rho U_e^2 \frac{\mathrm{d}\delta}{\mathrm{d}x} + \rho U_e V_e = -\delta\frac{\mathrm{d}p}{\mathrm{d}x} - \tau_w \tag{7.58}$$

The Bernoulli equation can be used at the edge of the boundary layer so that

$$p + \rho U_e^2 = \text{const.}, \qquad \frac{\mathrm{d}p}{\mathrm{d}x} = -\rho U_e \frac{\mathrm{d}U_e}{\mathrm{d}x}$$

After substituting Eqn (7.56) for V_e, introducing the definition, Eqn (7.17), of momentum thickness and using the result given above, Eqn (7.58) reduces to

$$\frac{\mathrm{d}}{\mathrm{d}x}(U_e^2\theta) + \delta^* U_e \frac{\mathrm{d}U_e}{\mathrm{d}x} - \rho U_e V_s = \frac{\tau_w}{\rho} \qquad (7.59)$$

This is the momentum integral equation first derived by von Kármán. Since no assumption is made at this stage about the relationship between τ_w and the velocity gradient at the wall, the momentum integral equation applies equally well to laminar or turbulent flow.

When suitable forms are chosen for the velocity profile the momentum integral equation can be solved to provide the variations of δ, δ^*, θ and C_f along the surface. A suitable approximate form for the velocity profile in the laminar boundary layer is derived in Section 7.6.1. In order to solve Eqn (7.59) in the turbulent case additional semi-empirical relationships must be introduced. In the simple case of the flat plate the solution to Eqn (7.59) can be found in closed form, as shown in Section 7.7. In the general case with a non-zero pressure gradient it is necessary to resort to computational methods to solve Eqn (7.59). Such methods are discussed in Section 7.11.

7.6.1   An approximate velocity profile for the laminar boundary layer

As explained in the previous subsection an approximate expression is required for the velocity profile in order to use the momentum integral equation. A reasonably accurate approximation can be obtained by using a cubic polynomial in the form:

$$\bar{u}(\equiv u/U_e) = a + b\bar{y} + c\bar{y}^2 + d\bar{y}^3 \qquad (7.60)$$

where $\bar{y} = y/\delta$. In order to evaluate the coefficients a, b, c and d four conditions are required, two at $\bar{y} = 0$ and two at $\bar{y} = 1$. Two of these conditions are readily available, namely

$$\bar{u} = 0 \quad \text{at} \quad \bar{y} = 0 \qquad (7.61a)$$

$$\bar{u} = 1 \quad \text{at} \quad \bar{y} = 1 \qquad (7.61b)$$

In real boundary-layer velocity profiles – see Fig. 7.10 – the velocity varies smoothly to reach U_e; there is no kink at the edge of the boundary layer. Accordingly, it follows that the velocity gradient is zero at $y = \delta$ giving a third condition, namely

$$\frac{\partial \bar{u}}{\partial \bar{y}} = 0 \quad \text{at} \quad \bar{y} = 1 \qquad (7.61c)$$

To obtain the fourth and final condition it is necessary to return to the boundary-layer equation (7.14). At the wall $y = 0$, $u = v = 0$, so both terms on the left-hand side are zero at $y = 0$. Thus, noting that $\tau = \mu\, \partial u/\partial y$, the required condition is given by

$$\frac{\mathrm{d}p}{\mathrm{d}x} = \frac{\partial \tau}{\partial y} \quad \text{at} \quad y = 0$$

Since $y = \bar{y}\delta$ and $p + \rho U_e^2 = \text{const.}$ the above equation can be rearranged to read

$$\frac{\partial^2 \bar{u}}{\partial \bar{y}^2} = -\frac{\delta^2}{v}\frac{\mathrm{d}U_e}{\mathrm{d}x} \quad \text{at} \quad \bar{y} = 0 \tag{7.61d}$$

In terms of the coefficients a, b, c and d the four conditions (7.61a, b, c, d) become

$$a = 0 \tag{7.62a}$$
$$b + c + d = 1 \tag{7.62b}$$
$$b + 2c + 3d = 0 \tag{7.62c}$$

$$2c = -\Lambda \quad \text{where} \quad \Lambda \equiv \frac{\delta^2}{v}\frac{\mathrm{d}U_e}{\mathrm{d}x} \tag{7.62d}$$

Equations (7.62b, c, d) can be readily solved for b, c and d to give the following approximate velocity profile

$$\bar{u} = \frac{3}{2}\bar{y} - \frac{1}{2}\bar{y}^3 + \frac{\Lambda}{4}(\bar{y} - 2\bar{y}^2 + \bar{y}^3) \tag{7.63}$$

The parameter Λ in Eqn (7.63) is often called the Pohlhausen parameter. It determines the effect of an external pressure gradient on the shape of the velocity profile. $\Lambda > 0$ and <0 correspond respectively to favourable and unfavourable pressure gradients. For $\Lambda = -6$ the wall shear stress $\tau_w = 0$ and for more negative values of Λ flow reversal at the wall develops. Thus $\Lambda = -6$ corresponds to boundary-layer separation. Velocity profiles corresponding to various values of Λ are plotted in Fig. 7.6. In this figure, the flat-plate profile corresponds to $\Lambda = 0$; $\Lambda = 6$ for the favourable pressure gradient; $\Lambda = -4$ for the mild adverse pressure gradient; $\Lambda = -6$ for the strong adverse pressure gradient; and $\Lambda = -9$ for the reversed-flow profile.

For the flat-plate case $\Lambda = 0$, the approximate velocity profile of Eqn (7.63) is compared with two other approximate profiles in Fig. 7.22. The velocity profile labelled Blasius is the accurate solution of the differential equations of motion given in Section 7.3.4 and Fig. 7.10.

The various quantities introduced in Sections 7.3.2 and 7.3.3 can readily be evaluated using the approximate velocity profile (7.63). For example, if Eqn (7.63) is substituted in turn into Eqns (7.16), (7.17) and (7.19), with use of Eqn (7.20), the following are obtained.

$$I_1 = \frac{\delta^*}{\delta} = \int_0^1 \left[1 - \left(\frac{3}{2} + \frac{\Lambda}{4}\right)\bar{y} + \frac{\Lambda}{2}\bar{y}^2 + \left(\frac{1}{2} - \frac{\Lambda}{4}\right)\bar{y}^3 \right] \mathrm{d}\bar{y}$$

$$= \left[\bar{y} - \left(\frac{3}{2} + \frac{\Lambda}{4}\right)\frac{\bar{y}^2}{2} + \frac{\Lambda}{2}\frac{\bar{y}^3}{3} + \left(\frac{1}{2} - \frac{\Lambda}{4}\right)\frac{\bar{y}^4}{4} \right]_0^1$$

$$= 1 - \left(\frac{3}{4} + \frac{\Lambda}{8}\right) + \frac{\Lambda}{6} + \left(\frac{1}{8} - \frac{\Lambda}{16}\right)$$

$$= \frac{3}{8} - \frac{\Lambda}{48} \tag{7.64a}$$

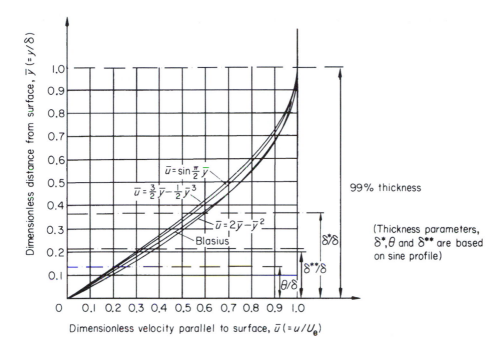

Fig. 7.22 Laminar velocity profile

$$I = \frac{\theta}{\delta} = \int_0^1 \left[\left(\frac{3}{2} + \frac{\Lambda}{4} \right) \bar{y} - \frac{\Lambda}{2} \bar{y}^2 - \left(\frac{1}{2} - \frac{\Lambda}{4} \right) \bar{y}^3 \right]$$

$$\times \left[1 - \left(\frac{3}{2} + \frac{\Lambda}{4} \right) \bar{y} + \frac{\Lambda}{2} \bar{y}^2 + \left(\frac{1}{2} - \frac{\Lambda}{4} \right) \bar{y}^3 \right] \mathrm{d}\bar{y}$$

$$= \left[\left(\frac{3}{2} + \frac{\Lambda}{4} \right) \frac{\bar{y}^2}{2} - \left\{ \left(\frac{3}{2} + \frac{\Lambda}{4} \right)^2 + \frac{\Lambda}{2} \right\} \frac{\bar{y}^3}{3} \right.$$

$$+ \left\{ 2 \left(\frac{3}{2} + \frac{\Lambda}{4} \right) \frac{\Lambda}{2} - \left(\frac{1}{2} - \frac{\Lambda}{4} \right) \right\} \frac{\bar{y}^4}{4} + \left\{ 2 \left(\frac{3}{2} + \frac{\Lambda}{4} \right) \left(\frac{1}{2} - \frac{\Lambda}{4} \right) - \frac{\Lambda^2}{4} \right\} \frac{\bar{y}^5}{5}$$

$$\left. - \left\{ 2 \frac{\Lambda}{2} \left(\frac{1}{2} - \frac{\Lambda}{4} \right) \right\} \frac{\bar{y}^6}{6} - \left(\frac{1}{2} - \frac{\Lambda}{4} \right)^2 \frac{\bar{y}^7}{7} \right]_0^1$$

$$= \frac{1}{280} \left(39 - \frac{\Lambda}{2} - \frac{1}{6} \Lambda^2 \right) \tag{7.64b}$$

$$C_{\mathrm{f}} = \frac{\tau_{\mathrm{w}}}{\frac{1}{2} \rho U_{\mathrm{e}}^2} = \frac{2\mu}{\rho U_{\mathrm{e}}^2} \left(\frac{\partial u}{\partial y} \right)_{\mathrm{w}} = \frac{2\mu}{\rho U_{\mathrm{e}} \delta} \left(\frac{\mathrm{d}\bar{u}}{\mathrm{d}\bar{y}} \right)_{\bar{y}=0}$$

$$= \frac{\mu}{\rho U_{\mathrm{e}} \delta} \left(3 + \frac{\Lambda}{2} \right) \tag{7.64c}$$

The quantities I_1 and I depend only on the shape of the velocity profile, and for this reason they are usually known as *shape parameters*. If the more accurate differential form of the boundary-layer equations were to be used, rather than the momentum-integral equation with approximate velocity profiles, the boundary-layer thickness δ would become a rather less precise quantity. For this reason, it is more common to use the shape parameter $H = \delta^*/\theta$. Frequently, H is referred to simply as the shape parameter.

For the numerical methods discussed in Section 7.11 below for use in the general case with an external pressure gradient, it is preferable to use a some-what more accurate quartic polynomial as the approximate velocity profile. This is particularly important for predicting the transition point. This quartic velocity profile is derived in a very similar way to that given above, the main differences are the addition of another term, $e\bar{y}^4$, on the right-hand side of Eqn (7.60), and the need for an additional condition at the edge of the boundary layer. This latter requires that

$$\frac{d^2\bar{u}}{d\bar{y}^2} = 0 \quad \text{at} \quad \bar{y} = 1$$

This has the effect of making the velocity profile even smoother at the edge of the boundary layer, and thereby improves the approximation. The resulting quartic velocity profile takes the form

$$\bar{u} = 2\bar{y} - 2\bar{y}^3 + \bar{y}^4 + \frac{\Lambda}{6}(\bar{y} - 3\bar{y}^2 + 3\bar{y}^3 - \bar{y}^4) \tag{7.65}$$

Using this velocity profile and following similar procedures to those outlined above leads to the following expressions:

$$I_1 = \frac{3}{10} - \frac{\Lambda}{120} \tag{7.64a'}$$

$$I = \frac{1}{63}\left(\frac{37}{5} - \frac{\Lambda}{15} - \frac{\Lambda^2}{144}\right) \tag{7.64b'}$$

$$C_f = \frac{\mu}{\rho U_e \delta}\left(4 + \frac{\Lambda}{3}\right) \tag{7.64c'}$$

Note that it follows from Eqn (7.64c') that with the quartic velocity profile the separation point where $\tau_w = 0$ now corresponds to $\Lambda = -12$.

7.7 Approximate methods for a boundary layer on a flat plate with zero pressure gradient

In this section, the momentum integral equation (7.59) will be solved to give approximate expressions for the skin-friction drag and for the variation of δ, δ^*, θ and C_f along a flat plate with laminar, turbulent and mixed laminar/turbulent boundary layers. This may seem a rather artificial and restrictive case to study in depth. It should be noted, however, that these results can be used to provide rough, but reasonable, estimates for any streamlined body. The equivalent flat plate for a specific streamlined body would have the same surface area and total streamwise length as the body. In this way reasonable estimates can be obtained, especially for the skin-friction drag, provided that the transition point is correctly located using the guidelines given at the end of Section 7.9.

7.7.1 Simplified form of the momentum integral equation

For the flat plate $dp/dx = 0$ and $U_e = U_\infty = \text{const.}$ so that $dU_e/dx = 0$. Accordingly, the momentum integral equation (7.59) reduces to the simple form

$$\tau_w = \rho U_e^2 \frac{d\theta}{dx} \tag{7.66}$$

Now the shape factor $I = \theta/\delta$ is simply a numerical quantity which depends only on the shape of the velocity profile.

So Eqn (7.66) may then be expressed in the alternative form

$$C_f = 2I \frac{d\delta}{dx} \tag{7.67}$$

where I has been assumed to be independent of x. Equations (7.66) and (7.67) are forms of the simple momentum integral equation.

7.7.2 Rate of growth of a laminar boundary layer on a flat plate

The rate of increase of the boundary-layer thickness δ may be found by integrating Eqn (7.67), after setting $\Lambda = 0$ in Eqns (7.64b and c) and substituting for I and C_f.

Thus Eqn (7.67) becomes

$$\frac{d\delta}{dx} = \frac{C_f}{2I} = \frac{140}{13} \frac{\mu}{\rho U_\infty \delta}$$

Therefore

$$\delta d\delta = \frac{140}{13} \frac{\mu}{\rho U_\infty}$$

whence

$$\frac{\delta^2}{2} = \frac{140}{13} \frac{\mu x}{\rho U_\infty}$$

The integration constant is zero if x is measured from the fore stagnation point where $\delta = 0$, i.e.

$$\delta = 4.64x/(Re_x)^{1/2} \tag{7.68}$$

The other thickness quantities may now be evaluated using Eqns (7.64a,b) with $\Lambda = 0$. Thus

$$\delta^* = 0.375\delta = 1.74x/(Re_x)^{1/2} \tag{7.69}$$

$$\theta = 0.139\delta = 0.646x/(Re_x)^{1/2} \tag{7.70}$$

7.7.3   Drag coefficient for a flat plate of streamwise length L with wholly laminar boundary layer

Note that

$$C_F = \frac{1}{L}\int_0^L C_f dx = \frac{2}{L}\int_0^L \frac{d\theta}{dx} = \frac{2\theta(L)}{L} \tag{7.71}$$

where $\theta(L)$ is the value of the momentum thickness at $x = L$. Thus using Eqn (7.70) in Eqn (7.71) gives

$$C_F = 1.293/Re^{1/2}$$

and

$$C_{D_F} = 2.586/Re^{1/2} \tag{7.72}$$

These expressions are shown plotted in Fig. 7.25 (lower curve).

Example 7.3   A flat plate of 0.6 m chord at zero incidence in a uniform airstream of $45\,\text{m}\,\text{s}^{-1}$. Estimate (i) the displacement thickness at the trailing edge, and (ii) the overall drag coefficient of the plate.

At the trailing edge, $x = 0.6\,\text{m}$ and

$$Re_x = \frac{45 \times 0.6}{14.6 \times 10^{-6}} = 1.85 \times 10^6$$

Therefore, using Eqn (7.69),

$$\delta^* = \frac{1.74 \times 0.6}{\sqrt{1.85 \times 10^3}} = 0.765 \times 10^{-3}\,\text{m} = 0.8\,\text{mm} \tag{i}$$

Re has the same value as Re_x at the trailing edge. So Eqn (7.72) gives

$$C_{D_F} = \frac{2.54}{\sqrt{1.85 \times 10^3}} = 0.0019 \tag{ii}$$

7.7.4   Turbulent velocity profile

A commonly employed, turbulent-boundary-layer profile is the seventh-root profile, which was proposed by Prandtl on the basis of friction-loss experiments with turbulent flow in circular pipes correlated by Blasius. The latter investigated experimental results on the resistance to flow and proposed the following empirical relationships between the local skin friction coefficient at the walls, $\bar{C}_f(= \tau_w/\frac{1}{2}\rho\bar{U}^2)$ and the Reynolds number of the flow $\overline{Re}$ (based on the average flow velocity $\bar{U}$ in the pipe and the diameter D). Blasius proposed the relationship

$$\bar{C}_f = \frac{0.0791}{\overline{Re}^{1/4}} \tag{7.73}$$

This expression gives reasonably good agreement with experiment for values of $\overline{Re}$ up to about 2.5×10^5.

Assuming that the velocity profile in the pipe may be written in the form

$$\frac{u}{U_m} = \left(\frac{y}{D/2}\right)^n = \left(\frac{y}{a}\right)^n \tag{7.74}$$

where u is the velocity at distance y from the wall and a = pipe radius $D/2$, it remains to determine the value of n. From Eqn (7.74), writing $U_m = C\bar{U}$, where C is a constant to be determined:

$$u = C\bar{U}\left(\frac{y}{a}\right)^n$$

i.e.

$$\bar{U} = \frac{u}{C}\left(\frac{a}{y}\right)^n \tag{7.75}$$

Substituting for $\bar{C}_f$ in the expression above for surface friction stress at the wall,

$$\tau_w = \bar{C}_f\frac{1}{2}\rho\bar{U}^2 = \frac{0.0791\nu^{1/4}}{D^{1/4}\bar{U}^{1/4}}\frac{1}{2}\rho\bar{U}^2 = 0.039\,55\rho\bar{U}^{7/4}\left(\frac{\nu}{D}\right)^{1/4} \tag{7.76}$$

From Eqn (7.75)

$$\bar{U}^{7/4} = \frac{u^{7/4}}{C^{7/4}}\left(\frac{a}{y}\right)^{7n/4}$$

so that Eqn (7.76) becomes

$$\tau_w = \frac{0.039\,55}{C^{7/4}}\rho u^{7/4}\frac{a^{[(7n/4)-(1/4)]}}{y^{7n/4}}\nu^{1/4}\left(\frac{1}{2}\right)^{1/4}$$

i.e.

$$\tau_w = \frac{0.0333}{C^{7/4}}\rho u^{7/4}\frac{\nu^{1/4}}{y^{7n/4}}a^{[(7n/4)-(1/4)]} \tag{7.77}$$

It may now be argued that very close to the wall, in the viscous sublayer ($u \neq 0$), the velocity u will not depend on the overall size of the pipe, i.e. that $u \neq f(a)$. If this is so, then it immediately follows that τ_w, which is $\mu(\partial u/\partial y)_w$, cannot depend on the pipe diameter and therefore the term $a^{[(7n/4)-(1/4)]}$ in Eqn (7.77) must be unity in order not to affect the expression for τ_w. For this to be so, $7n/4 - 1/4 = 0$ which immediately gives $n = \frac{1}{7}$. Substituting this back into Eqn (7.74) gives $u/U_m = (y/a)^{1/7}$. This expression thus relates the velocity u at distance y from the surface to the centre-line velocity U_m at distance a from the surface. Assuming that this will hold for very large pipes, it may be argued that the flow at a section along a flat, two-dimensional plate is similar to that along a small peripheral length of pipe, so that replacing a by δ will give the profile for the free boundary layer on the flat plate. Thus

$$\frac{u}{U_\infty} = \left(\frac{y}{\delta}\right)^{1/7} \quad\text{or}\quad \bar{u} = \bar{y}^{1/7} \tag{7.78}$$

This is Prandtl's seventh-root law and is found to give surprisingly good overall agreement with practice for moderate Reynolds numbers ($Re_x < 10^7$). It does, however, break down at the wall where the profile is tangential to the surface and gives an infinite value of $(\partial\bar{u}/\partial\bar{y})_w$. In order to find the wall shear stress, Eqn (7.77) must be used. The constant C may be evaluated by equating expressions for the total volume flow through the pipe, i.e. (using Eqns (7.75) and (7.78)),

$$\pi a^2 \bar{U} = 2\pi \int_0^a ur\,dr = 2\pi \bar{U} C \int_0^a \left(\frac{y}{a}\right)^{1/7} (a - y)\,dy = \frac{49}{60}\pi \bar{U} C a^2$$

giving $C = \dfrac{60}{49} = 1.224$. Substituting for C and n in Eqn (7.77) then gives

$$\tau_w = 0.0234\rho u^{7/4} \left(\frac{v}{y}\right)^{1/4}$$

that, on substituting for u from Eqn (7.78), gives

$$\tau_w = 0.0234\rho U_\infty^{7/4} \left(\frac{v}{\delta}\right)^{1/4} \tag{7.79}$$

Finally, since

$$C_f = \frac{\tau_w}{\frac{1}{2}\rho U_\infty^2}$$

for a free boundary layer:

$$C_f = 0.0468 \left(\frac{v}{U_\infty \delta}\right)^{1/4} = \frac{0.0468}{Re_\delta^{1/4}} \tag{7.80}$$

Using Eqns (7.78) and (7.80) in the momentum integral equation enables the growth of the turbulent boundary layer on a flat plate to be investigated.

7.7.5 Rate of growth of a turbulent boundary layer on a flat plate

$$d\delta/dx = C_f/2I$$

where

$$C_f = 0.0468(v/U_\infty \delta)^{1/4}$$

and

$$I = \int_0^1 \bar{u}(1 - \bar{u})\,d\bar{y} = \int_0^1 \bar{y}^{1/7}(1 - \bar{y}^{1/7})\,d\bar{y}$$

$$= \left[\frac{7}{8}\bar{y}^{8/7} - \frac{7}{9}\bar{y}^{9/7}\right]_0^1 = \frac{63 - 56}{72} = \frac{7}{72} \tag{a}$$

Therefore

$$\frac{d\delta}{dx} = \frac{72 \times 0.0468 v^{1/4}}{2 \times 7 \times (U_\infty \delta)^{1/4}}$$

i.e.

$$\delta^{1/4}d\delta = 0.241 \left(\frac{v}{U_\infty}\right)^{1/4} dx$$

Therefore

$$\frac{4}{5}\delta^{5/4} = 0.241\left(\frac{v}{U_\infty}\right)^{1/4} x$$

$$\delta = \left(\frac{5 \times 0.241}{4}\right)^{4/5}\left(\frac{v}{U_\infty}\right)^{1/5} x^{4/5}$$

or, in terms of Reynolds number Re_x, this becomes

$$\delta = 0.383\frac{x}{(Re_x)^{1/5}} \tag{7.81}$$

The developments of laminar and turbulent layers for a given stream velocity are shown plotted in Fig. 7.23.

In order to estimate the other thickness quantities for the turbulent layer, the following integrals must be evaluated:

$$\int_0^1 (1 - \bar{u})d\bar{y} = \int_0^1 (1 - \bar{y}^{1/7})d\bar{y} = \left[\bar{y} - \frac{7}{8}\bar{y}^{8/7}\right]_0^1 = 1 - \frac{7}{8} = 0.125 \tag{b}$$

$$\int_0^1 \bar{u}(1 - \bar{u}^2)d\bar{y} = \int_0^1 (\bar{y}^{1/7} - \bar{y}^{3/7})d\bar{y} = \left[\frac{7}{8}\bar{y}^{8/7} - \frac{7}{10}\bar{y}^{10/7}\right]_0^1$$

$$= \frac{7}{8} - \frac{7}{10} = 0.175 \tag{c}$$

Using the value for I in Eqn (a) above ($I = \frac{7}{72} = 0.0973$) and substituting appropriately for δ, from Eqn (7.81) and for the integral values, from Eqns (b) and (c), in Eqns (7.16), (7.17) and (7.18), leads to

$$\delta^* = 0.125\delta = \frac{0.0479x}{(Re_x)^{1/5}} \tag{7.82}$$

$$\theta = 0.0973\delta = \frac{0.0372x}{(Re_x)^{1/5}} \tag{7.83}$$

$$\delta^{**} = 0.175\delta = \frac{0.0761x}{(Re_x)^{1/5}} \tag{7.84}$$

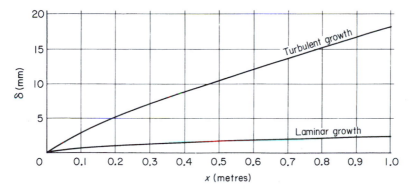

Fig. 7.23 Boundary layer growths on flat plate at free stream speed of $60 \, \mathrm{m/s}^{-1}$

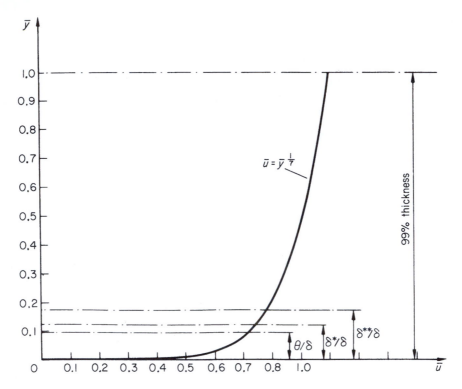

Fig. 7.24 Turbulent velocity profile

The seventh-root profile with the above thickness quantities indicated is plotted in Fig. 7.24.

Example 7.4 A wind-tunnel working section is to be designed to work with no streamwise pressure gradient when running empty at an airspeed of 60 m s^{-1}. The working section is 3.6 m long and has a rectangular cross-section which is 1.2 m wide by 0.9 m high. An approximate allowance for boundary-layer growth is to be made by allowing the side walls of the working section to diverge slightly. It is to be assumed that, at the upstream end of the working section, the turbulent boundary layer is equivalent to one that has grown from zero thickness over a length of 2.5 m; the wall divergence is to be determined on the assumption that the net area of flow is correct at the entry and exit sections of the working section. What must be the width between the walls at the exit section if the width at the entry section is exactly 1.2 m?

For the seventh-root profile:

$$\delta^* = \frac{0.0479x}{(Re_x)^{1/5}} \quad \text{(Eqn (7.82))}$$

At entry, $x = 2.5$ m. Therefore

$$Re_x = \frac{U_\infty x}{\nu} = \frac{60 \times 2.5}{14.6 \times 10^{-6}} = 102.7 \times 10^5$$

$$Re_x^{1/5} = 25.2$$

i.e.

$$\delta^* = \frac{0.0479 \times 2.5}{25.2} = 0.00475 \, \text{m}$$

At exit, $x = 6.1$ m. Therefore

$$Re_x = \frac{60 \times 6.1}{14.6 \times 10^{-6}} = 251 \times 10^5$$

$$Re_x^{1/5} = 30.2$$

i.e.

$$\delta^* = \frac{0.0479 \times 6.1}{30.2} = 0.00968 \, \text{m}$$

Thus δ^* increases by $(0.00968 - 0.00475) = 0.00493$ m. This increase in displacement thickness occurs on all four walls, i.e. total displacement area at exit (relative to entry) = $0.00493 \times 2(1.2 + 0.9) = 0.0207 \, \text{m}^2$.

The allowance is to be made on the two side walls only so that the displacement area on side walls $= 2 \times 0.9 \times \Delta^* = 1.8\Delta^* \, \text{m}^2$, where Δ^* is the exit displacement per wall. Therefore

$$\Delta^* = \frac{0.0207}{1.8} = 0.0115 \, \text{m}$$

This is the displacement for each wall, so that the total width between side walls at the exit section $= 1.2 + 2 \times 0.0115 = 1.223$ m.

7.7.6 Drag coefficient for a flat plate with wholly turbulent boundary layer

The local friction coefficient C_f may now be expressed in terms of x by substituting from Eqn (7.81) in Eqn (7.80). Thus

$$C_f = 0.0468 \left(\frac{v}{U_\infty}\right)^{1/4} \frac{(Re_x)^{1/20}}{(0.383x)^{1/4}} = \frac{0.0595}{(Re_x)^{1/5}} \tag{7.85}$$

whence

$$\tau_w = \frac{0.0595}{2(U_\infty x)^{1/5}} v^{1/5} \rho U_\infty^2 = 0.02975 \rho v^{1/5} U_\infty^{9/5} x^{-1/5} \tag{7.86}$$

The total surface friction force and drag coefficient for a wholly turbulent boundary layer on a flat plate follow as

$$C_F = \int_0^1 C_f \mathrm{d}\left(\frac{x}{L}\right) = \int_0^1 0.0595 \left(\frac{v}{U_\infty}\right)^{1/5} x^{-1/5} \mathrm{d}\left(\frac{x}{L}\right)$$

$$= \left(\frac{v}{U_\infty L}\right)^{1/5} \times 0.0595 \left[\frac{5}{4}\left(\frac{x}{L}\right)^{4/5}\right]_0^1 = 0.0744 Re^{-1/5} \tag{7.87}$$

and

$$C_{D_F} = 0.1488 Re^{-1/5} \tag{7.88}$$

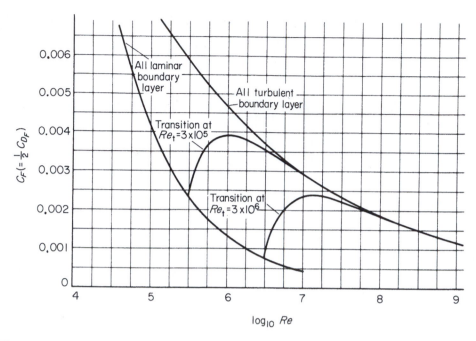

Fig. 7.25 Two-dimensional surface friction drag coefficients for a flat plate. Here Re = plate Reynolds number, i.e. $U_\infty L/\nu$; Re_t = transition Reynolds number, i.e. $U_\infty x_t/\nu_1$, $C_F = F/\frac{1}{2}\rho U_\infty^2 L$; F = skin friction force per surface (unit width)

These expressions are shown plotted in Fig. 7.25 (upper curve). It should be clearly understood that these last two coefficients refer to the case of a flat plate for which the boundary layer is turbulent over the entire streamwise length.

In practice, for Reynolds numbers (Re) up to at least 3×10^5, the boundary layer will be entirely laminar. If the Reynolds number is increased further (by increasing the flow speed) transition to turbulence in the boundary layer may be initiated (depending on free-stream and surface conditions) at the trailing edge, the transition point moving forward with increasing Re (such that Re_x at transition remains approximately constant at a specific value, Re_t, say). However large the value of Re there will inevitably be a short length of boundary layer near the leading edge that will remain laminar to as far back on the plate as the point corresponding to $Re_x = Re_t$. Thus, for a large range of practical Reynolds numbers, the boundary-layer flow on the plate will be partly laminar and partly turbulent. The next stage is to investigate the conditions at transition in order to evaluate the overall drag coefficient for the plate with mixed boundary layers.

7.7.7 Conditions at transition

It is usually assumed for boundary-layer calculations that the transition from laminar to turbulent flow within the boundary layer occurs instantaneously. This is obviously not exactly true, but observations of the transition process do indicate that the transition region (streamwise distance) is fairly small, so that as a first approximation the assumption is reasonably justified. An abrupt change in momentum thickness at the transition point would imply that $d\theta/dx$ is infinite. The simplified momentum integral equation (7.66) shows that this in turn implies that the local skin-friction coefficient C_f would be infinite. This is plainly unacceptable on physical grounds, so it follows that the momentum thickness will remain constant across the transition position. Thus

$$\theta_{L_t} = \theta_{T_t} \tag{7.89}$$

where the suffices L and T refer to laminar and turbulent boundary layer flows respectively and t indicates that these are particular values at transition. Thus

$$\delta_{L_t}\left(\int_0^1 \bar{u}(1-\bar{u})d\bar{y}\right)_L = \delta_{T_t}\left(\int_0^1 \bar{u}(1-\bar{u})d\bar{y}\right)_T$$

The integration being performed in each case using the appropriate laminar or turbulent profile. The ratio of the turbulent to the laminar boundary-layer thicknesses is then given directly by

$$\frac{\delta_{T_t}}{\delta_{L_t}} = \frac{\left(\int_0^1 \bar{u}(1-\bar{u})d\bar{y}\right)_L}{\left(\int_0^1 \bar{u}(1-\bar{u})d\bar{y}\right)_T} = \frac{I_L}{I_T} \tag{7.90}$$

Using the values of I previously evaluated for the cubic and seventh-root profiles (Eqns (ii), Sections 7.6.1 and 7.7.3):

$$\frac{\delta_{T_t}}{\delta_{L_t}} = \frac{0.139}{0.0973} = 1.43 \tag{7.91}$$

This indicates that on a flat plate the boundary layer increases in thickness by about 40% at transition.

It is then assumed that the turbulent layer, downstream of transition, will grow as if it had started from zero thickness at some point ahead of transition and developed along the surface so that its thickness reached the value δ_{T_t} at the transition position.

7.7.8 Mixed boundary layer flow on a flat plate with zero pressure gradient

Figure 7.26 indicates the symbols employed to denote the various physical dimensions used. At the leading edge, a laminar layer will begin to develop, thickening with distance downstream, until transition to turbulence occurs at some Reynolds number $Re_t = U_\infty x_t/\nu$. At transition the thickness increases suddenly from δ_{L_t} in the laminar layer to δ_{T_t} in the turbulent layer, and the latter then continues to grow as if it had started from some point on the surface distant x_{T_t} ahead of transition, this distance being given by the relationship

$$\delta_{T_t} = \frac{0.383 x_{T_t}}{(Re_x)_{T_t}^{1/5}}$$

for the seventh-root profile.

The total skin-friction force coefficient C_F for one side of the plate of length L may be found by adding the skin-friction force per unit width for the laminar boundary layer of length x_t to that for the turbulent boundary layer of length $(L - x_t)$, and

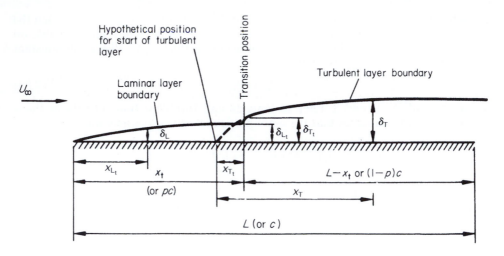

Fig. 7.26

dividing by $\frac{1}{2}\rho U_\infty^2 L$, where L is here the wetted surface area per unit width. Working in terms of Re_t, the transition position is given by

$$x_t = \frac{v}{U_\infty} Re_t \tag{7.92}$$

The laminar boundary-layer momentum thickness at transition is then obtained from Eqn (7.70):

$$\theta_{L_t} = \frac{0.646 x_t}{(Re_t)^{1/2}} = 0.646 \left(\frac{v}{U_\infty}\right)^{1/2} x_t^{1/2}$$

that, on substituting for x_t from Eqn (7.92), gives

$$\theta_{L_t} = 0.646 \frac{v}{U_\infty} (Re_t)^{1/2} \tag{7.93}$$

The corresponding turbulent boundary-layer momentum thickness at transition then follows directly from Eqn (7.83):

$$\theta_{T_t} = \frac{0.037 x_{T_t}}{(Re_x)_{T_t}^{1/5}} \tag{7.94}$$

The equivalent length of turbulent layer (x_{T_t}) to give this thickness is obtained from setting $\theta_{L_t} = \theta_{T_t}$; using Eqn (7.93) and (7.94) this gives

$$0.646 x_t \left(\frac{v}{U_\infty x_t}\right)^{1/2} = 0.037 x_{T_t} \left(\frac{v}{U_\infty x_{T_t}}\right)^{1/5}$$

leading to

$$x_{T_t}^{4/5} = \frac{0.646}{0.037} \left(\frac{v}{U_\infty}\right)^{4/5} Re_t^{1/2}$$

Thus

$$x_{T_t} = 35.5 \frac{v}{U_\infty} Re_t^{5/8} \tag{7.95}$$

Now, on a flat plate with no pressure gradient, the momentum thickness at transition is a measure of the momentum defect produced in the laminar boundary layer between the leading edge and the transition position by the surface friction stresses only. As it is also being assumed here that the momentum thickness through transition is constant, it is clear that the actual surface friction force under the laminar boundary layer of length x_t must be the same as the force that would exist under a turbulent boundary layer of length x_{T_t}. It then follows that the total skin-friction force for the whole plate may be found simply by calculating the skin-friction force under a turbulent boundary layer acting over a length from the point at a distance x_{T_t} ahead of transition, to the trailing edge. Reference to Fig. 7.26 shows that the total effective length of turbulent boundary layer is, therefore, $L - x_t + x_{T_t}$.

Now, from Eqn (7.21),

$$F = \int_0^{L-x_t+x_{T_t}} \tau_w dx = \frac{1}{2} \rho U_\infty^2 \int_0^{L-x_t+x_{T_t}} C_f dx$$

where C_f is given from Eqn (7.85) as

$$\frac{0.0595}{(Re_x)^{1/5}} = 0.0595 \left(\frac{v}{U_\infty}\right)^{1/5} x^{-1/5}$$

Thus

$$F = \frac{1}{2} \rho U_\infty^2 \times 0.0595 \left(\frac{v}{U_\infty}\right)^{1/5} \frac{5}{4} \left[x^{4/5}\right]_0^{L-xt+xT_t}$$

Now, $C_F = F / \frac{1}{2} \rho U_\infty^2 L$, where L is the total chordwise length of the plate, so that

$$C_F = 0.0744 \left(\frac{v}{U_\infty}\right)^{1/5} \frac{(L - x_t + x_{T_t})^{4/5}}{L}$$

$$= 0.0744 \left(\frac{v}{U_\infty L}\right) \left(\frac{U_\infty L}{v} - \frac{U_\infty x}{v} + \frac{U_\infty x_{T_t}}{v}\right)^{4/5}$$

i.e.

$$C_F = \frac{0.0744}{Re} (Re - Re_t + 35.5 Re_t^{5/8})^{4/5} \tag{7.96}$$

This result could have been obtained, alternatively, by direct substitution of the appropriate value of Re in Eqn (7.87), making the necessary correction for effective chord length (see Example 7.5).

The expression enables the curve of either C_F or C_{D_F}, for the flat plate, to be plotted against plate Reynolds number $Re = (U_\infty L/v)$ for a known value of the transition Reynolds number Re_t. Two such curves for extreme values of Re_t of 3×10^5 and 3×10^6 are plotted in Fig. 7.25.

It should be noted that Eqn (7.96) is not applicable for values of Re less than Re_t, when Eqns (7.71) and (7.72) should be used. For large values of Re, greater than about 10^8, the appropriate all-turbulent expressions should be used. However,

Eqns (7.85) and (7.88) become inaccurate for $Re > 10^7$. At higher Reynolds numbers the semi-empirical expressions due to Prandtl and Schlichting should be used, i.e.

$$C_f = [2\log_{10}(Re_x) - 0.65]^{-2.3} \tag{7.97a}$$

$$C_F = \frac{0.455}{(\log_{10} Re)^{2.58}} \tag{7.97b}$$

For the lower transition Reynolds number of 3×10^5 the corresponding value of Re, above which the all-turbulent expressions are reasonably accurate, is 10^7.

Example 7.5   (1) Develop an expression for the drag coefficient of a flat plate of chord c and infinite span at zero incidence in a uniform stream of air, when transition occurs at a distance pc from the leading edge. Assume the following relationships for laminar and turbulent boundary layer velocity profiles, respectively:

$$\bar{u}_L = \frac{3}{2}\bar{y} - \frac{1}{2}\bar{y}^3, \qquad \bar{u}_T = \bar{y}^{1/7}$$

(2) On a thin two-dimensional aerofoil of 1.8 m chord in an airstream of $45\,\text{m s}^{-1}$, estimate the required position of transition to give a drag per metre span that is 4.5 N less than that for transition at the leading edge.

(1) Refer to Fig. 7.26 for notation.
From Eqn (7.95), setting $x_t = pc$

$$x_{T_t} = 35.5(pc)^{(5/8)} \left(\frac{\nu}{U_\infty}\right)^{(3/8)} \tag{i}$$

Equation (7.88) gives the drag coefficient for an all-turbulent boundary layer as $C_{D_F} = 0.1488/Re^{1/5}$. For the mixed boundary layer, the drag is obtained as for an all-turbulent layer of length $[x_{T_t} + (1-p)c]$. The corresponding drag coefficient (defined with reference to length $[x_{T_t} + (1-p)c]$) is then obtained directly from the all-turbulent expression where Re is based on the same length $[x_{T_t} + (1-p)c]$. To relate the coefficient to the whole plate length c then requires that the quantity obtained should now be factored by the ratio

$$\frac{[x_{T_t} + (1-p)c]}{c}$$

Thus

$$C_{D_F} = \frac{D_F}{\frac{1}{2}\rho U_\infty^2 c} = \frac{0.1488}{Re_{[xT_t+(1-p)c]}^{1/5}} \times \frac{[x_{T_t} + (1-p)c]}{c} = \frac{0.1488}{\left(\frac{U_\infty}{\nu}\right)^{1/5}} \frac{[x_{T_t} + (1-p)c]^{4/5}}{c}$$

$$= \frac{0.1488[x_{T_t} + (1-p)c]^{4/5}}{\frac{U_\infty}{\nu} c \left(\frac{\nu}{U_\infty}\right)^{4/5}} = \frac{0.1488}{Re}\left[\frac{U_\infty}{\nu}x_{T_t} + (1-p)Re\right]^{4/5}$$

N.B. Re is here based on total plate length c. Substituting from Eqn (i) for x_{T_t}, then gives

$$C_{D_F} = \frac{0.1488}{Re}[35.5p^{5/8}Re^{5/8} + (1-p)Re]^{4/5}$$

This form of expression (as an alternative to Eqn (7.96)) is convenient for enabling a quick approximation to skin-friction drag to be obtained when the position of transition is likely to be fixed, rather than the transition Reynolds number, e.g. by position of maximum thickness, although strictly the profile shapes will not be unchanged with length under these conditions and neither will U_e over the length.

(2) With transition at the leading edge:

$$C_{D_F} = \frac{0.1488}{Re^{1/5}}$$

In this case

$$Re = \frac{Uc}{v} = \frac{45 \times 1.8}{14.6 \times 10^{-6}} = 55.5 \times 10^5$$

$$Re^{1/5} = 22.34$$

and

$$C_{D_F} = \frac{0.1488}{22.34} = 0.006\,67$$

The corresponding aerofoil drag is then $D_F = 0.006\,67 \times 0.6125 \times (45)^2 \times 1.8 = 14.88$ N. With transition at pc, $D_F = 14.86 - 4.5 = 10.36$ N, i.e.

$$C_{D_F} = \frac{10.36}{14.88} \times 0.006\,67 = 0.004\,65$$

Using this value in (i), with $Re^{5/8} = 16\,480$, gives

$$0.004\,65 = \frac{0.1488}{55.8 \times 10^5} [35.5 p^{5/8} \times 16\,480 + 55.8 \times 10^5 - 55.8 \times 10^5 p]^{4/5}$$

i.e.

$$5.84 \times 10^5 p^{5/8} - 55.8 \times 10^5 p = \left(\frac{55.8 \times 465}{0.1488}\right)^{5/4} - 55.8 \times 10^5 = (35.6 - 55.8)10^5$$

or

$$55.8p - 5.84p^{5/8} = 20.2$$

The solution to this (by successive approximation) is $p = 0.423$, i.e.

$$pc = 0.423 \times 1.8 = 0.671 \text{ m behind leading edge}$$

Example 7.6   A light aircraft has a tapered wing with root and tip chord-lengths of 2.2 m and 1.8 m respectively and a wingspan of 16 m. Estimate the skin-friction drag of the wing when the aircraft is travelling at 55 m/s. On the upper surface the point of minimum pressure is located at 0.375 chord-length from the leading edge. The dynamic viscosity and density of air may be taken as 1.8×10^{-5} kg s/m and 1.2 kg/m^3 respectively.

The average wing chord is given by $\bar{c} = 0.5(2.2 + 1.8) = 2.0$ m, so the wing is taken to be equivalent to a flat plate measuring 2.0 m $\times$ 16 m. The overall Reynolds number based on average chord is given by

$$Re = \frac{1.2 \times 55 \times 2.0}{1.8 \times 10^{-5}} = 7.33 \times 10^6$$

Since this is below 10^7 the guidelines at the end of Section 7.9 suggest that the transition point will be very shortly after the point of minimum pressure, so $x_t \simeq 0.375 \times 2.0 = 0.75$ m; also Eqn (7.96) may be used.

$$Re_t = 0.375 \times Re = 2.75 \times 10^6$$

So Eqn (7.96) gives

$$C_F = \frac{0.0744}{7.33 \times 10^6}\{7.33 \times 10^6 - 2.75 \times 10^6 + 35.5(2.75 \times 10^6)^{5/8}\}^{4/5} = 0.0023$$

Therefore the skin-friction drag of the upper surface is given by

$$D = \frac{1}{2}\rho U_\infty^2 \bar{c}sC_F = 0.5 \times 1.2 \times 55^2 \times 2.0 \times 16 \times 0.0023 = 133.8\,\text{N}$$

Finally, assuming that the drag of the lower surface is similar, the estimate for the total skin-friction drag for the wing is $2 \times 133.8 \simeq 270\,\text{N}$.

7.8 Additional examples of the application of the momentum integral equation

For the general solution of the momentum integral equation it is necessary to resort to computational methods, as described in Section 7.11. It is possible, however, in certain cases with external pressure gradients to find engineering solutions using the momentum integral equation without resorting to a computer. Two examples are given here. One involves the use of suction to control the boundary layer. The other concerns determining the boundary-layer properties at the leading-edge stagnation point of an aerofoil. For such applications Eqn (7.59) can be written in the alternative form with $H = \delta^*/\theta$:

$$\frac{C_f}{2} = \frac{V_s}{U_e} + \frac{\theta}{U_e}\frac{\mathrm{d}U_e}{\mathrm{d}x}(H+2) + \frac{\mathrm{d}\theta}{\mathrm{d}x} \tag{7.98}$$

When, in addition, there is no pressure gradient and no suction, this further reduces to the simple momentum integral equation previously obtained (Section 7.7.1, Eqn (7.66)), i.e. $C_f = 2(\mathrm{d}\theta/\mathrm{d}x)$.

Example 7.7 A two-dimensional divergent duct has a total included angle, between the plane diverging walls, of $20°$. In order to prevent separation from these walls and also to maintain a laminar boundary-layer flow, it is proposed to construct them of porous material so that suction may be applied to them. At entry to the diffuser duct, where the flow velocity is $48\,\text{m s}^{-1}$ the section is square with a side length of $0.3\,\text{m}$ and the laminar boundary layers have a general thickness (δ) of $3\,\text{mm}$. If the boundary-layer thickness is to be maintained constant at this value, obtain an expression in terms of x for the value of the suction velocity required, along the diverging walls. It may be assumed that for the diverging walls the laminar velocity profile remains constant and is given approximately by $\bar{u} = 1.65\bar{y}^3 - 4.30\bar{y}^2 + 3.65\bar{y}$.

The momentum equation for steady flow along the porous walls is given by Eqn (7.98) as

$$\frac{V_s}{U_e} = \frac{C_f}{2} - \frac{1}{U_e}\frac{\mathrm{d}U_e}{\mathrm{d}x}(H+2)\theta - \frac{\mathrm{d}\theta}{\mathrm{d}x}$$

If the thickness δ is to remain constant and the profile also, then $\theta = $ constant and $\mathrm{d}\theta/\mathrm{d}x = 0$. Also

$$\frac{C_f}{2} = \frac{\tau_w}{\rho U_e^2} = \frac{\mu\dfrac{U_e}{\delta}\left(\dfrac{\partial\bar{u}}{\partial\bar{y}}\right)_w}{\rho U_e^2} = \frac{v}{U_e\delta}\left(\frac{\partial\bar{u}}{\partial\bar{y}}\right)_w$$

i.e.

$$V_s = \frac{v}{\delta}\left(\frac{\partial \bar{u}}{\partial \bar{y}}\right)_w - \frac{dU_e}{dx}(H+2) \times \delta \frac{\theta}{\delta}$$

$$\frac{\partial \bar{u}}{\partial \bar{y}} = 4.95\bar{y}^2 - 8.60\bar{y} + 3.65$$

$$\left(\frac{\partial \bar{u}}{\partial \bar{y}}\right)_w = 3.65$$

Equation (7.16) gives

$$\frac{\delta^*}{\delta} = \int_0^1 (1 - \bar{u})d\bar{y} = \int_0^1 (1 - 1.65\bar{y}^3 + 4.30\bar{y}^2 - 3.65\bar{y})d\bar{y} = 0.1955$$

Equation (7.17) gives

$$\frac{\theta}{\delta} = \int_0^1 \bar{u}(1 - \bar{u})d\bar{y} = \int_0^1 (3.65\bar{y} - 17.65\bar{y}^2 + 33.05\bar{y}^3 - 30.55\bar{y}^4 + 14.2\bar{y}^5 - 2.75\bar{y}^6)d\bar{y} = 0.069$$

$$H = \frac{\delta^*}{\theta} = \frac{0.1955}{0.069} = 2.83$$

Also $\delta = 0.003\,\text{m}$

Diffuser duct cross-sectional area $= 0.09 + 0.06x \tan 10°$ where $x = $ distance from entry section, i.e.

$$A = 0.09 + 0.106x$$

and

$$A/A_e = 1 + 1.178x$$

where suffix i denotes the value at the entry section. Also

$$A_e U_{e_i} = A U_e$$

$$U_e = \frac{A_e}{A} U_{e_i} = \frac{48}{1 + 1.178x}$$

Then

$$\frac{dU_e}{dx} = -48 \times 1.178(1 + 1.178x)^{-2}$$

Finally

$$V_s = \frac{14.6 \times 10^{-6}}{0.003} \times 3.65 + \frac{48 \times 1.178 \times 4.83 \times 0.003 \times 0.069}{(1 + 1.178x)^2}$$

$$= 0.0178 + \frac{0.0565}{(1 + 1.178x)^2}\,\text{m s}^{-1}$$

Thus the maximum suction is required at entry, where $V_s = 0.0743\,\text{m s}^{-1}$.

For bodies with sharp leading edges such as flat plates the boundary layer grows from zero thickness. But in most engineering applications, e.g. conventional aerofoils, the leading edge is rounded. Under these circumstances the boundary layer has a finite thickness at the leading edge, as shown in Fig. 7.27a. In order to estimate the

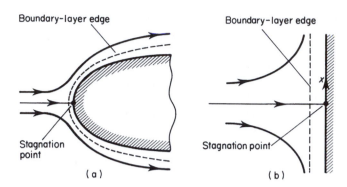

Fig. 7.27 Boundary-layer flow in the vicinity of the fore stagnation point

initial boundary-layer thickness it can be assumed that the flow in the vicinity of the stagnation point is similar to that approaching a flat plate oriented perpendicularly to the free-stream, as shown in Fig. 7.27b. For this flow $U_e = cx$ (where c is a constant) and the boundary-layer thickness does not change with x. In the example given below the momentum integral equation will be used to estimate the initial boundary-layer thickness for the flow depicted in Fig. 7.27b. An exact solution to the Navier–Stokes equations can be found for this stagnation-point flow (see Section 2.10.3). Here the momentum integral equation is used to obtain an approximate solution.

Example 7.8   Use the momentum integral equation (7.59) and the results (7.64a', b', c') to obtain expressions for δ, δ^*, θ and C_f. It may be assumed that the boundary-layer thickness does not vary with x and that $U_e = cx$.

$$\Lambda = \frac{\delta^2}{\nu}\frac{\mathrm{d}U_e}{\mathrm{d}x} = \frac{\delta^2}{\nu}c = \text{const.}$$

Hence $\theta = \text{const.}$ also and Eqn (7.59) becomes

$$\theta\frac{\mathrm{d}}{\mathrm{d}x}(c^2x^2) + \delta^*c^2x = \frac{\tau_w}{\rho}$$

Substituting Eqns (7.64a', b', c') leads to

$$2\delta c^2 x\frac{1}{63}\left(\frac{37}{5} - \frac{\Lambda}{15} - \frac{\Lambda^2}{144}\right) + c^2x\delta\left(\frac{3}{10} - \frac{\Lambda}{210}\right) = \nu cx\frac{1}{\delta}\left(2 + \frac{\Lambda}{6}\right)$$

Multiplying both sides by $\delta/\nu cx$ and using the above result for Λ, gives

$$\frac{2}{63}\left(\frac{37}{5} - \frac{\Lambda}{15} - \frac{\Lambda^2}{144}\right)\Lambda + \left(\frac{3}{10} - \frac{\Lambda}{120}\right)\Lambda = 2 + \frac{\Lambda}{6}$$

After rearrangement this equation simplifies to

$$\frac{1}{4536}\Lambda^3 + \left(\frac{1}{120} + \frac{2}{945}\right)\Lambda^2 + \left(\frac{1}{6} - \frac{3}{10} - \frac{74}{315}\right)\Lambda + 2 = 0$$

or

$$0.00022\Lambda^3 + 0.01045\Lambda^2 - 0.3683\Lambda + 2 = 0$$

It is known that Λ lies somewhere between 0 and 12 so it is relatively easy to solve this equation by trial and error to obtain

$$\Lambda = 7.052 \quad \Rightarrow \quad \delta = \sqrt{\frac{v\Lambda}{c}} = 2.655\sqrt{\frac{v}{c}}$$

Using Eqns (7.64a', b', c') then gives

$$\delta^* = 0.641\sqrt{\frac{v}{c}}, \qquad \theta = 0.278\sqrt{\frac{v}{c}}, \qquad C_f = 2.392\sqrt{\frac{v}{c}}\frac{1}{x}$$

Once the value of $c = (\mathrm{d}U_e/\mathrm{d}x)_{x=0}$ is specified (see Example 2.4) the results given above can be used to supply initial conditions for boundary-layer calculations over aerofoils.

7.9 Laminar-turbulent transition

It was mentioned in Section 7.2.5 above that transition from laminar to turbulent flow usually occurs at some point along the surface. This process is exceedingly complex and remains an active area of research. Owing to the very rapid changes in both space and time the simulation of transition is, arguably, the most challenging problem in computational fluid dynamics. Despite the formidable difficulties however, considerable progress has been made and transition can now be reliably predicted in simple engineering applications. The theoretical treatment of transition is beyond the scope of the present work. Nevertheless, a physical understanding of transition is vital for many engineering applications of aerodynamics, and accordingly a brief account of the underlying physics of transition in a boundary layer on a flat plate is given below.

Transition occurs because of the growth of small disturbances in the boundary layer. In many respects, the boundary layer can be regarded as a complex nonlinear oscillator that under certain circumstances has an initially linear wave-like response to external stimuli (or inputs). This is illustrated schematically in Fig. 7.28. In free flight or in high-quality wind-tunnel experiments several stages in the process can be discerned. The first stage is the conversion of external stimuli or disturbances into low-amplitude waves. The external disturbances may arise from a variety of different sources, e.g. free-stream turbulence, sound waves, surface roughness and vibration. The conversion process is still not well understood. One of the main difficulties is that the wave-length of a typical external disturbance is invariably very much larger than that of the wave-like response of the boundary layer. Once the low-amplitude wave is generated it will propagate downstream in the boundary layer and, depending on the local conditions, grow or decay. If the wave-like disturbance grows it will eventually develop into turbulent flow.

While their amplitude remains small the waves are predominantly two-dimensional (see Figs 7.28 and 7.29). This phase of transition is well understood and was first explained theoretically by Tollmien[*] with later extensions by Schlichting[†] and many others. For this reason the growing waves in the early so-called linear phase of transition are known as *Tollmien–Schlichting waves*. This linear phase extends for some 80% of the total transition region. The more advanced engineering predictions

---

[*] W. Tollmien (1929) Über die Entstehung der Turbulenz. *1. Mitt. Nachr. Ges. Wiss. Göttingen, Math. Phys. Klasse*, pp. 21–44.

[†] H. Schlichting (1933) Zur Entstehung der Turbulenz bei der Plattenströmung. *Z. angew. Math. Mech.*, **13**, 171–174.

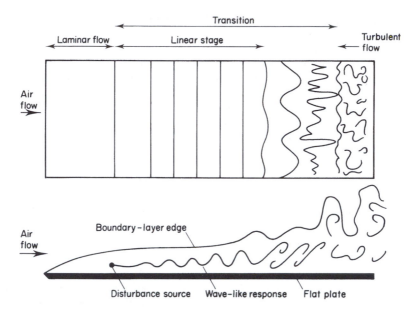

Fig. 7.28 Schematic of transition in a boundary layer over a flat plate, with disturbances generated by a harmonic line source

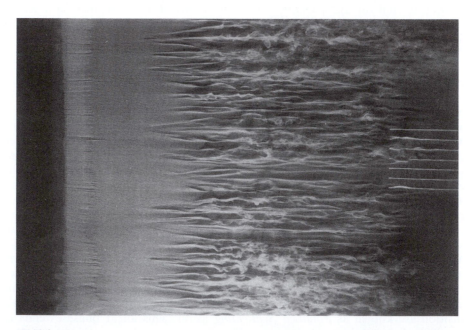

Fig. 7.29 Laminar-turbulent transition in a flat-plate boundary layer: This is a planform view of a dye sheet emitted upstream parallel to the wall into water flowing from left to right. Successive stages of transition are revealed, i.e. laminar flow on the upstream side, then the appearance of the two-dimensional Tollmien–Schlichting waves, followed by the formation of turbulent spots and finally fully developed turbulent flow. The Reynolds number based on distance along the wall is about 75 000. See Fig. 7.28 above for a schematic side-view of the transition process. (*The photograph was taken by H. Werlé at ONERA, France.*)

are, in fact, based on modern versions of Tollmien's linear theory. The theory is linear because it assumes the wave amplitudes are so small that their products can be neglected. In the later nonlinear stages of transition the disturbances become increasingly three-dimensional and develop very rapidly. In other words as the amplitude of the disturbance increases the response of the boundary layer becomes more and more complex.

This view of transition originated with Prandtl* and his research team at Göttingen, Germany, which included Tollmien and Schlichting. Earlier theories, based on neglecting viscosity, seemed to suggest that small disturbances could not grow in the boundary layer. One effect of viscosity was well known. Its so-called dissipative action in removing energy from a disturbance, thereby causing it to decay. Prandtl realized that, in addition to its dissipative effect, viscosity also played a subtle but essential role in promoting the growth of wave-like disturbances by causing energy to be transferred to the disturbance. His explanation is illustrated in Fig. 7.30. Consider a small-amplitude wave passing through a small element of fluid within the boundary

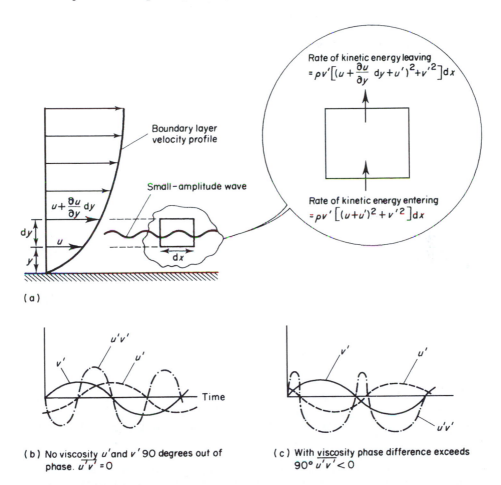

Boundary layer velocity profile

Rate of kinetic energy leaving
$$= \rho v' \left[\left(u + \frac{\partial u}{\partial y} \, dy + u' \right)^2 + v'^2 \right] dx$$

Rate of kinetic energy entering
$$= \rho v' \left[(u + u')^2 + v'^2 \right] dx$$

Small-amplitude wave

$u + \frac{\partial u}{\partial y} \, dy$

u

(a)

Time

(b) No viscosity u' and v' 90 degrees out of phase. $\overline{u'v'} = 0$

(c) With viscosity phase difference exceeds 90° $\overline{u'v'} < 0$

Fig. 7.30 Prandtl's explanation for disturbance growth

* L. Prandtl (1921) Bermerkungen über die Enstehung der Turbulenz, *Z. angew. Math. Mech.*, **1**, 431–436.

layer, as shown in Fig. 7.30a. The instantaneous velocity components of the wave are (u', v') in the (x, y) directions, u' and v' are very much smaller than u, the velocity in the boundary layer in the absence of the wave. The instantaneous rate of increase in kinetic energy within the small element is given by the difference between the rates at which kinetic energy leaves the top of the element and enters the bottom, i.e.

$$-\rho u' v' \frac{\partial u}{\partial y} + \text{higher order terms}$$

In the absence of viscosity u' and v' are exactly 90 degrees out of phase and the average of their product over a wave period, denoted by $\overline{u'v'}$, is zero, see Fig. 7.30b. However, as realized by Prandtl, the effects of viscosity are to increase the phase difference between u' and v' to slightly more than 90 degrees. Consequently, as shown in Fig. 7.30c, $\overline{u'v'}$ is now negative, resulting in a net energy transfer to the disturbance. The quantity $-\rho \overline{u'v'}$ is, in fact, the Reynolds stress referred to earlier in Section 7.2.4. Accordingly, the energy transfer process is usually referred to as *energy production by the Reynolds stress*. This mechanism is active throughout the transition process and, in fact, plays a key role in sustaining the fully turbulent flow (see Section 7.10).

Tollmien was able to verify Prandtl's hypothesis theoretically, thereby laying the foundations of the modern theory for transition. It was some time, however, before the ideas of the Göttingen group were accepted by the aeronautical community. In part this was because experimental corroboration was lacking. No sign of Tollmien–Schlichting waves could at first be found in experiments on natural transition. Schubauer and Skramstadt* did succeed in seeing them but realized that in order to study such waves systematically they would have to be created artificially in a controlled manner. So they placed a vibrating ribbon having a controlled frequency, ω, within the boundary layer to act as a wave-maker, rather than relying on natural sources of disturbance. Their results are illustrated schematically in Fig. 7.31. They found that for high ribbon frequencies, see Case (a), the waves always decayed. For intermediate frequencies (Case (b)) the waves were attenuated just downstream of the ribbon, then at a greater distance downstream they began to grow, and finally at still greater distances downstream decay resumed. For low frequencies the waves grew until their amplitude was sufficiently large for the nonlinear effects, alluded to above, to set in, with complete transition to turbulence occurring shortly afterwards. Thus, as shown in Fig. 7.31, Schubauer and Skramstadt were able to map out a curve of non-dimensional frequency versus $Re_x (= U_\infty x / \nu)$ separating the disturbance frequencies that will grow at a given position along the plate from those that decay. When disturbances grow the boundary-layer flow is said to be *unstable* to small disturbances, conversely when they decay it is said to be *stable*, and when the disturbances neither grow nor decay it is in a state of *neutral stability*. Thus the curve shown in Fig. 7.31 is known as the *neutral-stability boundary* or curve. Inside the neutral-stability curve, production of energy by the Reynolds stress exceeds viscous dissipation, and vice versa outside. Note that a *critical Reynolds number* Re_c and *critical frequency* ω_c exist. The Tollmien–Schlichting waves cannot grow at Reynolds numbers below Re_c or at frequencies above ω_c. However, since the disturbances leading to transition to turbulence are considerably lower than the critical frequency, the transitional Reynolds number is generally considerably greater than Re_c.

The shape of the neutral-stability curve obtained by Schubauer and Skramstadt agreed well with Tollmien's theory, especially at the lower frequencies of interest for

---

* G.B. Schubauer and H.K. Skramstadt (1948) Laminar boundary layer oscillations and transition on a flat plate. *NACA Rep., 909*.

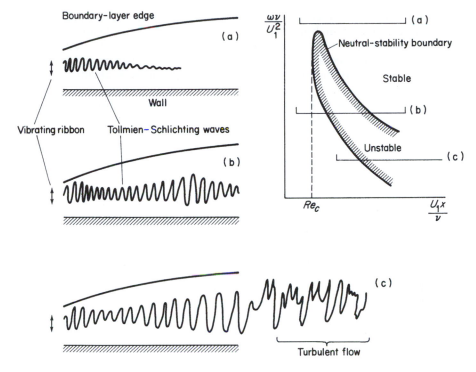

Fig. 7.31 Schematic of Schubauer and Skramstadt's experiment

transition. Moreover Schubauer and Skramstadt were also able to measure the growth rates of the waves and these too agreed well with Tollmien and Schlichting's theoretical calculations. Publication of Schubauer and Skramstadt's results finally led to the Göttingen 'small disturbance' theory of transition becoming generally accepted.

It was mentioned above that Tollmien–Schlichting waves could not be easily observed in experiments on natural transition. This is because the natural sources of disturbance tend to generate wave packets in an almost random fashion in time and space. Thus at any given instant there is a great deal of 'noise', tending to obscure the wave-like response of the boundary layer, and also disturbances having a wide range of frequencies are continually being generated. In contrast, the Tollmien–Schlichting theory is based on disturbances with a single frequency. Nevertheless, providing the initial level of the disturbances is low, what seems to happen is that the boundary layer responds preferentially, so that waves of a certain frequency grow most rapidly and are primarily responsible for transition. These most rapidly growing waves are those predicted by the modern versions of the Tollmien–Schlichting theory, thereby allowing the theory to predict, approximately at least, the onset of natural transition.

It has been explained above that provided the initial level of the external disturbances is low, as in typical free-flight conditions, there is a considerable difference between the critical and transitional Reynolds number. In fact, the latter is about 3×10^6 whereas $Re_c \simeq 3 \times 10^5$. However, if the initial level of the disturbances rises, for example because of increased free-stream turbulence or surface roughness, the downstream distance required for the disturbance amplitude to grow sufficiently for nonlinear effects to set in becomes shorter. Therefore, the transitional Reynolds number is reduced to a value closer to Re_c. In fact, for high-disturbance environments, such as those encountered in turbomachinery, the linear phase of transition is by-passed completely and laminar flow breaks down very abruptly into fully developed turbulence.

The Tollmien–Schlichting theory can also predict very successfully how transition will be affected by an external pressure gradient. The neutral-stability boundaries for the flat plate and for typical adverse and favourable pressure gradients are plotted schematically in Fig. 7.32. In accordance with the theoretical treatment Re_δ is used as the abscissa in place of Re_x. However, since the boundary layer grows with passage downstream Re_δ can still be regarded as a measure of distance along the surface. From Fig. 7.32 it can be readily seen that for adverse pressure gradients not only is $(Re_\delta)_c$ smaller than for a flat plate, but a much wider band of disturbance frequencies are unstable and will grow. When it is recalled that the boundary-layer thickness also grows more rapidly in an adverse pressure gradient, thereby reaching a given critical value of Re_δ sooner, it can readily be seen that transition is promoted under these circumstances. Exactly the converse is found for the favourable pressure gradient. This circumstance allows rough and ready predictions to be made for the transition

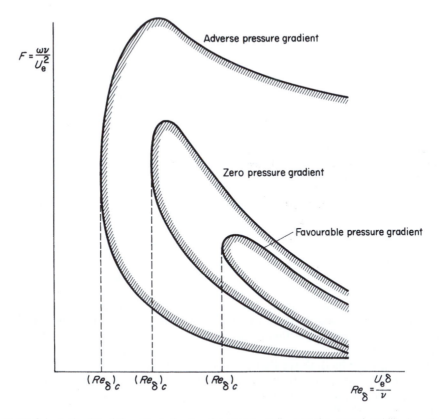

Fig. 7.32 Schematic plot of the effect of external pressure gradient on the neutral stability boundaries

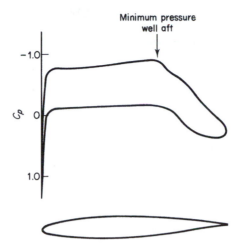

Fig. 7.33 Modern laminar-flow aerofoil and its pressure distribution

point on bodies and wings, especially in the case of the more classic streamlined shapes. These guidelines may be summarized as follows:

(i) If $10^5 < Re_L < 10^7$ (where $Re_L = U_\infty L/v$ is based on the total length or chord of the body or wing) then transition will occur very shortly downstream of the point of minimum pressure. For aerofoils at zero incidence or for streamlined bodies of revolution, the point of minimum pressure often, but not invariably, coincides with the point of maximum thickness.

(ii) If for an aerofoil Re_L is kept constant increasing the angle of incidence advances the point of minimum pressure towards the leading edge on the upper surface, causing transition to move forward. The opposite occurs on the lower surface.

(iii) At constant incidence an increase in Re_L tends to advance transition.

(iv) For $Re_L > 10^7$ the transition point may slightly precede the point of minimum pressure.

The effects of external pressure gradient on transition also explain how it may be postponed by designing aerofoils with points of minimum pressure further aft. A typical modern aerofoil of this type is shown in Fig. 7.33. The problem with this type of aerofoil is that, although the onset of the adverse pressure gradient is postponed, it tends to be correspondingly more severe, thereby giving rise to boundary-layer separation. This necessitates the use of boundary-layer suction aft of the point of minimum pressure in order to prevent separation and to maintain laminar flow. See Section 7.4 and 8.4.1 below.

7.10 The physics of turbulent boundary layers

In this section, a brief account is given of the physics of turbulent boundary layers. This is still very much a developing subject and an active research topic. But some classic empirical knowledge, results and methods have stood the test of time and are worth describing in a general textbook on aerodynamics. Moreover, turbulent flows are so important for engineering applications that some understanding of the relevant flow physics is essential for predicting and controlling flows.

7.10.1 Reynolds averaging and turbulent stress

Turbulent flow is a complex motion that is fundamentally three-dimensional and highly unsteady. Figure 7.34a depicts a typical variation of a flow variable, f, such as velocity or pressure, with time at a fixed point in a turbulent flow. The usual approach in engineering, originating with Reynolds*, is to take a time average. Thus the instantaneous velocity is given by

$$f = \bar{f} + f' \tag{7.99}$$

where the time average is denoted by ($\bar{}$) and ( )$'$ denotes the fluctuation (or deviation from the time average). The strict mathematical definition of the time average is

$$\bar{f} \equiv \lim_{T \to \infty} \frac{1}{T} \int_0^T f(x, y, z, t = t_0 + t')\mathrm{d}t' \tag{7.100}$$

where t_0 is the time at which measurement is notionally begun. For practical measurements T is merely taken as suitably large rather than infinite. The basic approach is often known as *Reynolds averaging*.

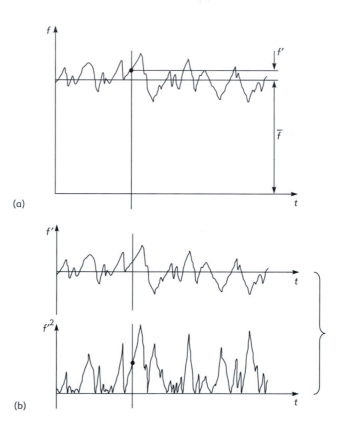

(a)

(b)

Fig. 7.34

* Reynolds, O. (1895) ' On the dynamical theory of incompressible viscous fluids and the determination of the criterion', *Philosophical Transactions of the Royal Society of London, Series A*, **186**, 123.

We will now use the Reynolds averaging approach on the continuity equation (2.94) and x-momentum Navier–Stokes equation (2.95a). When Eqn (7.99) with u for f and similar expressions for v and w are substituted into Eqn (2.94) we obtain

$$\frac{\partial \bar{u}}{\partial x} + \frac{\partial \bar{v}}{\partial y} + \frac{\partial \bar{w}}{\partial z} + \frac{\partial u'}{\partial x} + \frac{\partial v'}{\partial y} + \frac{\partial w'}{\partial z} = 0 \tag{7.101}$$

Taking a time average of a fluctuation gives zero by definition, so taking a time average of Eqn (7.101) gives

$$\frac{\partial \bar{u}}{\partial x} + \frac{\partial \bar{v}}{\partial y} + \frac{\partial \bar{w}}{\partial z} = 0 \tag{7.102}$$

Subtracting Eqn (7.102) from Eqn (7.101) gives

$$\frac{\partial u'}{\partial x} + \frac{\partial v'}{\partial y} + \frac{\partial w'}{\partial z} = 0 \tag{7.103}$$

This result will be used below.

We now substitute Eqn (7.99) to give expressions for u, v, w and p into Eqn (2.95a) to obtain

$$\rho \left(\frac{\partial(\bar{u} + u')}{\partial t} + (\bar{u} + u')\frac{\partial(\bar{u} + u')}{\partial x} + (\bar{v} + v')\frac{\partial(\bar{u} + u')}{\partial y} + (\bar{w} + w')\frac{\partial(\bar{u} + u')}{\partial z} \right)$$
$$= -\frac{\partial(\bar{p} + p')}{\partial x} + \mu \left(\frac{\partial^2(\bar{u} + u')}{\partial x^2} + \frac{\partial^2(\bar{u} + u')}{\partial y^2} + \frac{\partial^2(\bar{u} + u')}{\partial z^2} \right) \tag{7.104}$$

We now take a time average of each term, noting that although the time average of a fluctuation is zero by definition (see Fig. 7.34b), the time average of a product of fluctuations is not, in general, equal to zero (e.g. plainly $\overline{u'u'} = \overline{u'^2} > 0$, see Fig. 7.34b). Let us also assume that the turbulent boundary-layer flow is two-dimensional when time-averaged, so that no time-averaged quantities vary with z and $\bar{w} = 0$. Thus if we take the time average of each term of Eqn (7.104), it simplifies to

$$\rho \left(\bar{u}\frac{\partial \bar{u}}{\partial x} + \bar{v}\frac{\partial \bar{u}}{\partial y} \right) + \overline{\left(u'\frac{\partial u'}{\partial y} + v'\frac{\partial u'}{\partial y} + w'\frac{\partial u'}{\partial z} \right)}$$
$$\underbrace{\phantom{u'\frac{\partial u'}{\partial y} + v'\frac{\partial u'}{\partial y} + w'\frac{\partial u'}{\partial z}}}_{*}$$
$$= -\frac{\partial \bar{p}}{\partial x} + \mu \left(\frac{\partial^2 \bar{u}}{\partial x^2} + \frac{\partial^2 \bar{u}}{\partial y^2} \right) \tag{7.105}$$

The term marked with $*$ can be written as

$$\overline{\left(\frac{\partial u'^2}{\partial x} - u'\frac{\partial u'}{\partial y} + \frac{\partial(u'v')}{\partial y} - u'\frac{\partial v'}{\partial y} + \frac{\partial(u'w')}{\partial z} - u'\frac{\partial w'}{\partial z} \right)}$$

$$= \frac{\partial \overline{u'^2}}{\partial x} + \frac{\partial \overline{u'v'}}{\partial y} + \frac{\partial \overline{u'w'}}{\partial z} - \overline{u'\left(\frac{\partial u'}{\partial x} + \frac{\partial v'}{\partial y} + \frac{\partial w'}{\partial z} \right)}$$
$$\underbrace{\phantom{\frac{\partial u'}{\partial x} + \frac{\partial v'}{\partial y} + \frac{\partial w'}{\partial z}}}_{=0 \text{ from Eqn (7.103)}}$$

$$= \frac{\partial \overline{u'^2}}{\partial x} + \frac{\partial \overline{u'v'}}{\partial y} + \underbrace{\frac{\partial \overline{u'w'}}{\partial z}}_{=0 \text{ no variation with } z}$$

So that Eqn (7.105) becomes

$$\rho\left(\bar{u}\frac{\partial\bar{u}}{\partial x} + \bar{v}\frac{\partial\bar{u}}{\partial y}\right) = -\frac{\partial\bar{p}}{\partial x} + \mu\left(\frac{\partial\bar{\sigma}_{xx}}{\partial x} + \frac{\partial\bar{\sigma}_{xy}}{\partial y}\right) \tag{7.106}$$

where we have written

$$\bar{\sigma}_{xx} = \mu\frac{\partial\bar{u}}{\partial x} - \rho\overline{u'^2}; \qquad \bar{\sigma}_{xy} = \mu\frac{\partial\bar{u}}{\partial y} - \rho\overline{u'v'}$$

This notation makes it evident that when the turbulent flow is time-averaged $-\rho\overline{u'^2}$ and $-\rho\overline{u'v'}$ take on the character of a direct and shear stress respectively. For this reason, the quantities are known as *Reynolds stresses* or *turbulent stresses*. In fully turbulent flows, the Reynolds stresses are usually very much greater than the viscous stresses. If the time-averaging procedure is applied to the full three-dimensional Navier–Stokes equations (2.95), a Reynolds stress tensor is generated with the form

$$-\rho\begin{pmatrix} \overline{u'^2} & \overline{u'v'} & \overline{u'w'} \\ \overline{u'v'} & \overline{v'^2} & \overline{v'w'} \\ \overline{u'w'} & \overline{v'w'} & \overline{w'^2} \end{pmatrix} \tag{7.107}$$

It can be seen that, in general, there are nine components of the Reynolds stress comprising six distinct quantities.

7.10.2   Boundary-layer equations for turbulent flows

For the applications considered here, namely two-dimensional boundary layers (more generally, two-dimensional shear layers), only one of the Reynolds stresses is significant, namely the Reynolds shear stress, $-\rho\overline{u'v'}$. Thus for two-dimensional turbulent boundary layers the time-averaged boundary-layer equations (c.f. Eqns 7.7 and 7.14), can be written in the form

$$\frac{\partial\bar{u}}{\partial x} + \frac{\partial\bar{v}}{\partial y} = 0 \tag{7.108a}$$

$$\bar{u}\frac{\partial\bar{u}}{\partial x} + \bar{v}\frac{\partial\bar{u}}{\partial y} = -\frac{d\bar{p}}{dx} + \frac{\partial\bar{\tau}}{\partial y} \tag{7.108b}$$

The chief difficulty of turbulence is that there is no way of determining the Reynolds stresses from first principles, apart from solving the unsteady three-dimensional Navier–Stokes equations. It is necessary to formulate semi-empirical approaches for modelling the Reynolds shear stress before one can begin the process of solving Eqns (7.108a,b).

The momentum integral form of the boundary-layer equations derived in Section 7.6.1 is equally applicable to laminar or turbulent boundary layers, providing it is recognized that the time-averaged velocity should be used in the definition of momentum and displacement thicknesses. This is the basis of the approximate methods described in Section 7.7 that are based on assuming a 1/7th. power velocity profile and using semi-empirical formulae for the local skin-friction coefficient.

7.10.3   Eddy viscosity

Away from the immediate influence of the wall which has a damping effect on the turbulent fluctuations, the Reynolds shear stress can be expected to be very much greater than the viscous shear stress. This can be seen by comparing rough order-of-magnitude estimates of the Reynolds shear stress and the viscous shear stress, i.e.

$$-\rho \overline{u'v'} \quad \text{c.f.} \quad \mu \frac{\partial \bar{u}}{\partial y}$$

Assume that $\overline{u'v'} \sim CU_\infty^2$ (where C is a constant), then

$$\mu \frac{\partial \bar{u}}{\partial y} = \mathcal{O}(\frac{\mu U_\infty}{\delta}) = \mathcal{O}\Big(\frac{1}{C^2} \underbrace{\frac{\mu}{\rho U_\infty \delta}}_{1/Re}\Big) \rho \overline{u'v'}$$

where δ is the shear-layer width. So provided $C = O(1)$ then

$$\frac{\mu \partial \bar{u}/\partial y}{-\rho \overline{u'v'}} = \mathcal{O}\Big(\frac{1}{Re}\Big)$$

showing that for large values of Re (recall that turbulence is a phenomenon that only occurs at large Reynolds numbers) the viscous shear stress will be negligible compared with the Reynolds shear stress. Boussinesq* drew an analogy between viscous and Reynolds shear stresses by introducing the concept of the *eddy viscosity* ε_T:

$$\underbrace{\tau = \mu \frac{\partial \bar{u}}{\partial y}}_{\text{viscous shear stress}} \quad \text{cf.} \quad \underbrace{-\rho \overline{u'v'} = \rho \varepsilon_T \frac{\partial \bar{u}}{\partial y}}_{\text{Reynolds shear stress}}: \quad \varepsilon_T \gg \nu (= \mu/\rho) \qquad (7.109)$$

Boussinesq, himself, merely assumed that eddy viscosity was constant everywhere in the flow field, like molecular viscosity but very much larger. Until comparatively recently, his approach was still widely used by oceanographers for modelling turbulent flows. In fact, though, a constant eddy viscosity is a very poor approximation for wall shear flows like boundary layers and pipe flows. For simple turbulent free shear layers, such as the mixing layer and jet (see Fig. 7.35), and wake it is a reasonable assumption to assume that the eddy viscosity varies in the streamwise direction but not across a particular cross section. Thus, using simple dimensional analysis Prandtl[†] and Reichardt[‡] proposed that

$$\varepsilon_T = \underbrace{\kappa}_{\text{const.}} \times \underbrace{\Delta U}_{\text{Velocity difference across shear layer}} \times \underbrace{\delta}_{\text{shear−layer width}} \qquad (7.110)$$

κ is often called the *exchange coefficient* and it varies somewhat from one type of flow to another. Equation (7.110) gives excellent results and can be used to determine the variation of the overall flow characteristics in the streamwise direction (see Example 7.9).

The outer 80% or so of the turbulent boundary layer is largely free from the effects of the wall. In this respect it is quite similar to a free turbulent shear layer. In this

* J. Boussinesq (1872) Essai sur la théorie des eaux courantes. *Mémoires Acad. des Science*, Vol. 23, No. 1, Paris.

[†] L. Prandtl (1942) Bemerkungen zur Theorie der freien Turbulenz, *ZAMM*, **22**, 241–243.

[‡] H. Reichardt (1942) Gesetzmässigkeiten der freien Turbulenz, *VDI-Forschungsheft*, **414**, 1st Ed., Berlin.

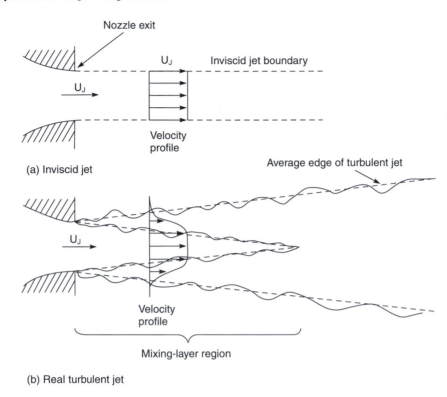

Fig. 7.35 An ideal inviscid jet compared with a real turbulent jet near the nozzle exit

outer region it is commonly assumed, following Laufer (1954), that the eddy viscosity can be determined by a version of Eqn (7.110) whereby

$$\varepsilon_T = \kappa U_e \delta^* \tag{7.111}$$

Example 7.9 The spreading rate of a mixing layer
Figure 7.35 shows the mixing layer in the intial region of a jet. To a good approximation the external mean pressure field for a free shear layer is atmospheric and therefore constant. Furthermore, the Reynolds shear stress is very much larger than the viscous stress, so that, after substituting Eqns (7.109) and (7.110), the turbulent boundary-layer equation (7.108b) becomes

$$\bar{u}\frac{\partial \bar{u}}{\partial x} + \bar{v}\frac{\partial \bar{u}}{\partial y} = \varepsilon_T \frac{\partial^2 \bar{u}}{\partial y^2} \qquad \left(= \kappa U_J \delta \frac{\partial^2 \bar{u}}{\partial y^2} \right)$$

The only length scale is the mixing-layer width, $\delta(x)$, which increases with x, so dimensional arguments suggest that the velocity profile does not change shape when expressed in terms of dimensionless y, i.e.

$$\frac{\bar{u}}{U_J} = F\left(\underbrace{\frac{y}{\delta}}_{\eta,\,\text{say}}\right)$$

This is known as making a *similarity* assumption. The assumed form of the velocity profile implies that

$$\frac{\partial \bar{u}}{\partial x} = \frac{\partial}{\partial x}\left(U_J F(\eta)\right) = \underbrace{\frac{\mathrm{d}U_J}{\mathrm{d}x}}_{=0} F(\eta) + U_J F'(\eta)\underbrace{\left(-\frac{\eta}{\delta}\frac{\mathrm{d}\delta}{\mathrm{d}x}\right)}_{\partial \eta/\partial x}$$

where $F'(\eta) \equiv \mathrm{d}F/\mathrm{d}\eta$.

Integrate Eqn (7.108a) to get

$$\bar{v} = -\int \frac{\partial \bar{u}}{\partial x}\mathrm{d}y = U_J \frac{\partial \delta}{\partial x}\int \eta F'(\eta)\mathrm{d}\eta$$

so

$$\bar{v} = U_J \frac{\partial \delta}{\partial x}G(\eta) \quad \text{where} \quad G = \int \eta F'(\eta)\mathrm{d}\eta$$

The derivatives with respect to y are given by

$$\frac{\partial \bar{u}}{\partial y} = \frac{\partial \eta}{\partial y}\frac{\mathrm{d}\bar{u}}{\mathrm{d}\eta} = \frac{U_J}{\delta}F'(\eta)$$

$$\frac{\partial^2 \bar{u}}{\partial y^2} = \frac{\partial \eta}{\partial y}\frac{\mathrm{d}}{\mathrm{d}\eta}\left(\frac{\partial \bar{u}}{\partial y}\right) = \frac{U_J}{\delta^2}F''(\eta)$$

The results given above are substituted into the reduced boundary-layer equation to obtain, after removing common factors,

$$-\underbrace{\frac{1}{\delta}\frac{\mathrm{d}\delta}{\mathrm{d}x}}_{\text{Fn. of } x \text{ only}}\underbrace{(\eta FF' + GF')}_{\text{Fn. of } \eta \text{ only}} = \underbrace{\kappa\frac{1}{\delta}}_{\text{Fn. of } x \text{ only}}\underbrace{F''}_{\text{Fn. of } \eta \text{ only}}$$

The braces indicate which terms are functions of x only or η only. So, we separate the variables and thereby see that, in order for the similarity form of the velocity to be a viable solution, we must require

$$\frac{\frac{1}{\delta}\frac{\mathrm{d}\delta}{\mathrm{d}x}}{\frac{\kappa}{\delta}} = -\frac{F''}{\eta FF' + GF'} = \text{const.}$$

After simplification the term on the left-hand side implies

$$\frac{\mathrm{d}\delta}{\mathrm{d}x} = \text{const.} \quad \text{or} \quad \delta \propto x$$

Setting the term, depending on η, with F'' as numerator, equal to a constant leads to a differential equation for F that could be solved to give the velocity profile. In fact, it is easy to derive a good approximation to the velocity profile, so this is a less valuable result.

When a turbulent (or laminar) flow is characterized by only one length scale – as in the present case – the term *self-similarity* is commonly used and solutions found this way are called *similarity* solutions. Similar methods can be used to determine the overall flow characteristics of other turbulent free shear layers.

7.10.4 Prandtl's mixing-length theory of turbulence

Equation (7.111) is not a good approximation in the region of the turbulent boundary layer or pipe flow near the wall. The eddy viscosity varies with distance from the wall in this region. A commonly used approach in this near-wall region is based on

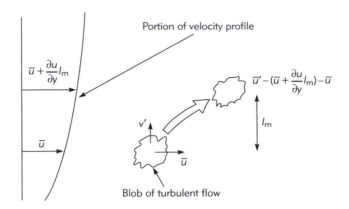

Fig. 7.36

Prandtl's *mixing-length* theory.* This approach to modelling turbulence is loosely based on the kinematic theory of gases. A brief account is given below and illustrated in Fig. 7.36.

Imagine a blob of fluid is transported upward by a fluctuating turbulent velocity v' through an average distance ℓ_m – *the mixing length* – (analogous to the mean free path in molecular dynamics). In the new position, assuming the streamwise velocity of the blob remains unchanged at the value in its original position, the fluctuation in velocity can be thought to be generated by the difference in the blob's velocity and that of its new surroundings. Thus

$$u' \simeq \underbrace{\left(\bar{u} + \frac{\partial \bar{u}}{\partial y}\ell_m\right)}_{\text{(i)}} - \underbrace{\bar{u}}_{\text{(ii)}} = \frac{\partial \bar{u}}{\partial y}\ell_m$$

Term (i) is the mean flow speed in the new environment. In writing the term in this form it is assumed that $\ell_m \ll \delta$, so that, in effect, it is the first two terms in a Taylor's series expansion.

Term (ii) is the mean velocity of blob.

If it is also assumed that $v' \sim (\partial \bar{u}/\partial y)\ell_m$, then

$$-\overline{u'v'} \sim \ell_m^2 \left(\frac{\partial \bar{u}}{\partial y}\right) \underbrace{\left|\frac{\partial \bar{u}}{\partial y}\right|}_{\text{(iii)}} \quad \text{implying} \quad \varepsilon_T = \text{const} \times \ell_m^2 \left|\frac{\partial \bar{u}}{\partial y}\right| \tag{7.112}$$

Term (iii) is written with an absolute value sign so that the Reynolds stress changes sign with $\partial \bar{u}/\partial y$, just as the viscous shear stress would.

7.10.5   Regimes of turbulent wall flow

As the wall is aproached it has a damping effect on the turbulence, so that very close to the wall the viscous shear stress greatly exceeds the Reynolds shear stress. This region right next to the wall where viscous effects dominate is usually known as the *viscous sub-layer*. Beyond the viscous sub-layer is a *transition* or *buffer* layer

---

* L. Prandtl (1925) Bericht über Untersuchunger zur ausgebildeten Turbulenz, *ZAMM*, **5**, 136–139.

where the viscous and Reynolds shear stresses are roughly equal in magnitude. This region blends into the fully turbulent region where the Reynolds shear stress is very much larger than the viscous shear stress. It is in this fully turbulent near-wall region that the mixing-length theory can be used. The outer part of the boundary layer is more like a free shear layer and there the Reynolds shear stress is given by Eqn (7.111).

A major assumption is that the fully turbulent layer begins at a height above the wall of $y \ll \delta$, so that

$$\tau = \tau_w + \underbrace{\frac{d\tau}{dy}y}_{\ll \tau_w} + \cdots \simeq \tau_w \tag{7.113}$$

Near the wall in the viscous sub-layer the turbulence is almost completely damped, so only molecular viscosity is important, thus

$$\tau = \mu \frac{d\bar{u}}{dy} = \tau_w \quad \text{therefore} \quad \bar{u} = \frac{\tau_w}{\mu}y \tag{7.114}$$

In the fully turbulent region the Reynolds shear stress is much greater than the viscous shear stress, so:

$$\tau = -\rho \overline{u'v'} = \tau_w$$

So if Eqn (7.112) is used and it is assumed that $\ell_m \propto y$, then

$$\left(\frac{d\bar{u}}{dy}\right)^2 \propto \frac{\tau_w}{\rho}\frac{1}{y^2} \quad \text{implying} \quad \frac{d\bar{u}}{dy} \propto \frac{V_*}{y} \tag{7.115}$$

where we have introduced the *friction velocity*:

$$V_* = \sqrt{\tau_w/\rho} \tag{7.116}$$

as the reference velocity that is subsequently used to render the velocity in the near-wall region non-dimensional.

Integrate Eqn (7.115) and divide by V_* to obtain the non-dimensional velocity profile in the fully turbulent region, and also re-write (7.114) to obtain the same in the viscous sub-layer. Thus

Fully turbulent flow:

$$\frac{\bar{u}}{V_*} = C_1 \ell n\left(\frac{yV_*}{\nu}\right) + C_2 \tag{7.117}$$

Viscous sub-layer:

$$\frac{\bar{u}}{V_*} = \frac{yV_*}{\nu} \tag{7.118}$$

where C_1 and C_2 are constants of integration to be determined by comparison with experimental data; and η or $y^+ = yV_*/\nu$ is the dimensionless distance from the wall; the length $\ell^+ = \nu/V_*$ is usually known as the *wall unit*.

Figure 7.37 compares (7.117) and (7.118) with experimental data for a turbulent boundary layer and we can thereby deduce that

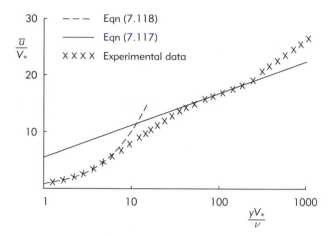

Fig. 7.37

Viscous sub-layer:

$$\frac{yV_*}{\nu} < 5 \qquad \mu\frac{\partial \bar{u}}{\partial y} \gg -\rho\overline{u'v'}$$

Buffer layer:

$$5 < \frac{yV_*}{\nu} < 50 \qquad \mu\frac{\partial \bar{u}}{\partial y} \simeq -\rho\overline{u'v'} \tag{7.119}$$

Fully developed turbulence:

$$\frac{yV_*}{\nu} > 50 \qquad \mu\frac{\partial \bar{u}}{\partial y} \ll -\rho\overline{u'v'}$$

The constants C_1 and C_2 can be determined from comparison with the experimental data so that (7.117) becomes:

Logarithmic velocity profile:

$$\frac{\bar{u}}{V_*} = 2.54\,\ell n\left(\frac{yV_*}{\nu}\right) + 5.56 \tag{7.120}$$

C_1 is often written as $1/\kappa$ where $\kappa = 0.41$ is known as the von Kármán constant* because he was the first to derive the logarithmic velocity profile. Equation (7.117) is often known as the *Law of the wall*. It applies equally well to the near-wall region of turbulent pipe and channel flows for which better agreement with experimental data is found for slightly different values of the constants. It is worth noting that it is not essential to evoke Prandtl's mixing-length theory to derive the law of the wall. The logarithmic form of the velocity profile can also be derived purely by means of dimensional analysis.[†]

---

*Th. von Kármán (1930) Mechanische Ähnlichkeit und Turbulenz, *Nachrichten der Akademie der Wissenschaften Göttingen*, Math.-Phys. Klasse, p. 58.

[†]G.I. Barenblatt and V.M. Prostokishin (1993) Scaling laws for fully-developed turbulent shear flows, *J. Fluid Mech.*, **248**, 513–529.

The outer boundary layer

The outer part of the boundary layer that extends for 70 or 80% of the total thickness is unaffected by the direct effect of the wall. It can be seen in Fig. 7.37 that the velocity profile deviates considerably from the logarithmic form in this outer part of the boundary layer. In many respects it is analogous to a free shear layer, especially a wake. It is sometimes referred to as the *defect layer* or *wake region*. Here inertial effects dominate and viscous effects are negligible, so the appropriate reference velocity and length scales to use for non-dimensionalization are U_e (the streamwise flow speed at the boundary-layer edge) and δ (the boundary-layer thickness) or some similar length scale. Thus the so-called *outer variables* are:

$$\frac{\bar{u}}{U_e} \quad \text{and} \quad \frac{y}{\delta}$$

7.10.6 Formulae for local skin-friction coefficient and drag

Although it is not valid in the outer part of the boundary layer, Eqn (7.117) can be used to obtain the following more accurate semi-empirical formulae for the local skin-friction coefficient and the corresponding drag coefficient for turbulent boundary layers over flat plates.

$$c_f \equiv \frac{\tau_w}{\frac{1}{2}\rho U_\infty^2} = (2\log_{10} Re_x - 0.65)^{-2.3} \tag{7.121}$$

$$C_{Df} \equiv \frac{D_f}{\frac{1}{2}\rho U_\infty^2 BL} = \frac{0.455}{(\log_{10} Re_L)^{2.58}} \tag{7.122}$$

where B and L are the breadth and length of the flat plate. The *Prandtl–Schlichting formula* (7.122) is more accurate than Eqn (7.88) when $Re_L > 10^7$.

Effects of wall roughness

Turbulent boundary layers, especially at high Reynolds numbers, are very sensitive to wall roughness. This is because any roughness element that protrudes through the viscous sub-layer will modify the law of the wall. The effect of wall roughness on the boundary layer depends on the size, shape and spacing of the elements. To bring a semblance of order Nikuradze matched each 'type' of roughness against an *equivalent sand-grain roughness* having roughness of height, k_s. Three regimes of wall roughness, corresponding to the three regions of the near-wall region, can be defined as follows:

Hydraulically smooth   If $k_s V_*/\nu \leq 5$ the roughness elements lie wholly within the viscous sublayer, the roughness therefore has no effect on the velocity profile or on the value of skin friction or drag.

Completely rough   If $k_s V_*/\nu \geq 50$ the roughness elements protrude into the region of fully developed turbulence. This has the effect of displacing the logarithmic profile downwards, i.e. reducing the value of C_2 in Eqn (7.117). In such cases the local skin-friction and drag coefficients are independent of Reynolds number and are given by

$$c_f = [2.87 + 1.58\log_{10}(x/k_s)]^{-2.5} \tag{7.123}$$

$$C_{Df} = [1.89 + 1.62\log_{10}(L/k_s)]^{-2.5} \tag{7.124}$$

Transitional roughness   If $5 \leq k_s V_*/\nu \leq 50$ the effect of roughness is more complex and the local skin-friction and drag coefficients depend both on Reynolds number and relative roughness, k_s/δ.

The relative roughness plainly varies along the surface. But the viscous sub-layer increases slowly and, although its maximum thickness is located at the trailing edge, the trailing-edge value is representative of most of the rest of the surface. The degree of roughness that is considered *admissible* in engineering practice is one for which the surface remains hydraulically smooth throughout, i.e. the roughness elements remain within the viscous sub-layer all the way to the trailing edge. Thus

$$k_{\mathrm{adm}} = 5\nu/(V_*)_{TE} \tag{7.125}$$

In the case of a flat plate it is found that Eqn (7.125) is approximately equivalent to

$$k_{\mathrm{adm}} \propto \frac{\nu}{V_\infty} \propto \frac{L}{Re_L} \tag{7.126}$$

Thus for plates of similar length the admissible roughness diminishes with increasing Re_L. In the case of ships' hulls admissible roughness ranges from $7\,\mu\mathrm{m}$ (large fast ships) to $20\,\mu\mathrm{m}$ (small slow ships); such values are utterly impossible to achieve in practice, and it is always neccessary to allow for a considerable increase in drag due to roughness. For aircraft admissible roughness ranges from $10\,\mu\mathrm{m}$ to $25\,\mu\mathrm{m}$ and that is just about attainable in practice. Model aircraft and compressor blades require the same order of admissible roughness and hydraulically smooth surfaces can be obtained without undue difficulty. At the other extreme there are steam-turbine blades that combine a small chord (L) with a fairly high Reynolds number (5×10^6) owing to the high velocities involved and to the comparatively high pressures. In this cases admissible roughness values are consequently very small, ranging from $0.2\,\mu\mathrm{m}$ to $2\,\mu\mathrm{m}$. This degree of smoothness can barely be achieved on newly manufactured blades and certainly the admissible roughness would be exceeded after a period of operation owing to corrosion and the formation of scaling.

The description of the aerodynamic effects of surface roughness given above has been in terms of equivalent sand-grain roughness. It is important to remember that the aerodynamic effects of a particular type of roughness may differ greatly from that of sand-grain roughness of the same size. It is even possible (see Section 8.5.3) for special forms of wall 'roughness', such as riblets, to lead to a reduction in drag.[*]

7.10.7 Distribution of Reynolds stresses and turbulent kinetic energy across the boundary layer

Figure 7.38 plots the variation of Reynolds shear stress and kinetic energy (per unit mass), $k = (\overline{u'^2} + \overline{v'^2} + \overline{w'^2})/2$ across the boundary layer. What is immediately striking is how comparatively high the levels are in the near-wall region. The Reynolds shear stress reaches a maximum at about $y^+ \simeq 100$ while the turbulence kinetic energy appears to reach its maximum not far above the edge of the viscous sub-layer.

Figure 7.39 plots the distributions of the so-called turbulence intensities of the velocity components, i.e. the square-roots of the direct Reynolds stresses, $\overline{u'^2}$, $\overline{v'^2}$ and $\overline{w'^2}$. Note that in the outer part of the boundary layer the three turbulent intensities tend to be the same (they are 'isotropic'), but they diverge widely as the wall is approached (i.e. they become 'anisotropic'). The distribution of eddy viscosity across

[*] P.W. Carpenter (1997) The right sort of roughness, *Nature*, **388**, 713–714.

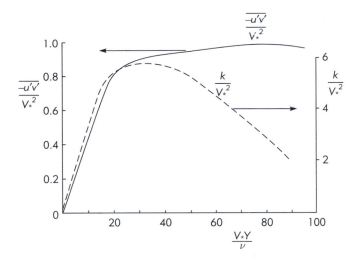

Fig. 7.38 Variation of Reynolds shear stress and turbulence kinetic energy across the near-wall region of the turbulent boundary layer

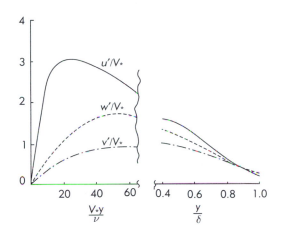

Fig. 7.39 Variations of the root mean squares of $\overline{u'^2}$, $\overline{v'^2}$ and $\overline{w'^2}$ across a turbulent boundary layer

the turbulent boundary layer is plotted in Fig. 7.40. This quantity is important for engineering calculations of turbulent boundary layers. Note that the form adopted in Eqn (7.112) for the near-wall region according to the mixing-length theory with $\ell_m \propto y$ is borne out by the behaviour shown in the figure.

If a probe were placed in the outer region of a boundary layer it would show that the flow is only turbulent for part of the time. The proportion of the time that the flow is turbulent is called the *intermittency* (γ). The intermittency distribution is also plotted in Fig. 7.40.

7.10.8   Turbulence structure in the near-wall region

The dominance of the near-wall region in terms of turbulence kinetic energy and Reynolds shear stress motivated engineers to study it in more detail with a view to

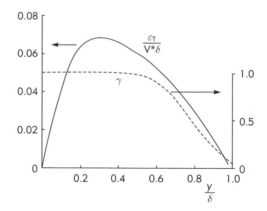

Fig. 7.40 Distributions of eddy viscosity and intermittency across a turbulent boundary layer

identifying the time-varying flow structures there. Kline *et al.** carried out a seminal study of this kind. They obtained hydrogen-bubble flow visualizations for the turbulent boundary layer. These revealed that streak-like structures develop within the viscous sub-layer. These are depicted schematically in Fig. 7.41. The streak-like structures are continuously changing with time. Observations over a period of time reveal that there are low- and high-speed streaks. The streak structures become less noticeable further away from the wall and apparently disappear in the law-of-the-wall region. In the outer region experiments reveal the turbulence to be intermittent and of larger scale (Fig. 7.41).

The conventional view is that the streaks are a manifestation of the existence of developing hair-pin vortices. See Figs 7.41 and 7.42 for a schematic illustration of the conceptual burst cycle of these structures, that is responsible for generating transient high levels of wall shear stress. The development of the 'hair pin vortices' tends to be quasi-periodic with the following sequence of events:

(i) *Formation of the low-speed streaks*: During this process the legs of the vortices lie close to the wall.
(ii) *Lift-up* or *ejection*: Stage E in Fig. 7.41. The velocity induced by these vortex legs tends to cause the vortex head to lift off away from the wall.
(iii) *Oscillation* or *instability*: The first part of Stage B in Fig. 7.41. A local point of inflexion develops in the velocity profile and the flow becomes susceptible to Helmholtz instability locally causing the head of the vortex to oscillate fairly violently.
(iv) *Bursting* or *break-up*: The latter part of Stage B in Fig. 7.41. The oscillation culminates in the vortex head bursting.
(v) *High-speed sweep*: Stage S in Fig. 7.41. After a period of quiescence the bursting event is followed by a high-speed sweep towards the wall. It is during this process that the shear stress at the wall is greatest and the new hairpin-vortex structures are generated.

It should be understood that Fig. 7.41 is drawn to correspond to a frame moving downstream with the evolving vortex structure. So a constant streamwise velocity is superimposed on the ejections and sweeps.

---

* S.J. Kline, W.J. Reynolds, F.A. Schraub and P.W. Runstadler (1967) The structure of turbulent boundary layers, *J. Fluid Mech.*, **30**, 741–773.

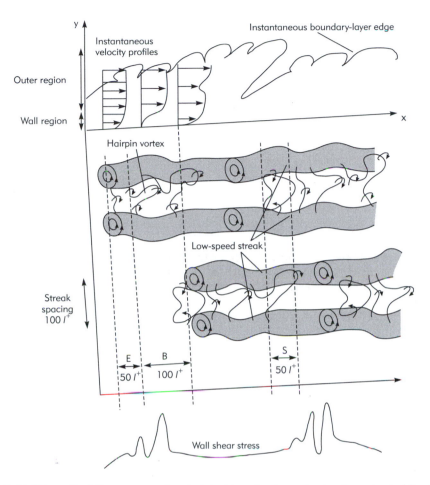

Fig. 7.41 Schematic of flow structures in a turbulent boundary layer showing the conceptual burst cycle *E*, ejection stage; *B*, break-up stage; *S*, sweep stage.

The events described above are quasi-periodic in a statistical sense. The mean values of their characteristics are as follows: spanwise spacing of streaks = $100\nu/V_*$; they reach a vertical height of $50\nu/V_*$; their streamwise extent is about $1000\nu/V_*$; and the bursting frequency is about $0.004V_*^2/\nu$. This estimate for bursting frequency is still a matter of some controversy. Some experts[*] think that the bursting frequency does not scale with the wall units, but other investigators have suggested that this result is an artefact of the measurement system.[†] Much greater detail on these near-wall structures together with the various concepts and theories advanced to explain their formation and regeneration can be found in Panton.[‡]

[*] K. Narahari Rao, R. Narasimha and M.A. Badri Narayanan (1971) The 'bursting' phenomenon in a turbulent boundary layer, *J. Fluid Mech.*, **48**, 339.

[†] Blackwelder, R.F. and Haritonidis, J.H. (1983) Scaling of the bursting frequency in turbulent boundary layers, *J. Fluid Mech.*, **132**, 87–103.

[‡] R.L. Panton (ed.) (1997) *Self-Sustaining Mechanisms of Wall Turbulence*, Computational Mechanics Publications, Southampton.

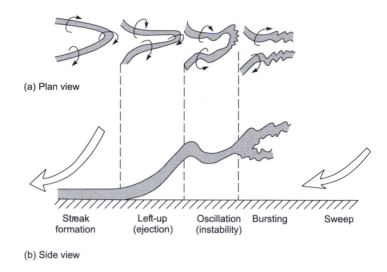

(a) Plan view

| Streak | Left-up | Oscillation | Bursting | Sweep |
| formation | (ejection) | (instability) | | |

(b) Side view

Fig. 7.42 Schematic of the evolution of a hairpin vortex in the near-wall region

Example 7.10   Determining specifications of a MEMS actuator for flow control
Modern technology is rapidly developing the capability of making very small machines that, among many other applications, can be used for actuation and sensing in flow-control systems. Such machines are collectively known as MEMS (Micro-electro-mechanical Systems). The term is usually used to refer to devices that have characteristic overall dimensions of less than 1 mm but more than 1 μm. Such devices combine electrical and mechanical components that are manufactured by means of integrated circuit batch-processing techniques.*

A conceptual MEMS actuator for use as part of a flow-control system is depicted schematically in Fig. 7.43. This device consists of a diaphragm located at the bottom of a buried cavity that connects to the boundary layer via an exit orifice. The diaphragm would be made of silicon or a suitable polymeric material and driven by a piezoceramic driver. When a voltage is applied to the driver, depending on the sign of the electrical signal, it either contracts or expands, thereby displacing the diaphragm up or down. Thus, if an alternating voltage is applied to the diaphragm, it will be periodically driven up and down. This in turn will alternately reduce and increase the volume of the cavity, thereby raising then reducing the air pressure there. An elevated cavity pressure will drive air through the exit orifice into the boundary layer followed by air returning to the cavity when the cavity pressure falls. This periodic outflow and inflow creates what is termed a *synthetic jet*. Despite the fact that over a cycle there is no net air leaving the cavity, vortical structures propagate into the boundary layer, much as they would from a steady micro-jet. It is also possible to drive the diaphragm with short-duration steady voltage thereby displacing the diaphragm suddenly upward and driving a 'puff' of air into the boundary layer.[†]

---

* For a general description of MEMS technology see Maluf, N. (2000) *An Introduction to Microelectro-mechanical Systems Engineering*. Artech House; Boston/London. And for a description of their potential use for flow control see Gad-el-Hak, M. (2000) *Flow Control. Passive, Active, and Reactive Flow Management*, Cambridge University Press.

[†] The concept of the synthetic jet was introduced by Smith, B.L. and Glezer, A. (1998) The formation and evolution of synthetic jets, *Phys. Fluids*, **10**, 2281–2297. For its use for flow control, see: Crook, A., Sadri, A.M. and Wood, N.J. (1999) The development and implementation of synthetic jets for the control of separated flow, *AIAA Paper 99–3176*; and Amitay, M. *et al.* (2001) Aerodynamics flow control over an unconventional airfoil using synthetic jet actuators, *AIAA Journal*, **39**, 361–370. A study of the use of these actuators for boundary-layer control based on numerical simulation is described by Lockerby, D.A., Carpenter, P.W. and Davies, C. (2002) Numerical simulation of the interaction of MEMS actuators and boundary layers, *AIAA Journal*, **39**, 67–73.

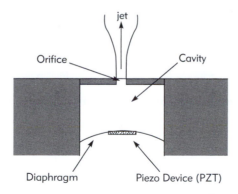

Fig. 7.43 Schematic sketch of a jet-type actuator

It has been proposed that synthetic-jet actuators be used to control the near-wall, streak-like, structures in the turbulent boundary layer over the flap of a large airliner. In particular, the aim is to increase the so-called bursting frequency in order to increase the level of turbulence. The increased turbulence level would be expected to lead to a delay of boundary-layer separation, allowing the flap to be deployed at a larger angle of incidence, thereby increasing its performance. The basic concept is illustrated in Fig. 7.44. The question to be addressed here is what dimensions and specifications should be chosen for the MEMS actuator in this application.

Consider an aircraft similar to the Airbus A340 with a mean wing chord of 6 m and a flap chord of 1.2 m. Assume that the approach speed is about 100 m/s and assume standard sea-level conditions so that the kinematic viscosity of the air is around 15×10^{-6} m^2/s. It is proposed to locate the array of MEMS actuators near the point of minimum pressure 200 mm from the leading edge of the flap. A new boundary layer will develop on the flap underneath the separated boundary layer from the main wing. This will ensure that the flap boundary layer is strongly disturbed provoking early transition to turbulence.

Thus

$$Re_x = \frac{xU_e}{\nu} \simeq \frac{0.2 \times 100}{15 \times 10^{-6}} = 1.33 \times 10^6.$$

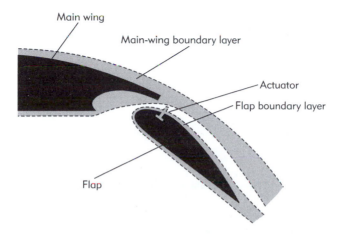

Fig. 7.44

In a low-disturbance environment we would normally expect the boundary layer to be laminar at this Reynolds number (see Section 7.9), but the flap boundary layer is highly disturbed by the separated boundary layer from the main wing, ensuring early transition. Using Eqn (7.121) we can estimate the local skin-friction coefficient, i.e.

$$c_f = (2\log_{10}(1.33 \times 10^6) - 0.65)^{-2.3} = 0.00356.$$

From this we can then estimate the wall shear stress and friction velocity, thus

$$\tau_w = \frac{1}{2}\rho U_e^2 c_f \simeq \frac{1}{2} \times 1.2 \times 100^2 \times 0.00356 = 21.36\,\text{Pa}, \qquad V_* = \sqrt{\frac{\tau_w}{\rho}} = \sqrt{\frac{21.36}{1.2}} = 4.22\,\text{m/s}.$$

The boundary-layer thickness is not strictly needed for determining the actuator specifications, but it is instructive to determine it also. In Eqn (7.71) it was shown that $C_{Df} = 2\Theta(L)/L$, i.e. there is a relationship between the coefficient of skin-friction drag and the momentum thickness at the trailing edge. This relationship can be exploited to estimate the momentum thickness in the present application. We merely assume for this purpose that the flap boundary layer terminates at the point in question, i.e. at $x = 200\,\text{mm}$, so that

$$\Theta(x) = \frac{x}{2}C_{Df}(x).$$

Using Eqn (7.122)

$$C_{Df}(x) = \frac{0.455}{(\log_{10}Re_x)^{2.58}} = \frac{0.455}{\{\log_{10}(1.33 \times 10^6)\}^{2.58}} = 0.00424.$$

Thus

$$\Theta(x) = \frac{x}{2}C_{Df}(x) = \frac{0.2}{2} \times 0.00424 \simeq 425\,\mu\text{m}.$$

From Eqn (7.83)

$$\Theta = 0.0973\delta \quad \text{giving} \quad \delta = \frac{0.424}{0.0973} = 4.36\,\text{mm}.$$

This illustrates yet again just how thin the boundary layer actually is in aeronautical applications.

We can now calculate the wall unit and thence the other dimensions of interest.
Wall unit:

$$\ell^+ \equiv \frac{\nu}{V_*} = \frac{15 \times 10^{-6}}{4.22} \simeq 3.5\,\mu\text{m};$$

Viscous sub-layer thickness:

$$5\,\ell^+ \simeq 17.5\,\mu\text{m};$$

Average spanwise spacing of streaks:

$$100\,\ell^+ \simeq 350\,\mu\text{m};$$

Bursting frequency:

$$0.004\frac{V_*^2}{\nu} = 0.004 \times \frac{4.22}{3.5 \times 10^{-6}} \simeq 4.8\,\text{kHz}.$$

This suggests that the spanwise dimensions of the MEMS actuators should not exceed about $100\,\mu\text{m}$; i.e. about 30 per cent of average spanwise streak spacing. They also need to be able to effect control at frequencies of at least ten times the bursting frequency, say $50\,\text{kHz}$.

7.11 Computational methods

7.11.1 Methods based on the momentum integral equation

In the general case with an external pressure gradient the momentum integral equation must be solved numerically. There are a number of ways in which this can be done. One method for laminar boundary layers is to use the approximate expressions (7.64a',b',c') with Eqn (7.59) (or 7.98) rewritten as

$$\frac{d\theta}{dx} = \frac{C_f}{2} - \frac{V_s}{U_e} - \frac{\theta}{U_e}\frac{dU_e}{dx}(H+2) \tag{7.127}$$

$d\theta/dx$ can be related to $d\delta/dx$ as follows:

$$\frac{d\theta}{dx} = \frac{d(\delta I)}{dx} = I\frac{d\delta}{dx} + \delta\frac{dI}{dx} = I\frac{d\delta}{dx} + \delta\frac{dI}{d\Lambda}\frac{d\Lambda}{d\delta}\frac{d\delta}{dx}$$

$$= \left(I + \delta\frac{dI}{d\Lambda}\frac{d\Lambda}{d\delta}\right)\frac{d\delta}{dx}$$

It follows from Eqns (7.62d) and (7.64b') respectively that

$$\delta\frac{d\Lambda}{d\delta} = \frac{2\delta^2}{\nu}\frac{dU_e}{dx} = 2\Lambda$$

$$\frac{dI}{d\Lambda} = -\frac{1}{63}\left(\frac{1}{15} + \frac{\Lambda}{72}\right)$$

So

$$\frac{d\theta}{dx} = F_1(\Lambda)\frac{d\delta}{dx} \tag{7.128}$$

where

$$F_1(\Lambda) = I - \frac{2\Lambda}{63}\left(\frac{1}{15} + \frac{\Lambda}{72}\right)$$

Thus Eqn (7.127) could be readily converted into an equation for $d\delta/dx$ by dividing both sides by $F_1(\Lambda)$. The problem with doing this is that it follows from Eqn (7.64c') that

$$C_f \propto \frac{1}{U_e\delta}$$

Thus in cases where either $\delta = 0$ or $U_e = 0$ at $x = 0$, the initial value of C_f, and therefore the right-hand side of Eqn (7.127) will be infinite. These two cases are in fact the two most common in practice. The former corresponds to sharp leading edges and the latter to blunt ones. To deal with the problem identified above both sides of Eqn (7.127) are multiplied by $2U_e\delta/\nu$, whereby with use of Eqn (7.128) it becomes

$$F_1(\Lambda)\frac{U_e}{\nu}\frac{d\delta^2}{dx} = F_2(\Lambda) - \frac{2V_s\delta}{\nu}$$

where use of Eqns (7.64a′,b′,c′) gives

$$F_2(\Lambda) = 4 + \frac{\Lambda}{3} - \frac{3}{10}\Lambda + \frac{\Lambda^2}{60} + \frac{4}{63}\left(\frac{37}{5} - \frac{\Lambda}{15} - \frac{\Lambda^2}{144}\right)$$

To obtain the final form of Eqn (7.127) for computational purposes $F_1(\Lambda)\delta^2 dU_e/dx$ is added to both sides and then both sides are divided by $F_1(\Lambda)$; thus

$$\frac{dZ}{dx} = \frac{F_2(\Lambda)}{F_1(\Lambda)} + \Lambda - \frac{2}{F_1(\Lambda)}\frac{V_s\delta}{v} \qquad (7.129)$$

where the dependent variable has been changed from δ to $Z = \delta^2 U_e/v$. In the usual case when $V_s = 0$ the right-hand side of Eqn (7.129) is purely a function of Λ. Note that

$$\Lambda = \frac{Z}{U_e}\frac{dU_e}{dx}, \qquad \text{and} \qquad \delta = \sqrt{\frac{Zv}{U_e}}$$

Since U_e is a prescribed function of x this allows both Λ and δ to be obtained from a value of Z. The other quantities of interest can be obtained from Eqn (7.64a′, b′, c′).

With the momentum-integral equation in the form (7.129) it is suitable for the direct application of standard methods for numerical integration of ordinary differential equations.* It is recommended that the fourth-order Runge–Kutta method be used. It could be used with an adaptive stepsize control, the advantage of which is that small steps would be chosen in regions of rapid change, e.g. near the leading edge, while larger steps would be taken elsewhere.

In order to begin the calculation it is necessary to supply initial values for Z and Λ at $x = 0$, say. For a sharp leading edge $\delta = 0$ giving $Z = \Lambda = 0$ at $x = 0$, whereas for a round leading edge $U_e = 0$ and $Z = 0$ but $\Lambda = 7.052$, see Example 7.8 above, this value of Λ should be used to evaluate the right-hand side of (7.129) at $x = 0$.

Boundary-layer computations using the method described above have been carried out for the case of a circular cylinder of radius 1 m in air flowing at 20 m/s ($v = 1.5 \times 10^{-5}$ m²/s). In Fig. 7.45, the computed values of Λ and momentum

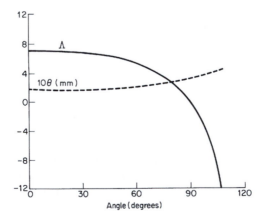

Fig. 7.45

* Complete FORTRAN subroutines for this are to be found in W.H. Press *et al.* (1992) *loc. cit.*

thickness are plotted against angle around the cylinder's surface measured from the fore stagnation point. A fourth-order Runge–Kutta integration scheme was used with 200 fixed steps between $x = 0$ and $x = 3$. This gave acceptable accuracy. According to this approximate calculation the separation point, corresponding to $\Lambda = -12$, occurs at 106.7°. This should be compared with the accurately computed value of 104.5° obtained using the differential form of the equations of motion. Thus it can be seen that the Pohlhausen method gives reasonably acceptable results in terms of a comparison with more accurate methods. In point of fact neither value given above for the separation point is close to the actual value found experimentally for a laminar boundary layer. Experimentally, separation is found to occur just ahead of the apex of the cylinder. The reason for the large discrepancy between theoretical and observed separation points is that the large wake substantially alters the flow outside the boundary layer. This main-stream flow, accordingly, departs markedly from the potential-flow solution assumed for the boundary-layer calculations. Boundary-layer theory only predicts the separation point accurately in the case of streamlined bodies with relatively small wakes. Nevertheless, the circular cylinder is a good test case for checking the accuracy of boundary-layer computations.

Numerical solutions of the momentum integral equation can also be found using Thwaites' method.* This method does not make use of the Pohlhausen approximate velocity profile Eqn (7.63) or Eqn (7.65). It is very simple to use and for some applications it is more accurate than the Pohlhausen method described above. A suitable FORTRAN program for Thwaites method is given by Cebeci and Bradshaw (1977).

Computational methods based on the momentum integral equation are also available for the turbulent boundary layer. In this case, one or more semi-empirical relationships are also required in addition to the momentum integral equation. For example, most methods make use of the formula for C_f due to Ludwieg and Tillmann (1949), namely

$$C_f = 0.246 \times 10^{-0.678H} \left(\frac{U_e \theta}{\nu} \right)^{-0.268}$$

A very good method of this type is due to Head (1958).† This method is relatively simple to use but, nevertheless, performs better than many of the much more complex methods based on the differential equations of motion. A FORTRAN program based on Head's method is also given by Cebeci and Bradshaw *loc. cit.*

In order to begin the computation of the turbulent boundary layer using Head's method it is necessary to locate the transition point and to supply initial values of θ and $H = \delta^*/\theta$. In Section 7.7.7, it was shown that for the boundary layer on a flat plate Eqn (7.89) held at the transition point, i.e. there is not a discontinuous change in momentum thickness. This applies equally well to the more general case. So once the transition point is located the starting value for θ in the turbulent boundary layer is given by the final value in the laminar part. However, since transition is assumed to occur instantaneously at a specific location along the surface, there will be a discontinuous change in velocity profile shape at the transition point. This implies a discontinuous change in the shape factor H. To a reasonable approximation

$$H_{Tt} = H_{Lt} - \Delta H$$

---

* B. Thwaites (1949) 'Approximate calculation of the laminar boundary layer', *Aero. Quart.*, **1**, 245.

† M.R. Head (1958) 'Entrainment in the turbulent boundary layers', *Aero. Res. Council, Rep. & Mem.*, 3152.

where

$$\Delta H = 0.821 + 0.114 \log_{10}(Re_{\theta t}) \quad Re_{\theta t} < 5 \times 10^4$$
$$\Delta H = 1.357 \quad\quad\quad\quad\quad\quad\quad Re_{\theta t} > 5 \times 10^4$$

7.11.2 Transition prediction

The usual methods of determining the transition point are based on the so-called e^n method developed by A.M.O. Smith and H. Gamberoni (1956) at Douglas Aircraft Co.* These methods are rather complex and involve specialized computational techniques. Fortunately, Smith and his colleagues have devised a simple, but highly satisfactory alternative.[†] They have found on a basis of predictions using the e^n method that the transition Reynolds number Re_{xt} (where $Re_x = U_e x/v$) is related to the shape factor H by the following semi-empirical formula:

$$\log_{10}(Re_{xt}) = -40.4557 + 64.8066H - 26.7538H^2 + 3.3819H^3 \quad 2.1 < H < 2.8$$
$$(7.130)$$

To use this method $\log_{10}(Re_x)$ is plotted against x (distance along the surface from the fore stagnation point on the leading edge). A laminar boundary-layer calculation is carried out and the right-hand side of Eqn (7.130) is also plotted against x. Initially, the former curve will lie below the latter; the transition point is located at the value of x (or Re_x) where the two curves *first* cross. This is illustrated in Fig. 7.46

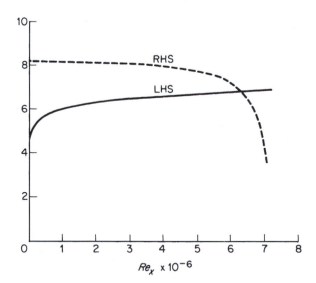

Fig. 7.46

* See N.A. Jaffe, T.T. Okamura and A.M.O. Smith (1970) 'Determination of spatial amplification factors and their application to predicting transition', *AIAA J.*, **8**, 301–308.

[†] A.R. Wazzan, C. Gazley and A.M.O. Smith (1981) 'H-R_x method for predicting transition', *AIAA J.*, **19**, 810–812.

where the left-hand side (denoted by LHS) and the right-hand side (denoted by RHS) of Eqn (7.130), calculated for the case of the circular cylinder illustrated in Fig. 7.45, are plotted against Re_x. In this case, the two curves cross at a value of $Re_x \simeq 6.2 \times 10^6$ and this is taken to correspond to the transition point.

7.11.3 Computational solution of the laminar boundary-layer equations

Nowadays, with the ready availability of powerful desktop computers, it is perfectly feasible to solve the boundary-layer equations computationally in their original form as partial differential equations. Consequently, methods based on the momentum integral equation are now less widely used. Computational solution of the boundary-layer equations is now a routine matter for industry and university researchers alike, and suitable computer programs are widely available. Nevertheless, specialized numerical techniques are necessary and difficulties are commonly encountered. Good expositions of the required techniques and the pitfalls to be avoided are available in several textbooks.* All-purpose commercial software packages for computational fluid dynamics are also widely available. As a general rule such software does not handle boundary layers well owing to the fine resolution required near the wall[†]. Only a brief introduction to the computational methods used for boundary layers will be given here. For full details the recommended texts should be consulted.

It is clear from the examples given that in aeronautical applications boundary layers are typically very thin compared with the streamwise dimensions of a body like a wing. This in itself would pose difficulties for computational solution. For this reason equations (7.7) and (7.14) are usually used in the following non-dimensional form

$$\frac{\partial U}{\partial X} + \frac{\partial V}{\partial Y} = 0 \tag{7.131}$$

$$U\frac{\partial U}{\partial X} + V\frac{\partial U}{\partial Y} = -\frac{\mathrm{d}P}{\mathrm{d}X} + \frac{\partial^2 U}{\partial Y^2} \tag{7.132}$$

where $U = u/U_\infty$, $V = v/(Re_L U_\infty)$, $P = p/(\rho U_\infty^2)$, $X = x/L$, and $Y = y/(LRe_L)$. In many respects this form was suggested by the method of derivation given in Section 7.3.1. What this form of the equations achieves is to make the effective range of both independent variables X and Y similar in size, i.e. $\mathcal{O}(1)$. The two dependent variables U and V are also both $\mathcal{O}(1)$. Thus the grid used to discretize the equations for computational solution can be taken as rectangular and rectilinear, as depicted in Fig. 7.47.

Mathematically, the boundary-layer equations are what are termed *parabolic*. What this means for computational purposes is that one starts at some initial point $X = X_0$, say, where U is known as a function of Y, e.g. the stagnation-point solution (see Section 2.10.3) should be used in the vicinity of the fore stagnation point. The object of the computational scheme is then to compute the solution at $X_0 + \Delta X$ where ΔX is a small step around the body surface. One then repeats this procedure until the trailing edge or the boundary-layer separation point is reached. For obvious

---

*For example, see Fersiger, J.H. (1998) *Numerical Methods for Engineering Application*, 2nd Ed., Wiley; Fersiger, J.H. and Peric, M. (1999) *Computational Methods for Fluid Dynamics*, 2nd Ed., Springer; Schlichting, H. and Gersten, K. (2000) *Boundary Layer Theory*, Chap. 23, 8th Ed., McGraw-Hill.

[†]A guide on how to use commercial CFD is given by Shaw, C.T. (1992) *Using Computational Fluid Mechanics*, Prentice Hall.

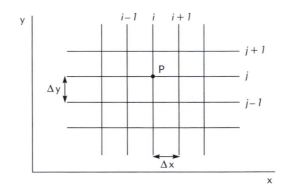

Fig. 7.47

reasons such a procedure is called a *marching* scheme. The fact that the equations are parabolic allows one to use such a marching scheme. It would not work for the Laplace equation, for example. In this case, quite different methods must be used (see Section 3.5).

The simplest approach is based on a so-called *explicit* finite-difference scheme. This will be briefly explained below.

It is assumed that the values of U are known along the line $X = X_i$, i.e. the discrete values at the grid points (like P) are known. The object is to devise a scheme for calculating the values of U at the grid points along $X = X_{i+1}$. To do this we rewrite Eqn (7.132) in a form for determining $\partial U/\partial X$ at the point P in Fig. 7.47. Thus

$$\left(\frac{\partial U}{\partial X}\right)_{i,j} = \frac{1}{U_{i,j}}\left\{-V_{i,j}\left(\frac{\partial U}{\partial Y}\right)_{i,j} - \left(\frac{dP}{dX}\right)_i + \left(\frac{\partial^2 U}{\partial Y^2}\right)_{i,j}\right\}. \tag{7.133}$$

Values at $X = X_{i+1}$ can then be estimated by writing

$$U_{i+1,j} = U_{i,j} + \left(\frac{\partial U}{\partial X}\right)_{i,j}\Delta X + \mathcal{O}((\Delta X)^2). \tag{7.134}$$

The last term on the right-hand side indicates the size of the error involved in using this approximation. The pressure gradient term in Eqn (7.133) is a known function of X obtained either from experimental data or from the solution for the potential flow around the body. But the other terms have to be estimated using finite differences. To obtain estimates for the first and second derivatives with respect to Y we start with Taylor expansions about the point P in the positive and negative Y directions. Thus we obtain

$$U_{i,j+1} = U_{i,j} + \left(\frac{\partial U}{\partial Y}\right)_{i,j}\Delta Y + \left(\frac{\partial^2 U}{\partial Y^2}\right)_{i,j}\frac{(\Delta Y)^2}{2} + \mathcal{O}((\Delta Y)^3) \tag{7.135a}$$

$$U_{i,j-1} = U_{i,j} - \left(\frac{\partial U}{\partial Y}\right)_{i,j}\Delta Y + \left(\frac{\partial^2 U}{\partial Y^2}\right)_{i,j}\frac{(\Delta Y)^2}{2} - \mathcal{O}((\Delta Y)^3) \tag{7.135b}$$

First we subtract Eqn (7.135b) from (7.135a) and rearrange to obtain the estimate

$$\left(\frac{\partial U}{\partial Y}\right)_{i,j} = \frac{U_{i,j+1} - U_{i,j-1}}{2\Delta Y} + \mathcal{O}((\Delta Y)^2). \qquad (7.136)$$

We then add Eqns (7.135a) and (7.135b) and rearrange to obtain the estimate

$$\left(\frac{\partial^2 U}{\partial Y^2}\right)_{i,j} = \frac{U_{i,j+1} - 2U_{i,j} + U_{i,j-1}}{(\Delta Y)^2} + \mathcal{O}((\Delta Y)^2). \qquad (7.137)$$

The results given in Eqns (7.136) and (7.137) are usually referred as *centred* finite differences. Finally, the value of $V_{i,j}$ also has to be estimated. This has to be obtained from Eqn (7.131) which can be rewritten as

$$\left(\frac{\partial V}{\partial Y}\right)_{i,j} = -\left(\frac{\partial U}{\partial X}\right)_{i,j}$$

Using a result analogous to Eqn (7.136) we get

$$\frac{V_{i,j+1} - V_{i,j-1}}{2\Delta Y} = -\left(\frac{\partial U}{\partial X}\right)_{i,j} \quad \text{i.e.} \quad V_{i,j+1} = V_{i,j-1} - 2\Delta Y \left(\frac{\partial U}{\partial X}\right)_{i,j}$$

There are two problems with this result. First it gives $V_{i,j+1}$ rather than $V_{i,j}$; this is easily remedied by replacing j by $j-1$ to obtain

$$V_{i,j} = V_{i,j-2} - 2\Delta Y \left(\frac{\partial U}{\partial X}\right)_{i,j-1} \qquad (7.138)$$

The other problem is that it requires a value of $(\partial U/\partial X)_{i,j-1}$. But the very reason we needed to estimate $V_{i,j}$ in the first place was to obtain an estimate for $(\partial U/\partial X)_{i,j}$! Fortunately, this does not represent a problem provided the calculations are done in the right order.

At the wall, say $j = 0$, $U = V = 0$ (assuming it to be impermeable), also from the continuity equation (7.131), $\partial V/\partial Y = 0$. Thus, using an equation analogous to Eqn (7.135a),

$$V_{i,1} = \underbrace{V_{i,0}}_{=0} + \underbrace{\left(\frac{\partial V}{\partial Y}\right)}_{=0} + \mathcal{O}((\Delta Y)^2) \simeq 0.$$

With this estimate of $V_{i,1}$, $(\partial U/\partial X)_{i,1}$ can now be estimated from Eqn (7.133), the first term on the right-hand side being equivalently zero. $U_{i+1,1}$ is then estimated from Eqn (7.134). All the values are now known for $j = 1$. The next step is to set $j = 2$ in Eqn (7.138), so that

$$V_{i,2} = \underbrace{V_{i,0}}_{=0} - 2\Delta Y \left(\frac{\partial U}{\partial X}\right)_{i,1}$$

$(\partial U/\partial X)_{i,2}$ can now be estimated from Eqn (7.133) and so on right across the boundary layer. It is necessary, of course, to choose an upper boundary to the computational domain at a finite, but suitably large, height, $y = Y_J$, corresponding to $j = J$, say.

This explicit finite-difference scheme is relatively simple to execute. But explicit schemes are far from ideal for boundary-layer calculations. Their main drawback is that their use leads to numerical instability that is characterized by the solution displaying increasingly large oscillations as the calculation marches downstream leading to unacceptedly large errors. In order to ensure that this does not happen, it is necessary to impose the following condition on the streamwise step length in order to ensure numerical stability:

$$\Delta X \leq \frac{1}{2} |\min(U_{i,j})(\Delta Y)^2| \tag{7.139}$$

Since U is very small at the first grid point near the wall, very small values of ΔX will be required to ensure numerical stability.

Because of the problems with numerical instability, so-called *implicit* schemes are much to be preferred, because these permit step-size to be determined by considerations of accuracy rather than numerical stability. Here a scheme based on the Crank–Nicholson method will be briefly described. The essential idea is to rewrite Eqn (7.134) in the form

$$U_{i+1,j} = U_{i,j} + \frac{1}{2}\left[\left(\frac{\partial U}{\partial X}\right)_{i,j} + \left(\frac{\partial U}{\partial X}\right)_{i+1,j}\right]\Delta X + \mathcal{O}((\Delta X)^3). \tag{7.140}$$

So that the derivative at $x = x_i$ is replaced by the average of that and the derivative at $x = x_{i+1}$. Formally, this is more accurate and numerically much more stable. The problem is that Eqn (7.133) implies that

$$\left(\frac{\partial U}{\partial X}\right)_{i+1,j} = \frac{1}{U_{i+1,j}}\left\{-V_{i+1,j}\left(\frac{\partial U}{\partial Y}\right)_{i+1,j} - \left(\frac{dP}{dX}\right)_{i+1} + \left(\frac{\partial^2 U}{\partial Y^2}\right)_{i+1,j}\right\}. \tag{7.141}$$

and Eqns (7.136) and (7.137) imply that

$$\left(\frac{\partial U}{\partial Y}\right)_{i+1,j} = \frac{U_{i+1,j+1} - U_{i+1,j-1}}{2\Delta Y} + \mathcal{O}((\Delta Y)^2) \tag{7.142}$$

$$\left(\frac{\partial^2 U}{\partial Y^2}\right)_{i+1,j} = \frac{U_{i+1,j+1} - 2U_{i+1,j} + U_{i+1,j-1}}{(\Delta Y)^2} + \mathcal{O}((\Delta Y)^2). \tag{7.143}$$

Thus the unknown values of U at X_{i+1}, i.e. $U_{i+1,j}(j = 1, \ldots, J)$, appear on both sides of Eqn (7.140). This is what is meant by the term *implicit*. In order to solve Eqn (7.140) for these unknowns it must be rearranged as a matrix equation of the form

$$\mathbf{AU} = \mathbf{R}$$

where $\mathbf{A}$ is the coefficient matrix, $\mathbf{U}$ is a column matrix containing the unknowns $U_{i+1,j}(j = 1, \ldots, J)$, and $\mathbf{R}$ is a column matrix of containing known quantities. Fortunately $\mathbf{A}$ has a tridiagonal form, i.e. only the main diagonal and the two diagonals either side of it are non-zero. Tridiagonal matrix equations can be solved very efficiently using the Thomas (or Tridiagonal) algorithm, versions of which can be found in Press *et al.* (*loc. cit.*) and on the Internet site associated with Ferziger (1998) (*loc. cit.*).

One of the most popular and widely used implicit schemes for computational solution of the boundary-layer equations is the Keller* *box scheme* which is slightly more accurate than the Crank–Nicholson method.

---

* Keller, H.B. (1978) Numerical methods in boundary-layer theory, *Annual Review Fluid Mech.*, **10**, 417–433.

7.11.4 Computational solution of turbulent boundary layers

The simplest approach is based on the turbulent boundary-layer equations (7.108a,b) written in the form:

$$\frac{\partial \bar{u}}{\partial x} + \frac{\partial \bar{v}}{\partial y} = 0 \tag{7.144}$$

$$\bar{u}\frac{\partial \bar{u}}{\partial x} + \bar{v}\frac{\partial \bar{u}}{\partial y} = -\frac{\mathrm{d}\bar{p}}{\mathrm{d}x} + \frac{\partial}{\partial y}\left(\mu\frac{\partial \bar{u}}{\partial y} - \rho\overline{u'v'}\right) \tag{7.145}$$

No attempt is made to write these equations in terms of non-dimensional variables as was done for Eqns (7.131) and (7.132). A similar procedure would be advantageous for computational solutions, but it is not necessary for the account given here.

The primary problem with solving Eqns (7.131) and (7.132) is not computational. Rather, it is that there are only two equations, but three dependent variables to determine by calculation, namely $\bar{u}$, $\bar{v}$ and $\overline{u'v'}$. The 'solution' described in Section 7.10 is to introduce an eddy-viscosity model (see Eqn 7.109) whereby

$$-\rho\overline{u'v'} = \rho\varepsilon_T\frac{\partial \bar{u}}{\partial y} \tag{7.146}$$

In order to solve Eqns (7.144) and (7.145), ε_T must be expressed in terms of $\bar{u}$ (and, possibly, $\bar{v}$). This can only be done semi-empirically. Below in Section 7.11.5 it will be explained how a suitable semi-empirical model for the eddy viscosity can be developed for computational calculation of turbulent boundary layers. First, a brief exposition of the wider aspects of this so-called *turbulence modelling* approach is given.

From the time in the 1950s when computers first began to be used by engineers, there has been a quest to develop increasingly more effective methods for computational calculation of turbulent flows. For the past two decades it has even been possible to carry out *direct numerical simulations* (*DNS*) of the full unsteady, three-dimensional, Navier–Stokes equations for relatively simple turbulent flows at comparatively low Reynolds numbers.[*] Despite the enormous advances in computer power, however, it is unlikely that DNS will be feasible, or even possible, for most engineering applications within the foreseeable future.[†] All alternative computational methods rely heavily on semi-empirical approaches known collectively as *turbulent modelling*. The modern methods are based on deriving additional transport equations from the Navier–Stokes equations for quantities like the various components of the Reynolds stress tensor (see Eqn 7.107), the turbulence kinetic energy, and the viscous dissipation rate. In a sense, such approaches are based on an unattainable goal, because each new equation that is derived contains ever more unknown quantities, so that the number of dependent variables always grows faster than the number of equations. As a consequence, an increasing number of semi-empirical formulae is required. Nevertheless, despite their evident drawbacks, the computational methods based on turbulence modelling have become an indispensable tool in modern

---

[*] Spalart, P.R. and Watmuff, J.H. (1993) Direct simulation of a turbulent boundary layer up to $R_\theta = 1410$, *J. Fluid Mech.*, **249**, 337–371; Moin, P. and Mahesh, K. (1998) Direct numerical simulation: A tool in turbulence research, *Annual Review Fluid Mech.*, **30**, 539–578; Friedrich, R. *et al.* (2001) Direct numerical simulation of incompressible turbulent flows, *Computers & Fluids*, **30**, 555–579.

[†] Agarwal, R. (1999) Computational fluid dynamics of whole-body aircraft, *Annual Review Fluid Mech.*, **31**, 125–170.

engineering. A brief account of one of the most widely used of these methods will be given in Section 7.11.6.

Increasingly, an alternative approach to this type of turbulence modelling is becoming a viable computational tool for engineering applications. This is *large-eddy simulation* (*LES*) that was first developed by meteorologists. It still relies on semi-empirical turbulence modelling, however. A brief exposition will be given in Section 7.11.7.

7.11.5 Zero-equation methods

Computational methods that are based on Eqns (7.144) and (7.145) plus semi-empirical formulae for eddy viscosity are often known as *zero-equation* methods. This terminology reflects the fact that no additional partial differential equations, derived from the Navier–Stokes equations, have been used. Here the Cebeci–Smith method[*] will be described. It was one of the most successful zero-equation methods developed in the 1970s.

Most of the zero-equation models are based on extensions of Prandtl's mixing-length concept (see Sections 7.10.4 and 7.10.5), namely:

$$-\overline{u'v'} = \varepsilon_T \frac{\partial \bar{u}}{\partial y} \quad \varepsilon_T = \kappa y^2 \left| \frac{\partial \bar{u}}{\partial y} \right| \quad (\text{i.e. } \ell_m = \kappa y)$$

The constant κ is often known as the von Kármán constant.

Three key modifications were introduced in the mid 1950s:

(1) *Damping near the wall*: Van Driest[†]

An exponential damping function was proposed that comes into play as $y \to 0$. This reflects the reduction in turbulence level as the wall is approached and extends the mixing-length model into the buffer layer and viscous sub-layer:

$$\ell_m = \kappa y [1 - \exp(-y^+ / A_0^+)] \quad A_0^+ = 26$$

(2) *Outer wake-like flow*: Clauser[‡]

It was recognized that the outer part of a boundary layer is like a free shear layer (specifically, like a wake flow), so there the Prandtl–Görtler eddy-viscosity model, see Eqns (7.110) and (7.111), is more appropriate:

$$\varepsilon = \underbrace{\alpha}_{\text{const.}} \times U_e \delta^*$$

where U_e is the flow speed at the edge of the boundary layer and δ^* is the boundary-layer displacement thickness.

(3) *Intermittency*: Corrsin and Kistler, and Klebannoff[§]

It was recognized that the outer part of the boundary layer is only intermittently turbulent (see Section 7.10.7 and Fig. 7.40). To allow for this it was proposed that ε_T be multiplied by the following semi-empirical intermittency factor:

[*] Cebeci, T. and Smith, A.M.O. (1974) *Analysis of Turbulent Boundary Layers*, Academic Press.

[†] Van Driest, E.R. (1956) On the turbulent flow near a wall, *J. Aeronautical Sciences*, **23**, 1007–1011.

[‡] Clauser, F.H. (1956) The turbulent boundary layer, *Adv. in Applied Mech.*, **4**, 1–51.

[§] Corrsin, S. and Kistler, A.L. (1954) The free-stream boundaries of turbulent flows, *NACA Tech. Note 3133*. Klebannoff, P.S. (1954) Characteristics of influence in a boundary layer with zero pressure gradient, *NACA Tech. Note 3178* and *NACA Rep.1247*.

$$\gamma_{tr} = \left[1 + \left(\frac{y}{\delta}\right)^6\right]^{-1}$$

Cebeci–Smith method

The Cebeci–Smith method incorporates versions of these three key modifications. For the *inner region* of the turbulent boundary layer:

$$(\varepsilon_T)_i = \ell^2 \left|\frac{\partial \bar{u}}{\partial y}\right| \gamma_{tr} \quad 0 \le y \le y_c \tag{7.147}$$

where the mixing length

$$\ell = \kappa y \underbrace{\left[1 - \exp\left(-\frac{y}{A}\right)\right]}_{i} \tag{7.148}$$

Term (i) is a semi-empirical modification of Van Driest's damping model that takes into account the effects of the streamwise pressure gradient; $\kappa = 0.4$ and
  Damping Length:

$$A = \frac{26\nu}{V_* \sqrt{1 - 11.8(\nu U_e / V_*^3) \mathrm{d}U_e/\mathrm{d}x}} \tag{7.149}$$

For the *outer region* of the turbulent boundary layer:

$$(\varepsilon_T)_o = \alpha U_e \delta^* \gamma_{tr} \quad y_c \le y \le \delta \tag{7.150}$$

where $\alpha = 0.0168$ when $Re_\theta \ge 5000$. y_c is determined by requiring

$$(\varepsilon_T)_i = (\varepsilon_T)_o \quad \text{at} \quad y = y_c \tag{7.151}$$

See Cebeci and Bradshaw (1977) for further details of the Cebeci–Smith method.
  It does a reasonably good job in calculating conventional turbulent boundary-layer flows. For applications involving separated flows, it is less successful and one-equation methods like that due to Baldwin and Lomax* are preferred.
  For further details on the Baldwin–Lomax and other one-equation methods, including computer codes, Wilcox (1993) and other specialist texts should be consulted.

7.11.6 The k–ε method – A typical two-equation method

Probably the most widely used method for calculating turbulent flows is the k–ε model which is incorporated into most commercial CFD software. It was independently developed at Los Alamos[†] and at Imperial College London.[‡]

---

[*] Baldwin, B.S. and Lomax, H. (1978) Thin layer approximation and algebraic model for separated turbulent flows. *AIAA Paper 78-257*.

[†] Harlow, F.H. and Nakayama, P.I. (1968) Transport of turbulence energy decay rate. Univ. of California, Los Alamos Science Lab. Rep. LA-3854.

[‡] Jones, W.P. and Launder, B.E. (1972) The prediction of laminarization with a two-equation model of turbulence, *Int. J. Heat Mass Transfer*, **15**, 301–314.

The basis of the $k-\varepsilon$ and most other two-equation models is an eddy-viscosity formula based on dimensional reasoning and taking the form:

$$\varepsilon_T = C_\mu k^{1/2} \ell \quad C_\mu \text{ is an empirical const.} \qquad (7.152)$$

Note that the kinetic energy per unit mass, $k \equiv (\overline{u'^2} + \overline{v'^2} + \overline{w'^2})/2$. Some previous two-equation models derived a transport equation for the length-scale ℓ. This seemed rather unphysical so, based on dimensional reasoning, the $k-\varepsilon$ model took

$$\ell = k^{3/2}/\varepsilon \qquad (7.153)$$

where ε is the viscous dissipation rate per unit mass and should not be confused with the eddy viscosity, ε_T. A transport equation for ε was then derived from the Navier–Stokes equations.

Both this equations for ε and the turbulence kinetic energy k contain terms involving additional unknown dependent variables. These terms must be modelled semi-empirically. For flows at high Reynolds number the transport equations for k and ε are modelled as follows:

Turbulence energy:

$$\rho \frac{\mathrm{D}k}{\mathrm{D}t} = \frac{\partial}{\partial y}\left(\frac{\rho \varepsilon_T}{\sigma_k}\frac{\partial k}{\partial y}\right) + \rho \varepsilon_T \left(\frac{\partial \bar{u}}{\partial y}\right)^2 - \rho\varepsilon \qquad (7.154)$$

Energy dissipation:

$$\rho \frac{\mathrm{D}\varepsilon}{\mathrm{D}t} = \frac{\partial}{\partial y}\left(\frac{\rho \varepsilon_T}{\sigma_\varepsilon}\frac{\partial \varepsilon}{\partial y}\right) + C_1 \frac{\varepsilon}{k}\rho \varepsilon_T \left(\frac{\partial \bar{u}}{\partial y}\right)^2 - C_2 \frac{\rho\varepsilon^2}{k} \qquad (7.155)$$

These equations contain 5 empirical constants that are usually assigned the following values:

C_μ	C_1	C_2	σ_k	σ_ε
0.09	1.55	2.0	1.0	1.3

where σ_k and σ_ε are often termed *effective turbulence Prandtl numbers*. Further modification of Eqns (7.154 and 7.155) is required to deal with relatively low Reynolds numbers. See Wilcox (1993) for details of this and the choice of wall boundary conditions.

The $k-\varepsilon$ model is intended for computational calculations of general turbulent flows. It is questionable whether it performs any better than, or even as well as, the zero-equation models described in Section 7.11.5 for boundary layers. But it can be used for more complex flows, although the results should be viewed with caution. A common misconception amongst practising engineers who use commercial CFD packages containing the $k-\varepsilon$ model is that they are solving the exact Navier–Stokes equations. They are, in fact, solving a system of equations that contains several approximate semi-empirical formulae, including the eddy-viscosity model described above. Real turbulent flows are highly unsteady and three-dimensional. The best one can expect when using the $k-\varepsilon$, or any other similar turbulence model, is an approximate result that gives guidance to some of the features of the real turbulent flow. At worst, the results can be completely misleading, for an example, see the discussion in Wilcox (1993) of the round-jet/plane-jet anomaly.

For a full description and discussion of two-equation turbulence models and other more advanced turbulence models see Wilcox (1993), Pope (2000) and other specialized books.

7.11.7   Large-eddy simulation

This approach to computational calculation of turbulent flow originated with meteorologists. In a sense LES is half way between the turbulence modelling based on Reynolds averaging and direct numerical simulations. LES is motivated by the view that the larger-scale motions are likely to vary profoundly between one type of turbulent flow and another, but that the small-scale turbulence is likely to be much more universal in character. Accordingly any semi-empirical turbulence modelling should be confined to the small-scale turbulence. With this in mind the flow variables are partitioned into

$$\{u, v, w, p\} = \underbrace{\{\tilde{u}, \tilde{v}, \tilde{w}, \tilde{p}\}}_{\text{Resolved field}} + \underbrace{\{u', v', w', p'\}}_{\text{Subgrid-scale field}} \tag{7.156}$$

The resolved or large-scale field is computed directly while the subgrid-scale field is modelled semi-empirically.

The resolved field is obtained by applying a *filter* to the flow variables, e.g.

$$\tilde{u}(\vec{x}) = \int \underbrace{G(\vec{x} - \vec{\xi})}_{\text{Filter function}} u(\vec{\xi}) \mathrm{d}\vec{\xi} \tag{7.157}$$

If the filter function is chosen appropriately this has the effect of 'averaging' over the sub-grid scales.

Two common choices of filter function

(1)   Box Filter *:

$$G(\vec{x} - \vec{\xi}) = \begin{cases} 1 & |\vec{x} - \vec{\xi}| < \Delta/2 \\ 0 & \text{Otherwise} \end{cases} \tag{7.158}$$

(2)   Gaussian Filter [†]:

$$G(\vec{x} - \vec{\xi}) = \left[\left(\frac{6}{\pi}\right)^{1/2} \frac{1}{\Delta_a} \right] \exp\left[-6(\vec{x} - \vec{\xi})^2 / \Delta_a^2 \right] \tag{7.159}$$

The choice made for the size of Δ or Δ_a in Eqns (7.158) or (7.159) determines the sub-grid scale. Filtering the Navier–Stokes equations gives:

$$\frac{\partial \tilde{u}_i}{\partial t} + \frac{\partial}{\partial x_1} \widetilde{u_i u_1} + \frac{\partial}{\partial x_2} \widetilde{u_i u_2} + \frac{\partial}{\partial x_3} \widetilde{u_i u_3} = -\frac{1}{\rho} \frac{\partial \tilde{p}}{\partial x_i} + \nu \nabla^2 \tilde{u}_i \quad i = 1, 2, 3 \tag{7.160}$$

$$\frac{\partial \tilde{u}_1}{\partial x_1} + \frac{\partial \tilde{u}_2}{\partial x_2} + \frac{\partial \tilde{u}_3}{\partial x_3} = 0 \tag{7.161}$$

where u_1, u_2 and u_3 denote u, v and w, and x_1, x_2 and x_3 denote x, y and z; and where

$$\widetilde{u_i u_j} = \widetilde{\tilde{u}_i \tilde{u}_j} + \widetilde{u'_i \tilde{u}_j} + \widetilde{u'_j \tilde{u}_i} + \widetilde{u'_i u'_j} \tag{7.162}$$

* Deardorff, J.W. (1970) A numerical study of three-dimensional turbulent channel flow at large Reynolds numbers, *J. Fluid Mech.*, **41**, 453–480.

[†] Leonard, A. (1974) Energy cascade in large eddy simulations of turbulent fluid flow, *Adv. Geophys.*, **18A**, 237–248.

Usually the following approximation suggested by Leonard (1974) is made

$$\widetilde{u_i u_j} \simeq \tilde{u}_i \tilde{u}_j + \underbrace{\frac{\Delta^2}{12} \frac{\partial \tilde{u}_i}{\partial u_\ell} \frac{\partial \tilde{u}_j}{\partial u_\ell}}_{\text{Leonard stress}} + \underbrace{\widetilde{u'_i u'_j}}_{\text{modelled semi-empirically}} \tag{7.163}$$

Sub-grid scale modelling

A common approach, originating with Smagorinsky* is to use an eddy viscosity, so that

$$\widetilde{u'_i u'_j} \simeq -\varepsilon_T \frac{1}{2} \left(\frac{\partial \tilde{u}_i}{\partial x_j} + \frac{\partial \tilde{u}_j}{\partial x_i} \right) \tag{7.164}$$

A common way of modelling ε_T is also due to Smagorinsky (1963):

$$\varepsilon_T = \frac{1}{2} (c\Delta)^2 \left[\sum_{i=1}^{3} \left(\frac{\partial \tilde{u}_i}{\partial x_j} + \frac{\partial \tilde{u}_j}{\partial x_i} \right) \left(\frac{\partial \tilde{u}_i}{\partial x_j} + \frac{\partial \tilde{u}_j}{\partial x_i} \right) \right]^{1/2} \tag{7.165}$$

where c is a semi-empirically determined constant.

For more information on LES see Wilcox (1993) and Pope (2000). LES is very demanding in terms of computational resources but with the rapid increase in computer power it is becoming more and more feasible for engineering calculations. An alternative that is less demanding on computational resources is to use conventional turbulence modelling based on Reynolds averaging, but to include the time derivatives of the mean velocity components in the Reynolds-averaged Navier–Stokes equations. This approach is sometimes known as very-large-eddy simulation (VLES). See Tucker (2001)[†] for a specialized treatment.

7.12 Estimation of profile drag from velocity profile in wake

At the trailing edge of a body immersed in a fluid flow, there will exist the boundary layers from the surfaces on either side. These boundary layers will join up and move downstream in the form of a wake of retarded velocity. The velocity profile will change with distance downstream, the wake cross-section increasing in size as the magnitude of its mean velocity defect, relative to free stream, decreases. At a sufficient distance downstream, the streamlines will all be parallel and the static pressure across the wake will be constant and equal to the free-stream value. If conditions at this station are compared with those in the undisturbed stream ahead of the body, then the rate at which momentum has been lost, while passing the body, will equate to the drag force on the body. The drag force so obtained will include both skin-friction and form-drag components, since these together will produce the overall momentum change. A method of calculating the drag of a two-dimensional body using the momentum loss in the wake is given below. The method depends on conditions remaining steady with time.

---

* Smagorinsky, J. (1963) General circulation experiments with the primitive equations: I. The basic equations, *Mon. Weather Rev.*, **91**, 99–164.

[†] Tucker, P.G. (2001) *Computation of Unsteady Internal Flows*. Kluwer Academic Publishers, Norwell, MA, U.S.A.

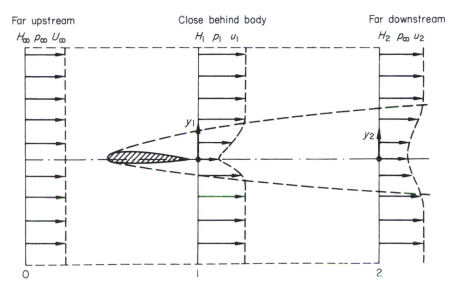

Fig. 7.48

7.12.1 The momentum integral expression for the drag of a two-dimensional body

Consider a two-dimensional control volume fixed in space (see Fig. 7.48) of unit width, with two faces (planes 0 and 2) perpendicular to the free stream, far ahead of and far behind the body respectively, the other two lying parallel to the undisturbed flow direction, and situated respectively far above and far below the body. For any stream tube (of vertical height δy) that is contained within the wake at the downstream boundary, the mass flow per unit time is $\rho u_2 \delta y_2$ and the velocity reduction between upstream and downstream is $U_\infty - u_2$. The loss of momentum per unit time in the stream tube $= \rho u_2(U_\infty - u_2)\delta y_2$ and, for the whole field of flow:

$$\text{Total loss of momentum per unit time} = \int_{-\infty}^{\infty} \rho u_2(U_\infty - u_2)\mathrm{d}y_2$$

In fact, the limits of this integration need only extend across the wake because the term $U_\infty - u_2$ becomes zero outside it.

This rate of loss of momentum in the wake is brought about by the reaction on the fluid of the profile drag force per unit span D, acting on the body. Thus

$$D = \int_w \rho u_2(U_\infty - u_2)\mathrm{d}y_2 \qquad (7.166)$$

This expression enables the drag to be calculated from an experiment arranged to determine the velocity profile at some considerable distance downstream of the body, i.e. where $p = p_\infty$.

For practical use it is often inconvenient, or impossible, to arrange for measurement so far away from the body, and methods that allow measurements to be made close behind the body (plane 1 in Fig. 7.48) have been developed by Betz* and B.M. Jones.[†] The latter's method is considerably the simpler and is reasonably accurate for most purposes.

7.12.2   B.M. Jones' wake traverse method for determining profile drag

In the wake close behind a body the static pressure, as well as the velocity, will vary from the value in the free stream outside the wake. B. Melville Jones allowed for this fact by assuming that, in any given stream tube between planes 1 (close to the body) and 2 (far downstream), the stagnation pressure could be considered to remain constant. This is very nearly the case in practice, even in turbulent wakes.

Let H_∞ be the stagnation pressure in any stream tube at plane 0 and $H_1 = H_2$ be its value in the same stream tube at planes 1 and 2. Then

$$H_\infty = p_\infty + \frac{1}{2}\rho U_\infty^2$$

$$H_1 = p_1 + \frac{1}{2}\rho u_1^2 = p_\infty + \frac{1}{2}\rho u_2^2$$

The velocities are given by

$$U_\infty = \sqrt{\frac{2}{\rho}(H_\infty - p_\infty)}, \quad u_1 = \sqrt{\frac{2}{\rho}(H_1 - p_1)} \quad \text{and} \quad u_2 = \sqrt{\frac{2}{\rho}(H_1 - p_\infty)}$$

Substituting the values for U_∞ and u_2 into Eqn (7.166) gives

$$D = 2 \int_{w_2} \sqrt{H_1 - p_\infty}(\sqrt{H_\infty - p_\infty} - \sqrt{H_1 - p_\infty})dy_2 \tag{7.167}$$

To refer this to plane 1, the equation of continuity in the stream tube must be used, i.e.

$$u_1\,\delta y_1 = u_2\,\delta y_2$$

or

$$\delta y_2 = \frac{u_1}{u_2}\delta y_1 = \sqrt{\frac{H_1 - p_1}{H_1 - p_\infty}}\delta y_1 \tag{7.168}$$

Referred to the wake at plane 1, Eqn (7.167) then becomes

$$D = 2 \int_{w_1} \sqrt{H_1 - p_1}(\sqrt{H_\infty - p_\infty} - \sqrt{H_1 - p_\infty})\,dy_1 \tag{7.169}$$

In order to express Eqn (7.169) non-dimensionally, the profile-drag coefficient C_{D_P} is used. For unit span:

$$C_{D_P} = \frac{D}{\frac{1}{2}\rho U_\infty^2 c} = \frac{D}{c(H_\infty - p_\infty)}$$

---

* A. Betz, *ZFM*, **16**, 42, 1925.
[†] B. Melville Jones, *ARCR and M*, 1688, 1936.

so that Eqn (7.169) becomes

$$C_{D_P} = 2 \int_{w_1} \sqrt{\frac{H_1 - p_1}{H_\infty - p_\infty}} \left[1 - \sqrt{\frac{H_1 - p_\infty}{H_\infty - p_\infty}} \right] d\left(\frac{y_1}{c}\right) \qquad (7.170)$$

It will be noticed again that this integral needs to be evaluated only across the wake, because beyond the wake boundary the stagnation pressure H_1 becomes equal to H_∞ so that the second term in the bracket becomes unity and the integrand becomes zero.

Equation (7.170) may be conveniently used in the experimental determination of profile drag of a two-dimensional body when it is inconvenient, or impracticable, to use a wind-tunnel balance to obtain direct measurement. It can, in fact, be used to determine the drag of aircraft in free flight. All that is required is a traversing mechanism for a pitot-static tube, to enable the stagnation and static pressures H_1 and p_1 to be recorded at a series of positions across the wake, ensuring that measurements are taken as far as the undisturbed stream on either side, and preferably an additional measurement made of the dynamic pressure, $H_\infty - p_\infty$, in the incoming stream ahead of the body. In the absence of the latter it can be assumed, with reasonable accuracy, that $H_\infty - p_\infty$ will be the same as the value of $H_1 - p_1$ outside the wake.

Using the recordings obtained from the traverse, values of $H_1 - p_1$ and $H_1 - p_\infty$ may be evaluated for a series of values of y_1/c across the wake, and hence a corresponding series of values of the quantity

$$\sqrt{\frac{H_1 - p_1}{H_\infty - p_\infty}} \left(1 - \sqrt{\frac{H_1 - p_\infty}{H_\infty - p_\infty}} \right)$$

By plotting a curve of this function against the variable y_1/c a closed area will be obtained (because the integral becomes zero at each edge of the wake). The magnitude of this area is the value of the integral, so that the coefficient C_{D_P} is given directly by twice the area under the curve.

In order to facilitate the actual experimental procedure, it is often more convenient to construct a comb or rake of pitot and static tubes, set up at suitable spacings. The comb is then positioned across the wake (it must be wide enough to read into the free stream on either side) and the pitot and static readings recorded.

The method can be extended to measure the drag of three-dimensional bodies, by making a series of traverses at suitable lateral (or spanwise) displacements. Each individual traverse gives the drag force per unit span, so that summation of these in a spanwise direction will give the total three-dimensional drag.

7.12.3   Growth rate of two-dimensional wake, using the general momentum integral equation

As explained the two boundary layers at the trailing edge of a body will join up and form a wake of retarded flow. The velocity profile across this wake will vary appreciably with distance behind the trailing edge. Some simple calculations can be made that will relate the rate of growth of the wake thickness to distance downstream, provided the wake profile shape and external mainstream conditions can be specified.

The momentum integral equation for steady incompressible flow, Eqn (7.98) may be reduced to

$$\frac{d\theta}{dx} = \frac{C_f}{2} - \frac{\theta}{U_e}\frac{dU_e}{dx}(2 + H) \qquad (7.171)$$

Now C_f is the local surface shear stress coefficient at the base of the boundary layer, and at the wake centre, where the two boundary layers join, there is no relative velocity and therefore no shearing traction. Thus, for each half of the wake, C_f is zero and Eqn (7.171) becomes

$$\frac{d\theta}{dx} = -\frac{\theta}{U_e}\frac{dU_e}{dx}(2 + H) \qquad (7.172)$$

It is clear from this that if the mainstream velocity outside the wake is constant, then $dU_e/dx = 0$ and the right-hand side becomes zero, i.e. the momentum thickness of the wake is constant. This would be expected from the direct physical argument that there are no overall shearing tractions at the wake edges under these conditions, so that the total wake momentum will remain unaltered with distance downstream. θ may represent the momentum thickness for each half of the wake, considered separately if it is unsymmetrical, or of the whole wake if it is symmetrical.

The general thickness δ of the wake is then obtainable from the relationship

$$\theta = \delta \int_0^1 \bar{u}(1 - \bar{u})\,d\bar{y} = I\delta$$

so that

$$\frac{\delta_b}{\delta_a} = \frac{\theta_b}{\theta_a}\frac{I_2}{I_b} \qquad (7.173)$$

where suffices a and b refer to two streamwise stations in the wake. Knowledge of the velocity profiles at stations a and b is necessary before the integrals I_a and I_b can be evaluated and used in this equation.

Example 7.11   A two-dimensional symmetrical aerofoil model of 0.3 m chord with a roughened surface is immersed, at zero incidence, in a uniform airstream flowing at $30\,\mathrm{m\,s^{-1}}$. The minimum velocity in the wake at a station 2.4 m downstream from the trailing edge is $27\,\mathrm{m\,s^{-1}}$. Estimate the general thickness of the whole wake at this station. Assume that each boundary layer at the trailing edge has a 'seventh root' profile and a thickness corresponding to a turbulent flat-plate growth from a point at 10% chord, and that each half-wake profile at the downstream station may be represented by a cubic curve of the form $\bar{u} = a\bar{y}^3 + b\bar{y}^2 + c\bar{y} + d$.

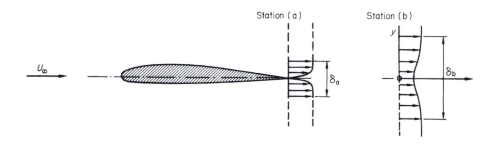

At the trailing edge, where $x = 0.3\,\text{m}$:

$$\frac{1}{2}\delta_a = \frac{0.383x}{Re_x^{1/5}} \quad \text{(Eqn (7.81))}$$

$$Re_x = \frac{30 \times 0.3}{14.6} \times 10^6 = 6.16 \times 10^5$$

$$Re_x^{1/5} = 14.39$$

$$\frac{1}{2}\delta_a = \frac{0.383 \times 0.27}{14.39} = 0.007\,19\,\text{m}$$

Also

$$I_a = 0.0973 \quad \text{Eqn (a) (Section 7.7.5)}$$

At the wake station:

$$\bar{u} = a\bar{y}^3 + b\bar{y}^2 + c\bar{y} + d, \qquad \frac{\partial \bar{u}}{\partial \bar{y}} = 3a\bar{y}^2 + 2b\bar{y} + c$$

The conditions to be satisfied are that (i) $\bar{u} = 0.9$ and (ii) $\partial \bar{u}/\partial \bar{y} = 0$, when $\bar{y} = 0$; and (iii) $\bar{u} = 1.0$ and (iv) $\partial \bar{u}/\partial \bar{y} = 0$, when $\bar{y} = 1$. (Condition (ii) follows because, once the wake is established at a short distance behind the trailing edge, the profile discontinuity at the centre-line disappears.) Thus:

$$d = 0.9, \qquad c = 0, \qquad 1 = a + b + 0.9 \qquad \text{and} \qquad 0 = 3a + 2b$$

i.e.

$$a = \frac{2}{3}b$$

$$1 = \left(1 - \frac{2}{3}\right)b + 0.9 \qquad \text{or} \qquad b = 0.3 \qquad \text{and} \qquad a = -0.2$$

$$\bar{u} = -0.2\bar{y}^3 + 0.3\bar{y}^2 + 0.9$$

$$I_b = \int_0^1 \bar{u}(1 - \bar{u})\,d\bar{y} = \int_0^1 (-0.2\bar{y}^3 + 0.3\bar{y}^2 + 0.9)(1 + 0.2\bar{y}^3 - 0.3\bar{y}^2 - 0.9)\,d\bar{y} = 0.0463$$

$$\frac{\delta_b}{\delta_a} = \frac{I_a}{I_b} = \frac{0.0973}{0.0463} = 2.1$$

i.e.

$$\delta_b = 2.1 \times 2 \times 0.007\,19 = 0.0302\,\text{m} \quad (30.2\,\text{mm})$$

7.13 Some boundary-layer effects in supersonic flow

A few comments may now be made about the qualitative effects on boundary-layer flow of shock waves that may be generated in the mainstream adjacent to the surface of a body. A normal shock in a supersonic stream invariably reduces the Mach number to a subsonic value and this speed reduction is associated with a very rapid increase in pressure, density and temperature.

For an aerofoil operating in a transonic regime, the mainstream flow just outside the boundary layer accelerates from a subsonic speed near the leading edge to sonic speed at some point near the subsonic-peak-suction position. At this point, the streamlines in the local mainstream will be parallel and the effect of the aerofoil surface curvature will be to cause the streamlines to begin to diverge downstream. Now the characteristics of a supersonic stream are such that this divergence is accompanied by an increase in Mach number, with a consequent decrease in pressure. Clearly, this state of affairs cannot be maintained, because the local mainstream flow must become subsonic again at a higher pressure by the time it reaches the undisturbed free-stream conditions downstream of the trailing edge. The only mechanism available for producing the necessary retardation of the flow is a shock wave, which will set itself up approximately normal to the flow in the supersonic region of the mainstream; the streamwise position and intensity (which will vary with distance from the surface) of the shock must be such that just the right conditions are established behind it, so that the resulting mainstream approaches ambient conditions far downstream. However, this simple picture of a near-normal shock requirement is complicated by the presence of the aerofoil boundary layer, an appreciable thickness of which must be flowing at subsonic speed regardless of the mainstream flow speed. Because of this, the rapid pressure rise at the shock, which cannot be propagated upstream in the supersonic regions of flow, can be so propagated in the subsonic region of the boundary layer. As a result, the rapid pressure rise associated with the shock becomes diffused near the base of the boundary layer and appears in the form of a progressive pressure rise starting at some appreciable distance upstream of the incident shock. The length of this upstream diffusion depends on whether the boundary layer is laminar or turbulent. In a laminar boundary layer the length may be as much as one hundred times the nominal general thickness (δ) at the shock, but for a turbulent layer it is usually nearer ten times the boundary-layer thickness. This difference can be explained by the fact that, compared with a turbulent boundary layer, a larger part of the laminar boundary-layer flow near the surface is at relatively low speed, so that the pressure disturbance can propagate upstream more rapidly and over a greater depth.

It has already been pointed out in Sections 7.2.6 and 7.4 that an adverse pressure gradient in the boundary layer will at least cause thickening of the layer and may well cause separation. The latter effect is more probable in the laminar boundary layer and an additional possibility in this type of boundary layer is that transition to turbulence may be provoked. There are thus several possibilities, each of which may affect the external flow in different ways.

7.13.1 Near-normal shock interaction with laminar boundary layer

There appear to be three general possibilities when a near-normal shock interacts with a laminar boundary layer. With a relatively weak shock, corresponding to an upstream Mach number just greater than unity, the diffused pressure rise may simply cause a gradual thickening of the boundary layer ahead of the shock with no transition and no separation. The gradual thickening causes a family of weak compression waves to develop ahead of the main shock (these are required to produce the supersonic mainstream curvature) and the latter sets itself up at an angle, between itself and the upstream surface, of rather less than 90° (see Fig. 7.49). The compression waves join the main shock at some small distance from the surface, giving a diffused base to the shock.

Immediately behind the shock, the boundary layer tends to thin out again and a local expansion takes place which brings a small region of the mainstream up to a slightly supersonic speed again and this is followed by another weak near-normal

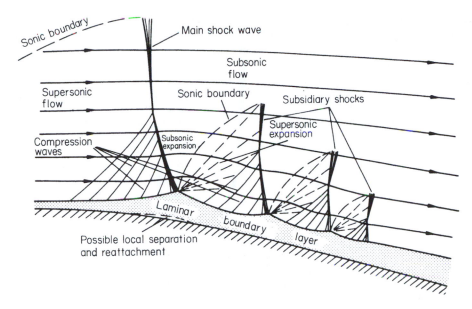

Fig. 7.49

shock which develops in the same way as the initial shock. This process may be repeated several times before the mainstream flow settles down to become entirely subsonic. Generally speaking this condition is not associated with boundary-layer separation, although there may possibly be a very limited region of separation near the base of the main shock wave.

As the mainstream speed increases so that the supersonic region is at a higher Mach number, the above pattern tends to change, the first shock becoming very much stronger than the subsequent ones and all but one of the latter may well not occur at all for local upstream Mach numbers much above 1.3. This is to be expected, because a strong first shock will produce a lower Mach number in the mainstream behind it. This means that there is less likelihood of the stream regaining supersonic speed. Concurrently with this pattern change, the rate of thickening of the boundary layer, upstream of the first and major shock, becomes greater and the boundary layer at the base of the normal part of the shock will generally separate locally before reattaching. There is a considerable possibility that transition to turbulence will occur behind the single subsidiary shock. This type of flow is indicated in Fig. 7.50.

With still greater local supersonic Mach numbers, the pressure rise at the shock may be sufficient to cause separation of the laminar boundary layer well ahead of the main shock position. This will result in a sharp change in direction of the mainstream flow just outside the boundary layer and this will be accompanied by a well-defined oblique shock which joins the main shock at some distance from the surface. This type of shock configuration is called a lambda-shock, for obvious reasons. It is unlikely that the boundary layer will reattach under these conditions, and the secondary shock, which normally appears as the result of reattachment or boundary-layer thinning, will not develop. This type of flow is indicated in Fig. 7.51. This sudden separation of the upper-surface boundary layer on an aerofoil, as Mach number increases, is usually associated with a sudden decrease in lift coefficient and the phenomenon is known as a shock stall.

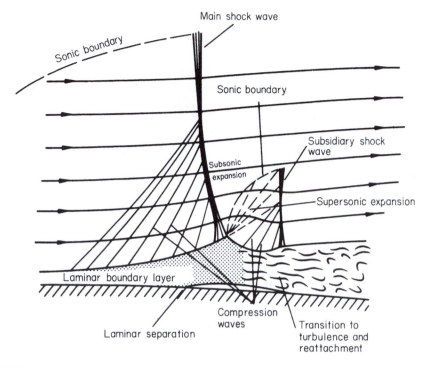

Fig. 7.50

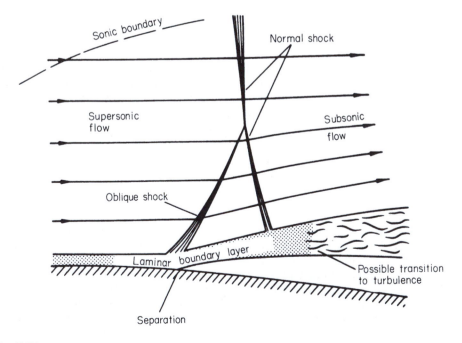

Fig. 7.51

7.13.2   Near-normal shock interaction with turbulent boundary layer

Because the turbulent boundary layer is far less susceptible to disturbance by an adverse pressure gradient than is a laminar one, separation is not likely to occur for local mainstream Mach numbers, ahead of the shock, of less than about 1.3 (this corresponds to a downstream-to-upstream pressure ratio of about 1.8). When no separation occurs, the thickening of the boundary layer ahead of the shock is rapid and the compression wavelets near the base of the main shock are very localized, so that the base of the shock appears to be slightly diffused, although no lambda formation is apparent. Behind the shock no subsequent thinning of the boundary layer appears to occur and the secondary shocks, typical of laminar boundary-layer interaction, do not develop.

If separation of the turbulent layer occurs ahead of the main shock a lambda shock develops and the mainstream flow looks much like that for a fully separated laminar boundary layer.

7.13.3   Shock-wave/boundary-layer interaction in supersonic flow

One of the main differences between subsonic and supersonic flows, as far as boundary-layer behaviour is concerned, is that the pressure gradient along the flow is of opposite sign with respect to cross-sectional area change. Thus in a converging supersonic flow the pressure rises and in a diverging flow the pressure falls in the stream direction (see Section 6.2). As a result the pressure gradient at a convex corner is negative and the boundary layer will generally negotiate the corner without separating, and the effect of the boundary layer on the external or mainstream flow will be negligible (Fig. 7.52a). Conversely, at a concave corner an oblique shock wave is generated and the corresponding pressure rise will cause boundary-layer thickening ahead of the shock, and in the case of a laminar boundary layer will probably cause local separation at the corner (see Figs 7.52b and 7.53). The resultant curvature of the flow just outside the boundary layer causes a wedge of compression wavelets to develop which, in effect, diffuse the base of the shock wave as shown in Fig. 7.52b.

At the nose of a wedge, the oblique nose shock will be affected by the boundary-layer growth; the presence of the rapidly thickening boundary layer near the leading edge produces an effective curvature of the nose of the wedge and a small region of expansive (Prandtl–Meyer) flow will develop locally behind the nose shock, which will now be curved and slightly detached from the nose (Fig. 7.54a). A similar effect will occur at the leading edge of a flat plate where a small detached curved local shock will develop. This shock will rapidly degenerate into a very weak shock approximating to a Mach wave at a small distance from the leading edge (Fig. 7.54b).

In some cases, an oblique shock that has been generated at some other point in the mainstream may be incident on the surface and boundary layer. Such a shock will be at an angle, between the upstream surface and itself, of considerably less than 90°. The general reaction of the boundary layer to this condition is similar to that already discussed in the transonic case, except that the oblique shock does not, in general, reduce the mainstream flow to a subsonic speed.

If the boundary layer is turbulent, it appears to reflect the shock wave as another shock wave in much the same way as would the solid surface in the absence of the boundary layer, although some thickening of the boundary layer occurs. There may also be local separation and reattachment, in which case the reflected shock originates just

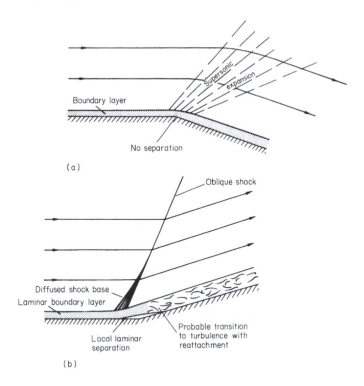

Fig. 7.52

Fig. 7.53 Supersonic flow through a sharp concave corner: The flow is from left to right at a downstream Mach number of 2.5. The holographic interferogram shows flow turning through an angle of 11° thereby forming an oblique shock wave that interacts with the turbulent boundary layer present on the wall. Each fringe corresponds to constant density. The boundary layer transmits the effect of the shock wave a short distance upstream but there is no flow separation. Compare with Fig. 7.52b above. (*The photograph was taken by P.J. Bryanston-Cross in the Engineering Department, University of Warwick, UK.*)

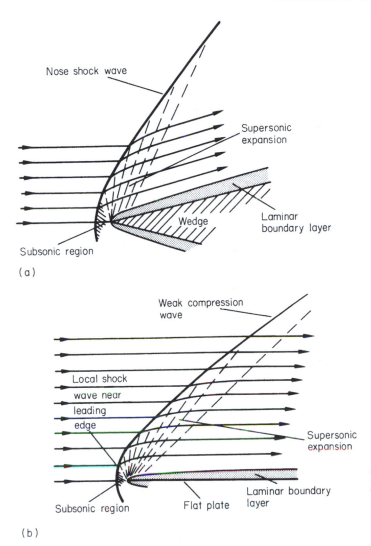

Nose shock wave

Supersonic expansion

Wedge

Laminar boundary layer

Subsonic region

(a)

Weak compression wave

Local shock wave near leading edge

Supersonic expansion

Subsonic region

Flat plate

Laminar boundary layer

(b)

Fig. 7.54

ahead of the point of incidence. A laminar boundary layer, however, thickens gradually up to the point of incidence, and may separate locally in this region, and then rapidly become thinner again. The shock then reflects as a fan of expansion waves, followed by a diffused shock a little farther downstream. There is also a set of weak compression waves set up ahead of the incident shock, owing to the boundary-layer thickening, but these do not usually set up a lambda configuration as with a near-normal incident shock.

Approximate representations of the above cases are shown in Fig. 7.55.

One other condition of interest that occurs in a closed uniform duct (two-dimensional or circular) when a supersonic stream is being retarded by setting up a back pressure in the duct. In the absence of boundary layers, the retardation would normally occur through a plane normal shock across the duct, reducing the flow, in one jump, from supersonic to subsonic speed. However, because of the presence of

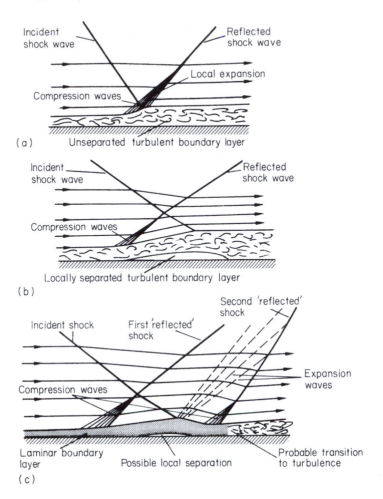

Fig. 7.55

the boundary layer, which thickens ahead of the shock, causing the base of the shock to thicken or bifurcate (lambda shock) depending on the nature and thickness of the boundary layer, in much the same way as for the transonic single-surface case.

Because of the considerable thickening of the boundary layers, the net flow area is reduced and may reaccelerate the subsonic flow to supersonic, causing another normal shock to be set up to re-establish the subsonic condition. This situation may be repeated several times until the flow reduces to what is effectively fully-developed subsonic boundary-layer flow. If the boundary layers are initially thick, the first shock may show a large degree of bifurcation owing to the large change of flow direction well ahead of the normal part of the shock. In some cases, the extent of the normal shock may be reduced almost to zero and a diamond pattern of shocks develops in the duct. Several typical configurations of this sort are depicted in Fig. 7.56.

To sum up the last two sections, it can be stated that, in contrast to the case of most subsonic mainstream flows, interaction between the viscous boundary layer and the effectively inviscid, supersonic, mainstream flow is likely to be appreciable. In the subsonic case, unless complete separation takes place, the effect of the boundary-layer

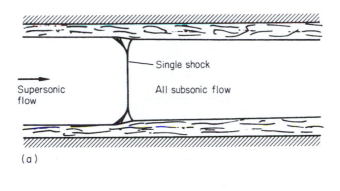

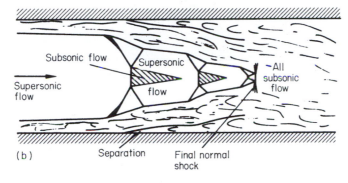

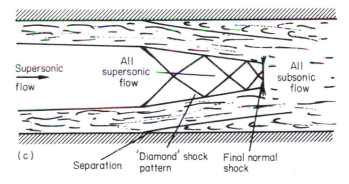

Fig. 7.56 (a) Low upstream Mach number; thin boundary layer; relatively small pressure rise; no separation. (b) Higher upstream Mach number; thin boundary layer; larger overall pressure rise; separation at first shock. (c) Moderately high upstream Mach number; thick boundary layer; large overall pressure rise; separation at first shock

development on the mainstream can usually be neglected, so that an inviscid main-stream-flow theory can be developed independently of conditions in the boundary layer. The growth of the latter can then be investigated in terms of the velocities and streamwise pressure gradients that exist in the previously determined mainstream flow.

In the supersonic (and transonic) case, the very large pressure gradients that exist across an incident shock wave are propagated both up- and downstream in the boundary layer. The rapid thickening and possible local separation that result

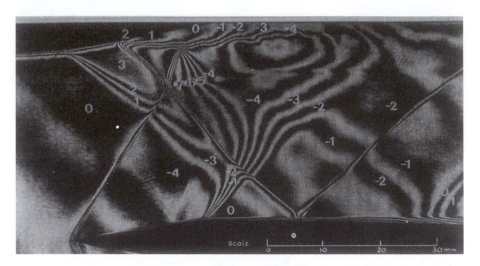

Fig. 7.57 Complex wave interactions in supersonic flow: The flow is from left to right for this holographic interferogram. Complex interactions occur between shock waves, expansion waves and boundary layers on the upper and lower walls. An oblique shock wave runs up and to the right from the leading edge of the wedge. This interacts with a fan of expansion waves running downwards and to the right from a sharp turn in the upper surface. A subsequent compression turn in the upper surface located at the top of the photograph generates a second shock wave running downward and to the right which interacts first with the leading-edge shock wave, then with an expansion wave emanating from the lower surface, and finally with the boundary layer on the lower surface. The pressure rise associated with this second shock wave has led to boundary-layer separation on the upper surface close to the label 2. See Fig. 7.55 on page 480. Interferograms can supply quantitative data in form of density or Mach number values. The Mach numbers corresponding to the numerical labels (given in parentheses) appearing in the photograph are as follows: $0.92(-7)$, $0.98(-6)$, $1.05(-5)$, $1.13(-4)$, $1.21(-3)$, $1.28(-2)$, $1.36(-1)$, $1.44(0)$, $1.53(1)$, $1.64(2)$, $1.75(3)$. (*The photograph was taken by P.J. Bryanston-Cross in the Engineering Department, University of Warwick, UK.*)

frequently have considerable effects on the way in which the shock is reflected by the boundary layer (see Fig. 7.57). In this way, the whole character of the mainstream flow may frequently be changed. It follows from this that a supersonic mainstream flow is much more dependent on Reynolds number than a subsonic one because of the appreciably different effects of an incident shock on laminar and turbulent boundary layers. The Reynolds number, of course, has a strong influence on the type of boundary layer that will occur. The theoretical quantitative prediction of supersonic stream behaviour in the presence of boundary layers is, consequently, extremely difficult.

Exercises

1 A thin plate of length 50 cm is held in uniform water flow such that the length of the plate is parallel to the flow direction. The flow speed is 10 m/s, the viscosity, $\mu = 1.0 \times 10^{-3}$ Pa s, and the water density is 998 kg/m³.

  (i) What is the Reynolds number based on plate length. What can be deduced from its value?

  (ii) On the assumption that the boundary layer is laminar over the whole surface, use the approximate theory based on the momentum-integral equation to find:

    (a) The boundary-layer thickness at the trailing edge of the plate; and

    (b) The skin-friction drag coefficient.

(iii) Repeat (ii), but now assume that the boundary layer is turbulent over the whole surface. (Use the formulae derived from the 1/7-power-law velocity profile.)
(*Answer*: $Re_L = 500\,000$; $\delta \simeq 3.3$ mm, $C_{Df} \simeq 0.0018$; $\delta \simeq 13.4$ mm, $C_{Df} \simeq 0.0052$)

2 A thin plate of length 1.0 m is held in a uniform air flow such that the length of the plate is parallel to the flow direction. The flow speed is 25 m/s, the viscosity, $\mu = 14.96 \times 10^{-6}$ Pa s, and the air density is 1.203 kg/m^3.

(i) If it is known that transition from laminar to turbulent flow occurs when the Reynolds number based on x reaches 500 000, find the transition point.
(ii) Calculate the equivalent plate length for an all-turbulent boundary layer with the same momentum thickness at the trailing edge as the actual boundary layer.
(iii) Calculate the coefficient of skin-friction drag per unit breadth for the part of the plate with
(a) A laminar boundary layer; and
(b) A turbulent boundary layer. (Use the formulae based on the 1/7-power-law velocity profile.)
(iv) Calculate the total drag per unit breadth.
(v) Estimate the percentage of drag due to the turbulent boundary layer alone.
(*Answer*: 300 mm from the leading edge; 782 mm; 0.00214, 0.00431; 1.44 N/m per side; 88%)

3 The geometric and aerodynamic data for a wing of a large white butterfly is as follows: Flight speed, $U_\infty = 1.35$ m/s; Average chord, $c = 25$ mm; Average span, $s = 50$ mm; air density $= 1.2$ kg/m^3; air viscosity, $\mu = 18 \times 10^{-6}$ Pa s; Drag at zero lift $= 120\,\mu$N (measured on a miniature wind-tunnel balance). Estimate the boundary-layer thickness at the trailing edge. Also compare the measured drag with the estimated skin-friction drag. How would you account for any difference in value? (*Answer*: 2.5 mm; 75 μN)

4 A submarine is 130 m long and has a mean perimeter of 50 m. Assume its wetted surface area is hydraulically smooth and is equivalent to a flat plate measuring 130 m $\times$ 50 m. Calculate the power required to maintain a cruising speed of 16 m/s when submerged in a polar sea at 0 °C. If the engines develop the same power as before, at what speed would the submarine be able to cruise in a tropical sea at 20 °C?

Take the water density to be 1000 kg/m^3, and its kinematic viscosity to be 1.79×10^{-6} m^2/s at 0 °C and 1.01×10^{-6} m^2/s at 20 °C.
(*Answer*: 20.5 MW; 16.37 m/s)

5 A sailing vessel is 64 m long and its hull has a wetted surface area of 560 m^2. Its top speed is about 9 m/s. Assume that normally the equivalent sand-grain roughness, k_s, of the hull is about 0.2 mm.

The total resistance of the hull is composed of wave drag plus skin-friction drag. Assume that the latter can be estimated by assuming it to be the same as the equivalent flat plate. The skin-friction drag is exactly half the total drag when sailing at the top speed under normal conditions. Assuming that the water density and kinematic viscosity are 1000 kg/m^3 and 1.2×10^{-6} m^2/s respectively, estimate:

(a) The admissible roughness for the vessel;
(b) The power required to maintain the vessel at its top speed when the hull is unfouled (having its original sand-grain roughness);

(c) The amount by which the vessel's top speed would be reduced if barnacles and seaweed were allowed to remain adhered to the hull, thereby raising the equivalent sand-grain roughness, k_s, to about 5 mm.

(*Answer*: 2.2 μm; 1.06 MW; 8.04 m/s, i.e. a reduction of 10.7%)

6 Suppose that the top surface of a light-aircraft wing travelling at an air speed of 55 m/s were assumed to be equivalent to a flat plate of length 2 m. Laminar-turbulent transition is known to occur at a distance of 0.75 m from the leading edge. Given that the kinematic viscosity of air is 15×10^{-6} m²/s, estimate the coefficient of skin-friction drag.

(*Answer*: 0.00223)

7 Dolphins have been observd to swim at sustained speeds up to 11 m/s. According to the distinguished zoologist Sir James Gray, this speed could only be achieved, assuming normal hydrodynamic conditions prevail, if the power produced per unit mass of muscle far exceeds that produced by other mammalian muscles. This result is known as Gray's paradox. The object of this exercise is to carry out revised estimates of the power required in order to check the soundness of Gray's calculations.

Assume that the dolphin's body is hydrodynamically equivalent to a prolate spheroid (formed by an ellipse rotated about its major axis) of 2 m length with a maximum thickness-to-length ratio of 1:6.

$$\text{Volume of a prolate spheroid} = \frac{4}{3}\pi a b^2$$

$$\text{Surface/area} = 2\pi b^2 + \frac{2\pi a^2 b}{\sqrt{a^2 - b^2}} \text{arc sin}\left(\frac{\sqrt{a^2 - b^2}}{a}\right)$$

where $2a$ is the length and $2b$ the maximum thickness.

Calculate the dimensions of the equivalent flat plate and estimate the power required to overcome the hydrodynamic drag (assuming it to be solely due to skin friction) at 11 m/s for the following two cases:

(a) Assuming that the transitional Reynolds number takes the same value as the maximum found for a flat plate, i.e. 2×10^6, say;
(b) Assuming that transition occurs at the point of maximum thickness (i.e. at the point of minimum pressure), which is located half-way along the body.

The propulsive power is supplied by a large group of muscles arranged around the spine, typically their total mass is about 36 kg, the total mass of the dolphin being typically about 90 kg. Assuming that the propulsive efficiency of the dolphin's tail unit is about 75%, estimate the power required per unit mass of muscle for the two cases above. Compare the results with the values given below.
Running man, 40 W/kg; Hovering humming-bird, 65 W/kg.

(*Answer*: 2 m × 0.832 m; 2.87 kW, 1.75 kW; 106 W/kg, 65 W/kg)

8 Many years ago the magazine *The Scientific American* published a letter concerned with the aerodynamics of pollen spores. A photograph accompanied the letter showing a spore having a diameter of about 20 μm and looking remarkably like a golf ball. The gist of the letter was that nature had discovered the principle of golf-ball aerodynamics millions of years before man. Explain why the letter-writer's logic is faulty.

8

Flow control and wing design

Preamble

This chapter deals with some of the fundamental principles of wing design for maximizing lift and minimizing drag. In this regard the behaviour of the boundary layer is critical and many design techniques and flow-control methods are available to counter adverse developments. Ways of identifying the most advantageous pressure distributions over the wing are given for low- and high-speed flows. Lift augmentation at low speeds by the use of multi-element aerofoils and various types of flap is described, along with several methods of direct boundary-layer control. The chapter closes with descriptions of the methods for reducing drag in its various forms (skin-friction, form, induced, and wave drag).

8.1 Introduction

Wing design is an exceedingly complex and multi-faceted subject. It is not possible to do justice to all that it involves in the present text. It is possible, however, to cover some of the fundamental principles that underly design for high lift and low drag.

For fixed air properties and freestream speed, lift can be augmented in four main ways, namely:

  (i) Increase in wing area;
 (ii) Rise in angle of incidence;
(iii) Increased camber; or
(iv) Increased circulation by the judicious application of high-momentum fluid.

The extent to which (ii) and (iii) can be exploited is governed by the behaviour of the boundary layer. A wing can only continue to generate lift successfully if boundary-layer separation is either avoided or closely controlled. Lift augmentation is usually accomplished by deploying various high-lift devices, such as flaps and multi-element aerofoils. Such devices lead to increased drag, so they are generally used only at the low speeds encountered during take-off and landing. Nevertheless, it is instructive to examine the factors governing the maximum lift achievable with an unmodified single-element aerofoil before passing to a consideration of the various high-lift devices. Accordingly, in what follows the maximization of lift for single-element aerofoils is considered in Section 8.2, followed by Section 8.3 on multi-element aerofoils and various types of flap, and Section 8.4 on other methods of boundary-layer control. Finally, the various methods used for drag reduction are described in Sections 8.5 to 8.8.

8.2   Maximizing lift for single-element aerofoils

This section addresses the question of how to choose the pressure distribution, particularly that on the upper wing surface, to maximize the lift. Even when a completely satisfactory answer is found to this rather difficult question, it still remains to determine the appropriate shape the aerofoil should assume in order to produce the specified pressure distribution. This second step in the process is the so called inverse problem of aerofoil design. It is very much more demanding than the direct problem, discussed in Chapter 4, of determining the pressure distribution for a given shape of aerofoil. Nevertheless, satisfactory inverse design methods are available. They will not, however, be discussed any further here. Only the more fundamental question of choosing the pressure distribution will be considered.

In broad terms the maximum lift achievable is limited by two factors, namely:

(i) Boundary-layer separation; and
(ii) The onset of supersonic flow.

In both cases it is usually the upper wing surface that is the more critical. Boundary-layer separation is the more fundamental of the two factors, since supercritical wings are routinely used even for subsonic aircraft, despite the substantial drag penalty in the form of wave drag that will result if there are regions of supersonic flow over the wing. However, no conventional wing can operate at peak efficiency with significant boundary-layer separation.

In two-dimensional flow boundary-layer separation is governed by:

(a) The severity and quality of the adverse pressure gradient; and
(b) The kinetic-energy defect in the boundary layer at the start of the adverse pressure gradient.

This latter quantity can be measured by the kinetic-energy thickness, δ^{**}, introduced in Section 7.3.2. Factor (a) is more vague. Precisely how is the severity of an adverse pressure gradient assessed? What is the optimum variation of adverse pressure distribution along the wing? Plainly when seeking an answer to the first of these questions a suitable non-dimensional local pressure must be used in order to remove, as far as possible, the effects of scale. What soon becomes clear is that the conventional definition of coefficient of pressure, namely

$$C_p = \frac{p - p_\infty}{\frac{1}{2} \rho_\infty V_\infty^2}$$

is not at all satisfactory. Use of this non-dimensional quantity invariably makes pressure distributions with high negative values of C_p appear to be the most severe. It is difficult to tell from the variation of C_p along an aerofoil whether or not the boundary layer has a satisfactory margin of safety against separation. Yet it is known from elementary dimensional analysis that if the Reynolds number is the same for two aerofoils of the same shape, but different size and freestream speed, the boundary layers will behave in an identical manner. Furthermore, Reynolds-number effects, although very important, are relatively weak.

There is a more satisfactory definition of pressure coefficient for characterizing the adverse pressure gradient. This is the *canonical* pressure coefficient, $\bar{C}_p$, introduced by A.M.O. Smith.[*] The definition of $\bar{C}_p$ is illustrated in Fig. 8.1. Note that local pressure is measured as a departure from the value of pressure, p_m, (the corresponding local velocity at the edge of the boundary layer is U_m) at the start of the pressure rise. Also note that the local dynamic pressure at the start of the pressure rise is now used to make the pressure difference non-dimensional. When the canonical representation is used, $\bar{C}_p = 0$ at the start of the adverse pressure gradient and $\bar{C}_p = 1$, corresponding to the stagnation point where $U = 0$, is the maximum possible value. Furthermore, if two pressure distributions have the same shape a boundary layer experiencing a deceleration of $(U/U_\infty)^2$ from 20 to 10 is no more or less likely to separate than one experiencing a deceleration of $(U/U_\infty)^2$ from 0.2 to 0.1. With the pressure-magnitude effects scaled out it is much easier to assess the effect of the adverse pressure gradient by simple inspection than when a conventional C_p distribution is used.

How are the two forms of pressure coefficient related? From the Bernoulli equation it follows that

$$C_p = 1 - \left(\frac{U}{U_\infty}\right)^2 \quad \text{and} \quad \bar{C}_p = 1 - \left(\frac{U}{U_m}\right)^2$$

Therefore it follows that

$$C_p = 1 - \left(\frac{U}{U_\infty}\right)^2 = 1 - \left(\frac{U}{U_m}\right)^2 \left(\frac{U_m}{U_\infty}\right)^2$$

$$= 1 - (1 - \bar{C}_p)\left(\frac{U_m}{U_\infty}\right)^2$$

The factor $(U_m/U_\infty)^2$ is just a constant for a given pressure distribution or aerofoil shape.

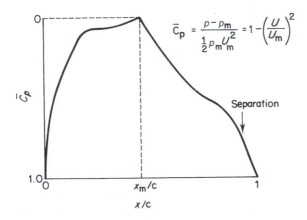

$$\bar{C}_p = \frac{p - p_m}{\frac{1}{2}p_m U_m^2} = 1 - \left(\frac{U}{U_m}\right)^2$$

Separation

Fig. 8.1 Smith's canonical pressure distribution

[*] A.M.O. Smith (1975) 'High-Lift Aerodynamics', *J. Aircraft*, **12**, 501–530. Many of the topics discussed in Sections 8.1 and 8.2 are covered in greater depth by Smith.

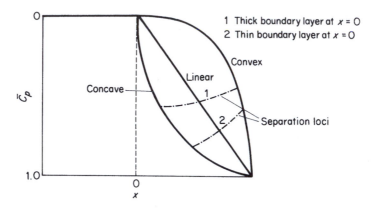

Fig. 8.2 Effects of different types of adverse pressure variation on separation

Figure 8.2 gives some idea of how the quality of the adverse pressure distribution affects boundary-layer separation. For this figure it is assumed that a length of constant pressure is followed by various types of adverse pressure gradient. Suppose that from the point $x = 0$ onwards $\bar{C}_p \propto x^m$. For the curve labelled *convex*, $m \simeq 4$, say; for that labelled *linear*, $m = 1$; and for that labelled *concave*, $m \simeq 1/4$. One would not normally design a wing for which the flow separates before the trailing edge is reached, so ideally the separation loci should coincide with the trailing edge. The separation loci in Fig. 8.2 depend on two additional factors, namely the thickness of the boundary layer at the start of the adverse pressure gradient, as shown in Fig. 8.2; and also the Reynolds number per unit length in the form of U_m/ν. This latter effect is not illustrated, but as a general rule the higher the value of U_m/ν the higher the value of $\bar{C}_p$ that the boundary layer can sustain before separating.

It is mentioned above that the separation point is affected by the energy defect in the boundary layer at the start of the adverse pressure gradient, $x = 0$. Other things being equal this implies that the thinner the boundary layer is at $x = 0$, the farther the boundary layer can develop in the adverse pressure gradient before separating. This point is illustrated in Fig. 8.3. This figure is based on calculations (using Head's method) of a turbulent boundary layer in an adverse pressure gradient with a preliminary constant-pressure region of variable length, x_0. It is shown very clearly that the shorter x_0 is, the longer the distance Δx_s from $x = 0$ to the separation point. It may be deduced from this result that it is best to keep the boundary layer laminar, and therefore thin, up to the start of the adverse pressure gradient. Ideally, transition should occur at or shortly after $x = 0$, since turbulent boundary layers can withstand adverse pressure gradients much better than laminar ones. Fortunately the physics of transition, see Section 7.9, ensures that this desirable state of affairs can easily be achieved.

The canonical plot in Fig. 8.2 contains much information of practical value. For example, suppose that at typical cruise conditions the value of $(U/U_\infty)^2$ at the trailing edge is 0.8 corresponding to $C_p = 0.2$, and typically $\bar{C}_p = 0.4$ (say) there. In this case any of the $\bar{C}_p$ curves in Fig. 8.2 would be able to sustain the pressure rise without leading to separation. Therefore, suitable aerofoils with a wide variety of pressure distributions could be designed to meet the specification. If, on the other hand, the goal is to achieve the maximum possible lift, then a highly concave pressure-rise curve with $m \simeq 1/4$ would be the best choice. This is because, assuming that separation

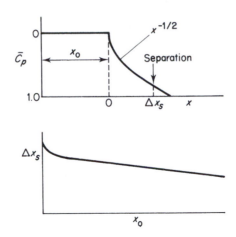

Fig. 8.3 Variation of location of separation with length of initial flat plate for a turbulent boundary layer in a specified adverse pressure variation

occurs at the trailing edge, the highly concave distribution not only gives the largest possible value of $(\bar{C}_p)_{TE}$ and therefore the largest possible value of U_m/U_{TE}; but also because the pressure rises to its value at the trailing edge the most rapidly. This latter attribute is of great advantage because it allows the region of constant pressure to be maintained over as much of the aerofoil surface as possible, leading to the greatest possible average value of $|C_p|$ on the upper surface and, therefore, the greatest possible lift. For many people this conclusion is counter-intuitive, since it seems to violate the classic rules of streamlining that seek to make the adverse pressure gradient as gentle as possible. Nevertheless, the conclusions based on Fig. 8.2 are practically sound.

The results depicted in Fig. 8.2 naturally suggest an important practical question. Is there, for a given situation, a best choice of adverse pressure distribution? The desired goals would be as above, namely to maximize U_m/U_{TE} and to maximize the rate of pressure rise. This question, or others very similar, have been considered by many researchers and designers. A widely quoted method of determining the optimum adverse pressure distribution is due to Stratford.* His theoretically derived pressure distributions lead to a turbulent boundary layer that is on the verge of separation, but remains under control, for much of the adverse pressure gradient. It is quite similar qualitatively to the concave distribution in Fig. 8.2. Two prominent features of Stratford's pressure distribution are:

(a) The initial pressure gradient $d\bar{C}_p/dx$ is infinite, so that small pressure rises can be accomplished in very short distances.
(b) It can be shown that in the early stages $C_p \propto x^{1/3}$.

If compressible effects are taken into account and it is considered desirable to avoid supersonic flow on the upper wing surface, the minimum pressure must correspond to sonic conditions. The consequences of this requirement are illustrated in Fig. 8.4. Here it can be seen that at comparatively low speeds very high values of suction pressure can be sustained before sonic conditions are reached, resulting in a pronounced peaky pressure distribution. For high subsonic Mach numbers, on the

---

*B.S. Stratford (1959) The prediction of separation of the turbulent boundary layer. *J. Fluid Mech.*, **5**, 1–16.

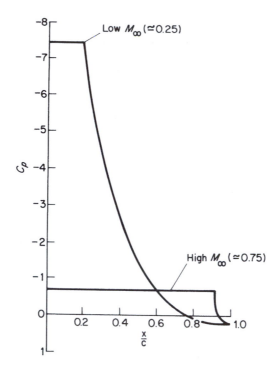

Fig. 8.4 Upper-wing-surface pressure distributions with laminar rooftop

other hand, only modest maximum suction pressures are permissible before sonic conditions are reached. In this case, therefore, the pressure distribution is very flat. An example of the practical application of these ideas for low flight speeds is illustrated schematically in Fig. 8.5. This shows a Liebeck* aerofoil. This sort of aerofoil was used as a basis for the aerofoil designed by Lissaman[†] specially for the successful man-powered aircraft *Gossamer Albatross* and *Condor*. In this application high lift and low drag were paramount. Note that there is a substantial fore-portion of the aerofoil with a favourable pressure gradient, rather than a very rapid initial acceleration up to a constant-pressure region. The favourable pressure gradient ensures that the boundary layer remains laminar until the onset of the adverse pressure gradient, thereby minimizing the boundary-layer thickness at the start of the pressure rise. Incidentally, note that the maximum suction pressure in Fig. 8.5 is considerably less than that in Fig. 8.4 for the low-speed case. But, it is not, of course, suggested here that at the speeds encountered in man-powered flight the flow over the upper wing surface is close to sonic conditions.

There is some practical disadvantage with aerofoils designed for concave pressure–recovery distributions. This is illustrated in Fig. 8.6 which compares the variations of lift coefficient with angle of incidence for typical aerofoils with convex and concave pressure distributions. It is immediately plain that the concave distribution leads to much higher values of $(C_L)_{max}$. But the trailing-edge stall is much more gentle, initially at least, for the aerofoil with the convex distribution. This is a desirable

* R.H. Liebeck (1973) A class of aerofoils designed for high lift in incompressible flow. *J. of Aircraft*, **10**, 610–617.

[†] P.B.S. Lissaman (1983) 'Low-Reynolds-number airfoils', *Annual Review of Fluid Mechanics*, **15**, 223–239.

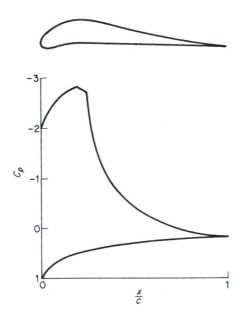

Fig. 8.5 Typical low-speed high-lift aerofoil – schematic representation of a Liebeck aerofoil

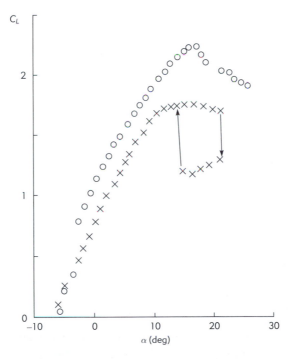

Fig. 8.6 Comparison of the variations of lift coefficient versus angle of incidence for aerofoils with concave and convex pressure–recovery distributions. $Re = 2 \times 10^5$. ×, Wortmann FX-137 aerofoil (convex); ○, Selig-Guglielmo S1223 aerofoil (concave)
Source: Based on Figs 7 and 14 of M.S. Selig and J.J. Guglielmo (1997) 'High-lift low Reynolds number airfoil design', *AIAA Journal of Aircraft*, **34**(1), 72–79

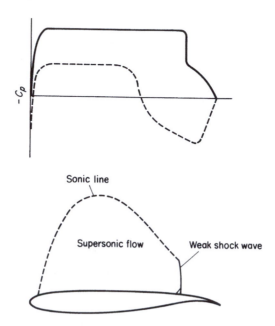

Fig. 8.7 Schematic figure illustrating a modern supercritical aerofoil

feature from the viewpoint of safety. The much sharper fall in C_L seen in the case of the aerofoil with the concave pressure distribution is explained by the fact that the boundary layer is close to separation for most of the aerofoil aft of the point of minimum pressure. (Recall that the ideal Stratford distribution aims for the boundary layer to be on the verge of separation throughout the pressure recovery.) Consequently, when the angle of incidence that provokes separation is reached, any further rise in incidence sees the separation point move rapidly forward.

As indicated above, it is not really feasible to design efficient wings for aircraft cruising at high subsonic speeds without permitting a substantial region of supersonic flow to form over the upper surface. However, it is still important to minimize the wave drag as much as possible. This is achieved by tailoring the pressure distribution so as to minimize the strength of the shock-wave system that forms at the end of the supersonic-flow region. A schematic figure illustrating the main principles of modern supercritical aerofoils is shown in Fig. 8.7. This sort of aerofoil would be designed for M_∞ in the range of 0.75–0.80. The principles behind this design are not very dissimilar from those exemplified by the high-speed case in Fig. 8.4, in the sense that a constant pressure is maintained over as much of the upper surface as possible.

8.3   Multi-element aerofoils

At the low speeds encountered during landing and take-off, lift needs to be greatly augmented and stall avoided. Lift augmentation is usually achieved by means of flaps* of various kinds – see Fig. 8.8. The plain flap shown in Fig. 8.8a increases the camber and angle of incidence; the Fowler flap (Fig. 8.8b) increases camber, angle of

---

* The most complete account is given by A.D. Young (1953) 'The aerodynamic characteristics of flaps', *Aero. Res. Council, Rep. & Mem. No. 2622.*

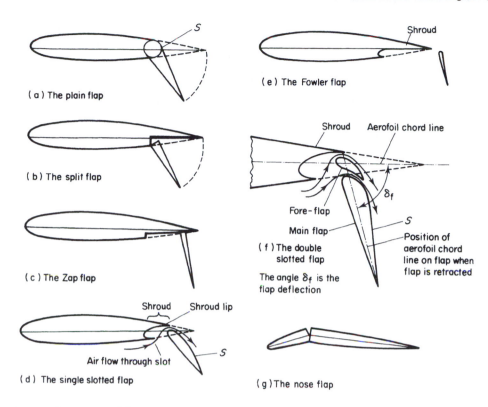

(a) The plain flap

(b) The split flap

(c) The Zap flap

Shroud   Shroud lip

Air flow through slot

(d) The single slotted flap

Shroud

(e) The Fowler flap

Shroud   Aerofoil chord line

Fore-flap

Main flap

(f) The double
slotted flap

The angle δ_f is the
flap deflection

δ_f

S

Position of
aerofoil chord
line on flap when
flap is retracted

S

(g) The nose flap

Fig. 8.8 Some types of flaps

incidence and wing area; and the nose flap (Fig. 8.8g) increases camber. The flaps
shown in Fig. 8.8 are relatively crude devices and are likely to lead to boundary-layer
separation when deployed. Modern aircraft use combinations of these devices in the
form of multi-element wings – Fig. 8.9. The slots between the elements of these wings
effectively suppress the adverse effects of boundary-layer separation, providing that
they are appropriately designed. Multi-element aerofoils are not a new idea. The
basic concept dates back to the early days of aviation with the work of Handley Page
in Britain and Lachmann in Germany. Nature also exploits the concept in the wings
of birds. In many species a group of small feathers, attached to the thumb-bone and
known as the alula, acts as a slat.

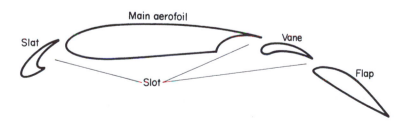

Main aerofoil

Slat

Vane

Slot

Flap

Fig. 8.9 Schematic sketch of a four-element aerofoil

How do multi-element aerofoils greatly augment lift without suffering the adverse effects of boundary-layer separation? The conventional explanation is that, since a slot connects the high-pressure region on the lower surface of a wing to the relatively low-pressure region on the top surface, it therefore acts as a blowing type of boundary-layer control (see Section 8.4.2). This explanation is to be found in a large number of technical reports and textbooks, and as such is one of the most widespread misconceptions in aerodynamics. It can be traced back to no less an authority than Prandtl* who wrote:

> The air coming out of a slot blows into the boundary layer on the top of the wing and imparts fresh momentum to the particles in it, which have been slowed down by the action of viscosity. Owing to this help the particles are able to reach the sharp rear edge without breaking away.

This conventional view of how slots work is mistaken for two reasons. Firstly, since the stagnation pressure in the air flowing over the lower surface of a wing is exactly the same as for that over the upper surface, the air passing through a slot cannot really be said to be high-energy air, nor can the slot act like a kind of nozzle. Secondly, the slat does not give the air in the slot a high velocity compared to that over the upper surface of the unmodified single-element wing. This is readily apparent from the accurate and comprehensive measurements of the flow field around a realistic multi-element aerofoil reported by Nakayama *et al.*[†] In fact, as will be explained below, the slat and slot usually act to reduce the flow speed over the main aerofoil.

The flow field associated with a typical multi-element aerofoil is highly complex. Its boundary-layer system is illustrated schematically in Fig. 8.10 based on the measurements of Nakayama *et al.* It is noteworthy that the wake from the slot does not interact strongly with the boundary layer on the main aerofoil before reaching the trailing edge of the latter. The wake from the main aerofoil and boundary layer from the flap also remain separate entities. As might well be expected, given the complexity of the flow field, the true explanation of how multi-element aerofoils augment lift, while avoiding the detrimental effects of boundary-layer separation, is multifaceted. And, the beneficial aerodynamic action of a well-designed multi-element aerofoil is due to a number of different primary effects, that will be described in turn.[‡]

Fig. 8.10 Typical boundary-layer behaviour for a three-element aerofoil

---

* L. Prandtl and O.G. Tietjens *Applied Hydro- and Aeromechanics*, Dover, New York, p. 227.

[†] A. Nakayama, H.-P. Kreplin and H.L. Morgan (1990) 'Experimental investigation of flowfield about a multielement airfoil', *AIAA J.*, **26**, 14–21.

[‡] Many of the ideas described in the following passages are due to A.M.O. Smith (1975) ibid.

8.3.1 The slat effect

To appreciate qualitatively the effect of the upstream element (e.g. the slat) on the immediate downstream element (e.g. the main aerofoil) the former can be modelled by a vortex. The effect is illustrated in Fig. 8.11. When one considers the component of the velocity induced by the vortex in the direction of the local tangent to the aerofoil contour in the vicinity of the leading edge (see inset in Fig. 8.11), it can be seen that the slat (vortex) acts to reduce the velocity along the edge of the boundary layer on the upper surface and has the opposite effect on the lower surface. Thus the effect of the slat is to reduce the severity of the adverse pressure gradient on the main aerofoil. In the case illustrated schematically in Fig. 8.11 it can be seen that the consequent reduction in pressure over the upper surface is counter-balanced by the rise in pressure on the lower surface. For a well-designed slat/main-wing combination it can be arranged that the latter effect predominates resulting in a slight rise in lift coefficient.

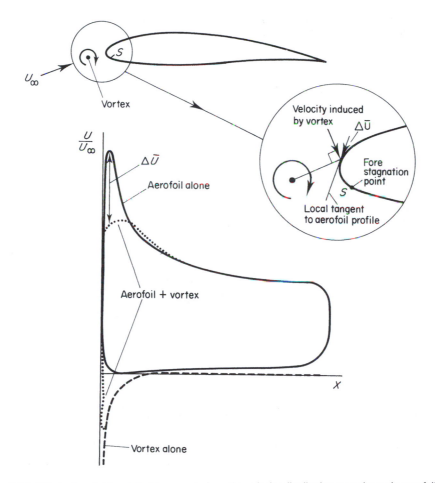

Fig. 8.11 Effect of a slat (modelled by a vortex) on the velocity distribution over the main aerofoil

8.3.2 The vane effect

In a similar way the effect of the downstream element (e.g. the vane) on the immediate upstream element (e.g. the main aerofoil) can also be modelled approximately by placing a vortex near the trailing edge of the latter. This effect is illustrated in Fig. 8.12. This time the vane (vortex) near the trailing edge induces a velocity over the main aerofoil surface that leads to a rise in velocity on both upper and lower surfaces. In the case of the upper surface this is beneficial because it raises the velocity at the trailing edge, thereby reducing the severity of the adverse pressure gradient. In addition to this, the vane has a second beneficial effect. This can be understood from the inset in Fig. 8.12. Note that owing to the velocity induced by the vane at the trailing edge, the effective angle of attack has been increased. If matters were left unchanged the streamline would not now leave smoothly from the trailing edge of the main aerofoil. This would violate the Kutta condition – see Section 4.1.1. What must happen is that viscous effects generate additional circulation in order that the Kutta condition be satisfied once again. Thus the presence of the vane leads to enhanced circulation and, therefore, higher lift.

8.3.3 Off-the-surface recovery

What happens with a typical multi-element aerofoil, as shown in Figs 8.9 and 8.13, is that the boundary layer develops in the adverse pressure gradient of the slat,

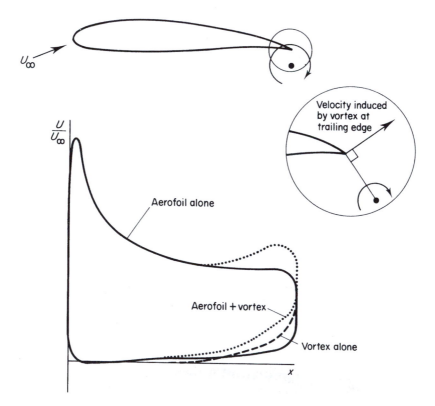

Fig. 8.12 Effect of a vane (modelled by a vortex) on the velocity distribution over the main wing

reaches the trailing edge in an unseparated state, and then leaves the trailing edge forming a wake. The slat wake continues to develop in the adverse pressure gradient over the main aerofoil; but for well-designed multi-element aerofoils the slot is sufficiently wide for the slat wake and main-aerofoil boundary layer to remain separate, likewise the wake of the main aerofoil and flap boundary layer. It is perfectly possible for the flow within the wakes to decelerate to such an extent in the downstream adverse pressure gradient that reversed flow occurs in the wake. This would give rise to stall, immediately destroying any beneficial effect. For well-designed cases it appears that the wake flows can withstand adverse pressure gradients to a far greater degree than attached boundary layers. Accordingly, flow reversal and wake breakdown are usually avoided. Consequently, for a multi-element aerofoil the total deceleration (or recovery, as it is often called) of the velocity along the edge of the boundary layer can take place in stages, as illustrated schematically in Fig. 8.13. In terms of the canonical pressure coefficient, U/U_m takes approximately the same value at the trailing edge of each element and, moreover, the boundary layer is on the verge of separation at the trailing edge of each element. (In fact, owing to the vane effect, described above, the value of $(U/U_m)_{TE}$ for the flap will be lower than that for the main aerofoil.) It is then evident that the overall reduction in (U/U_∞) from $(U_m/U_\infty)_{slat}$ to $(U_{TE}/U_\infty)_{flap}$ will be very much greater than the overall reduction for a single-element aerofoil. In this way the multi-element aerofoil can withstand a

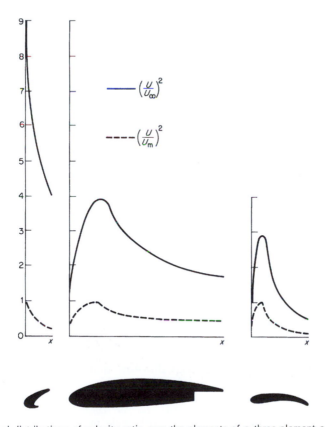

Fig. 8.13 Typical distributions of velocity ratio over the elements of a three-element aerofoil

very much greater overall velocity ratio or pressure difference than a comparable single-element aerofoil.

8.3.4 Fresh boundary-layer effect

It is evident from Fig. 8.10 that the boundary layer on each element develops largely independently from those on the others. This has the advantage of ensuring a fresh thin boundary layer, and therefore small kinetic-energy defect, at the start of the adverse pressure gradient on each element. The length of pressure rise that the boundary layer on each element can withstand before separating is thereby maximized – c.f. Fig. 8.3.

8.3.5 Use of multi-element aerofoils on racing cars

In the 1960s and early 1970s several catastrophic accidents occurred in which racing cars became airborne. In some cases aerodynamic interference from nearby competing vehicles was undoubtedly a factor. Nevertheless, these accidents are a grim reminder of what can happen to a racing car if insufficient aerodynamic downforce is generated. Modern Grand Prix cars generate their prodigious aerodynamic downforces from two main sources, namely 'ground effect' and inverted wings. Under current Formula-One rules the undertray of the car must be completely flat between the front and rear wheels. This severely limits the ability of the racing-car designer to exploit ground effect for generating downforce.*

Inverted wings, mounted in general above the front and rear axles (Fig. 8.14), first began to appear on Formula-One cars in 1968. The resultant increase in the downward force between the tyre and road immediately brought big improvements in cornering, braking and traction performance. The front wing is the most efficient aerodynamic device on the car. Except when closely following another car, this wing operates in undisturbed airflow, so there is nothing preventing the use of conventional aerofoils to generate high downforce (negative lift) with a relatively small drag. If the wing is located close to the ground the negative lift is further enhanced owing to increased acceleration of the air between the bottom of the wing and the ground, leading to lower suction pressure. (Fig. 8.15.) However, if the ground clearance is too small, the adverse pressure gradient over the rear of the wing becomes more severe, resulting in stall. Even if stall is avoided, too close a proximity to the ground may result in large and uncontrollable variations in downforce when there are unavoidable small changes in ride height due to track undulations or to roll and pitch of the vehicle. Sudden large changes in downward force that are inevitably accompanied by sudden changes to the vehicle's centre of pressure could make the car extremely difficult to drive. Racing-car designers must therefore compromise between optimum aerodynamic efficiency and controllability.

Under Formula-One rules the span of each wing is limited, so that the adverse three-dimensional effects found with wings of low aspect ratio are relatively severe. One of these adverse effects is the strong reduction in the spanwise lift distribution from root to tip. A common solution to this problem is to use plane end-plates, as illustrated in Fig. 8.14; these help keep the flow quasi-two-dimensional over the

*The information for this section comes from two main sources, namely, R.G. Dominy (1992) 'Aerodynamics of Grand Prix Cars', *Proc. I. Mech. E., Part D: J. of Automobile Engineering*, **206**, 267–274; and P.G. Wright (1982) 'The influence of aerodynamics on the design of Formula One racing cars', *Int. J. of Vehicle Design*, **3**(4), 383–397.

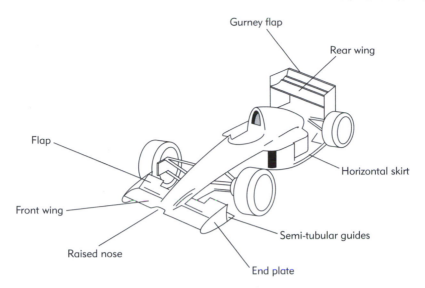

Fig. 8.14 Main aerodynamic features of a Grand Prix car
Source: Based on Fig. 1 of R.G. Dominy (1992) 'Aerodynamics of Grand Prix Cars', *Proc. I. Mech. E., Part D: J. of Automobile Engineering*, **206**, 267–274

entire span. End-plates do not eliminate the generation of strong wing-tip vortices which have other undesirable effects. Consequently, semi-tubular guides along the lower edges of the end-plates are often used in an attempt to control these vortices (see Fig. 8.14). It can also be seen in Fig. 8.14 that the front wing comprises a main wing and a flap. The chord and camber of the flap are very much greater over its outer section compared with inboard. This arrangement is adopted in order to reduce

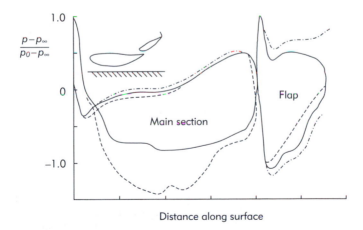

Fig. 8.15 Effects of ground proximity and a Gurney flap on the pressure distribution over a two-element front wing – schematic only. Key: ——, wing in free flow; - - - -, wing in close proximity to the ground; - · - · -, wing fitted with a Gurney flap and in close proximity to the ground
Source: Based on Figs 5 and 6 of R.G. Dominy (1992) 'Aerodynamics of Grand Prix Cars', *Proc. I. Mech. E., Part D: J. of Automobile Engineering*, **206**, 267–274

the adverse effects of the front wing's wake on the cooling air entering the radiator intakes.

The rear wing has to operate in the vehicle's wake. So the generation of high downforce by the rear wing is inevitably much less efficient than for the front wing. The car's wake is a highly unsteady, turbulent flow containing complex vortical flow structures. As a consequence, the effective angle of incidence along the leading edge of the rear wing may vary by up to 20°. Also the effective onset speeds may be much reduced compared with the front wings, further impairing aerodynamic efficiency. Despite all these problems, in order to maintain the required position for the centre of pressure, the design engineers have to ensure that the rear wing generates more than twice the downforce of the front wings. This is achieved by resorting to the sort of highly cambered, multi-element, aerofoils deployed by aircraft wings for landing. The high drag associated with the rear wing places severe limits on the top speed of the cars. But the drag penalty is more than offset by the much higher cornering speeds enabled by the increased downforce.

8.3.6   Gurney flaps

As well as being a great racing-car driver, Dan Gurney is also well-known for his technical innovations. His most widely emulated innovation is probably the now-obligatory practice of winning drivers spraying their supporters with champagne from vigorously shaken bottles. But it is for the Gurney flap that he is known in aerodynamics. This is a deceptively simple device consisting merely of a small plate fixed to and perpendicular to the trailing edge of a wing. It can be seen attached to the trailing edge of the multi-element rear wing in Figs 8.14 and 8.15.

Gurney first started fitting these 'spoilers' pointing upwards at the end of the rear deck of his Indy 500 cars in the late 1960s in order to enhance the generation of the downforce. The idea was completely contrary to the classic concepts of aerodynamics. Consequently, he was able to disguise his true motives very effectively by telling his competitors that the devices were intended to prevent cut hands when the cars were pushed out. So successful was this deception that some of his competitors attached the tabs projecting downwards in order to better protect the hands. Although this 'improved' arrangement undoubtedly impaired, rather than enhanced, the generation of a downforce, it was several years before they eventually realized the truth.

Gurney flaps became known in aerodynamics after Dan Gurney discussed his ideas with the aerodynamicist and wing designer, Bob Liebeck of Douglas Aircraft. They reasoned that if the tabs worked at the rear end of a car, they should be capable of enhancing the lift generated by conventional wings. This was confirmed experimentally by Liebeck.* The beneficial effects of a Gurney flap in generating an enhanced downforce is illustrated by the pressure distribution over the flap of the two-element aerofoil shown in Fig. 8.15. The direct effects of Gurney flaps of various heights on the lift and drag of wings were demonstrated by other experimental studies, see Fig. 8.16. It can be seen that the maximum lift rises as the height of the flap is increased from 0.005 to 0.02 chord. It is plain, though, that further improvement to aerodynamic performance diminishes rapidly with increased flap height. The drag polars plotted in Fig. 8.16b show that for a lift coefficient less than unity the drag is generally greater with a Gurney flap attached. They are really only an advantage for generating high lift.

---

* R.H. Liebeck (1978) 'Design of subsonic airfoils for high lift', *AIAA J. of Aircraft*, **15**(9), 547–561.

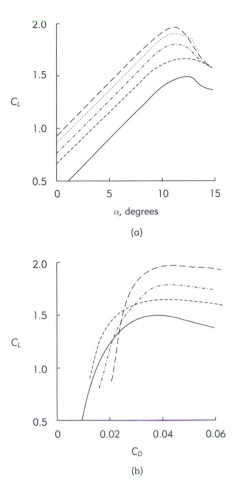

Fig. 8.16 The effects of Gurney flaps placed at the trailing edge of a NACA 4412 wing on the variation of lift and drag with angle of incidence. The flap height varies from 0.005 to 0.02 times the chord, c. ——, baseline without flap; - - - -, $0.005c$; - · - · -, $0.01c$; ······, $0.015c$; – – –, $0.02c$
Source: Based on Fig. 7 of B.L Storms and C.S. Jang (1994) 'Lift enhancement of an airfoil using a Gurney flap and vortex generators,' *AIAA J. of Aircraft*, **31**(3), 542–547

Why do Gurney flaps generate extra lift? The answer is to be found in the twin-vortex flow field depicted in Fig. 8.17. Something like this was hypothesized by Liebeck (1978).* However, it has only been confirmed comparatively recently by the detailed laser-Doppler measurements carried out at Southampton University (England)[†] of the flow fields created by Gurney flaps. As can be seen in Fig. 8.17, two contra-rotating vortices are created behind the flap. A trapped vortex is also included immediately ahead of the flap even though this is not shown clearly in the

* R.H. Liebeck (1978) 'Design of subsonic airfoils for high lift', *AIAA J. of Aircraft*, **15**(9), 547–561.

[†] D. Jeffrey, X. Zhang and D.W. Hurst (2000) 'Aerodynamics of Gurney flaps on a single-element high-lift wing', *AIAA J. of Aircraft*, **37**(2), 295–301; D. Jeffrey, X. Zhang and D.W. Hurst (2001) 'Some aspects of the aerodynamics of Gurney flaps on a double-element wing', *Trans. of ASME, J. of Fluids Engineering*, **123**, 99–104.

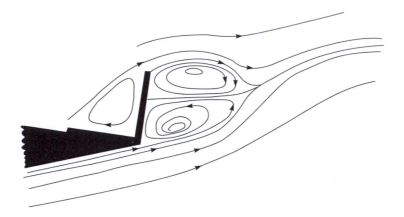

Fig. 8.17 Flow pattern downstream of a Gurney flap
Source: Based on figures in D. Jeffrey, X. Zhang and D.W. Hurst (2000) 'Aerodynamics of Gurney flaps on a single-element high-lift wing', *AIAA J. of Aircraft*, **37**(2), 295–301

measurements. This must be present, as was originally suggested by Liebeck. In an important respect, however, Fig. 8.17 is misleading. This is because it cannot depict the unsteady nature of the flow field. The vortices are, in fact, shed alternately in a similar fashion to the von Kármán vortex street behind a circular cylinder (see Section 7.5). It can be also seen in Fig. 8.17 (showing the configuration for enhancing downforce) that the vortices behind the Gurney flap deflect the flow downstream upwards. In some respects the vortices have a similar circulation-enhancing effect as the downstream flap in a multi-element aerofoil (see Section 8.3.2).

The principle of the Gurney flap was probably exploited in aeronautics almost by accident many years before its invention. Similar strips had been in use for many years, but were intended to reduce control-surface oscillations caused by patterns of flow separation changing unpredictably. It is also likely that the split and Zap flaps, shown in Fig. 8.8b and c, that date back to the early 1930s, produced similar flow fields to the Gurney flap. Nevertheless, it is certainly fair to claim that the Gurney flap is unique as the only aerodynamic innovation made in automobile engineering that has been transferred to aeronautical engineering. Today Gurney flaps are widely used to increase the effectiveness of the helicopter stabilizers.* They were first used in helicopters on the trailing edge of the tail on the Sikorsky S-76B because the first flight tests had revealed insufficient maximum (upwards) lift. This problem was overcome by fitting a Gurney flap to the inverted NACA 2412 aerofoil used for the horizontal tail. Similar circumstances led to the use of a Gurney flap on the horizontal stabilizer of the Bell JetRanger (Fig. 8.18.). Apparently, in this case the design engineers had difficulty estimating the required incidence of the stabilizer. Flight tests indicated that they had not guessed it quite correctly. This was remedied by adding a Gurney flap.

Another example is the double-sided Gurney flap installed on the trailing edge of the vertical stabilizer of the Eurocopter AS-355 TwinStar. This is used to cure a problem on thick surfaces with large trailing-edge angles. In such a case lift reversal

---

* The information on helicopter aerodynamics used here is based on an article by R.W. Prouty, 'The Gurney Flap, Part 2' in the March 2000 issue of *Rotor & Wing* (http://www.aviationtoday.com/reports/rotorwing/).

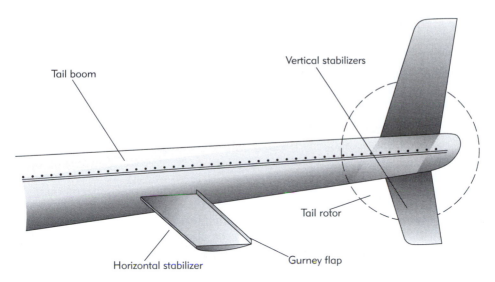

Fig. 8.18 The Gurney flap installed on the horizontal stabilizer of a Bell 206 JetRanger

can occur for small angles of attack, as shown in Fig. 8.19, thereby making the stabilizer a 'destabilizer'! The explanation for this behaviour is that at small positive angle of attack, the boundary layer separates near to the trailing edge on the upper (suction) side of the aerofoil. On the lower side the boundary layer remains attached. Consequently the pressure is lower there than over the top surface. The addition of a double Gurney flap stabilizes the boundary-layer separation and eliminates the lift reversal.

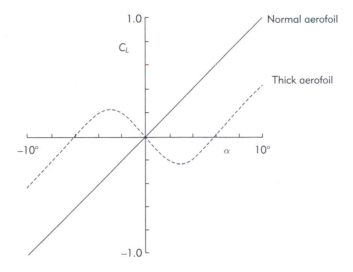

Fig. 8.19 Lift reversal for thick aerofoils

8.3.7   Movable flaps: artificial bird feathers*

This concept is illustrated in Fig. 8.20. Superficially it appears similar to the Gurney flap. However, the mode of operation is quite different. And, in any case, for *positive* high lift the Gurney flap would be attached to the trailing edge pointing downwards. The basic idea here is that at high angles of attack when flow separation starts to occur near the trailing edge, the associated reversed flow causes the movable flap to be raised. This then acts as a barrier to the further migration of reversed flow towards the leading edge, thereby controlling flow separation.

The movable flap concept originated with Liebe[†] who was the inventor of the boundary-layer fence (see Section 8.4.3). He observed that during the landing approach or in gusty winds, the feathers on the upper surface of many bird wings tend to be raised near the trailing edge. (Photographs of the phenomenon on a skua wing are to be found in Bechert *et al.* 1997.) Liebe interpreted this behaviour as a form of biological high-lift device and his ideas led to some flight tests on a Messerchmitt Me 109 in 1938. The device led to the development of asymmetric lift distributions that made the aircraft difficult to control and the project was abandoned. Many years later a few preliminary flight tests were carried out in Aachen on a glider.[‡] In this case small movable plastic sheets were installed on the upper surface of the wing. Apparently it improved the glider's handling qualities at high angles of attack.

There are problems with movable flaps. Firstly, they have a tendency to flip over at high angles of attack when the reversed flow becomes too strong. Secondly, they tend not to lie flat at low angles of attack, leading to a deterioration in aerodynamic performance. This is because when the boundary layer is attached the pressure rises towards the trailing edge, so the space under the flap connects with a region of slightly higher pressure that tends to lift it from the surface. These problems were largely overcome owing to three features of the design depicted in Fig. 8.21 which was fitted to a laminar glider aerofoil (see Bechert *et al.* 1997). Ties limited the maximum deflection of the flaps. And making the flap porous and the trailing edge jagged both helped to equalize the static pressure on either side of the flap during attached-flow conditions. These last two features are also seen in birds' feathers. The improvement in the aerodynamic characteristics can also be seen in Fig. 8.21.

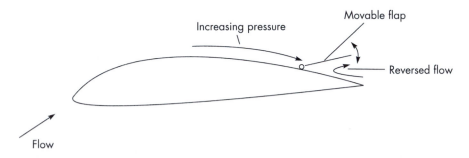

Fig. 8.20 Schematic illustrating the basic concept of the movable flap

---

* The account given here is based on a more detailed treatment by D.W. Bechert, M. Bruse, W. Hage and R. Meyer (1997) 'Biological surfaces and their technological application – Laboratory and flight experiments on drag reduction and separation control', *AIAA Paper 97-1960*.

[†] W. Liebe (1975) 'Der Auftrieb am Tragflügel: Enstehung and Zusammenbruch', *Aerokurier*, Heft **12**, 1520–1523.

[‡] B. Malzbender (1984) 'Projekte der FV Aachen, Erfolge im Motor- und Segelflug', *Aerokurier*, Heft **1**, 4.

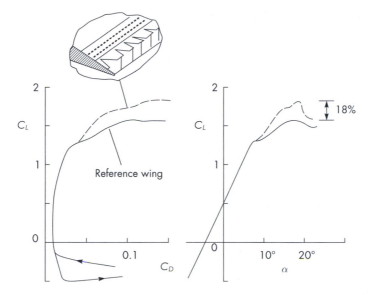

Fig. 8.21 Improved design of the movable flap and resulting improvement in aerodynamic characteristics for a laminar glider aerofoil
Source: Based on Fig. 25 of Bechert *et al.* (1997)

Successful flight tests on similar movable flaps were carried out later on a motor glider.

8.4 Boundary layer control for the prevention of separation

Many of the widely used techniques have already been described in Section 8.3. But there are various other methods of flow-separation control that are used on aircraft and in other engineering applications. These are described here.* Some of the devices used are active, i.e. they require the expenditure of additional power from the propulsion units; others are passive and require no additional power. As a general rule, however, the passive devices usually lead to increased drag at cruise when they are not required. The active techniques are discussed first.

8.4.1 Boundary-layer suction

The basic principle was demonstrated experimentally in Prandtl's paper that introduced the boundary-layer concept to the world.[†] He showed that the suction through a slot could be used to prevent flow separation from the surface of a cylinder. The basic principle is illustrated in Fig. 8.22. The layer of low-energy ('tired') air near the surface approaching the separation point is removed through a suction slot.

---

*A more complete recent account is to be found in M. Gad-el-Hak (2000) *Flow Control: Passive, Active and Reactive Flow Management*, Cambridge University Press.

[†]L. Prandtl (1904) 'Über Flüssigkeitsbewegung bei sehr kleiner Reibung', in *Proc. 3rd Int. Math. Mech.*, **5**, 484–491, Heidelberg, Germany.

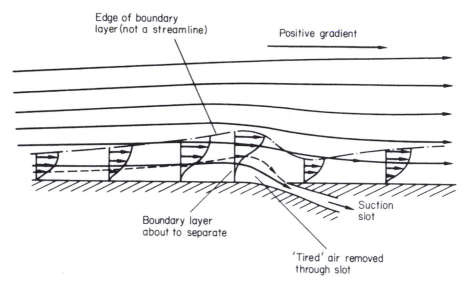

Fig. 8.22

The result is a much thinner, more vigorous, boundary layer that is able to progress further along the surface against the adverse pressure gradient without separating.

Suction can be used to suppress separation at high angles of incidence, thereby obtaining very high lift coefficients. In such applications the trailing edge may be permitted to have an appreciable radius instead of being sharp. The circulation is then adjusted by means of a small spanwise flap, as depicted in Fig. 8.23. If sufficient boundary layer is removed by suction, then a flow regime, that is virtually a potential flow, may be set up and, on the basis of the Kutta–Zhukovsky hypothesis, the sharp-edged flap will locate the rear stagnation point. In this way aerofoils with elliptic, or even circular, cross-sections can generate very high-lift coefficients.

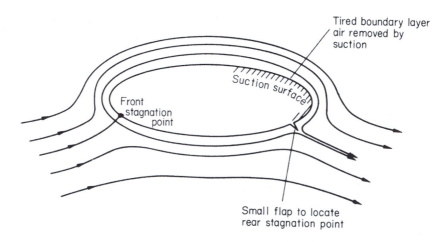

Fig. 8.23

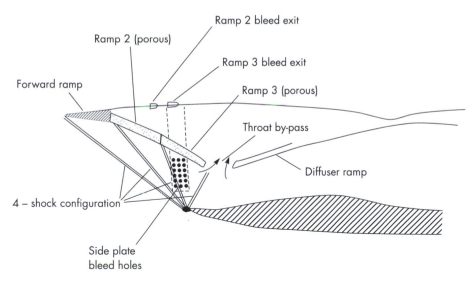

Fig. 8.24 Features of the F-15 engine-inlet flow management

There are great practical disadvantages for this type of high-lift device. First of all it is very vulnerable to dust blocking the suction slots. Secondly, it is entirely reliant on the necessary engine power being available for suction. Either blockage or engine failure would lead to catastrophic failure. For these reasons suction has not been used in this way for separation control in production aircraft. But it has been tested on rotors in prototype helicopters.

Many supersonic aircraft feature forms of suction in the intakes to their engines in order to counter the effects of shock-wave/boundary-layer interaction. Without such measures the boundary layers in the inlets would certainly thicken and be likely to separate. And some form of shock-wave system is indispensible because the air needs to be slowed down from the supersonic flight speed to about a Mach number of 0.4 at entry to the compressor. Two commonly used methods of implementing boundary-layer suction (or bleed) are porous surfaces and a throat slot by-pass. Both were used for the first time in a production aircraft on the McDonnell Douglas F-4 Phantom. Another example is the wide slot at the throat that acts as an effective and sophisticated form of boundary-layer bleed on the Concorde, thereby making the intake tolerant of changes in engine demand or the amount of bleed. The McDonnell Douglas F-15 Eagle also incorporates a variety of such boundary-layer control methods, as illustrated in Fig. 8.24. This aircraft has porous areas on the second and third engine-inlet ramps, plus a throat by-pass in the form of a slot and a porous region on the sideplates in the vicinity of the terminal shock wave. All the porous areas together account for about 30% of the boundary-layer removal with the throat by-pass accounting for the remainder.

8.4.2 Control by tangential blowing

Since flow separation is due to the complete loss of kinetic energy in the boundary layer immediately adjacent to the wall, another method of preventing it is to re-energize the 'tired' air by blowing a thin, high-speed jet into it. This method is often used with trailing-edge flaps (Fig. 8.25). To obtain reasonable results with this

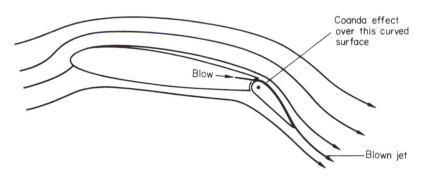

Fig. 8.25 A blown trailing-edge flap

method, great care must be taken with the design of the blowing duct. It is essential that good mixing takes place between the blown air and the boundary layer.

Most applications of tangential blowing for flow control exploit the so-called *Coanda effect*. This name is used for the tendency of a fluid jet issuing tangentially on to a curved or angled solid surface to adhere to it, as illustrated in Fig. 8.26. The name derives from the Franco-Romanian engineer, Henri Coanda, who filed a French patent in 1932 for a propulsive device exploiting the phenomenon. The explanation for the phenomenon can be understood by considering the radial equilibrium of the fluid element depicted in Fig. 8.26a. This can be expressed in simple terms as follows:

$$\frac{\partial p}{\partial r} = \frac{\rho V^2}{r} \tag{8.1}$$

where p is the pressure within the jet boundary layer (strictly, the wall jet) issuing from the nozzle exit slot, r is the radial distance from the centre of curvature of the surface, ρ is the fluid density, and V is the local flow speed. It is easy to see that the pressure field thereby created forces the flow issuing from the nozzle to adhere to the surface. But this does not explain why the equally valid flow solution shown in Fig. 8.26b is only found in practice when the Coanda effect breaks down. Presumably the slightly enhanced viscous drag, experienced by the jet on its surface side as it emerges from the nozzle, tends to deflect it towards the surface. Thereafter, the pressure field set up by the requirements of radial equilibrium will tend to force the jet towards the surface. Another viscous effect, namely entrainment of the fluid between the jet and the surface, may also help pull the jet towards the surface.

The practical limits on the use of the Coanda effect can also be understood to a certain extent by considering the radial equilibrium of the fluid element depicted in Fig. 8.26a. Initially we will assume that the flow around the curved surface is inviscid so that it obeys Bernoulli's equation

$$p = p_0 - \frac{1}{2}\rho V^2 \tag{8.2}$$

where p_0 is the stagnation pressure of the flow issuing from the nozzle. Equation (8.2) may be substituted into Eqn (8.1) which is then rearranged to give

$$\frac{\mathrm{d}V}{V} = \mathrm{d}r, \quad \text{i.e.} \quad V = V_\mathrm{w} \exp\left(\frac{r}{R_\mathrm{c}}\right), \tag{8.3}$$

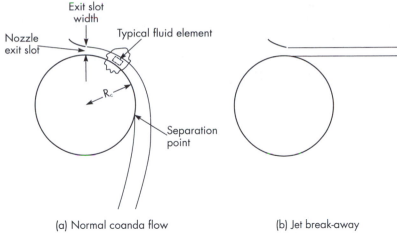

(a) Normal coanda flow (b) Jet break-away

Fig. 8.26 The Coanda effect – the flow of a jet around a circular cylinder
Source: Based on Fig. 1 of P.W. Carpenter and P.N. Green (1997) 'The aeroacoustics and aerodynamics of high-speed Coanda devices', *J. Sound & Vibration*, **208**(5), 777–801

where V_w is the (inviscid) flow speed along the wall and R_c is the radius of curvature of the surface. When the ratio of the exit-slot width, b, to the radius of curvature is small, $r \simeq R_c$ and $V \simeq V_w$. It then follows from Eqn (8.1) that near the exit slot the pressure at the wall is given by

$$p = \int \frac{\rho V^2}{r}\, dr \simeq \rho \int \frac{V_w^2}{R_c}\, dr = p_\infty - \rho \int_r^b \frac{V_w^2}{R_c}\, dr = p_\infty - \frac{\rho V^2 b}{R_c} \tag{8.4}$$

where p_∞ is the ambient pressure outside the Coanda flow.

It can be seen from Eqn (8.4) that the larger $\rho V^2 b/R_c$ is, the more the wall pressure falls below p_∞. In the actual viscous flow the average flow speed tends to fall with distance around the surface. As a consequence, the wall pressure rises with distance around the surface, thereby creating an adverse pressure gradient and eventual separation. This effect is intensified for large values of $\rho V^2 b/R_c$, so the nozzle exit-slot height, b, must be kept as small as possible. For small values of b/R_c the Coanda effect may still break down if the exit flow speed is high enough. But the simple analysis leading to Eqn (8.4) ignores compressible-flow effects. In fact, the blown air normally reaches supersonic speeds before the Coanda effect breaks down. At sufficiently high supersonic exit speeds shock-wave/boundary-layer interaction will provoke flow separation and cause the breakdown of the Coanda effect.* This places practical limits on the strength of blowing that can be employed.

The Coanda principle may be used to delay separation over the upper surface of a trailing-edge flap. The blowing is usually powered by air ducted from the engines. By careful positioning of the flap surface relative to the blown air jet and the main wing surface, advantage can be taken of the Coanda effect to make the blown jet adhere to the upper surface of the flap even when it is deflected downwards by as much as 60° (Fig. 8.25). In this way the circulation around the wing can be greatly enhanced.

---

* For a recent review on the aerodynamics of the Coanda effect, see P.W. Carpenter and P.N. Green (1997) 'The aeroacoustics and aerodynamics of high-speed Coanda devices', *J. Sound & Vibration*, **208**(5), 777–801.

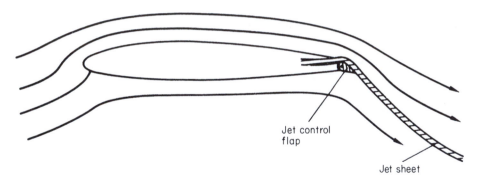

Fig. 8.27 A jet flap with a vestigial control flap

A more extreme version of the principle is depicted in Fig. 8.27 where only a vestigial flap is used. This arrangement is occasionally found at the trailing edge of a conventional blown flap. The term *jet flap* has sometimes been applied to this device, but the term is used rather imprecisely; it has even been applied to blown-flap systems in general. Here we will reserve the term for the case where the air is blown so strongly as to be supersonic. Such an arrangement is found on fighter aircraft with small wings, such as the Lockheed F-104 Starfighter, the Mig-21 PFM, and the McDonnell Douglas F-4 Phantom. This was done in order to increase lift at low speeds, thereby reducing the landing speed. The air is bled from the engine compressor and blown over the trailing-edge flaps. According to McCormick,* prior to 1951 it was thought that, if supersonic blown air were to be used, it would not only fail to adhere to the flap surface, but also lead to unacceptable losses due to the formation of shock waves. This view was dispelled by an undergraduate student, John Attinello, in his honours thesis at Lafayette College in the United States. Subsequently, his concept was subjected to more rigorous and sophisticated experimental studies before being flight tested and ultimately used on many aircraft, including the examples mentioned above.

Table 8.1 Aerodynamic performance of some high-lift systems

System	$C_{L_{max}}$
Internally blown flap	9
Upper surface blowing	8
Externally blown flap	7
Vectored thrust	3
Boeing 767 with slat + triple flap	2.8
Boeing 727 with slat + single flap	2.45

Source: Based on Tables 2 and 3 of A. Filippone 1999–2001 *Aerodynamics Database – Lift Coefficients* (http://aerodyn.org/HighLift/tables.html).

* B.W. McCormick (1979) *Aerodynamics, Aeronautics and Flight Mechanics*, Wiley.

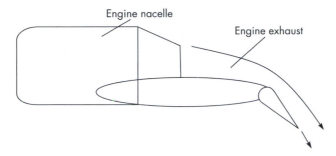

Fig. 8.28 Upper surface blowing

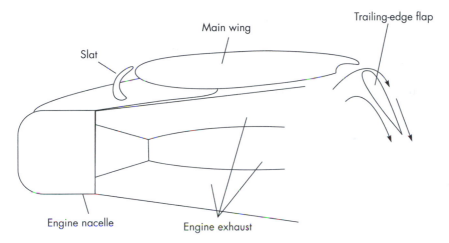

Fig. 8.29 An externally blown flap

Internally blown flaps give the best performance of any high-lift system, see Table 8.1, but upper surface blowing (Fig. 8.28) is also very effective. This arrangement is used on various versions of the Antonov An 72/74 transport aircraft. A slightly less efficient system is the externally blown flap (Fig. 8.29). A version of this is used on the Boeing C-17 Globemaster heavy transport aircraft. The engine exhaust flow is directed below and through slotted flaps to produce an additional lifting force. This allows the aircraft to make a steep, low-speed, final approach with a low landing speed for routine short-field landings. Many STOL (Short Take-Off and Landing) aircraft and fighter aircraft make use of thrust vectoring that also exploits the Coanda effect. One possible arrangement is depicted in Fig. 8.30.

Blown flaps and some other high-lift systems actually generate substantial additional circulation and do not just generate the required high lift owing to an increased angle of attack. For this reason in some applications the term *circulation-control* wings is often used. It is not necessary to install a flap on a circulation-control wing. For example, see the system depicted in Fig. 8.31. Rotors have been fitted with both suction-type circulation control (see Fig. 8.23) and the more common blown and jet flaps, and have been tested on a variety of helicopter prototypes.* But, as yet, circulation-control rotors

* See R.W. Prouty's articles on 'Aerodynamics' in *Aviation Today: Rotor & Wing* May, June and July, 2000 (http://www.aviationtoday.com/reports/rotorwing/).

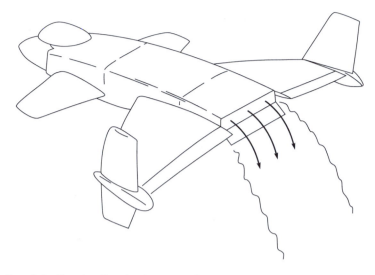

Fig. 8.30 Use of the Coanda effect for thrust vectoring
Source: Based on Fig. 1 of P.M. Bevilaqua and J.D. Lee (1987) 'Design of supersonic Coanda jet nozzles', *Proc. of the Circulation-Control Workshop 1986, NASA Conf. Pub. 2432*

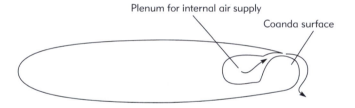

Fig. 8.31 A circulation-control wing

have not been used on any production aircraft. A recent research development, mainly in the last 10 years, is the use of periodic blowing for separation control.[*] Significant lift enhancement can be achieved efficiently with the use of very low flow rates. Almost all the experimental studies are at fairly low Reynolds number, but Seifert and Pack[†] have carried out wind-tunnel tests at Reynolds numbers typical of flight conditions.

Tangential blowing can only be applied to the prevention of separation, unlike suction that can be employed for this purpose or for laminar-flow control. The flow created by blowing tends to be very vulnerable to laminar-turbulent transition, so tangential blowing almost inevitably triggers transition.

8.4.3 Other methods of separation control

Passive flow control through the generation of streamwise vortices is frequently used on aircraft and other applications. Some of the devices commonly in use are depicted

---

[*] See the recent review by D. Greenblatt and I. Wygnanski (2000) 'The control of flow separation by periodic excitation', *Prog. in Aerospace Sciences*, **36**, 487–545.

[†] A. Seifert and L.G. Pack (1999) 'Oscillatory control of separation at high Reynolds number', *AIAA J.*, **37**(9), 1062–1071.

in Fig. 8.32. Figure 8.32a shows a row of vortex generators on the upper surface of a wing. These take a variety of forms and often two rows at two different chordwise locations are used. The basic principle is to generate an array of small streamwise vortices. These act to promote increased mixing between the high-speed air in the main stream and outer boundary layer with the relatively low-speed air nearer the surface. In this way the boundary layer is re-energized. Vortex generators promote the reattachment of separated boundary layers within separation bubbles, thereby postponing fully developed stall.

Fixed vortex generators are simple, cheap, and rugged. Their disadvantages are that they cannot be used for active stall control, a technology now being used for highly manœuvrable fighter aircraft; also they generate parasitic drag at cruise conditions where stall suppression is not required. These disadvantages have led to the development of vortex-generator jets (VGJ) whereby angled small jets are blown, either steadily or in a pulsatory mode, through orifices in the wing surface. The concept was first proposed by Wallis in Australia and Pearcey in the U.K.* primarily

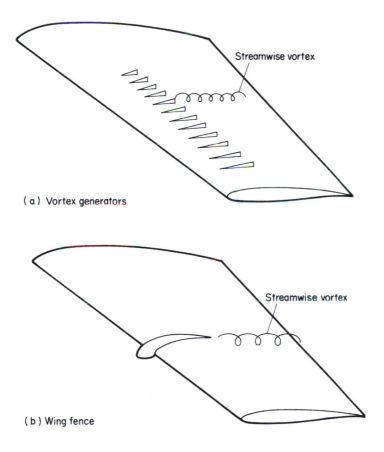

Streamwise vortex

(a) Vortex generators

Streamwise vortex

(b) Wing fence

Fig. 8.32

* R.A. Wallis (1952) 'The use of air jets for boundary layer control', *Aerodynamic Research Laboratories, Australia, Aero. Note 110 (N-34736)*; H.H. Pearcey (1961) 'Shock-induced separation and its prevention', in *Boundary Layer & Flow Control*, Vol. 2 (edited by G.V. Lachmann), Pergamon, pp. 1170–1344.

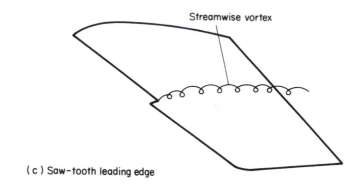

(c) Saw-tooth leading edge

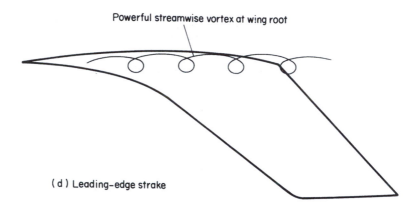

(d) Leading-edge strake

Fig. 8.32 (Continued)

for the control of shock-induced separation. More recently the concept has been reexamined as an alternative to conventional vortex generators*.

Wing fences (Fig. 8.32b) and 'vortilons' act as a barrier to tipward flow on swept-back wings. They also generate powerful streamwise vortices. The saw-tooth leading edge (Fig. 8.32c) is another common device for generating a powerful streamwise vortex; as is the leading-edge strake (Fig. 8.32d). In this last case the vortex reenergizes the complex, three-dimensional, boundary-layer flow that develops along the wing-body junction.

8.5 Reduction of skin-friction drag

Four main types of drag are found in aerodynamics – see Section 1.5.5 – namely: skin-friction drag, form drag, induced drag, and wave drag. The methods in use for

---

* For example, see J.P. Johnston and M. Nishi (1990) 'Vortex generator jets – means for flow separation control', *AIAA J.*, **28**(6), 989–994; see also the recent reviews by Greenblatt and Wygnanski (2000) referenced in Section 8.4.2 and Gad-el-Hak (2000) referenced at the beginning of Section 8.4., and J.C. Magill and K.R. McManus (2001) 'Exploring the feasibility of pulsed jet separation control for aircraft configurations', *AIAA J. of Aircraft*, **38**(1), 48–56.

the reduction of each type of drag are discussed in turn in the sections that follow. A more detailed recent account of drag reduction is given by Gad-el-Hak.*

In broad terms skin-friction drag[†] can be reduced in one of two ways. Either laminar flow can be maintained by postponing laminar-turbulent transition, this is the so-called laminar-flow technology, or ways are found to reduce the surface shear stress generated by the turbulent boundary layer. The maintenance of laminar-flow by prolonging a favourable or constant-pressure region over the wing surface is discussed briefly in Section 7.9. Active laminar-flow control requires the use of boundary-layer suction and this is described in Section 8.5.1. Another laminar-flow technique based on the use of compliant walls (artificial dolphin skin) is described in Section 8.5.2. Riblets are the main technique available for reducing turbulent skin-friction and their use is described in Section 8.5.3.

8.5.1 Laminar flow control by boundary-layer suction

Distributed suction acts in two main ways to suppress laminar-turbulent transition. First, it reduces the boundary-layer thickness. Recall from Section 7.9 that for a fixed pressure gradient a critical Reynolds number based on boundary-layer thickness must be reached before transition is possible. Second, it creates a much fuller velocity profile within the boundary layer, somewhat similar to the effect of a favourable pressure gradient. This makes the boundary layer much more stable with respect to the growth of small disturbances (e.g. Tollmien-Schlichting waves). In effect, this also greatly increases the critical Reynolds number. The earliest work on laminar-flow control (LFC) including the use of suction was carried out in Germany and Switzerland during the late 1930s in wind-tunnels.[‡] The first flight tests were carried out in the United States in 1941 using a B-18 bomber fitted with a wing glove. The maximum flight speed available and the chord of the wing glove limited the transitional Reynolds number achieved to a lower value than that obtained in wind-tunnel tests.

Research on suction-type LFC continued up to the 1960s in Great Britain and the United States. This included several flight tests using wing gloves on aircraft like the F-94 and Vampire. In such tests full-chord laminar flow was maintained on the wing's upper surface at Reynolds numbers up to 30×10^6. To achieve this transition delay exceptionally well-made smooth wings were required. Even very small surface roughness, due to insect impact, for example, caused wedges of turbulent flow to form behind each individual roughness element. Further flight tests in the United States and Great Britain (the latter used a vertically mounted test wing on a Lancaster bomber) revealed that it was much more difficult to maintain laminar flow over swept wings. This was because swept leading edges bring into play more powerful routes to transition than the amplification of Tollmien-Schlichting waves. First of all, turbulence propagates along the leading edge from the wing roots, this is termed *leading-edge contamination*. Secondly, completely different and more powerful

* M. Gad-el-Hak (2000) *Flow Control: Passive, Active and Reactive Flow Management*, Cambridge University Press.

† Reviews of many aspects of this subject are to be found in *Viscous Drag Reduction in Boundary Layers*, edited by D.M. Bushnell and J.N. Hefner, AIAA: Washington, D.C. (1990).

‡ H. Holstein (1940) 'Messungen zur Laminarhaltung der Grenzschicht an einem Flügel', *Lilienthal Bericht*, **S10**, 17–27; J. Ackeret, M. Ras, and W. Pfenninger (1941) 'Verhinderung des Turbulentwerdens einer Grenzschicht durch Absaugung', *Naturwissenschaften*, **29**, 622–623; and M. Ras and J. Ackeret (1941) 'Über Verhinderung der Grenzschicht-Turbulenz durch Absaugung', *Helv. Phys. Acta*, **14**, 323.

disturbances form in the boundary layer over the leading-edge region of swept wings. These are called *cross-flow vortices*.

Owing to the practical difficulties and to the relatively low price of aviation fuel, LFC research was discontinued at the end of the 1960s. More recently, with the growing awareness of the environmental requirements for fuel economy and limiting engine emissions, it has been revived. LFC is really the only technology currently available with the potential for very substantial improvement to fuel economy. For transport aircraft, the reduction in fuel burnt could exceed 30%. Recent technical advances have also helped to overcome some of the practical difficulties. The principal such advances are:

(i) Krueger (Fig. 8.33) flaps at the leading edge that increase lift and act to protect the leading-edge region from insect impact during take-off and climb-out;

(ii) Improved manufacturing techniques, such as laser drilling and electron-beam technology, that permit the leading edges to be smooth perforated titanium skins;

(iii) The use of hybrid LFC.

The application of the first two innovations is illustrated in Fig. 8.33. Perforated skins give distributed suction which is more effective than the use of discrete suction slots. Hybrid LFC would be particularly useful for swept-back wings because it is not possible to maintain laminar flow over them by means of natural LFC alone. This depends on shaping the wing section in order to postpone the onset of an adverse pressure gradient to as far aft as possible. Tollmien-Schlichting waves can be suppressed in this way, but not the more powerful transition mechanisms of leading-edge contamination and cross-flow vortices found in the leading-edge region on swept wings. With hybrid LFC, suction is used only in the leading-edge region in order to suppress the cross-flow vortices and leading-edge contamination. Over the remainder of the wing where amplification of Tollmien-Schlichting waves is the main route to

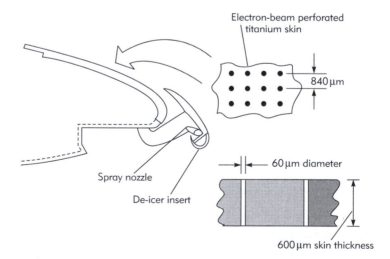

Fig. 8.33 Leading-edge arrangement for 1983–1987 flight tests conducted on a JetStar aircraft at NASA Dryden Flight Research Center. Important features were: (1) Suction on upper surface only; (2) Suction through electron-beam-perforated skin; (3) Leading-edge shield extended for insect protection; (4) De-icer insert on shield for ice protection; (5) Supplementary spray nozzles for protection from insects and ice
Source: Based on Fig. 12 of Braslow (2000) ibid

transition, wing-profile shaping can be used to reduce the effects of an adverse pressure gradient. In practice, it is easier to achieve this for the upper surface only. Owing to the higher flow speeds there, the upper surface produces most of the skin-friction drag. Hybrid LFC wings were extensively and successfully flight tested by Boeing on a modified 757 airliner during the early 1990s. Although, LFC based on the use of boundary-layer suction has yet to be used in any operational aircraft, the technology in the form of the less risky hybrid LFC has been established as practically realizable. In this way, based on proven current technology, a 10 to 20% improvement in fuel consumption could be achieved for moderate-sized subsonic commercial aircraft.

A detailed account of LFC technology and its history is given by Braslow.[*]

8.5.2   Compliant walls: artificial dolphin skins

It is widely thought that some dolphin species possess an extraordinary laminar-flow capability. Certainly mankind has long admired the swimming skills of these fleet creatures. Scientific interest in dolphin hydrodynamics dates back at least as far as 1936 when Gray[†] published his analysis of dolphin energetics. It is widely accepted that species like the bottle-nosed dolphin (*Tursiops truncatus*) can maintain a sustained swimming speed of up to 9 m/s. Gray followed the usual practice of marine engineers in modelling the dolphin's body as a one-sided flat plate of length 2 m. The corresponding value of Reynolds number based on overall body length was about 20×10^6. Even in a very-low-noise flow environment the Reynolds number, Re_{xt}, for transition from laminar to turbulent flow does not exceed 2 to 3×10^6 for flow over a flat plate. Accordingly, Gray assumed that if conventional hydrodynamics were involved, the flow would be mostly turbulent and the dolphin would experience a large drag force. So large, in fact, that at 9 m/s its muscles would have to deliver about seven times more power per unit mass than any other mammalian muscle. This led him and others to argue that the dolphin must be capable of maintaining laminar flow by some extraordinary means. This hypothesis has come to be known as *Gray's Paradox*.

Little in detail was known about laminar-turbulent transition in 1936 and Gray would have been unaware of the effects of the streamwise pressure gradient along the boundary layer (see Section 7.9). We now know that transition is delayed in favourable pressure gradients and promoted in adverse ones. Thus, for the dolphin, the transition point would be expected to occur near the point of minimum pressure. For *Tursiops truncatus* this occurs about half way along the body corresponding to $Re_{xt} = 10 \times 10^6$. When this is taken into account the estimated drag is very much less and the required power output from the muscles only exceeds the mammalian norm by no more than a factor of two. There is also some recent evidence that dolphin muscle is capable of a higher output. So on re-examination of Gray's paradox there is now much less of an anomaly to explain. Nevertheless, the dolphin may still find advantage in a laminar-flow capability. Moreover, there is ample evidence which will be briefly reviewed below, that the use of properly designed, passive, artificial dolphin skins, i.e. compliant walls, can maintain laminar flow at much higher Reynolds numbers than found for rigid surfaces.

In the late 1950s, Max Kramer,[‡] a German aeronautical engineer working in the United States, carried out a careful study of the dolphin epidermis and designed

---

[*] A.L. Braslow (2000) *Laminar-Flow Control*, NASA web-based publication on http://www/dfrc.nasa.gov/History/Publications/LFC.

[†] J. Gray (1936) 'Studies in animal locomotion. VI The propulsive powers of the dolphin', *J. Experimental Biology*, **13**, 192–199.

[‡] M.O. Kramer (1957, 1960) 'Boundary layer stabilization by distributed damping', *J. Aeronautical Sciences*, **24**, 459; and *J. American Society of Naval Engineers*, **74**, 341–348.

compliant coatings closely based on what he considered to be its key properties. Figure 8.34 shows his compliant coatings and test model. Certainly, his coatings bore a considerable resemblance to dolphin skin, particularly with respect to dimensions (see also Fig. 8.35). They were manufactured from soft natural rubber and he mimicked the effects of the fatty, more hydrated tissue, by introducing a layer of highly viscous silicone oil into the voids created by the short stubs. He achieved drag reductions of up to 60% for his best compliant coating compared with the rigid-walled control in sea-water at a maximum speed of 18 m/s. Three grades of rubber and various silicone oils with a range of viscosities were tested to obtain the largest drag reduction. The optimum viscosity was found to be about 200 times that of water.

Although no evidence existed beyond the drag reduction, Kramer believed that his compliant coatings acted as a form of laminar-flow control. His idea was that they reduced or suppressed the growth of the small-amplitude Tollmien-Schlichting waves, thereby postponing transition to a much higher Reynolds number or even eliminating it entirely. He believed that the fatty tissue in the upper dermal layer of

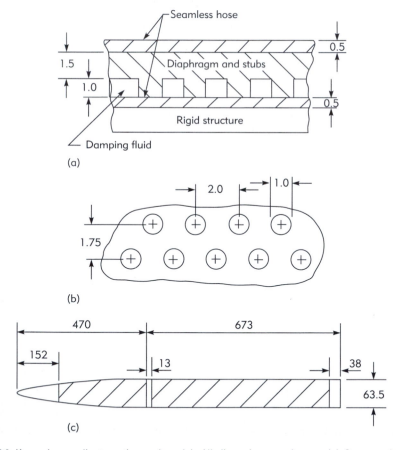

Fig. 8.34 Kramer's compliant coating and model. All dimensions are in mm. (a) Cross-section; (b) Cut through stubs; (c) Model: shaded regions were coated

Source: Based on Fig. 1 of P.W. Carpenter, C. Davies and A.D. Lucey (2000) 'Hydrodynamics and compliant walls: Does the dolphin have a secret?', *Current Science*, **79**(6), 758–765

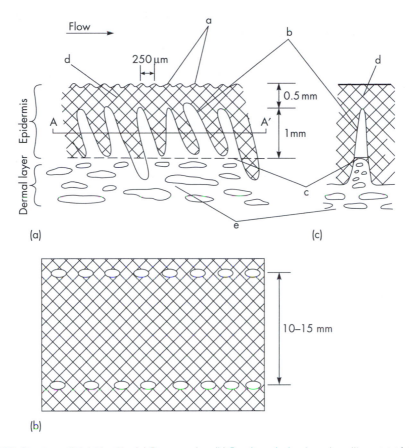

Fig. 8.35 Structure of dolphin skin. (a) Cross-section; (b) Cut through the dermal papillae at AA'; (c) Front view. Key: a, cutaneous ridges or microscales; b, dermal papillae; c, dermal ridge; d, upper epidermal layer; e, fatty tissue
Source: Based on Fig. 1 of P.W. Carpenter *et al.* (see Fig. 8.34)

the dolphin skin and, by analogy the silicone oil in his coatings, acted as damping to suppress the growth of the waves. This must have seemed eminently reasonable at the time. Surprisingly, however, the early theoretical work by Benjamin,[*] while showing that wall compliance can indeed suppress the growth of Tollmien-Schlichting waves, also showed that wall damping in itself promoted wave growth (i.e. the waves grew faster for a high level of damping than for a low level). This led to considerable scepticism about Kramer's claims. But the early theories, including that of Benjamin, were rather general in nature and made no attempt to model Kramer's coatings theoretically. A detailed theoretical assessment of the laminar-flow capabilities of his coatings was carried out much later by Carpenter and Garrad[†] who modelled the coatings as elastic plates supported on spring foundations with the effects of visco-elastic damping and the viscous damping fluid included. Their results broadly

---

[*] T.B. Benjamin (1960) 'Effects of a flexible boundary on hydrodynamic stability', *J. Fluid Mech.*, **9**, 513–532.

[†] P.W. Carpenter and A.D. Garrad (1985) 'The hydrodynamic stability of flows over Kramer-type compliant surfaces. Pt. 1. Tollmien-Schlichting instabilities', *J. Fluid Mech.*, **155**, 465–510.

confirm that the Kramer coatings were capable of substantially reducing the growth of Tollmien-Schlichting waves.

Experimental confirmation for the stabilizing effects of wall compliance on Tollmien-Schlichting waves was provided by Gaster* who found close agreement between the measured growth and the predictions of the theory. Subsequently many authors have used versions of this theory to show how suitably designed compliant walls can achieve a fivefold or greater increase in the transitional Reynolds number, Re_{xt}, as compared with the corresponding rigid surfaces. Although, compliant walls have yet to be used for laminar-flow control, there is little doubt that they have the potential for this in certain marine applications. In principle, they could also be used in aeronautical applications. But, in practice, owing to the need to match the inertias of the air and the wall, the wall structure would have to be impractically light and flimsy.[†]

8.5.3 Riblets

A moderately effective way of reducing turbulent skin friction involves surface modification in the form of *riblets*. These may take many forms, but essentially consist of minute streamwise ridges and valleys. One possible configuration is depicted in Fig. 8.36b. Similar triangular-shaped riblets are available in the form of polymeric film from the 3M Company. The optimum, non-dimensional, spanwise spacing between the riblets is given in wall units (see Section 7.10.5) by

$$s^+ = s\sqrt{\tau_w/\nu} = 10 \text{ to } 20$$

This corresponds to an actual spacing of 25 to 75 μm for flight conditions. (Note that the thickness of a human hair is approximately 70 μm). The 3M riblet film has been flight tested on an in-service Airbus A300–600 and on other aircraft. It is currently being used on regular commercial flights of the Airbus A340–300 aircraft by Cathay Pacific. The reduction in skin-friction drag observed was of the order of 5 to 8%. Skin-friction drag accounts for about 50% of the total drag for the Airbus A340–300 (a rather higher proportion than for many other types of airliner). Probably only about 70% of the surface of the aircraft is available to be covered with riblets leading to about 3% reduction in total drag.[‡] This is fairly modest but represents a worth-while savings in fuel and increase in payload. Riblets have also been used on Olympic-class rowing shells in the United States and on the hull of the *Stars and Stripes*, the winner of the 1987 America's Cup yacht race.

The basic concept behind the riblets had many origins, but it was probably the work at NASA Langley[§] in the United States that led to the present developments.

---

* M. Gaster (1987) 'Is the dolphin a red herring?', *Proc. of IUTAM Symp. on Turbulence Management and Relaminarisation*, edited by H.W. Liepmann and R. Narasimha, Springer, New York, pp. 285–204. See also A.D. Lucey and P.W. Carpenter (1995) 'Boundary layer instability over compliant walls: comparison between theory and experiment', *Physics of Fluids*, **7**(11), 2355–2363.

[†] Detailed reviews of recent progress can be found in: P.W. Carpenter (1990) 'Status of transition delay using compliant walls', *Viscous Drag Reduction in Boundary Layers*, edited by D.M. Bushnell and J.N. Hefner, Vol. 123, *Progress in Astronautics and Aeronautics*, AIAA, Washington, D.C., pp. 79–113; M. Gad-el-Hak (2000) *Flow control*, Cambridge University Press; and P.W. Carpenter, A.D. Lucey and C. Davies (2001) 'Progress on the use of compliant walls for laminar-flow control', *AIAA J. of Aircraft*, **38**(3), 504–512.

[‡] See D.W. Bechert, M. Bruse, W. Hage and R. Meyer (1997) 'Biological surfaces and their technological application – laboratory and flight experiments and separation control', *AIAA Paper 97-1960*.

[§] M.J. Walsh and L.M. Weinstein (1978) 'Drag and heat transfer on surfaces with longitudinal fins', *AIAA Paper No. 78-1161*.

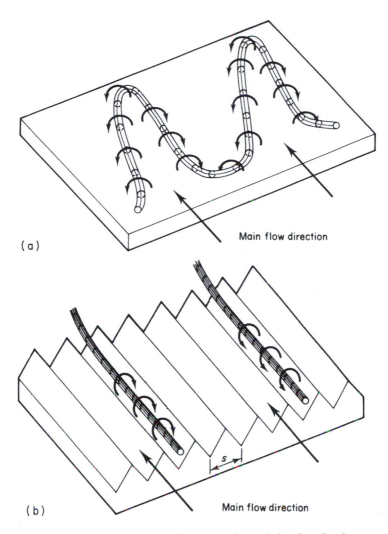

(a)

Main flow direction

(b)

Main flow direction

Fig. 8.36 The effect of riblets on the near-wall structures in a turbulent boundary layer

The concept was also discovered independently in Germany through the study of the hydrodynamics of riblet-like formations on shark scales.* The non-dimensional riblet spacings found on shark scales lie in the range $8 < s^+ < 18$, i.e. almost identical to the range of values given above for optimum drag reduction in the experiments of NASA and others on man-made riblets.

Given that the surface area is increased by a factor of 1.5 to 2.0, the actual reduction in mean surface shear stress achieved with riblets is some 12–16%. How do riblets produce a reduction in skin-friction drag? At first sight it is astonishing that such minute modifications to the surface should have such a large effect. The

* See W.-E. Reif and A. Dinkelacker (1982) 'Hydrodynamics of the squamation in fast swimming sharks', *Neues Jahrbuch für Geologie und Paleontologie*, **164**, 184–187; and D.W. Bechert, G. Hoppe and W.-E. Reif (1985) 'On the drag reduction of the shark skin', *AIAA Paper No. 85-0546*.

phenomenon is also in conflict with the classic view in aerodynamics and hydro-dynamics that surface roughness should lead to a drag increase. A plausible explanation for the effect of riblets is that they interfere with the development of the near-wall structures in the turbulent boundary that are mainly responsible for generating the wall shear stress. (See Section 7.10.8) These structures can be thought of as 'hairpin' vortices that form near the wall, as depicted in Fig. 8.36a. As these vortices grow and develop in time they reach a point where the head of the vortex is violently ejected away from the wall. Simultaneously the contra-rotating, streamwise-oriented, legs of the vortex move closer together, thereby inducing a powerful down-wash of high-momentum fluid between the vortex legs. This sequence of events is often termed a 'near-wall burst'. It is thought that riblets act to impede the close approach of the vortex legs, thereby weakening the bursting process.

8.6   Reduction of form drag

Form drag is kept to a minimum by avoiding flow separation and in this respect has already been discussed in the previous sections. Streamlining is vitally important for reducing form drag. It is worth noting that at high Reynolds numbers a circular cylinder has roughly the same overall drag as a classic streamlined aerofoil with a chord length equal to 100 cylinder radii. Form drag is overwhelmingly the main contribution to the overall drag for bluff bodies like the cylinder, whereas in the case of streamlined bodies skin-friction drag is predominant, form drag being less than ten per cent of the overall drag. For bluff bodies even minimal streamlining can be very effective.

8.7   Reduction of induced drag

Aspects of this topic have already been discussed in Chapter 5. There it was shown that, in accordance with the classic wing theory, induced drag falls as the aspect ratio of the wing is increased. It was also shown that, for a given aspect ratio, elliptic-shaped wings (strictly, wings with elliptic wing loading) have the lowest induced drag. Over the past 25 years the winglet has been developed as a device for reducing induced drag without increasing the aspect ratio. A typical example is depicted in Figs 8.37a and 8.40. Winglets of this and other types have been fitted to many different civil aircraft ranging from business jets to very large airliners.

The physical principle behind the winglet is illustrated in Figs. 8.37b and 8.37c. On all subsonic wings there is a tendency for a secondary flow to develop from the high-pressure region below the wing round the wing-tip to the relatively low-pressure region on the upper surface (Fig. 8.37b). This is part of the process of forming the trailing vortices. If a winglet of the appropriate design and orientation is fitted to the wing-tip, the secondary flow causes the winglet to be at an effective angle of incidence, giving rise to lift and drag components L_w and D_w relative to the winglet, as shown in Fig. 8.37c. Both L_w and D_w have components in the direction of flight. L_w provides a component to counter the aircraft drag, while D_w provides one that augments the aircraft drag. For a well-designed winglet the contribution of L_w predominates, resulting in a net reduction in overall drag, or a thrust, equal to ΔT (Fig. 8.37c). For example, data available for the Boeing 747–400 indicate that

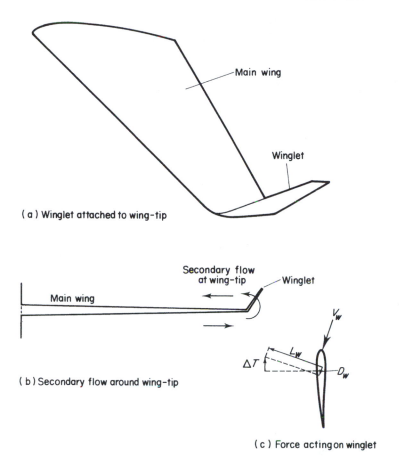

Fig. 8.37 Using winglets to reduce induced drag

winglets reduce drag by about 2.5% corresponding to a weight saving of 9.5 tons at take-off.*

The winglet shown in Fig. 8.37a has a sharp angle where it joins the main wing. This creates the sort of corner flow seen at wing-body junctions. Over the rear part of the wing the boundary layer in this junction is subject to an adverse streamwise pressure gradient from both the main wing and the winglet. This tends to intensify the effect of the adverse pressure gradient leading to a risk of flow separation and increased drag. This can be avoided by the use of blended winglets (Fig. 8.38a) or a winglet that is shifted downstream (Fig. 8.38b). Variants of both these designs are very common. The pressure distributions over the upper surface of the main wing close to the wing-tip are plotted in Fig. 8.39 for all three types of winglet and for the unmodified wing. The winglet with the sharp corner has a distribution with a narrow suction peak close to the leading edge that is followed by a steep adverse pressure gradient. This type of pressure distribution favours early laminar-turbulent transition and also risks flow separation. In contrast, the other two designs, especially the

* See A. Filippone (1999–2001) 'Wing-tip devices' (http://aerodyn.org/Drag/tip_devices.html).

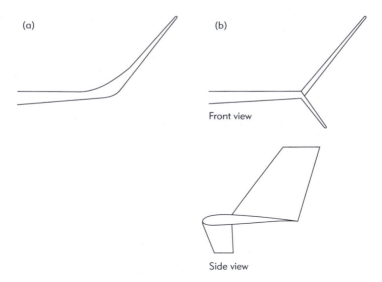

Fig. 8.38 Alternative winglet designs. (a) The blended winglet; (b) The winglet shifted downstream.

downstream-shifted winglet, have much more benign pressure distributions. Calculations using the panel method indicate that all three winglet types lead to a similar reduction in induced drag.* This suggests that the two winglet designs shown in Fig. 8.38 are to be preferred to the one with a sharp corner.

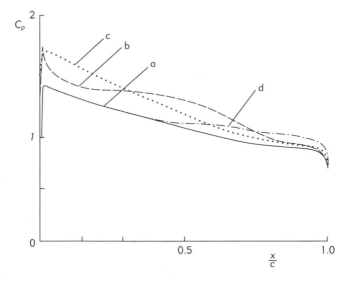

Fig. 8.39 Streamwise pressure distributions over the upper surface of the main wing close to the wing-tip for different winglet configurations (schematic only). a, wing-tip without winglet; b, winglet with sharp corner; c, blended winglet; d, winglet shifted downstream
Source: Based on a figure from 'Winglets, a close look' (http://beadec1.ea.bs.dlr.de/airfoils/winglt1.htm)

* 'Winglets, a close look' (http://beadec1.ea.bs.dlr.de/airfoils/winglt1.htm).

Fig. 8.40 A view of the Airbus A340 showing the winglets attached to the wing-tips. These devices are used in order to reduce induced drag. See Fig. 8.37, page 523. In the foreground is the wing of the Airbus A320-200 fitted with another wing-tip device known as a wing-tip fence. (*The photograph was provided by Gert Wunderlich.*)

8.8 Reduction of wave drag

Aspects of this have been covered in the discussion of swept wings in Section 5.7 and of supercritical aerofoils in Sections 7.9 and 8.2. In the latter case it was found that keeping the pressure uniform over the upper wing surface minimized the shock strength, thereby reducing wave drag. A somewhat similar principle holds for the whole wing-body combination of a transonic aircraft. This was encapsulated in the *area rule* formulated in 1952 by Richard Whitcomb* and his team at NACA Langley. It was known that as the wing-body configuration passed through the speed of sound, the conventional straight fuselage, shown in Fig. 8.41a, experienced a sharp rise in wave drag. Whitcomb's team showed that this rise in drag could be considerably reduced if the fuselage was waisted, as shown in Fig. 8.41b, in such a way as to keep the total cross-sectional area of the wing-body combination as uniform as possible. Waisted fuselages of this type became common features of aircraft designed for transonic operation.

The area rule was first applied to a production aircraft in the case of the Convair F-102A, the USAF's first supersonic interceptor. Emergency application of the area rule became necessary owing to a serious problem that was revealed during the flight tests of the prototype aircraft, the YF-102. Its transonic drag was found to exceed the thrust produced by the most powerful engine then available. This threatened to

* R.T. Whitcomb (1956) 'A study of the zero-lift drag-rise characteristics of wing-body combinations near the speed of sound', *NACA Rep. 1273*.

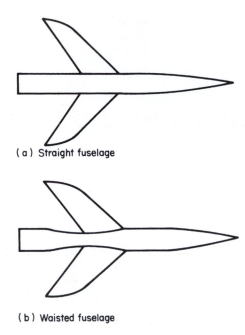

(a) Straight fuselage

(b) Waisted fuselage

Fig. 8.41 Application of the area rule for minimizing wave drag

jeopardize the whole programme because a supersonic flight speed was an essential USAF specification. The area rule was used to guide a major revised design of the fuselage. This reduced the drag sufficiently for supersonic Mach numbers to be achieved.

9

Propellers and propulsion

Preamble

Propulsive systems using atmospheric air include propellers, turbo-jets, ramjets, helicopters and hovercraft. Those which are independent of the atmosphere (if any) through which they move include rocket motors. In every case mentioned above the propulsive force is obtained by increasing the momentum of the working gas in the direction opposite to that of the force, assisted in the case of the hovercraft by a cushioning effect. A simple momentum theory of propulsion is applied to airscrews and rotors that permits performance criteria to be derived. A blade-element theory is also described. For the rocket motor and rocket-propelled body a similar momentum treatment is used. The hovercraft is briefly treated separately.

The forward propulsive force, or thrust, in aeronautics is invariably obtained by increasing the rearward momentum of a quantity of gas. Aircraft propulsion systems may be divided into two classes:

(I) those systems where the gas worked on is wholly or principally atmospheric air;
(II) other propulsive systems, in which the gas does not contain atmospheric air in any appreciable quantity.

Class I includes turbo-jets, ram-jets and all systems using airscrews or helicopter rotors. It also includes ornithopters (and, in nature, birds, flying insects, etc.). The only example of the Class II currently used in aviation is the rocket motors.

9.1 Froude's momentum theory of propulsion

This theory applies to propulsive systems of Class I. In this class, work is done on air from the atmosphere and its energy increased. This increase in energy is used to increase the rearwards momentum of the air, the reaction to which appears as a thrust on the engine or airscrew.

The theory is based on the concept of the ideal actuator disc or pure energy supplier. This is an infinitely thin disc of area S which offers no resistance to air

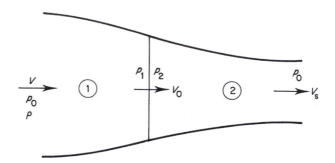

Fig. 9.1 The ideal actuator disc, and flow in the slipstream

passing through it. Air passing through the disc receives energy in the form of pressure energy from the disc, the energy being added uniformly over the whole area of the disc. It is assumed that the velocity of the air through the disc is constant over the whole area and that all the energy supplied to the disc is transferred to the air.

Consider the system shown in Fig. 9.1. This represents an actuator disc at rest in a fluid that, a long way ahead of the disc, is moving uniformly with a speed of V and has a pressure of p_0. The outer curved lines represent the streamlines that separate the fluid which passes through the disc from that which does not. As the fluid between these streamlines approaches the disc it accelerates to a speed V_0, its pressure decreasing to p_1. At the disc, the pressure is increased to p_2 but continuity prohibits a sudden change in speed. Behind the disc the air expands and accelerates until, well behind the disc, its pressure has returned to p_0, when its speed is V_s. The flow between the bounding streamlines behind the disc is known as the *slipstream*.

In unit time:

$$\text{mass of fluid passing through disc} = \rho S V_0 \tag{9.1}$$

Increase of rearward momentum of this mass of fluid

$$= \rho S V_0 (V_s - V) \tag{9.2}$$

and this is the thrust on the disc. Thus

$$T = \rho S V_0 (V_s - V) \tag{9.3}$$

The thrust can also be calculated from the pressures on the two sides of the disc as

$$T = S(p_2 - p_1) \tag{9.4}$$

The flow is seen to be divided into two regions 1 and 2, and Bernoulli's equation may be applied within each of these regions. Since the fluid receives energy at the disc Bernoulli's equation may not be applied through the disc. Then

$$p_0 + \frac{1}{2}\rho V^2 = p_1 + \frac{1}{2}\rho V_0^2 \tag{9.5}$$

and

$$p_2 + \frac{1}{2}\rho V_0^2 = p_0 + \frac{1}{2}\rho V_s^2 \tag{9.6}$$

From Eqns (9.5) and (9.6)

$$\left(p_2 + \frac{1}{2}\rho V_0^2\right) - \left(p_1 + \frac{1}{2}\rho V_0^2\right) = \left(p_0 + \frac{1}{2}\rho V_s^2\right) - \left(p_0 + \frac{1}{2}\rho V^2\right)$$

i.e.

$$p_2 - p_1 = \frac{1}{2}\rho(V_s^2 - V^2) \tag{9.7}$$

Substituting this in Eqn (9.4) and equating the result to Eqn (9.3), i.e. equating the two expressions for the thrust:

$$\frac{1}{2}\rho S(V_s^2 - V^2) = \rho S V_0(V_s - V)$$

and dividing this by $\rho S(V_s - V)$ gives

$$V_0 = \frac{1}{2}(V_s + V) \tag{9.8}$$

showing that the velocity through the disc is the arithmetic mean of the velocities well upstream, and in the fully developed slipstream. Further, if the velocity through the disc V_0 is written as

$$V_0 = V(1 + a) \tag{9.9}$$

it follows from Eqn (9.8) that

$$V_s + V = 2V_0 = 2V(1 + a)$$

whence

$$V_s = V(1 + 2a) \tag{9.10}$$

The quantity a is termed the inflow factor.

Now unit mass of the fluid upstream of the disc has kinetic energy of $\frac{1}{2}V^2$ and a pressure energy appropriate to the pressure p_0, whereas the same mass well behind the disc has, after passing through the disc, kinetic energy of $\frac{1}{2}V_s^2$ and pressure energy appropriate to the pressure p_0. Thus unit mass of the fluid receives an energy increase of $\frac{1}{2}(V_s^2 - V^2)$ on passing through the disc. Thus the rate of increase of energy of the fluid in the system, dE/dt, is given by

$$\frac{dE}{dt} = \rho S V_0 \frac{1}{2}(V_s^2 - V^2)$$

$$= \frac{1}{2}\rho S V_0(V_s^2 - V^2) \tag{9.11}$$

This rate of increase of energy of the fluid is, in fact, the power supplied to the actuator disc.

If it is now imagined that the disc is moving from right to left at speed V into initially stationary fluid, useful work is done at the rate TV. Thus the efficiency of the disc as a propulsive system is

$$\eta_i = \frac{TV}{\frac{1}{2}\rho S V_0(V_s^2 - V^2)}$$

Substituting for T from Eqn (9.3) gives

$$\eta_i = \frac{\rho S V_0 (V_s - V) V}{\frac{1}{2} \rho S V_0 (V_s^2 - V^2)}$$

$$= \frac{V}{\frac{1}{2}(V_s + V)} \tag{9.12}$$

This is the ideal propulsive efficiency or the Froude efficiency of the propulsive system.

In practice the part of the ideal actuator disc would be played by the airscrew or jet engine, which will violate some or all of the assumptions made. Each departure from the ideal will lead to a reduction in efficiency, and thus the efficiency of a practical propulsive system will always be less than the Froude efficiency as calculated for an ideal disc of the same area producing the same thrust under the same conditions.

Equation (9.12) may be treated to give several different expression for the efficiency, each of which has its own merit and use. Thus

$$\eta_i = \frac{V}{\frac{1}{2}(V_s + V)}$$

$$= \frac{2}{[1 + (V_s/V)]} \tag{9.12a}$$

$$= \frac{V}{V_0} \tag{9.12b}$$

$$= \frac{1}{(1 + a)} \tag{9.12c}$$

Also, since useful power $= TV$, and the efficiency is V/V_0, the power supplied is

$$P = \frac{TV}{V/V_0} = TV_0 \tag{9.13}$$

Of particular interest is Eqn (9.12a). This shows that, for a given flight speed V, the efficiency decreases with increasing V_s. Now the thrust is obtained by accelerating a mass of air. Consider two extreme cases. In the first, a large mass of air is affected, i.e. the diameter of the disc is large. Then the required increase in speed of the air is small, so V_s/V differs little from unity, and the efficiency is relatively high. In the second case, a disc of small diameter affects a small mass of air, requiring a large increase in speed to give the same thrust. Thus V_s/V is large, leading to a low efficiency. Therefore to achieve a given thrust at a high efficiency it is necessary to use the largest practicable actuator disc.

An airscrew does, in fact, affect a relatively large mass of air, and therefore has a high propulsive efficiency. A simple turbo-jet or ram-jet, on the other hand, is closer to the second extreme considered above, and consequently has a poor propulsive efficiency. However, at high forward speeds compressibility causes a marked reduction in the efficiency of a practical airscrew, when the advantage shifts to the jet engine. It was to improve the propulsive efficiency of the turbo-jet engine that the by-pass or turbo-fan type of engine was introduced. In this form of engine only part of the air taken is fully compressed and passed through the combustion chambers and turbines. The remainder is slightly compressed and ducted round the combustion chambers. It is then exhausted at a relatively low speed, producing thrust at a fairly high propulsive efficiency. The air that passed through the combustion chambers is ejected at high speed, producing thrust at a comparatively low efficiency. The overall propulsive efficiency is thus slightly greater than that of a simple turbo-jet engine giving the same thrust. The turbo-prop engine is, in effect, an extreme form of by-pass engine in which nearly all the thrust is obtained at high efficiency.

Another very useful equation in this theory may be obtained by expressing Eqn (9.3) in a different form. Since

$$V_0 = V(1+a) \quad \text{and} \quad V_s = V(1+2a)$$
$$T = \rho S V_0 (V_s - V) = \rho S V (1+a)[V(1+2a) - V]$$
$$= 2\rho S V^2 a (1+a) \tag{9.14}$$

Example 9.1 An airscrew is required to produce a thrust of 4000 N at a flight speed of 120 m s^{-1} at sea level. If the diameter is 2.5 m, estimate the minimum power that must be supplied, on the basis of Froude's theory.

$$T = 2\rho S V^2 a (1+a)$$

i.e.

$$a + a^2 = \frac{T}{2\rho S V^2}$$

Now $T = 4000$ N, $V = 120$ m s^{-1} and $S = 4.90$ m^2. Thus

$$a + a^2 = \frac{4000}{2 \times 1.226 \times 14\,400 \times 4.90} = 0.0232$$

whence

$$a = 0.0227$$

Then the ideal efficiency is

$$\eta_i = \frac{1}{1.0227}$$

$$\text{Useful power} = TV = 480\,000 \text{ W}$$

Therefore minimum power supplied, P, is given by

$$P = 480\,000 \times 1.0227 = 491 \text{ kW}$$

The actual power required by a practical airscrew would probably be about 15% greater than this, i.e. about 560 kW.

Example 9.2 A pair of airscrews are placed in tandem (Fig. 9.2), at a streamwise spacing sufficient to eliminate mutual interference. The rear airscrew is of such a diameter that it just fills the slipstream of the front airscrew. Using the simple momentum theory calculate: (i) the efficiency of the combination and (ii) the efficiency of the rear airscrew, if the front airscrew has a Froude efficiency of 90%, and if both airscrews deliver the same thrust. (U of L)

For the front airscrew, $\eta_i = 0.90 = \frac{9}{10}$. Therefore

$$1/(1+a) = \frac{9}{10}; \quad 1+a = \frac{10}{9}; \quad a = \frac{1}{9}$$

Thus

$$V_0 = V(1+a) = \frac{10}{9} V$$

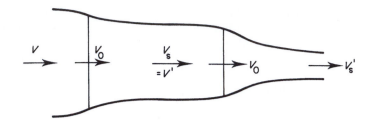

Fig. 9.2 Actuator discs in tandem

and

$$V_s = V(1 + 2a) = \frac{11}{9} V$$

The thrust of the front airscrew is

$$T = \rho S_1 V_0 (V_s - V) = \rho S_1 \left(\frac{10}{9} V \right) \left(\frac{2}{9} V \right) = \frac{20}{81} \rho S_1 V^2$$

The second airscrew is working entirely in the slipstream of the first. Therefore the speed of the approaching flow is V_s, i.e. $\frac{11}{9} V$. The thrust is

$$T = \rho S_2 V_0' (V_s' - V') = \rho S_2 V_0' (V_s' - V_s)$$

Now, by continuity:

$$\rho S_2 V_0' = \rho S_1 V_0$$

and also the thrusts from the two airscrews are equal. Therefore

$$T = \rho S_1 V_0 (V_s - V) = \rho S_2 V_0' (V_s' - V') = \rho S_1 V_0 (V_s' - V_s)$$

whence

$$V_s - V = V_s' - V_s$$

i.e.

$$V_s' = 2V_s - V = \left(\frac{22}{9} - 1 \right) V = \frac{13}{9} V$$

Then, if the rate of mass flow through the discs is $\dot{m}$:

$$\text{thrust of rear airscrew} = \dot{m}(V_s' - V_s) = \dot{m} \left(\frac{13}{9} - \frac{11}{9} \right) V = \frac{2}{9} \dot{m} V$$

The useful power given by the second airscrew is TV, not TV_s, and therefore:

$$\text{useful power from 2nd airscrew} = \frac{2}{9} \dot{m} V^2$$

Kinetic energy added per second by the second airscrew, which is the power supplied by (and to) the second disc, is

$$\frac{dE}{dt} = P = \frac{1}{2} \dot{m}(V_s'^2 - V_s^2) = \frac{1}{2} \dot{m} \left(\frac{169}{81} - \frac{121}{81} \right) V^2 = \frac{8}{27} \dot{m} V^2$$

Thus the efficiency of the rear components is

$$\frac{\frac{2}{9}\dot{m}V^2}{\frac{8}{27}\dot{m}V^2} = 0.75 \quad \text{or} \quad 75\%$$

$$\text{Power input to front airscrew} = \frac{TV}{0.90}$$

$$\text{Power input to rear airscrew} = \frac{TV}{0.75}$$

Therefore

$$\text{total power input} = \left(\frac{10}{9} + \frac{4}{3}\right)TV = \frac{22}{9}TV$$

$$\text{total useful power output} = 2TV$$

Therefore

$$\text{efficiency of combination} = \frac{2TV}{\frac{22}{9}TV} = \frac{9}{11} = 0.818 \quad \text{or} \quad 81.8\%.$$

9.2 Airscrew coefficients

The performance of an airscrew may be determined by model tests. As is the case with all model tests it is necessary to find some way of relating these to the full-scale performance, and dimensional analysis is used for this purpose. This leads to a number of coefficients, analogous to the lift and drag coefficients of a body. These coefficients also serve as a very convenient way of presenting airscrew performance data, which may be calculated by blade-element theory (Section 9.4), for use in aircraft design.

9.2.1 Thrust coefficient

Consider an airscrew of diameter D revolving at n revolutions per second, driven by a torque Q, and giving a thrust of T. The characteristics of the fluid are defined by its density, ρ, its kinematic viscosity, v, and its modulus of bulk elasticity, K. The forward speed of the airscrew is V. It is then assumed that

$$T = \mathrm{h}(D, n, \rho, v, K, V)$$
$$= CD^a n^b \rho^c v^d K^e V^f \tag{9.15}$$

Then, putting this in dimensional form,

$$[\mathrm{MLT}^{-2}] = [(\mathrm{L})^a (\mathrm{T})^{-b} (\mathrm{ML}^{-3})^c (\mathrm{L}^2 \mathrm{T}^{-1})^d (\mathrm{ML}^{-1} \mathrm{T}^{-2})^e (\mathrm{LT}^{-1})^f]$$

Separating this into the three fundamental equations gives

(M) $1 = c + e$

(L) $1 = a - 3c + 2d - e + f$

(T) $2 = b + d + 2e + f$

Solving these three equations for a, b and c in terms of d, e and f gives

$$a = 4 - 2e - 2d - f$$
$$b = 2 - d - 2e - f$$
$$c = 1 - e$$

Substituting these in Eqn (9.15) gives

$$T = CD^{4-2e-2d-f} n^{2-d-2e-f} \rho^{1-e} v^d K^e V^f$$
$$= C\rho n^2 D^4 \mathrm{f}\left[\left(\frac{v}{D^2 n}\right)^d \left(\frac{K}{\rho D^2 n^2}\right)^e \left(\frac{V}{nD}\right)^f\right] \tag{9.16}$$

Consider the three factors within the square brackets.

(i) $v/D^2 n$; the product Dn is a multiple of the rotational component of the blade tip speed, and thus the complete factor is of the form $v/(\text{length} \times \text{velocity})$, and is therefore of the form of the reciprocal of a Reynolds number. Thus ensuring equality of Reynolds numbers as between model and full scale will take care of this term.

(ii) $K/\rho D^2 n^2$; $K/\rho = a^2$, where a is the speed of sound in the fluid. As noted above, Dn is related to the blade tipspeed and therefore the complete factor is related to $(\text{speed of sound/velocity})^2$, i.e. it is related to the tip Mach number. Therefore care in matching the tip Mach number in model test and full-scale flight will allow for this factor.

(iii) V/nD; V is the forward speed of the airscrew, and therefore V/n is the distance advanced per revolution. Then V/nD is this advance per revolution expressed as a multiple of the airscrew diameter, and is known as the advance ratio, denoted by J.

Thus Eqn (9.16) may be written as

$$T = C\rho n^2 D^4 \mathrm{h}(Re, \, M, J) \tag{9.17}$$

The constant C and the function $\mathrm{h}(Re, M, J)$ are usually collected together, and denoted by k_T, the thrust coefficient. Thus, finally.

$$T = k_T \rho n^2 D^4 \tag{9.18}$$

k_T being a dimensionless quantity dependent on the airscrew design, and on Re, M and J. This dependence may be found experimentally, or by the blade-element theory.

9.2.2   Torque coefficient

The torque Q is a force multiplied by a length, and it follows that a rational expression for the torque is

$$Q = k_Q \rho n^2 D^5 \tag{9.19}$$

k_Q being the torque coefficient which, like k_T, depends on the airscrew design and on Re, M and J.

9.2.3 Efficiency

The power supplied to an airscrew is P_{in} where

$$P_{in} = 2\pi n Q$$

whereas the useful power output is P_{out} where

$$P_{out} = TV$$

Therefore, the airscrew efficiency, η, is given by

$$\eta = \frac{TV}{2\pi n Q} = \frac{k_T \rho n^2 D^4 V}{k_Q \rho n^2 D^5 2\pi n}$$

$$= \frac{1}{2\pi} \frac{k_T}{k_Q} \frac{V}{nD} = \frac{1}{2\pi} \frac{k_T}{k_Q} J \tag{9.20}$$

9.2.4 Power coefficient

The power required to drive an airscrew is

$$P = 2\pi n Q = 2\pi n (k_Q \rho n^2 D^5) = 2\pi k_Q \rho n^3 D^5 \tag{9.21}$$

The power coefficient, C_P, is then defined by

$$P = C_P \rho n^3 D^5 \tag{9.22}$$

i.e.

$$C_P = \frac{P}{\rho n^3 D^5} \tag{9.22a}$$

By comparison of Eqns (9.21) and (9.22) it is seen that

$$C_P = 2\pi k_Q \tag{9.22b}$$

Then, from Eqn (9.20), the efficiency of the airscrew is

$$\eta = J\left(\frac{k_T}{C_P}\right) \tag{9.23}$$

9.2.5 Activity factor

The activity factor is a measure of the power-absorbing capacity of the airscrew, which, for optimum performance, must be accurately matched to the power produced by the engine.

Consider an airscrew of diameter D rotating at n with zero forward speed, and consider in particular an element of the blade at a radius of r, the chord of the element being c. The airscrew will, in general, produce a thrust and therefore there will be a finite speed of flow through the disc. Let this inflow be ignored, however. Then the motion and forces on the element are as shown in Fig. 9.3.

$$\frac{\delta Q}{r} = C_D \frac{1}{2} \rho (2\pi r n)^2 c \delta r$$

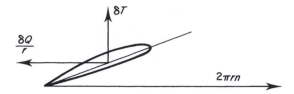

Fig. 9.3

and therefore the torque associated with the element is

$$\delta Q = 2\pi^2 \rho C_D n^2 (cr^3)\delta r$$

It is further assumed that C_D is constant for all blade sections. This will not normally be true, since much of the blade will be stalled. However, within the accuracy required by the concept of activity factor, this assumption is acceptable. Then the total torque required to drive an airscrew with B blades is

$$Q = 2\pi^2 \rho C_D B n^2 \int_{\text{root}}^{\text{tip}} cr^3 \, dr$$

Thus the power absorbed by the airscrew under static conditions is approximately

$$P = 2\pi n Q = 4\pi^3 \rho C_D B n^3 \int_{\text{root}}^{\text{tip}} cr^3 \, dr$$

In a practical airscrew the blade roots are usually shielded by a spinner, and the lower limit of the integral is, by convention, changed from zero (the root) to 0.1 D. Thus

$$P = 4\pi^3 \rho C_D B n^3 \int_{0.1D}^{0.5D} cr^3 \, dr$$

Defining the activity factor (AF) as

$$\text{AF} = \frac{10^5}{D^5} \int_{0.1D}^{0.5D} cr^3 \, dr$$

leads to

$$P = 4\pi^3 \rho C_D B n^3 \left(\frac{D}{10}\right)^5 \times (\text{AF})$$

Further work on the topic of airscrew coefficients is most conveniently done by means of examples.

Example 9.3 An airscrew of 3.4 m diameter has the following characteristics:

J	1.06	1.19	1.34	1.44
k_Q	0.0410	0.0400	0.0378	0.0355
η	0.76	0.80	0.84	0.86

Calculate the forward speed at which it will absorb 750 kW at 1250 rpm at 3660 m ($\sigma = 0.693$) and the thrust under these conditions. Compare the efficiency of the airscrew with that of the ideal actuator disc of the same area, giving the same thrust under the same conditions.

$$\text{Power} = 2\pi n Q$$

Therefore

$$\text{torque } Q = \frac{750\,000 \times 60}{2\pi \times 1250} = 5730 \,\text{N m}$$

$$n = \frac{1250}{60} = 20.83 \,\text{rps} \qquad n^2 = 435 \,(\text{rps})^2$$

Therefore

$$k_Q = \frac{Q}{\rho n^2 D^5} = \frac{5730}{0.639 \times 435 \times (3.4)^5 \times 1.226} = 0.0368$$

Plotting the given values of k_Q and η against J shows that, for $k_Q = 0.0368$, $J = 1.39$ and $\eta = 0.848$. Now $J = V/nD$, and therefore

$$V = JnD = 1.39 \times 20.83 \times 3.4 = 98.4 \,\text{m s}^{-1}$$

Since the efficiency is 0.848 (or 84.8%), the thrust power is

$$750 \times 0.848 = 635 \,\text{kW}$$

Therefore the thrust is

$$T = \frac{\text{Power}}{\text{Speed}} = \frac{635\,000}{98.4} = 6460 \,\text{N}$$

For the ideal actuator disc

$$a(1 + a) = \frac{T}{2\rho S V^2} = \frac{6460}{2 \times 0.693 \times \frac{\pi}{4}(3.4)^2 \times (98.4)^2 \times 1.226} = 0.0434$$

whence

$$a = 0.0417$$

Thus the ideal efficiency is

$$\eta_1 = \frac{1}{1.0417} = 0.958 \quad \text{or} \quad 95.8\%$$

Thus the efficiency of the practical airscrew is (0.848/0.958) of that of the ideal actuator disc. Therefore the relative efficiency of the practical airscrew is 0.885, or 88.5%.

Example 9.4 An aeroplane is powered by a single engine with speed–power characteristic:

Speed (rpm)	1800	1900	2000	2100
Power (kW)	1072	1113	1156	1189

The fixed-pitch airscrew of 3.05 m diameter has the following characteristics:

J	0.40	0.42	0.44	0.46	0.48	0.50
k_T	0.118	0.115	0.112	0.109	0.106	0.103
k_Q	0.0157	0.0154	0.0150	0.0145	0.0139	0.0132

and is directly coupled to the engine crankshaft. What will be the airscrew thrust and efficiency during the initial climb at sea level, when the aircraft speed is $45\,\mathrm{m\,s^{-1}}$?

Preliminary calculations required are:

$$Q = k_Q \rho n^2 D^5 = 324.2\,k_Q n^2$$

after using the appropriate values for ρ and D.

$$J = V/nD = 14.75/n$$

The power required to drive the airscrew, P_r, is

$$P_\mathrm{r} = 2\pi n Q$$

With these expressions, the following table may be calculated:

rpm	1800	1900	2000	2100
$P_\mathrm{a}(\mathrm{kW})$	1072	1113	1156	1189
n (rps)	30.00	31.67	33.33	35.00
$n^2(\mathrm{rps})^2$	900	1003	1115	1225
J	0.492	0.465	0.442	0.421
k_Q	0.013 52	0.014 36	0.014 94	0.015 38
$Q(\mathrm{Nm})$	3950	4675	5405	6100
$P_\mathrm{r}(\mathrm{kW})$	745	930	1132	1340

In this table, P_a is the brake power available from the engine, as given in the data, whereas the values of k_Q for the calculated values of J are read from a graph.

A graph is now plotted of P_a and P_r against rpm, the intersection of the two curves giving the equilibrium condition. This is found to be at a rotational speed of 2010 rpm, i.e. $n = 33.5$ rps. For this value of n, $J = 0.440$ giving $k_T = 0.112$ and $k_Q = 0.0150$. Then

$$T = 0.112 \times 1.226 \times (33.5)^2 \times (3.05)^4 = 13\,330\,\mathrm{N}$$

and

$$\eta = \frac{1}{2\pi}\frac{k_T}{k_Q}J = \frac{1}{2\pi}\frac{0.112}{0.0150} \times 0.440 = 0.523 \quad \text{or} \quad 52.3\%$$

As a check on the correctness and accuracy of this result, note that

$$\text{thrust power} = TV = 13\,300 \times 45 = 599\,\mathrm{kW}$$

At 2010 rpm the engine produces 1158 kW (from engine data), and therefore the efficiency is $599 \times 100/1158 = 51.6\%$, which is in satisfactory agreement with the earlier result.

9.3   Airscrew pitch

By analogy with screw threads, the pitch of an airscrew is the advance per revolution. This definition, as it stands, is of little use for airscrews. Consider two extreme cases. If the airscrew is turning at, say, 2000 rpm while the aircraft is stationary, the advance per revolution is zero. If, on the other hand, the aircraft is gliding with the engine stopped the advance per revolution is infinite. Thus the pitch of an airscrew can take any value and is therefore useless as a term describing the airscrew. To overcome this difficulty two more definite measures of airscrew pitch are accepted.

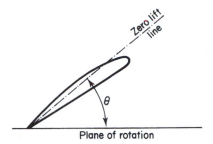

Fig. 9.4

9.3.1   Geometric pitch

Consider the blade section shown in Fig. 9.4, at radius r from the airscrew axis. The broken line is the zero-lift line of the section, i.e. the direction relative to the section of the undisturbed stream when the section gives no lift. Then the geometric pitch of the element is $2\pi r \tan\theta$. This is the pitch of a screw of radius r and helix angle $(90 - \theta)$ degrees. This geometric pitch is frequently constant for all sections of a given airscrew. In some cases, however, the geometric pitch varies from section to section of the blade. In such cases, the geometric pitch of that section at 70% of the airscrew radius is taken, and called the geometric mean pitch.

The geometric pitch is seen to depend solely on the geometry of the blades. It is thus a definite length for a given airscrew, and does not depend on the precise conditions of operation at any instant, although many airscrews are mechanically variable in pitch (see Section 9.3.3).

9.3.2   Experimental mean pitch

The experimental mean pitch is defined as the advance per revolution when the airscrew is producing zero net thrust. It is thus a suitable parameter for experimental measurement on an existing airscrew. Like the geometric pitch, it has a definite value for any given airscrew, provided the conditions of test approximate reasonably well to practical flight conditions.

9.3.3   Effect of geometric pitch on airscrew performance

Consider two airscrews differing only in the helix angles of the blades and let the blade sections at, say, 70% radius be as drawn in Fig. 9.5. That of Fig. 9.5a has a fine pitch, whereas that of Fig. 9.5b has a coarse pitch. When the aircraft is at rest, e.g. at the start of the take-off run, the air velocity relative to the blade section is the resultant V_R of the velocity due to rotation, $2\pi nr$, and the inflow velocity, V_{in}. The blade section of the fine-pitch airscrew is seen to be working at a reasonable incidence, the lift δL will be large, and the drag δD will be small. Thus the thrust δT will be large and the torque δQ small and the airscrew is working efficiently. The section of the coarse-pitch airscrew, on the other hand, is stalled and therefore gives little lift and much drag. Thus the thrust is small and the torque large, and the airscrew is inefficient. At high flight speeds the situation is much changed, shown in Fig. 9.5c,d. Here the section of the coarse-pitch airscrew is working efficiently, whereas the fine-pitch airscrew is now giving a negative thrust, a situation that might arise in a steep dive. Thus an airscrew that has

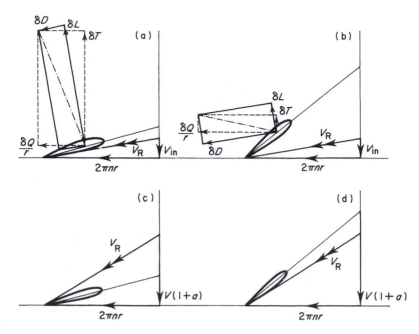

Fig. 9.5 Effect of geometric pitch on airscrew performance

a pitch suitable for low-speed flight and take-off is liable to have a poor performance at high forward speeds, and vice versa. This was the one factor that limited aircraft performance in the early days of powered flight.

A great advance was achieved consequent on the development of the two-pitch airscrew. This is an airscrew in which each blade may be rotated bodily, and set in either of two positions at will. One position gives a fine pitch for take-off and climb, whereas the other gives a coarse pitch for cruising and high-speed flight. Consider Fig. 9.6 which shows typical variations of efficiency η with J for (a) a fine-pitch and (b) a coarse-pitch airscrew.

For low advance ratios, corresponding to take-off and low-speed flight, the fine pitch is obviously better whereas for higher speeds the coarse pitch is preferable. If the pitch may be varied at will between these two values the overall performance

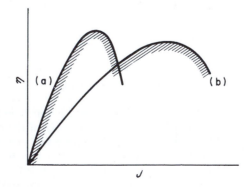

Fig. 9.6 Efficiency for a two-pitch airscrew

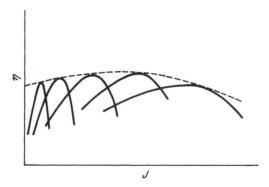

Fig. 9.7 Efficiency for a constant-speed airscrew

attainable is as given by the hatched line, which is clearly better than that attainable from either pitch separately.

Subsequent research led to the development of the constant-speed airscrew in which the blade pitch is infinitely variable between predetermined limits. A mechanism in the airscrew hub varies the pitch to keep the engine speed constant, permitting the engine to work at its most efficient speed. The pitch variations also result in the airscrew working close to its maximum efficiency at all times. Figure 9.7 shows the variation of efficiency with J for a number of the possible settings. Since the blade pitch may take any value between the curves drawn, the airscrew efficiency varies with J as shown by the dashed curve, which is the envelope of all the separate η, J curves. The requirement that the airscrew shall be always working at its optimum efficiency while absorbing the power produced by the engine at the predetermined constant speed calls for very skilful design in matching the airscrew with the engine.

The constant-speed airscrew, in turn, led to the provision of feathering and reverse-thrust facilities. In feathering, the geometric pitch is made so large that the blade sections are almost parallel to the direction of flight. This is used to reduce drag and to prevent the airscrew turning the engine (windmilling) in the event of engine failure. For reverse thrust, the geometric pitch is made negative, enabling the airscrew to give a negative thrust to supplement the brakes during the landing ground run, and also to assist in manoeuvring the aircraft on the ground.

9.4   Blade element theory

This theory permits direct calculation of the performance of an airscrew and the design of an airscrew to achieve a given performance.

9.4.1   The vortex system of an airscrew

An airscrew blade is a form of lifting aerofoil, and as such may be replaced by a hypothetical bound vortex. In addition, a trailing vortex is shed from the tip of each blade. Since the tip traces out a helix as the airscrew advances and rotates, the trailing vortex will itself be of helical form. A two-bladed airscrew may therefore be considered to be replaced by the vortex system of Fig. 9.8. Photographs have been taken of aircraft taking off in humid air that show very clearly the helical trailing vortices behind the airscrew.

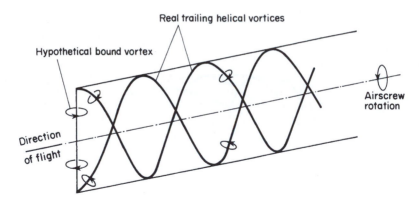

Fig. 9.8 Simplified vortex system for a two-bladed airscrew

Rotational interference The slipstream behind an airscrew is found to be rotating, in the same sense as the blades, about the airscrew axis. This rotation is due in part to the circulation round the blades (the hypothetical bound vortex) and the remainder is induced by the helical trailing vortices. Consider three planes: plane (i) immediately ahead of the airscrew blades; plane (ii), the plane of the airscrew blades; and plane (iii) immediately behind the blades. Ahead of the airscrew, in plane (i) the angular velocity of the flow is zero. Thus in this plane the effects of the bound and trailing vortices exactly cancel each other. In plane (ii) the angular velocity of the flow is due entirely to the trailing vortices, since the bound vortices cannot produce an angular velocity in their own plane. In plane (iii) the angular velocity due to the bound vortices is equal in magnitude and opposite in sense to that in plane (i), and the effects of the trailing and bound vortices are now additive.

Let the angular velocity of the airscrew blades be Ω, the angular velocity of the flow in the plane of the blades be $b\Omega$, and the angular velocity induced by the bound vortices in planes ahead of and behind the disc be $\pm\beta\Omega$. This assumes that these planes are equidistant from the airscrew disc. It is also assumed that the distance between these planes is small so that the effect of the trailing vortices at the three planes is practically constant. Then, ahead of the airscrew (plane (i)):

$$(b - \beta)\Omega = 0$$

i.e.

$$b = \beta$$

Behind the airscrew (plane (iii)), if ω is the angular velocity of the flow

$$\omega = (b + \beta)\Omega = 2b\Omega$$

Thus the angular velocity of the flow behind the airscrew is twice the angular velocity in the plane of the airscrew. The similarity between this result and that for the axial velocity in the simple momentum theory should be noted.

9.4.2 The performance of a blade element

Consider an element, of length δr and chord c, at radius r of an airscrew blade. This element has a speed in the plane of rotation of Ωr. The flow is itself rotating in the same plane and sense at $b\Omega$, and thus the speed of the element relative to the air in this plane is $\Omega r(1 - b)$. If the airscrew is advancing at a speed of V the velocity through the disc is $V(1 + a)$, a being the inflow at the radius r. Note that in this theory it is not necessary for a and b to be constant over the disc. Then the total velocity of the flow relative to the blade is V_R as shown in Fig. 9.9.

If the line CC′ represents the zero-lift line of the blade section then θ is, by definition, the geometric helix angle of the element, related to the geometric pitch, and α is the absolute angle of incidence of the section. The element will therefore experience lift and drag forces, respectively perpendicular and parallel to the relative velocity V_R, appropriate to the absolute incidence α. The values of C_L and C_D will be those for a two-dimensional aerofoil of the appropriate section at absolute incidence α, since three-dimensional effects have been allowed for in the rotational interference term, $b\Omega$. This lift and drag may be resolved into components of thrust and 'torque-force' as in Fig. 9.9. Here δL is the lift and δD is the drag on the element. δR is the resultant aerodynamic force, making the angle γ with the lift vector. δR is resolved into components of thrust δT and torque force $\delta Q/r$, where δQ is the torque required to rotate the element about the airscrew axis. Then

$$\tan \gamma = \delta D / \delta L = C_D / C_L \tag{9.24}$$

$$V_R = V(1 + a)\operatorname{cosec} \phi = \Omega r(1 - b) \sec \phi \tag{9.25}$$

$$\delta T = \delta R \cos(\phi + \gamma) \tag{9.26}$$

$$\frac{\delta Q}{r} = \delta R \sin(\phi + \gamma) \tag{9.27}$$

$$\tan \phi = \frac{V(1 + a)}{\Omega r(1 - b)} \tag{9.28}$$

The efficiency of the element, η_1, is the ratio, useful power out/power input, i.e.

$$\eta_1 = \frac{V}{\Omega} \frac{\delta T}{\delta Q} = \frac{V \cos(\phi + \gamma)}{\Omega r \sin(\phi + \gamma)} \tag{9.29}$$

Now from the triangle of velocities, and Eqn (9.28):

$$\frac{V}{\Omega r} = \frac{1 - b}{1 + a} \tan \phi$$

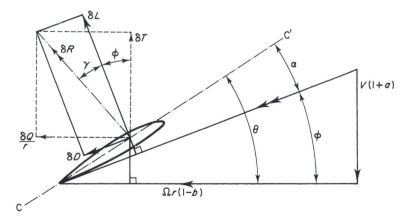

Fig. 9.9 The general blade element

whence, by Eqn (9.29):

$$\eta_1 = \frac{1-b}{1+a} \frac{\tan \phi}{\tan(\phi + \gamma)} \tag{9.30}$$

Let the solidity of the annulus, σ, be defined as the ratio of the total area of blade in annulus to the total area of annulus. Then

$$\sigma = \frac{Bc}{2\pi r} \frac{\delta r}{\delta r} = \frac{Bc}{2\pi r} \tag{9.31}$$

where B is the number of blades.
Now

$$\delta L = Bc\delta r \frac{1}{2} \rho V_R^2 C_L \tag{9.32a}$$

$$\delta D = Bc\delta r \frac{1}{2} \rho V_R^2 C_D \tag{9.32b}$$

From Fig. 9.9

$$\delta T = \delta L \cos \phi - \delta D \sin \phi$$

$$= Bc\delta r \frac{1}{2} \rho V_R^2 (C_L \cos \phi - C_D \sin \phi)$$

Therefore

$$\frac{\mathrm{d}T}{\mathrm{d}r} = Bc \frac{1}{2} \rho V_R^2 (C_L \cos \phi - C_D \sin \phi)$$

$$= 2\pi r \sigma \frac{1}{2} \rho V_R^2 (C_L \cos \phi - C_D \sin \phi) \tag{9.33}$$

Bearing in mind Eqn (9.24), Eqn (9.33) may be written as

$$\frac{\mathrm{d}T}{\mathrm{d}r} = \pi r \sigma \rho V_R^2 C_L (\cos \phi - \tan \gamma \sin \phi)$$

$$= \pi r \sigma \rho V_R^2 C_L \sec \gamma \, (\cos \phi \cos \gamma - \sin \phi \sin \gamma)$$

Now for moderate incidences of the blade section, $\tan \gamma$ is small, about 0.02 or so, i.e. $L/D \simeq 50$, and therefore $\sec \gamma \simeq 1$, when the above equation may be written as

$$\frac{\mathrm{d}T}{\mathrm{d}r} = \pi r \sigma \rho V_R^2 C_L \cos(\phi + \gamma)$$

Writing

$$t = C_L \cos(\phi + \gamma) \tag{9.34}$$

Then

$$\frac{\mathrm{d}T}{\mathrm{d}r} = \pi r \sigma t \rho V_R^2 \quad \text{for the airscrew} \tag{9.35a}$$

$$= Bc \frac{1}{2} \rho V_R^2 t \quad \text{for the airscrew} \tag{9.35b}$$

$$= c \frac{1}{2} \rho V_R^2 t \quad \text{per blade} \tag{9.35c}$$

Similarly

$$\frac{\delta Q}{r} = \delta L \sin \phi + \delta D \cos \phi$$

whence, using Eqn (9.32a and b)

$$\frac{dQ}{dr} = 2\pi r^2 \sigma \frac{1}{2} \rho V_R^2 (C_L \sin \phi + C_D \cos \phi)$$

Writing now

$$q = C_L \sin(\phi + \gamma) \qquad (9.36)$$

leads to

$$\frac{dQ}{dr} = \pi r^2 \sigma q \rho V_R^2 \quad \text{total} \qquad (9.37a)$$

$$= Bcr \frac{1}{2} \rho V_R^2 q \quad \text{total} \qquad (9.37b)$$

$$= cr \frac{1}{2} \rho V_R^2 q \quad \text{per blade} \qquad (9.37c)$$

The quantities dT/dr and dQ/dr are known as the thrust grading and the torque grading respectively.

Consider now the axial momentum of the flow through the annulus. The thrust δT is equal to the product of the rate of mass flow through the element with the change in the axial velocity, i.e. $\delta T = \dot{m} \delta V$. Now

$$\dot{m} = \text{area of annulus} \times \text{velocity through annulus} \times \text{density}$$
$$= (2\pi r \delta r)[V(1 + a)]\rho$$
$$= 2\pi r \rho \delta r \, V(1 + a)$$
$$\Delta V = V_s - V = V(1 + 2a) - V = 2aV$$

whence

$$\delta T = 2\pi r \rho \delta r V^2 2a(1 + a)$$

giving

$$\frac{dT}{dr} = 4\pi \rho r V^2 a(1 + a) \qquad (9.38)$$

Equating Eqn (9.38) and (9.35a) and using also Eqn (9.25), leads to:

$$4\pi \rho r V^2 a(1 + a) = \pi r \sigma t \rho V^2 (1 + a)^2 \text{cosec}^2 \phi$$

whence

$$\frac{a}{1 + a} = \frac{1}{4} \sigma t \, \text{cosec}^2 \phi \qquad (9.39)$$

In the same way, by considering the angular momentum

$$\delta Q = \dot{m}\Delta\omega r^2$$

where $\Delta\omega$ is the change in angular velocity of the air on passing through the airscrew. Then

$$\begin{aligned}\delta Q &= (2\pi r\delta r)[\rho V(1+a)](2b\Omega)r^2 \\ &= 4\pi r^3 \rho Vb(1+a)\Omega\delta r\end{aligned} \tag{9.40}$$

whence

$$\frac{\mathrm{d}Q}{\mathrm{d}r} = 4\pi r^3 \rho Vb(1+a)\Omega\delta \tag{9.41}$$

Now, as derived previously,

$$\frac{\mathrm{d}Q}{\mathrm{d}r} = \pi r^2 \sigma q\rho V_R^2 \qquad \text{(Eqn (9.37a))}$$

Substituting for V_R both expressions of Eqn (9.25), this becomes

$$\frac{\mathrm{d}Q}{\mathrm{d}r} = \pi r^2 \sigma\rho[V(1+a)\mathrm{cosec}\,\phi][\Omega r(1-b)\sec\phi]q$$

Equating this expression for $\mathrm{d}Q/\mathrm{d}r$ to that of Eqn (9.41) gives after manipulation

$$\begin{aligned}\frac{b}{1-b} &= \frac{1}{4}\sigma q\,\mathrm{cosec}\,\phi\sec\phi \\ &= \frac{1}{2}\sigma q\,\mathrm{cosec}\,2\phi\end{aligned} \tag{9.42}$$

The local efficiency of the blade at the element, η_1, is found as follows.

$$\text{Useful power output} = V\,\delta T = V\frac{\mathrm{d}T}{\mathrm{d}r}\delta r$$

$$\text{Power input} = 2\pi n\,\delta Q = 2\pi n\frac{\mathrm{d}Q}{\mathrm{d}r}\delta r$$

Therefore

$$\begin{aligned}\eta_1 &= \frac{V}{2\pi n}\frac{\mathrm{d}T/\mathrm{d}r}{\mathrm{d}Q/\mathrm{d}r} \\ &= \frac{V}{2\pi n}\frac{2\pi r\sigma\,\frac{1}{2}\rho V_R^2 t}{2\pi r^2\sigma\,\frac{1}{2}\rho V_R^2 q} \\ &= \frac{V}{2\pi nr}\frac{t}{q}\end{aligned} \tag{9.43}$$

which is an alternative expression to Eqn (9.30).

With the expressions given above, $\mathrm{d}T/\mathrm{d}r$ and $\mathrm{d}Q/\mathrm{d}r$ may be evaluated at several radii of an airscrew blade given the blade geometry and section characteristics, the forward and rotational speeds, and the air density. Then, by plotting $\mathrm{d}T/\mathrm{d}r$ and

dQ/dr against the radius r and measuring the areas under the curves, the total thrust and torque per blade and for the whole airscrew may be estimated. In the design of a blade this is the usual first step. With the thrust and torque gradings known, the deflection and twist of the blade under load can be calculated. This furnishes new values of θ along the blade, and the process is repeated with these new values of θ. The iteration may be repeated until the desired accuracy is attained.

A further point to be noted is that portions of the blade towards the tip may attain appreciable Mach numbers, large enough for the effects of compressibility to become important. The principal effect of compressibility in this connection is its effect on the lift-curve slope of the aerofoil section. Provided the Mach number of the relative flow does not exceed about 0.75, the effect on the lift-curve slope may be approximated by the Prandtl–Glauert correction (see Section 6.8.2). This correction states that, if the lift curve slope at zero Mach number, i.e. in incompressible flow, is a_0 the lift-curve slope at a subsonic Mach number M is a_M where

$$a_M = \frac{a_0}{\sqrt{1 - M^2}}$$

Provided the Mach number does not exceed about 0.75 as stated above, the effect of compressibility on the section drag is very small. If the Mach number of any part of the blade exceeds the value given above, although the exact value depends on the profile and thickness/chord ratio of the blade section, that part of the blade loses lift while its drag rises sharply, leading to a very marked loss in overall efficiency and increase in noise.

Example 9.5 At 1.25 m radius on a 4-bladed airscrew of 3.5 m diameter the local chord of each of the blades is 250 mm and the geometric pitch is 4.4 m. The lift-curve slope of the blade section in incompressible flow is 0.1 per degree, and the lift/drag ratio may, as an approximation, be taken to be constant at 50. Estimate the thrust and torque gradings and the local efficiency in flight at 4600 m ($\sigma = 0.629$, temperature $= -14.7\,°C$), at a flight speed of 67 m s^{-1} TAS and a rotational speed of 1500 rpm.

The solution of this problem is essentially a process of successive approximation to the values of a and b.

$$\text{solidity } \sigma = \frac{Bc}{2\pi r} = \frac{4 \times 0.25}{2\pi \times 1.25} = 0.1273$$

$$1500\,\text{rpm} = 25\,\text{rps} = n$$

$$\tan \gamma = \frac{1}{50} \quad \text{whence} \quad \gamma = 1.15°$$

$$\text{geometric pitch} = 2\pi r \tan \theta = 4.4$$

whence

$$\tan \theta = 0.560, \qquad \theta = 29.3°$$

$$\tan \phi = \frac{V(1 + a)}{\Omega r(1 - b)} = \frac{67(1 + a)}{62.5\pi(1 - b)} = 0.3418 \frac{1 + a}{1 - b}$$

$$\text{Speed of sound in atmosphere} = 20.05(273 - 14.7)^{1/2} = 325\,\text{m s}^{-1}$$

Suitable values for initial guesses for a and b are $a = 0.1$, $b = 0.02$. Then

$$\tan \phi = 0.3418 \frac{1.1}{0.98} = 0.383$$

$$\phi = 20.93°, \qquad \alpha = 29.3 - 20.93 = 8.37°$$
$$V_R = V(1 + a) \operatorname{cosec} \phi$$
$$= \frac{V(1 + a)}{\sin \phi} = \frac{67 \times 1.1}{0.357} = 206 \, \mathrm{m \, s^{-1}}$$

$$M = \frac{206}{325} = 0.635, \sqrt{1 - M^2} = 0.773$$

$$\frac{\mathrm{d}C_L}{\mathrm{d}\alpha} = \frac{0.1}{0.773} = 0.1295 \text{ per degree}$$

Since α is the absolute incidence, i.e. the incidence from zero lift:

$$C_L = \alpha \frac{\mathrm{d}C_L}{\mathrm{d}\alpha} = 0.1295 \times 8.37 = 1.083$$

Then

$$q = C_L \sin(\phi + \gamma) = 1.083 \sin(20.93 + 1.15)° = 0.408$$

and

$$t = C_L \cos(\phi + \gamma) = 1.083 \cos 22.08° = 1.004$$

$$\frac{b}{1 - b} = \frac{1}{2}\sigma q \operatorname{cosec} 2\phi = \frac{\sigma q}{2 \sin 2\phi} = \frac{0.1274 \times 0.408}{2 \times 0.675} = 0.0384$$

giving

$$b = \frac{0.0384}{1.0384} = 0.0371$$

$$\frac{a}{1 + a} = \frac{1}{4}\sigma t \operatorname{cosec}^2 \phi = \frac{0.1274 \times 1.004}{4 \times 0.357 \times 0.357} = 0.2515$$

giving

$$a = \frac{0.2515}{0.7485} = 0.336$$

Thus the assumed values $a = 0.1$ and $b = 0.02$ lead to the better approximations $a = 0.336$ and $b = 0.0371$, and a further iteration may be made using these values of a and b. A rather quicker approach to the final values of a and b may be made by using, as the initial values for an iteration, the arithmetic mean of the input and output values of the previous iteration. Thus, in the present example, the values for the next iteration would be $a = 0.218$ and $b = 0.0286$. The use of the arithmetic mean is particularly convenient when giving instructions to computers (whether human or electronic).

The iteration process is continued until agreement to the desired accuracy is obtained between the assumed and derived values of a and b. The results of the iterations were:

$$a = 0.1950 \qquad b = 0.0296$$

to four significant figures. With these values for a and b substituted in the appropriate equations, the following results are obtained:

$$\phi = 22°48'$$
$$\alpha = 6°28'$$

$$V_R = 207\,\text{m s}^{-1} \qquad M = 0.640$$

giving

$$\frac{dT}{dr} = \frac{1}{2}\rho V_R^2 ct = 3167\,\text{N m}^{-1} \text{ per blade}$$

and

$$\frac{dQ}{dr} = \frac{1}{2}\rho V_R^2 crq = 1758\,\text{N m m}^{-1} \text{ per blade}$$

Thus the thrust grading for the whole airscrew is $12\,670\,\text{N m}^{-1}$ and the torque grading is $7032\,\text{N m m}^{-1}$.

The local efficiency is

$$\eta_l = \frac{V}{2\pi nr}\frac{t}{q} = 0.768 \quad \text{or} \quad 76.8\%$$

9.5 The momentum theory applied to the helicopter rotor

In most, but not all, states of helicopter flight the effect of the rotor may be approximated by replacing it by an ideal actuator disc to which the simple momentum theory applies. More specifically, momentum theory may be used for translational, i.e. forward, sideways or rearwards, flight, climb, slow descent under power and hovering.

9.5.1 The actuator disc in hovering flight

In steady hovering flight the speed of the oncoming stream well ahead of (i.e. above) the disc is zero, while the thrust equals the helicopter weight, ignoring any downward force arising from the downflow from the rotor acting on the fuselage, etc. If the weight is W, the rotor area A, and using the normal notation of the momentum theory, with ρ as the air density

$$W = \rho A V_0(V_s - V) = \rho A V_0 V_s \tag{9.44}$$

since $V = 0$. V_s is the slipstream velocity and V_0 the velocity at the disc.

The general momentum theory shows that

$$V_0 = \frac{1}{2}(V_s + V) \quad (\text{Eqn}(9.8))$$
$$= \frac{1}{2}V_s \text{ in this case} \tag{9.45}$$

or

$$V_s = 2V_0$$

which, substituted in Eqn (9.44), gives

$$W = 2\rho A V_0^2 \tag{9.46}$$

i.e.

$$V_0 = \sqrt{W/2\rho A} \tag{9.47}$$

Defining the effective disc loading, l_{de}, as

$$l_{de} = W/A\sigma \tag{9.48}$$

where σ is the relative density of the atmosphere, then

$$\frac{W}{2\rho A} = \frac{W}{A\sigma}\frac{1}{2}\frac{\sigma}{\rho} = \frac{1}{2\rho_0}l_{de}$$

ρ_0 being sea-level standard density. Then

$$V_0 = \sqrt{l_{de}/2\rho_0} \tag{9.49}$$

The power supplied is equal to the rate of increase of kinetic energy of the air, i.e.

$$P = \frac{1}{2}\rho A V_0(V_s^2 - V^2)$$
$$= \frac{1}{2}\rho V_0 V_s^2 A = 2\rho A V_0^3 \tag{9.50}$$

Substituting for V_0 from Eqn (9.47) leads to

$$P = 2\rho A \left(\frac{W}{2\rho A}\right)^{3/2} = \sqrt{\frac{W^3}{2\rho A}} \tag{9.51a}$$

$$= W\sqrt{\frac{l_{de}}{2\rho_0}} \tag{9.51b}$$

This is the power that must be supplied to the ideal actuator disc. A real rotor would require a considerably greater power input.

9.5.2   Vertical climbing flight

The problem of vertical climbing flight is identical to that studied in Section 9.1, with the thrust equal to the helicopter weight plus the air resistance of the fuselage etc., to the vertical motion, and with the oncoming stream speed V equal to the rate of climb of the helicopter.

9.5.3   Slow, powered, descending flight

In this case, the air approaches the rotor from below and has its momentum decreased on passing through the disc. The associated loss of kinetic energy of the air appears as a power input to the ideal actuator, which therefore acts as a windmill. A real rotor will, however, still require to be driven by the engine, unless the rate of descent is large. This case, for the ideal actuator disc, may be treated by the methods of Section 9.1 with the appropriate changes in sign, i.e. V positive, $V_s < V_0 < V$, $p_1 > p_2$ and the thrust $T = -W$.

9.5.4 Translational helicopter flight

It is assumed that the effect of the actuator disc used to approximate the rotor is to add incremental velocities ν_v and ν_h, vertically and horizontally respectively, at the disc. It is further assumed, in accordance with the simple axial momentum theory of Section 9.1, that in the slipstream well behind the disc these incremental velocities increase to $2\nu_v$ and $2\nu_h$ respectively. The resultant speed through the disc is denoted by U and the resultant speed in the fully developed slipstream by U_1. Then, by considering vertical momentum:

$$W = \rho A U(2\nu_v) = 2\rho A U \nu_v \tag{9.52}$$

Also, from the vector addition of velocities:

$$U^2 = (V + \nu_h)^2 + (\nu_v)^2 \tag{9.53}$$

where V is the speed of horizontal flight. By consideration of horizontal momentum

$$\frac{1}{2}\rho V^2 A C_D = 2\rho A U \nu_h \tag{9.54}$$

where C_D is the drag coefficient of the fuselage, etc., based on the rotor area A.

Power input = rate of increase of KE, i.e.

$$P = \frac{1}{2}\rho A U(U_1^2 - V^2) \tag{9.55}$$

and from vector addition of velocities:

$$U_1^2 = (V + 2\nu_h)^2 + (2\nu_v)^2 \tag{9.56}$$

The most useful solution of the five equations Eqn (9.52) to Eqn (9.56) inclusive is obtained by eliminating U_1, ν_h and ν_v.

$$\nu_v = \frac{W}{2\rho A U} \tag{9.52a}$$

$$\nu_h = \frac{\frac{1}{2}\rho V^2 A C_D}{2\rho A U} = \frac{C_D}{4U}V^2 \tag{9.54a}$$

Then, from Eqn (9.53):

$$U^2 = V^2 + 2V\nu_h + \nu_h^2 + \nu_v^2$$

Substituting for ν_v and ν_h, and multiplying by U^2 gives

$$U^4 - U^2 V^2 = \frac{1}{2}C_D U V^3 + \frac{1}{16}C_D^2 V^4 + \left(\frac{W}{2\rho A}\right)^2$$

Introducing the effective disc loading, l_{de}, from Eqn (9.48) leads to

$$U^4 - U^2 V^2 - \frac{1}{2}C_D V^3 U = \frac{1}{16}C_D^2 V^4 + \left(\frac{l_{de}}{2\rho_0}\right)^2 \tag{9.57}$$

a quartic equation for U in terms of given quantities. Since, from Eqn (9.56),

$$U_1^2 = V^2 + 4V\nu_h + 4\nu_h^2 + 4\nu_v^2$$

Then

$$P = \frac{1}{2}\rho A U(U_1^2 - V^2) = \frac{1}{2}\rho A U[4V\nu_h + 4\nu_h^2 + 4\nu_v^2]$$

$$= 2\rho A\left[\frac{1}{4}C_D V^3 + \frac{1}{16}C_D^2\frac{V^4}{U} + \frac{1}{U}\left(\frac{l_{de}}{2\rho_0}\right)^2\right] \tag{9.58}$$

which, with the value of U calculated from Eqn (9.57) and the given quantities, may be used to calculate the power required.

Example 9.6 A helicopter weighs 24 000 N and has a single rotor of 15 m diameter. Using momentum theory, estimate the power required for level flight at a speed of 15 m s^{-1} at sea level. The drag coefficient, based on the rotor area, is 0.006.

$$A = \frac{\pi}{4}(15)^2 = 176.7\,\text{m}^2$$

$$l_{de} = \frac{W}{A\sigma} = \frac{24\,000}{176.7 \times 1} = 136\,\text{N}\,\text{m}^{-2}$$

$$\frac{l_{de}}{2\rho_0} = \frac{136}{2 \times 1.226} = 55.6\,\text{m}^2\,\text{s}^{-2}$$

With the above values, and with $V = 15\,\text{m}\,\text{s}^{-1}$, Eqn (9.57) is

$$U^4 - 225U^2 - \frac{1}{2}U(0.006)(3375) = (55.6)^2 + \frac{(0.006)^2}{16}(15)^4$$

i.e.

$$U^4 - 225U^2 - 10.125U = 3091$$

This quartic equation in U may be solved by any of the standard methods (e.g. Newton–Raphson), the solution being $U = 15.45\,\text{m}\,\text{s}^{-1}$ to four significant figures. Then

$$P = 2 \times 1.226 \times 176.7\left[\frac{0.006 \times (15)^3}{4} + \frac{(0.006)^2 \times (15)^4}{16 \times 15.45} + \frac{(55.6)^2}{15.45}\right]$$

$$= 88.9\,\text{kW}$$

This is the power required if the rotor behaves as an ideal actuator disc. A practical rotor would require considerably more power than this.

9.6 The rocket motor

As noted on page 527 the rocket motor is the only current example of aeronautical interest in Class II of propulsive systems. Since it does not work by accelerating atmospheric air, it cannot be treated by Froude's momentum theory. It is unique among current aircraft power plants in that it can operate independently of air from the atmosphere. The consequences of this are:

(i) it can operate in a rarefied atmosphere, or an atmosphere of inert gas
(ii) its maximum speed is not limited by the thermal barrier set up by the high ram-compression of the air in all air-breathing engines.

In a rocket, some form of chemical is converted in the combustion chamber into gas at high temperature and pressure, which is then exhausted at supersonic speed through a nozzle. Suppose a rocket to be travelling at a speed of V, and let the gas leave the nozzle with a speed of v relative to the rocket. Let the rate of mass flow of gas be $\dot{m}$.* This gas is produced by the consumption, at the same rate, of the chemicals in the rocket fuel tanks (or solid charge). Whilst in the tanks the mean m of fuel has a forward momentum of mV. After discharge from the nozzle the gas has a rearward momentum of $m(v - V)$. Thus the rate of increase of rearward momentum of the fuel/gas is

$$\dot{m}(v - V) - (-\dot{m}V) = \dot{m}v \tag{9.59}$$

and this rate of change of momentum is equal to the thrust on the rocket. Thus the thrust depends only on the rate of fuel consumption and the velocity of discharge relative to the rocket. The thrust does not depend on the speed of the rocket itself. In particular, the possibility exists that the speed of the rocket V can exceed the speed of the gas relative to both the rocket, v, and relative to the axes of reference, $v - V$.

When in the form of fuel in the rocket, the mass m of the fuel has a kinetic energy of $\frac{1}{2}mV^2$. After discharge it has a kinetic energy of $\frac{1}{2}m(v - V)^2$. Thus the rate of change of kinetic energy is

$$\frac{\mathrm{d}E}{\mathrm{d}t} = \frac{1}{2}\dot{m}[(v - V)^2 - V^2] = \frac{1}{2}\dot{m}(v^2 - 2vV) \tag{9.60}$$

the units being Watts.

Useful work is done at the rate TV, where $T = \dot{m}v$ is the thrust. Thus the propulsive efficiency of the rocket is

$$\eta_P = \frac{\text{rate of useful work}}{\text{rate of increase of KE of fuel}}$$
$$= \frac{2vV}{v^2 - 2vV} = \frac{2}{(v/V) - 2} \tag{9.61}$$

Now suppose $v/V = 4$. Then

$$\eta_P = \frac{2}{4 - 2} = 1 \quad \text{or} \quad 100\%$$

If $v/V < 4$, i.e. $V > v/4$, the propulsive efficiency exceeds 100%.

This derivation of the efficiency, while theoretically sound, is not normally accepted, since the engineer is unaccustomed to efficiencies in excess of 100%. Accordingly an alternative measure of the efficiency is used. In this the energy input is taken to be the energy liberated in the jet, plus the initial kinetic energy of the fuel while in the tanks. The total energy input is then

$$\frac{\mathrm{d}E}{\mathrm{d}t} = \frac{1}{2}\dot{m}v^2 + \frac{1}{2}\dot{m}V^2$$

---

* Some authors denote mass flow by m in rocketry, using the mass discharged (per second, understood) as the parameter.

giving for the efficiency

$$\eta_P = \frac{2(v/V)}{(v/V)^2 + 1} \tag{9.62}$$

By differentiating with respect to v/V, this is seen to be a maximum when $v/V = 1$, the propulsive efficiency then being 100%. Thus the definition of efficiency leads to a maximum efficiency of 100% when the speed of the rocket equals the speed of the exhaust gas relative to the rocket, i.e. when the exhaust is at rest relative to an observer past whom the rocket has the speed V.

If the speed of the rocket V is small compared with the exhaust speed v, as is the case for most aircraft applications, V^2 may be ignored compared with v^2 giving

$$\eta_P \simeq \frac{2vV}{v^2} = 2\frac{V}{v} \tag{9.63}$$

9.6.1 The free motion of a rocket-propelled body

Imagine a rocket-propelled body moving in a region where aerodynamic drag and lift and gravitational force may be neglected, i.e. in space remote from any planets, etc. At time t let the mass of the body plus unburnt fuel be M, and the speed of the body relative to some axes be V. Let the fuel be consumed at a rate of $\dot{m}$, the resultant gas being ejected at a speed of v relative to the body. Further, let the total rearwards momentum of the rocket exhaust, produced from the instant of firing to time t, be I relative to the axes. Then, at time t, the total forward momentum is

$$H_1 = MV - I \tag{9.64}$$

At time $(t + \delta t)$ the mass of the body plus unburnt fuel is $(M - \dot{m}\delta t)$ and its speed is $(V + \delta V)$, while a mass of fuel $\dot{m}\delta t$ has been ejected rearwards with a mean speed, relative to the axes, of $(v - V - \frac{1}{2}\delta V)$. The total forward momentum is then

$$H_2 = (M - \dot{m}\delta t)(V + \delta V) - \dot{m}\delta t(v - V - \frac{1}{2}\delta V) - I$$

Now, by the conservation of momentum of a closed system:

$$H_1 = H_2$$

i.e.

$$\begin{aligned} MV - I =& MV + M\delta V - \dot{m}V\delta t - \dot{m}\delta t\delta V - \dot{m}v\delta t + \dot{m}V\delta t \\ &+ \frac{1}{2}\dot{m}\delta t\delta V - I \end{aligned}$$

which reduces to

$$M\delta V - \frac{1}{2}\dot{m}\delta t\delta V - \dot{m}v\delta t = 0$$

Dividing by δt and taking the limit as $\delta t \to 0$, this becomes

$$M\frac{\mathrm{d}V}{\mathrm{d}t} - \dot{m}v = 0 \tag{9.65}$$

Note that this equation can be derived directly from Newton's second law, force = mass × acceleration, but it is not always immediately clear how to apply this law to bodies of variable mass. The fundamental appeal to momentum made above removes any doubts as to the legitimacy of such an application. Equation (9.65) may now be rearranged as

$$\frac{dV}{dt} = \frac{\dot{m}}{M}v$$

i.e.

$$\frac{1}{v}dV = \frac{\dot{m}}{M}dt$$

Now $\dot{m} = -dM/dt$, since $\dot{m}$ is the rate of which fuel is burnt, and therefore

$$\frac{1}{v}dV = -\frac{1}{M}\frac{dM}{dt}dt = -\frac{dM}{M}$$

Therefore

$$V/v = -M + \text{constant}$$

assuming v, but not necessarily $\dot{m}$, to be constant. If the rate of fuel injection into the combustion chamber is constant, and if the pressure into which the nozzle exhausts is also constant, e.g. the near-vacuum implicit in the initial assumptions, both $\dot{m}$ and v will be closely constant. If the initial conditions are $M = M_0$, $V = 0$ when $t = 0$ then

$$0 = -\ln M_0 + \text{constant}$$

i.e. the constant of integration is $\ln M_0$. With this

$$\frac{V}{v} = \ln M_0 - \ln M = \ln\left(\frac{M_0}{M}\right)$$

or, finally

$$V = v\ln(M_0/M) \tag{9.66}$$

The maximum speed of a rocket in free space will be reached when all the fuel is burnt, i.e. at the instant the motor ceases to produce thrust. Let the mass with all fuel burnt be M_1. Then, from Eqn (9.66)

$$V_{max} = v\ln(M_0/M_1) = v\ln R \tag{9.66a}$$

where R is the mass ratio M_0/M_1. Note that if the mass ratio exceeds $e = 2.718\ldots$, the base of natural logarithms, the speed of the rocket will exceed the speed of ejection of the exhaust relative to the rocket.

Distance travelled during firing

From Eqn (9.66),

$$V = v\ln(M_0/M) = v\ln M_0 - v\ln(M_0 - \dot{m}t)$$

Now if the distance travelled from the instant of firing is x in time t:

$$x = \int_0^t V \, dt$$

$$= v \int_0^t [\ln M_0 - \ln(M_0 - \dot{m}t)] dt$$

$$= vt \ln M_0 - v \int_0^t \ln(M_0 - \dot{m}t) dt \tag{9.67}$$

To solve the integral (G, say) in Eqn (9.67), let

$$y = \ln(M_0 - \dot{m}t)$$

Then

$$\exp(y) = M_0 - \dot{m}t \qquad \text{and} \qquad t = \frac{1}{\dot{m}}(M_0 - e^y)$$

whence

$$dt = -\frac{1}{\dot{m}} e^y dy$$

Then

$$G = \int_0^t \ln(M_0 - \dot{m}t) dt$$

$$= \int_{y_0}^{y_1} y \left(-\frac{1}{\dot{m}} e^y dy \right)$$

where

$$y_0 = \ln M_0 \qquad \text{and} \qquad y_1 = \ln(M_0 - \dot{m}t)$$

Therefore

$$G = -\frac{1}{\dot{m}} \int_{y_0}^{y_1} y e^y dy = \frac{1}{\dot{m}} \int_{y_0}^{y_1} y \, d(e^y)$$

which, on integrating by parts, gives

$$G = -\frac{1}{\dot{m}} [e^y(y - 1)]_{y_0}^{y_1} = \frac{1}{\dot{m}} [e^y(1 - y)]_{y_0}^{y_1}$$

Substituting back for y in terms of M_0, $\dot{m}$, and t gives

$$G = \frac{1}{\dot{m}} [(M_0 - \dot{m}t)(1 - \ln\{M_0 - \dot{m}t\})]_0^t$$

$$= \frac{1}{\dot{m}} [M(1 - \ln M) - M_0(1 - \ln M_0)]$$

where $M = M_0 - \dot{m}t$. Thus finally

$$G = \frac{1}{\dot{m}} [(M - M_0) - M \ln M + M_0 \ln M_0]$$

Substituting this value of the integral back into Eqn (9.67) gives, for the distance travelled

$$x = vt \ln M_0 - \frac{v}{\dot{m}} \{(M - M_0) - M \ln M + M_0 \ln M_0\} \qquad (9.68)$$

Now, if $\dot{m}$ is constant:

$$t = \frac{1}{\dot{m}} (M_0 - M) \qquad (9.69)$$

which, substituted into Eqn (9.68), gives

$$x = \frac{v}{\dot{m}} \{(M_0 - M) \ln M_0 - (M - M_0) + M \ln M - M_0 \ln M_0\}$$
$$= v \frac{M_0}{\dot{m}} \left[\left(1 - \frac{M}{M_0}\right) - \frac{M}{M_0} \ln \left(\frac{M_0}{M}\right) \right] \qquad (9.70)$$

For the distance at all-burnt, when $x = X$ and $M = M_1 = M_0/R$:

$$X = v \frac{M_0}{\dot{m}} \left(1 - \frac{1}{R}(1 + \ln R)\right) \qquad (9.71)$$

Alternatively, this may be written as

$$X = \frac{v}{\dot{m}} \frac{M_0}{R} [(R - 1) - \ln R]$$

i.e.

$$X = v \frac{M_1}{\dot{m}} [(R - 1) - \ln R] \qquad (9.72)$$

Example 9.7 A rocket-propelled missile has an initial total mass of 11 000 kg. Of this mass, 10 000 kg is fuel which is completely consumed in 5 minutes burning time. The exhaust is 1500 m s^{-1} relative to the rocket. Plot curves showing the variation of acceleration, speed and distance with time during the burning period, calculating these quantities at each half-minute.

For the acceleration

$$\frac{\mathrm{d}V}{\mathrm{d}t} = \frac{\dot{m}}{M} v \qquad (9.65a)$$

Now

$$\dot{m} = \frac{10\,000}{5} = 2000 \text{ kg min}^{-1} = 33.\dot{3} \text{ kg s}^{-1}$$

$$M = M_0 - \dot{m}t = 11\,000 - 33.\dot{3}t \text{ kg}$$

where t is the time from firing in seconds, or

$$M = 11\,000 - 1000\,N \text{ kg}$$

where N is the number of half-minute periods elapsed since firing.

$$\frac{M_0}{\dot{m}} = \frac{11\,000}{100/3} = 330 \text{ seconds}$$

Table 9.1

t (min)	M (1000 kg)	Acceleration (m s^{-2})	V (m s^{-1})	x (km)
0	11	4.55	0	0
0.5	10	5.00	143	2.18
1	9	5.55	300	9.05
1.5	8	6.25	478	19.8
2	7	7.15	679	37.6
2.5	6	8.33	919	61.4
3	5	10.0	1180	92.0
3.5	4	12.5	1520	133
4	3	16.7	1950	185
4.5	2	25.0	2560	256
5	1	50.0	3600	342
5.5	1	0	3600	450

Substituting the above values into the appropriate equations leads to the final results given in Table 9.1. The reader should plot the curves defined by the values in Table 9.1. It should be noted that, in the 5 minutes of burning time, the missile travels only 342 kilometres but, at the end of this time, it is travelling at 3600 m s^{-1} or 13 000 km h^{-1}. Another point to be noted is the rapid increase in acceleration towards the end of the burning time, consequent on the rapid percentage decrease of total mass. In Table 9.1, the results are given also for the first half-minute after all-burnt.

9.7 The hovercraft

In conventional winged aircraft lift, associated with circulation round the wings, is used to balance the weight, for helicopters the 'wings' rotate but the lift generation is the same. A radically different principle is used for sustaining of the hovercraft. In machines of this type, a more or less static region of air, at slightly more than atmospheric pressure, is formed and maintained below the craft. The difference between the pressure of the air on the lower side and the atmospheric pressure on the upper side produces a force tending to lift the craft. The trapped mass of air under the craft is formed by the effect of an annular jet of air, directed inwards and downwards from near the periphery of the underside. The downwards ejection of the annular jet produces an upwards reaction on the craft, tending to lift it. In steady hovering, the weight is balanced by the jet thrust and the force due to the cushion of air below the craft. The difference between the flight of hovercraft and normal jet-lift machines lies in the air cushion effect which amplifies the vertical force available, permitting the direct jet thrust to be only a small fraction of the weight of the craft. The cushion effect requires that the hovering height/diameter ratio of the craft be small, e.g. 1/50, and this imposes a severe limitation on the altitude attainable by the hovercraft.

Consider the simplified system of Fig. 9.10, showing a hovercraft with a circular planform of radius r, hovering a height h above a flat, rigid horizontal surface. An annular jet of radius r, thickness t, velocity V and density ρ is ejected at an angle θ to the horizontal surface. The jet is directed inwards but, in a steady, equilibrium state, must turn to flow outwards as shown. If it did not, there would be a continuous increase of mass within the region C, which is impossible. Note that such an increase of mass will occur for a short time immediately after starting, while the air cushion is being built up. The curvature of the path of the air jet shows that it possesses a centripetal acceleration and this

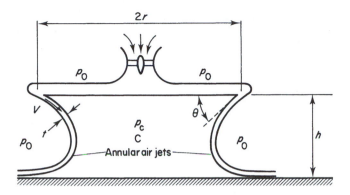

Fig. 9.10 The simplified hovercraft system

is produced by a difference between the pressure p_c within the air cushion and the atmospheric pressure p_0. Consider a short peripheral length δs of the annular jet and assume:

(i) that the pressure p_c is constant over the depth h of the air cushion
(ii) that the speed V of the annular jet is unchanged throughout the motion.

Then the rate of mass flow within the element of peripheral length δs is $\rho V t\, \delta s$ kg s^{-1}. This mass has an initial momentum parallel to the rigid surface (or ground) of $\rho V t\, \delta s V \cos\theta = \rho V^2 t \cos\theta\, \delta s$ inwards.

After turning to flow radially outwards, the air has a momentum parallel to the ground of $\rho V t\, \delta s V = \rho V^2 t\, \delta s$ outwards. Therefore there is a rate of change of momentum parallel to the ground of $\rho V^2 t(1 + \cos\theta)\, \delta s$. This rate of change of momentum is due to the pressure difference $(p_c - p_0)$ and must, indeed, be equal to the force exerted on the jet by this pressure difference, parallel to the ground, which is $(p_c - p_0)h\, \delta s$. Thus

$$(p_c - p_0)h\, \delta s = \rho V^2 t(1 + \cos\theta)\, \delta s$$

or

$$(p_c - p_0) = \frac{\rho V^2 t}{h}(1 + \cos\theta) \tag{9.73}$$

Thus the lift L_c due to the cushion of air on a circular body of radius r is

$$L_c = \pi r^2(p_c - p_0) = \frac{\pi \rho r^2 V^2 t}{h}(1 + \cos\theta) \tag{9.74}$$

The direct lift due to the downwards ejection of the jet is

$$L_j = \rho V t\, 2\pi r V \sin\theta = 2\pi r \rho V^2 t \sin\theta \tag{9.75}$$

and thus the total lift is

$$L = \pi r \rho V^2 t\left\{2\sin\theta + \frac{r}{h}(1 + \cos\theta)\right\} \tag{9.76}$$

If the craft were remote from any horizontal surface such as the ground or sea, so that the air cushion has negligible effect, the lift would be due only to the direct jet thrust, with the maximum value $L_{j_0} = 2\pi r \rho V^2 t$ when $\theta = 90°$. Thus the lift amplification factor, L/L_{j_0}, is

$$\frac{L}{L_{j_0}} = \sin\theta + \frac{r}{2h}(1 + \cos\theta) \tag{9.77}$$

Differentiation with respect to θ shows that this has a maximum value when

$$\tan\theta = \frac{2h}{r} \tag{9.78}$$

Since machines of this type are intended to operate under conditions such that h is very small compared to r, it follows that the maximum amplification is achieved when θ is close to zero, i.e. the jet is directed radially inwards. Then with the approximations $\sin\theta = 0$, $\cos\theta = 1$:

$$\frac{L}{L_{j_0}} = \frac{h}{r} \tag{9.79}$$

and

$$L = \frac{r}{h} L_{j_0} = \frac{r}{h} 2\pi r \rho V^2 t = \frac{2\pi r^2 \rho V^2 t}{h} \tag{9.80}$$

It will be noted that the direct jet lift is now, in fact, negligible.

The power supplied is equal to the kinetic energy contained in the jet per unit time,* which is

$$\frac{1}{2} 2\pi r \rho V t V^2 = \pi r \rho V^3 t \tag{9.81}$$

Denoting this by P, combining Eqns (9.80) and (9.81), and setting lift L equal to the weight W, leads to

$$\frac{P}{W} = \frac{Vh}{2r}$$

as the minimum power necessary for sustentation, while, if $\theta \neq 0$,

$$\frac{P}{W} \simeq \frac{Vh}{r(1 + \cos\theta)}$$

ignoring a term involving $\sin\theta$. Thus if V is small, and if h is small compared to r, it becomes possible to lift the craft with a comparatively small power.

The foregoing analysis applies to hovering flight and has, in addition, involved a number of simplifying assumptions. The first is the assumption of a level, rigid surface below the machine. This is reasonably accurate for operation over land but is not justified over water, when a depression will be formed in the water below the craft. It must be remembered that the weight of the craft will be reacted by a pressure distributed over the surface below the machine, and this will lead to deformation of a non-rigid surface.

---

* The power supplied to the jet will also contain a term relating to the increase in potential (pressure) energy, since the jet static pressure will be slightly greater than atmospheric. Since the jet pressure will be approximately equal to p_c, which is, typically, about $750\,\text{N m}^{-2}$ above atmospheric, the increase in pressure energy will be very small and has been neglected in this simplified analysis.

Another assumption is that the pressure p_c is constant throughout the air cushion. In fact, mixing between the annular jet and the air cushion will produce eddies leading to non-uniformity of the pressure within the cushion. The mixing referred to above, together with friction between the air jet and the ground (or water) will lead to a loss of kinetic energy and speed of the air jet, whereas it was assumed that the speed of the jet remained constant throughout the motion. These effects produce only small corrections to the results of the analysis above.

If the power available is greater than is necessary to sustain the craft at the selected height h, the excess may be used either to raise the machine to a greater height, or to propel the craft forwards.

Exercises

1 If an aircraft of wing area S and drag coefficient C_D is flying at a speed of V in air of density ρ and if its single airscrew, of disc area A, produces a thrust equal to the aircraft drag, show that the speed in the slipstream V_s, is, on the basis of Froude's momentum theory

$$V_s = V\sqrt{1 + \frac{S}{A}C_D}$$

2 A cooling fan is required to produce a stream of air, 0.5 m in diameter, with a speed of $3\,\mathrm{m\,s^{-1}}$ when operating in a region of otherwise stationary air of standard density. Assuming the stream of air to be the fully developed slipstream behind an ideal actuator disc, and ignoring mixing between the jet and the surrounding air, estimate the fan diameter and the power input required. (*Answer*: 0.707 m diameter; 3.24 W)

3 Repeat Example 9.2 in the text for the case where the two airscrews absorb equal powers, and finding (i) the thrust of the second airscrew as a percentage of the thrust of the first, (ii) the efficiency of the second and (iii) the efficiency of the combination. (*Answer*: 84%; 75.5%; 82.75%)

4 Calculate the flight speed at which the airscrew of Example 9.3 of the text will produce a thrust of 7500 N, and the power absorbed, at the same rotational speed. (*Answer*: $93\,\mathrm{m\,s^{-1}}$; 840 kW)

5 At 1.5 m radius, the thrust and torque gradings on each blade of a 3-bladed airscrew revolving at 1200 rpm at a flight speed of $90\,\mathrm{m\,s^{-1}}$ TAS at an altitude where $\sigma = 0.725$ are $300\,\mathrm{N\,m^{-1}}$ and $1800\,\mathrm{N\,m\,m^{-1}}$ respectively. If the blade angle is $28°$, find the blade section absolute incidence. Ignore compressibility. (*Answer*: $1°48'$) (CU)

6 At 1.25 m radius on a 3-bladed airscrew, the aerofoil section has the following characteristics:

solidity $= 0.1$; $\theta = 29°7'$; $\alpha = 4°7'$; $C_L = 0.49$; $L/D = 50$

Allowing for both axial and rotational interference find the local efficiency of the element. (*Answer*: 0.885) (CU)

7 The thrust and torque gradings at 1.22 m radius on each blade of a 2-bladed airscrew are $2120\,\mathrm{N\,m^{-1}}$ and $778\,\mathrm{N\,m\,m^{-1}}$ respectively. Find the speed of rotation (in $\mathrm{rad\,s^{-1}}$) of the airstream immediately behind the disc at 1.22 m radius. (*Answer*: $735\,\mathrm{rad\,s^{-1}}$)

8 A 4-bladed airscrew is required to propel an aircraft at $125\,\mathrm{m\,s^{-1}}$ at sea level, the rotational speed being 1200 rpm. The blade element at 1.25 m radius has an absolute incidence of 6° and the thrust grading is $2800\,\mathrm{N\,m^{-1}}$ per blade. Assuming a reasonable value for the sectional lift curve slope, calculate the blade chord at 1.25 m radius. Neglect rotational interference, sectional drag and compressibility.

(*Answer*: 240 mm)

9 A 3-bladed airscrew is driven at 1560 rpm at a flight speed of $110\,\mathrm{m\,s^{-1}}$ at sea level. At 1.25 m radius the local efficiency is estimated to be 87%, while the lift/drag ratio of the blade section is 57.3. Calculate the local thrust grading, ignoring rotational interference.

(*Answer*: $9000\,\mathrm{N\,m^{-1}}$ per blade)

10 Using simple momentum theory develop an expression for the thrust of a propeller in terms of its disc area, the air density and the axial velocities of the air a long way ahead, and in the plane, of the propeller disc. A helicopter has an engine developing 600 kW and a rotor of 16 m diameter with a disc loading of $170\,\mathrm{N\,m^{-2}}$. When ascending vertically with constant speed at low altitude, the product of the lift and the axial velocity of the air through the rotor disc is 53% of the power available. Estimate the velocity of ascent.   (*Answer*: $110\,\mathrm{m\,min^{-1}}$) (U of L)

Appendix 1: symbols and notation

A	Moment of inertia about OX. Aspect ratio (also (AR)). With suffices, coefficients in a Fourier series of sine terms, or a polynomial series in z
AF	Activity factor of an airscrew
AR	Aspect ratio (also A)
a	Speed of sound. Axial inflow factor in airscrew theory. Lift curve slope. $\mathrm{d}C_L/\mathrm{d}\alpha$ (suffices denote particular values). Radius of vortex core. Acceleration or deceleration
B	Number of blades on an airscrew
b	Rotational interference factor in airscrew theory. Total wing span ($=2s$). Hinge moment coefficient slope
CG	Centre of gravity
C_D	Total drag coefficient
C_{D_0}	Zero-lift drag coefficient
C_{D_v}	Trailing vortex drag coefficient
C_{D_L}	Lift-dependent drag coefficient. (Other suffices are used in particular cases.)
C_H	Hinge moment coefficient
C_L	Lift coefficient
C_M	Pitching moment coefficient
C_p	Pressure coefficient
C_p	Power coefficient for airscrews
C_R	Resultant force coefficient
c	Wing chord. A distance
$\bar{c}$	Standard or geometric mean chord
$\bar{c}$ or $\bar{c}_A$	Aerodynamic mean chord
c_0	Root chord
c_T	Tip chord
c_p, c_V	Specific heats at constant pressure and constant volume
CP	Centre of pressure
D	Drag (suffices denote particular values). Airscrew diameter. A length (occasionally)
d	Diameter, occasionally a length
d_v	Spanwise trailing vortex drag grading ($=\rho w\Gamma$)
E	Internal energy per unit mass. Kinetic energy
F	Fractional flap chord. Force
f( ), fn( )	Function of the stated variables
g	Acceleration due to gravity
g( )	Function of the stated variables
H	Hinge moment. Total pressure. Momentum. Shape factor, δ^*/θ

h	Fractional camber of a flapped plate aerofoil. Distance between plates in Newton's definition of viscosity. Enthalpy per unit mass
h( )	Function of the stated variables
h_0	Fractional position of the aerodynamic centre
I	Momentum of rocket exhaust
i	The imaginary operator, $\sqrt{-1}$
J	Advance ratio of an airscrew
K	Modulus of bulk elasticity
k	Chordwise variation of vorticity. Lift-dependent drag coefficient factor
k_{CP}	Centre of pressure coefficient
K_T, k_Q	Thrust and torque coefficients (airscrews)
L	Lift. Dimension of length. Temperature lapse rate in the atmosphere
l	Length. Lift per unit span
l_{de}	Effective disc loading of a helicopter
M	Dimension of mass. Mach number. Pitching moment about Oy
m	Mass. Strength of a source (-sink). An index
$\dot{m}$	Rate of mass flow
N	Rpm of an airscrew. Normal influence coefficient. Number of panel points
n	Revolutions per second of an airscrew. Frequency. An index
$\hat{n}$	Unit normal vector
O	Origin of coordinates
P	Power. The general point in space
p	Static pressure in a fluid
Q	Torque, or a general moment. Total velocity of a uniform stream
q	Angular velocity in pitch about Oy. Local resultant velocity. A coefficient in airscrew theory
q_n, q_t	Radial and tangential velocity components
Re	Real part of a complex number
Re	Reynolds number
R	Resultant force. Characteristic gas constant. Radius of a circle
r	Radius vector, or radius generally
S	Wing area. Vortex tube area. Area of actuator disc. Entropy
S'	Tailplane area
s	Semi-span ($= \frac{1}{2}b$). Distance. Specific entropy
s'	Spacing of each trailing vortex centre from aircraft centre-line
T	Dimension of time. Thrust. Temperature (suffices denote particular values). Tangential influence coefficient
t	Time. Aerofoil section thickness. A coefficient in airscrew theory
$\hat{t}$	Tangential unit vector
U	Velocity. Steady velocity parallel to Ox
U_∞	Freestream flow speed
U_e	Mainstream flow speed
u	Velocity component parallel to Ox
u'	Disturbance velocity parallel to Ox
V	Velocity. Volume. Steady velocity parallel to Oy
V_s	Stalling speed
V_E	Equivalent air speed
V_R	Resultant speed
v	Velocity component parallel to Oy. Velocity
v'	Disturbance velocity parallel to Oy
W	Weight. Steady velocity parallel to Oz
w	Wing loading. Downwash velocity. Velocity parallel to Oz
X, Y, Z	Components of aerodynamic or external force
x, y, z	Coordinates of the general point P
x, X	Distance
z	Distance. Spanwise coordinate

α	Angle of incidence or angle of attack. An angle, generally
β	A factor in airscrew theory. An angle generally
Γ	Circulation
Γ	Half the dihedral angle; the angle between each wing and the Oxy plane
γ	Ratio of specific heats, c_p/c_V. Shear strain
δ	Boundary layer thickness. A factor. Camber of an aerofoil section
ε	Downwash angle. Surface slope. Strain
ζ	Vorticity. Complex variable in transformed plane ($= \xi + i\eta$)
η	Efficiency. Ordinate in ζ-plane
θ	Dimension of temperature. Polar angular coordinate. Blade helix angle (airscrews). Momentum thickness
Λ	Angle of sweepback or sweep-forward. Pohlhausen pressure-gradient parameter
λ	Taper ratio ($= c_T/c_0$). A constant
μ	Strength of a doublet. Dynamic viscosity. Aerofoil parameter in lifting line theory
v	Kinematic viscosity. Prandtl–Meyer angle
ξ	Abscissa in ζ-plane
ρ	Density. Radius of curvature
Σ	Summation sign
σ	Relative density. Blade or annular solidity (airscrews). Stress
τ	Shear stress
ϕ	Sweepback angle. Velocity potential. A polar coordinate. Angle of relative wind to plane of airscrew disc
ψ	The stream function
Ω	Angular velocity of airscrew
ω	Angular velocity in general
∇^2	Laplace's operator $\nabla^2 (= \partial^2/\partial x^2 + \partial^2/\partial y^2)$

Suffices

0	No lift. Standard sea level. Straight and level flight. Undisturbed stream
$\frac{1}{4}$	Quarter chord point
1	A particular value
2	A particular value
3	A particular value
∞	Infinity or two-dimensional conditions
AC	Aerodynamic centre
a	available
c	Chord from Ox axis. Compressible
CP	Centre of pressure
f	Full scale or flight
g	Ground
h	Horizontal
i	Ideal, computation numbering sequence. Incompressible
in	Input. Computation numbering sequence. Length
L	Lower surface
LE	Leading edge
l	Local
m	Model
max	Maximum
min	Minimum
md	Minimum drag
n	normal
n	Denotes general term
opt	Optimum
out	Output
P	Prandtl–Meyer

p	Propulsive parallel
r	Required
s	Stagnation or reservoir conditions. Slipstream. Stratosphere. Surface
TE	Trailing edge
t	Thickness (aerofoil), panel identification in computation. Tangential
U	Upper surface
V	Vertical
W	Wall

Primes and superscripts

$'$	Perturbance or disturbance
$*$	Throat (locally sonic) conditions. Boundary-layer displacement thickness
$**$	Boundary-layer energy thickness
$\wedge$	Unit vector
$\rightarrow$	Vector

The dot notation is frequently used for differentials, e.g. $\dot{y} = dy/dx$, the rate of change of y with x.

Appendix 2: the international standard atmosphere

(1) Sea level conditions

$T_0 = +15\,°C = 288.16\,K$ $\qquad \mu_0 = 1.783 \times 10^{-5}\,kg\,m^{-1}s^{-1}$

$p_0 = 101325\,Nm^{-2}$ $\qquad \nu_0 = 1.455 \times 10^{-5}\,m^2s^{-1}$

$\rho = 1.2256\,kg\,m^{-3}$

(2) Relative values

Altitude (m)	Temperature $\theta = T/T_0$	Pressure $\delta = p/p_0$	Density		Viscosity	
			$\sigma = \rho/\rho_0$	$\sigma^{1/2}$	$\tilde{\mu} = \mu/\mu_0$	$\tilde{\nu} = \nu/\nu_0$
−0	1	1	1	1	1	1
250	0.9944	0.9707	0.9762	0.9880	0.9956	1.0198
500	0.9887	0.9421	0.9528	0.9761	0.9911	1.0402
750	0.9831	0.9142	0.9299	0.9643	0.9867	1.0610
1000	0.9774	0.8869	0.9074	0.9526	0.9822	1.0824
1250	0.9718	0.8604	0.8853	0.9409	0.9777	1.1044
1500	0.9661	0.8344	0.8637	0.9293	0.9733	1.1269
1750	0.9605	0.8091	0.8424	0.9178	0.9688	1.1500
2000	0.9549	0.7845	0.8215	0.9064	0.9642	1.1737
2250	0.9492	0.7604	0.8011	0.8950	0.9597	1.1980
2500	0.9436	0.7369	0.7810	0.8837	0.9552	1.2230
2750	0.9379	0.7141	0.7613	0.8725	0.9506	1.2487
3000	0.9323	0.6918	0.7420	0.8614	0.9461	1.2750
3250	0.9266	0.6701	0.7231	0.8503	0.9415	1.3020
3500	0.9210	0.6489	0.7045	0.8394	0.9369	1.3298
3750	0.9154	0.6283	0.6863	0.8285	0.9323	1.3584
4000	0.9097	0.6082	0.6685	0.8176	0.9277	1.3877
4250	0.9041	0.5886	0.6511	0.8069	0.9231	1.4178
4500	0.8984	0.5696	0.6339	0.7962	0.9184	1.4488
4750	0.8928	0.5510	0.6172	0.7856	0.9138	1.4806
5000	0.8872	0.5329	0.6007	0.7751	0.9091	1.5133
5250	0.8815	0.5154	0.5846	0.7646	0.9044	1.5470
5500	0.8759	0.4983	0.5689	0.7542	0.8997	1.5816
5750	0.8702	0.4816	0.5534	0.7439	0.8950	1.6172
6000	0.8646	0.4654	0.5383	0.7337	0.8903	1.6538
6250	0.8589	0.4497	0.5235	0.7236	0.8855	1.6915

Altitude (m)	Temperature $\theta = T/T_0$	Pressure $\delta = p/p_0$	Density		Viscosity	
			$\sigma = \rho/\rho_0$	$\sigma^{1/2}$	$\tilde{\mu} = \mu/\mu_0$	$\tilde{\nu} = \nu/\nu_0$
6500	0.8533	0.4344	0.5091	0.7135	0.8808	1.7303
6750	0.8477	0.4195	0.4949	0.7035	0.8760	1.7702
7000	0.8420	0.4050	0.4810	0.6936	0.8713	1.8113
7250	0.8364	0.3910	0.4674	0.6837	0.8665	1.8536
7500	0.8307	0.3773	0.4542	0.6739	0.8617	1.8972
7750	0.8251	0.3640	0.4412	0.6642	0.8568	1.9421
8000	0.8194	0.3511	0.4285	0.6546	0.8520	1.9884
8250	0.8138	0.3386	0.4161	0.6450	0.8471	2.0361
8500	0.8082	0.3264	0.4039	0.6356	0.8423	2.0852
8750	0.8025	0.3146	0.3921	0.6262	0.8374	2.1359
9000	0.7969	0.3032	0.3805	0.6168	0.8325	2.1881
9250	0.7912	0.2921	0.3691	0.6076	0.8276	2.2420
9500	0.7856	0.2813	0.3581	0.5984	0.8227	2.2976
9750	0.7799	0.2708	0.3472	0.5893	0.8177	2.3549
10 000	0.7743	0.2607	0.3367	0.5802	0.8128	2.4141
10 250	0.7687	0.2509	0.3264	0.5713	0.8078	2.4752
10 500	0.7630	0.2413	0.3163	0.5624	0.8028	2.5383
10 750	0.7574	0.2321	0.3064	0.5536	0.7978	2.6034
11 000	0.7517	0.2232	0.2968	0.5448	0.7928	2.6707
11 500	Constant	0.2062	0.2743	0.5238	Constant	2.8897
12 000	in	0.1906	0.2535	0.5035	in	3.1268
12 500	stratosphere	0.1761	0.2343	0.4841	stratosphere	3.3833
13 000		0.1628	0.2166	0.4654		3.6608
13 500		0.1505	0.2001	0.4474		3.9811
14 000		0.1390	0.1850	0.4301		4.2860
14 500		0.1285	0.1709	0.4135		4.6376
15 000		0.1188	0.1580	0.3975		5.0180
15 500		0.1098	0.1470	0.3821		5.4297
16 000		0.1014	0.1349	0.3673		5.8751
16 500		0.0937	0.1247	0.3531		6.3570
17 000		0.0866	0.1153	0.3395		6.8785
17 500		0.0801	0.1065	0.3264		7.4427
18 000		0.0740	0.0984	0.3138		8.0532
18 500		0.0684	0.0910	0.3016		8.7138
19 000		0.0632	0.0841	0.2900		9.4286
19 500		0.0584	0.0777	0.2788		10.202
20 000	Constant	0.0540	0.0718	0.2680	Constant	11.039
20 500	in	0.0499	0.0664	0.2576	in	11.945
21 000	stratosphere	0.0461	0.0613	0.2477	stratosphere	12.924
21 500		0.0426	0.0567	0.2381		13.985
22 000		0.0394	0.0524	0.2289		15.132
22 500		0.0364	0.0484	0.2200		16.373
23 000		0.0336	0.0447	0.2115		17.716
23 500		0.0311	0.0414	0.2034		19.169
24 000		0.0287	0.0382	0.1955		20.742
24 500		0.0266	0.0353	0.1879		22.443
25 000		0.0245	0.0326	0.1807		24.284

Appendix 3*: a solution of integrals of the type of Glauert's integral

Glauert's integral

$$G_n = \int_0^\pi \frac{\cos n\theta}{\cos \theta - \cos \theta_1} \, \mathrm{d}\theta$$

In Chapters 4 and 5 much use is made of the integral

$$G_n = \int_0^\pi \frac{\cos n\theta}{\cos \theta - \cos \theta_1} \, \mathrm{d}\theta$$

the result for which was quoted as

$$\pi \frac{\sin n\theta_1}{\sin \theta_1}$$

This may be proved, by contour integration, as follows.
In the complex plane, integrate the function

$$\mathrm{f}(z) = \frac{z^n}{z^2 - 2z \cos \theta_1 + 1}$$

with respect to z round the circle of unit radius centred at the origin. On this circle $z = \mathrm{e}^{\mathrm{i}\theta}$ and therefore

$$\int_c \frac{z^n \mathrm{d}z}{z^2 - 2z \cos \theta_1 + 1} = \int_{-\pi}^{+\pi} \frac{\mathrm{e}^{2\mathrm{i}n\theta} \mathrm{i} \mathrm{e}^{\mathrm{i}\theta} \mathrm{d}\theta}{\mathrm{e}^{2\mathrm{i}\theta} - 2\mathrm{e}^{\mathrm{i}\theta} \cos \theta_1 + 1}$$

which, cancelling $\mathrm{e}^{\mathrm{i}\theta}$ from numerator and denominator, putting

$$\mathrm{e}^{\mathrm{i}\theta} = \cos \theta + \mathrm{i} \sin \theta$$

and using De Moivre's theorem, reduces to

$$\frac{\mathrm{i}}{2} \int_{-\pi}^{+\pi} \frac{\cos n\theta + \mathrm{i} \sin n\theta}{\cos \theta - \cos \theta_1} \, \mathrm{d}\theta \tag{A3.1}$$

---

* This section may be omitted at a first reading.

The *poles* or *singularities* of the function f(z) are those points where f(z) is infinite, i.e. in this case where

$$z^2 - 2z \cos \theta_1 + 1 = 0$$

i.e.

$$z = \cos \theta_1 \pm \sin \theta_1 = e^{\pm i\theta_1} \tag{A3.2}$$

In general if a function f(z) has a simple pole at the point $z = c$, then

$$\lim_{z \to c} (z - c) f(z)$$

is finite and its value is called the *residue* at the pole. In this case

$$\lim_{z \to c} (z - c) f(z) = \lim_{z \to c} \left(\frac{(z - c) z^n}{z^2 - 2z \cos \theta_1 + 1} \right)$$

which, by L'Hôpital's theorem

$$= \left[\frac{\frac{d}{dz}((z - c) z^n)}{\frac{d}{dz}(z^2 - 2z \cos \theta_1 + 1)} \right]_{z=c}$$

Thus differentiating and reducing, and for this case putting $c = e^{\pm i\theta_1}$, from Eqn (A3.2) the residues at the two poles are $(\sin n\theta_1 \pm \cos n\theta)/2 \sin \theta_1$ and the sum of the residues is

$$\frac{\sin n\theta_1}{\sin \theta_1} \tag{A3.3}$$

Now for this case the poles (at the points $z = e^{\pm i\theta_1}$) are on the contour of integration and by Cauchy's residue theorem the value of the integral (Eqn (A3.1)) is equal to $\pi i \times$ (sum of the residues on the contour). Thus

$$\frac{i}{2} \int_{-\pi}^{+\pi} \frac{\cos n\theta + i \sin n\theta}{\cos \theta - \cos \theta_1} d\theta = \pi i \frac{\sin n\theta_1}{\sin \theta_1} \tag{A3.4}$$

Equating the imaginary parts of this equation,

$$\frac{i}{2} \int_{-\pi}^{+\pi} \frac{\cos n\theta}{\cos \theta - \cos \theta_1} d\theta = \pi \frac{\sin n\theta_1}{\sin \theta_1}$$

and by the symmetry of the integrand:

$$\int_0^\pi \frac{\cos n\theta}{\cos \theta - \cos \theta_1} d\theta = \frac{i}{2} \int_{-\pi}^{+\pi} \frac{\cos n\theta}{\cos \theta - \cos \theta_1} d\theta$$

i.e.

$$\int_0^\pi \frac{\cos n\theta}{\cos \theta - \cos \theta_1} d\theta = \pi \frac{\sin n\theta_1}{\sin \theta_1} \tag{A3.5}$$

Using the result that

$$\sin n\theta \sin \theta = \frac{1}{2} \cos(n - 1)\theta - \frac{1}{2} \cos(n + 1)\theta$$

it follows that

$$\int_0^\pi \frac{\sin n\theta \sin \theta}{\cos \theta - \cos \theta_1} d\theta = \frac{1}{2} \int_0^\pi \frac{\cos(n - 1)\theta}{\cos \theta - \cos \theta_1} d\theta - \frac{1}{2} \int_0^\pi \frac{\cos(n - 1)\theta}{\cos \theta - \cos \theta_1} d\theta$$

Then from Eqn (A3.5) it follows that

$$\int_0^\pi \frac{\sin n\theta \sin \theta}{\cos \theta - \cos \theta_1} \, d\theta = \frac{\pi}{2} \frac{\sin(n-1)\theta_1 - \sin(n+1)\theta_1}{\sin \theta_1}$$

and since

$$\sin(n-1)\theta_1 - \sin(n+1)\theta_1 = -2\cos n\theta_1 \sin \theta_1$$

the integral becomes

$$\int_0^\pi \frac{\sin n\theta \sin \theta}{\cos \theta - \cos \theta_1} \, d\theta = -\pi \cos n\theta_1 \qquad n = 0, 1, 2, \ldots \qquad (A3.6)$$

Appendix 4: conversion of imperial units to système international (SI) units

The conversion between Imperial units and SI units is based on the fact that the fundamental units (pound mass, foot, second and degree Centigrade) of the Imperial system have been defined in terms of the corresponding units of the SI. These definitions are as follows:

1 foot = 0.3048 m
1 pound = 0.453 592 27 kg

The second and the degree Celsius (degree Centigrade) are identical in the two systems.

Working from these definitions, the conversion factors given in Table A4.1 are calculated. This table covers the more common quantities encountered in aerodynamics. The conversion factors have been rounded to five significant figures where appropriate.

Table A4.1 Conversion factors between Imperial units and SI units

One of these	is equal to this number	of these
ft	0.3048	m
in	25.4	mm
statute mile	1609.3	m
nautical mile	1853.2	m
ft^2	0.0929	m^2
in^2	6.4516×10^{-4}	m^2
in^2	645.16	mm^2
ft^3	0.028 32	m^3
in^3	1.6387×10^{-5}	m^3
in^3	16 387	mm^3
slug	14.594	kg
slug ft^{-3}	515.38	$kg\ m^{-3}$
lbf	4.4482	N
lbf ft^{-2}	47.880	Nm^{-2}
lbf in^{-2}	6894 8	Nm^{-2}
ft lbf	1.3558	J
hp	745.70	W

lbf ft	1.3558	Nm
ft s^{-1}	0.3048	m s^{-1}
mile h^{-1}	0.447 04	m s^{-1}
knot	0.514 77	m s^{-1}

The knot Even with standardization on the SI, the knot continues to be used as a preferred non-metric unit in practical aeronautics. The knot is a unit of speed, and is defined as one nautical mile per hour, where one nautical mile is equal to 1853 m (6080 feet).

Bibliography

ABBOTT, I.H. and DOENHOFF, A.E. von (1958) *Theory of Wing Sections* (Dover Publications, Inc.)

ACKROYD, J.A.D., AXCELL, B.P. and RUBAN, A. (2001) *Early Developments of Modern Aerodynamics* (Butterworth-Heinemann)

ANDERSON, J.D. Jr. (1997) *A History of Aerodynamics* (Cambridge University Press)

ANDERSON, J.D. Jr. (2001) *Fundamentals of Aerodynamics*, 3rd Edition (McGraw-Hill)

ANDERSON, J.D. Jr. (1990) *Modern Compressible Flow: With Historical Perspective*, 2nd Edition (McGraw-Hill)

ASHLEY, H. and LANDAHL, M. (1985) *Aerodynamics of Wings and Bodies* (Dover Pub.)

BARNARD, R.H. (2001) *Road Vehicle Aerodynamic Design*, 2nd Edition (MechAero Pub.)

BARNARD, R.H. and PHILPOTT, D.R. (1994) *Aircraft Flight*, 2nd Edition (Prentice Hall)

BATCHELOR, G.B. (1967) *Introduction to Fluid Dynamics* (Cambridge University Press)

BRADSHAW, P. (1971) *An Introduction to Turbulence and its Measurement*, 2nd Edition (Pergamon Press)

BRADSHAW, P. (1970) *Experimental Fluid Mechanics* (Pergamon Press)

BRAMWELL, A.R.S., BALMFORD, D.E.H. and DONE, G.T.S. (2001) *Bramwell's Helicopter Dynamics* (Butterworth-Heinemann)

CARPENTER, C.J. (1996) *Flightwise Vol. 1: Principles of Aircraft Flight* (Airlife Publishing)

CEBECI, T. (1999) *An Engineering Approach to the Calculation of Aerodynamic Flows* (Springer)

CEBECI, T. and BRADSHAW, P. (1977) *Momentum Transfer in Boundary Layers* (Hemisphere)

CEBECI, T. and COUSTEIX, J. (1998) *Modeling and Computation of Boundary-Layer Flows* (Springer)

COX, R.N. and CRABTREE, L.F. (1965) *Elements of Hypersonic Aerodynamics* (English Universities Press)

CUMPSTY, N.A. (1989) *Compressor Aerodynamics* (Longmann)

DALTON, S. (1999) *The Miracle of Flight* (Firefly Books Ltd.)

DONALDSON, C. duP. (General Editor) (1954–64) *High Speed Aerodynamics and Jet Propulsion*, 12 vols (Princeton University Press; Oxford University Press)

DRAZIN, P.G. and REID, W.H. (1981) *Hydrodynamic Stability* (Cambridge University Press)

DUNCAN, W.J., THOM, A.S. and YOUNG, A.D. (1970) *Mechanics of Fluids*, 2nd Edition (Edward Arnold)

DURAND, W.J. (Editor-in-Chief) (1934–36) *Aerodynamic Theory*, 6 vols (Springer; Dover)

FERZIGER, J.H. (1998) *Numerical Methods for Engineering Applications*, 2nd Edition (Wiley)

FERZIGER, J.H. and PERIC, M. (1999) *Computational Methods for Fluid Dynamics*, 2nd Edition, revised (Springer)

FLETCHER, C.A.J. (1991) *Computational Techniques for Fluid Dynamics*, 2 vols, 2nd Edition (Springer)

GAD-EL-HAK, M. (2000) *Flow Control* (Cambridge University Press)

GLAUERT, H. (1947) *The Elements of Airscrew and Aerofoil Theory* (Cambridge University Press)

GOLDSTEIN, S. (1938) *Modern Developments in Fluid Dynamics*, 2 vols (Oxford University Press; Dover)

HAYES, W.D. and PROBSTEIN, R.F. (1966) *Hypersonic Flow Theory*, 2nd Edition (Academic Press)

HENNE, P.A. (Editor) (1990) *Applied Computational Aerodynamics* (AIAA)

HINZE, J.O. (1975) *Turbulence*, 2nd Edition (McGraw-Hill)

HOERNER, S.F. (1965) *Fluid Dynamic Drag* (Hoerner, New Jersey)

HOUGHTON, E.L. and BOSWELL, R.P. (1969) *Further Aerodynamics for Engineering Students* (Edward Arnold)

HOUGHTON, E.L. and BROCK, A.E. (1970) *Aerodynamics for Engineering Students*, 2nd Edition (Edward Arnold)

HOUGHTON, E.L. and BROCK, A.E. (1975) *Tables for the Compressible Flow of Dry Air*, 3rd Edition (Edward Arnold)

JONES, R.T. (1990) *Wing Theory* (Oxford: Princeton UniversityPress)

KARAMCHETI, K. (1966) *Principles of Ideal-Fluid Aerodynamics* (Wiley)

KÜCHEMANN, D. (1978) *The Aerodynamic Design of Aircraft* (Pergamon)

KATZ, J. (1995) *Race Car Aerodynamics* (Robert Bently)

KATZ, J. and PLOTKIN, A. (2001) *Low-speed Aerodynamics*, 2nd Edition (Cambridge University Press)

KUETHE, A.M. and CHOW, C.-Y. (1997) *Foundations of Aerodynamics*, 5th Edition (Wiley)

LANCHESTER, F.W. (1907) *Aerodynamics* (Constable)

LANCHESTER, F.W. (1908) *Aerodynetics* (Constable)

LIEPMANN, H.W. and ROSHKO, A. (1957) *Elements of Gasdynamics* (Wiley)

LIGHTHILL, J. (1986) *An Informal Introduction to Theoretical Fluid Mechanics* (Cambridge University Press)

McCORMICK, B.W. (1995) *Aerodynamics, Aeronautics, and Flight Mechanics*, 2nd Edition (Wiley)

MERZKIRCH, W. (1974) *Flow Visualisation* (Academic Press)

MILLIKAN, C.B. (1941) *Aerodynamics of the Airplane* (Wiley, New York; Chapman and Hall, London)

MILNE-THOMSON, L.M. (1948) *Theoretical Aerodynamics* (Macmillan)

NICKEL, K. and WOHLFAHRT, M. (1994) *Tailless Aircraft in Theory and Practice* (Edward Arnold)

OWCZAREK, J.A. (1964) *Fundamentals of Gas Dynamics* (International Textbook Co.)

OWER, E. and PANKHURST, R.C. (1966) *Measurement of Airflow*, 4th Edition (Pergamon)

PANKHURST, R.C. and HOLDER, D.W. (1952) *Wind-Tunnel Technique* (Pitman)

PANTON, R.L. (Editor) (1997) *Self-sustaining Mechanisms of Wall Turbulence* (Computational Mechanics Publications)

POPE, A. and RAE, W.H. (1984) *Low Speed Wind Tunnel Testing*, 2nd Edition (Wiley)

POPE, S.B. (2000) *Turbulent Flows* (Cambridge University Press)

PRANDTL, L. (1952) *Essentials of Fluid Dynamics* (Blackie)

PRANDTL, L. and TIETJENS, O.G. (1934) *Appled Hydro- and Aeromechanics* (McGraw-Hill; Dover)

PRANDTL, L. and TIETJENS, O.G. (1934) *Fundamentals of Hydro- and Aeromechanics* (McGraw-Hill; Dover)

PRESS, W.H. *et al.* (1997) *Numerical Recipes. The Art of Scientific Computing*, 2nd Edition (Cambridge University Press)

ROSENHEAD, L. (Editor) (1963) *Laminar Boundary Layers* (Clarendon Press Oxford)

SCHLICHTING, H. (1979) *Boundary-Layer Theory*, 7th Edition (McGraw-Hill)

SCHLICHTING, H. and GERSTEN, K. (2000) *Boundary Layer Theory*, 8th Edition (Springer)

SEDDON, J. and NEWMAN, S. (2001) *Basic Helicopter Aerodynamics*, 2nd Edition (Blackwell Science)

SHAPIRO, A.H. (1953) *The Dynamics and Thermodynamics of Compressible Fluid Flow*, 2 vols (Ronald Press Co.)

SHAW, C.T. (1992) *Using Computational Fluid Mechanics* (Prentice Hall)

SUMANTRAN, V. (1997) *Vehicle Aerodynamics* (SAE)

SUTTON, O.G. (1949) *The Science of Flight* (Pelican Books)

TENNEKES, H. (1997) *The Simple Science of Flight* (The MIT Press)

THWAITES, B. (Editor) (1960) *Incompressible Aerodynamics* (Oxford University Press)

TRITTON, D.J. (1988) *Physical Fluid Dynamics*, 2nd Edition (Clarendon)
TUCKER, P.G. (2001) *Computation of Unsteady Internal Flows* (Kluwer)
VAN DYKE, M. (1982) *An Album of Fluid Motion* (Parabolic Press)
WILCOX, D.C. (1993) *Turbulence Modeling for CFD* (DCW Industries)
WHITE, F.M. (1999) *Fluid Mechanics*, 4th Edition (McGraw-Hill)
WHITE, F.M. (1991) *Viscous Fluid Flow*, 2nd Edition (McGraw-Hill)
WHITFORD, R. (2000) *Fundamentals of Fighter Design* (Airlife Publishing Ltd.)
YOUNG, A.D. (1989) *Boundary Layers* (BSP Professional)
ZUCROW, M.J. and HOFFMAN, J.D. (1976) *Gas Dynamics*, 2 vols (Wiley)

Index